A NEW HISTORY OF THE HUMANITIES

Many histories of science have been written, but *A New History of the Humanities* offers the first overarching history of the humanities from Antiquity to the present. There are already historical studies of musicology, logic, art history, linguistics, and historiography, but this volume gathers these, and many other humanities disciplines, into a single coherent account.

Its central theme is the way in which scholars throughout the ages and in virtually all civilizations have sought to identify patterns in texts, art, music, languages, literature, and the past. What rules can we apply if we wish to determine whether a tale about the past is trustworthy? By what criteria are we to distinguish consonant from dissonant musical intervals? What rules jointly describe all possible grammatical sentences in a language? How can modern digital methods enhance pattern-seeking in the humanities? Rens Bod contends that the hallowed opposition between the sciences (mathematical, experimental, dominated by universal laws) and the humanities (allegedly concerned with unique events and hermeneutic methods) is a mistake born of a myopic failure to appreciate the pattern-seeking that lies at the heart of this inquiry. *A New History of the Humanities* amounts to a persuasive plea to give Panini, Valla, Bopp, and countless other often overlooked intellectual giants their rightful place next to the likes of Galileo, Newton, and Einstein.

Rens Bod is a professor at the Institute for Logic, Language and Computation at the University of Amsterdam. He has published extensively on linguistics, digital humanities and the history of the humanities.

'Bod takes the humanities back to their rightful place in the family tree of sciences'

Frederik Stjernfelt, *Weekendavisen*

'Bod's effort has become a reason for debate and interdisciplinary encounters among different scholars who agree that the "disunity" of the sciences, which is commonplace in the post-positivist epistemological era, does not necessarily mean disunity of culture'

Alessandro Pagnini, *Domenica24*

'Bod constantly makes comparisons between what he finds in the separate disciplines and in different times. With the analysis these comparisons yield he aims to create an integrated picture of the development of the humanities. The result is a truly impressive volume written in a highly accessible style'

Bart Karstens, *Beiträge zur Geschichte der Sprachwissenschaft*

A New History of the Humanities

The Search for Principles and Patterns from Antiquity to the Present

RENS BOD

Translated from the Dutch by

LYNN RICHARDS

OXFORD
UNIVERSITY PRESS

OXFORD
UNIVERSITY PRESS

Great Clarendon Street, Oxford, OX2 6DP,
United Kingdom

Oxford University Press is a department of the University of Oxford.
It furthers the University's objective of excellence in research, scholarship,
and education by publishing worldwide. Oxford is a registered trade mark of
Oxford University Press in the UK and in certain other countries

First published 2013
First published in paperback 2015

Published in the United States of America by Oxford University Press
198 Madison Avenue, New York, NY 10016, United States of America

British Library Cataloguing in Publication Data
Data available

Library of Congress Cataloging in Publication Data
Data available

ISBN 978–0–19–966521–1 (Hbk.)
ISBN 978–0–19–875839–6 (Pbk.)

The publisher gratefully acknowledges the support
of the Dutch Foundation for Literature.

If two rules conflict, the last rule prevails.

Panini, *c.*500 BCE

For many states that were once great have now become small, and in my lifetime those that are great used to be small.

Herodotus, *c.*440 BCE

The poet should prefer probable impossibilities to improbable possibilities.

Aristotle, *c.*335 BCE

No pyknon, neither a complete pyknon nor a part of one, is melodically adjacent to another pyknon.

Aristoxenus, *c.*310 BCE

When things reach their period of greatest flourishing, they begin to decay.

Sima Qian, *c.*100 BCE

His style was fine and subtle, his brush without a flaw, yet his workmanship was inferior to his ideas, and (in his case) fame surpassed reality.

Xie He, *c.*500

It is wrong when the beginning of a sentence is in conflict with the end of it.

Sibawayh, *c.*790

The tradition regarding an event which in itself does not contradict either logical or physical laws will invariably depend for its character as true or false upon the character of the reporters.

al-Biruni, 1030

History is composed of the news about the days, states, and previous centuries. It is a theory, an analysis, and justification of creatures and their principles, and a science of how events happened and their causes.

Ibn Khaldun, 1377

In every painting it is therefore important that all members do what they are supposed to do in such a way that even the smallest part does not fail to contribute to the subject.

Leon Battista Alberti, 1435

But at last he brought it thus far, that he could demonstrate the whole Trinity to be represented by these first rudiments of grammar, as clearly and plainly as it was possible for a mathematician to draw a triangle in the sand.

Desiderius Erasmus, 1511

I wish to be a good grammarian. Religious discord depends on nothing except ignorance of grammar.

Joseph Scaliger, 1603/7

Whenever I find an unknown source, I am so overjoyed that I cannot sleep.

Gu Yanwu, 1660

One can understand only what one has created.

Giambattista Vico, 1725

History is on every occasion the record of what one age finds worthy of note in another.

Jacob Burckhardt, 1885

A linguistic system is a series of differences of sound combined with a series of differences of ideas.

Ferdinand de Saussure, 1916

All fairy tales are of one type in regard to their structure.

Vladimir Propp, 1928

Music was destined to reach its culmination in the likeness of itself.

Heinrich Schenker, 1935

A fully adequate grammar must assign to each of an infinite range of sentences a structural description indicating how this sentence is understood by the ideal speaker-hearer.

Noam Chomsky, 1965

Every discourse, even a poetic or oracular sentence, carries with it a system of rules for producing analogous things and thus an outline of methodology.

Jacques Derrida, 1986

To understand what understanding a film means, we must determine what previous knowledge someone needs to have in order understand filmic 'language'.

Christian Metz, 1992

And I think to myself, what a wonderful world.

Bob Thiele, George Weiss, Louis Armstrong

Contents

List of Illustrations xi
List of Table xi
Preface to the Paperback Edition xii
Preface xiii

1. Introduction: The Quest for Principles and Patterns 1

2. Antiquity: The Dawn of the 'Humanities' 13
2.1 Linguistics: The Birth of Grammar 14
2.2 Historiography: The Source Problem and the Form of the Past 21
2.3 Philology: The Problem of Text Reconstruction 31
2.4 Musicology: The Laws of Harmony and Melody 37
2.5 Art Theory: The Visual Reproduction of the World 44
2.6 Logic: The Rules of Reasoning 52
2.7 Rhetoric: Oratory as a Discipline 58
2.8 Poetics: The Study of Literature and Theatre 65
Conclusion: Common Patterns in the Humanities of the Ancient World 71

3. The Middle Ages: The Universal and the Particular 74
3.1 Linguistics: From Rules to Examples 75
3.2 Historiography: Universal History and Formal Transmission Theory 84
3.3 Philology: Copyists, Encyclopaedists, and Translators 103
3.4 Musicology: The Formalization of Musical Practice 109
3.5 Art Theory: Following and Breaking Rules 118
3.6 Logic: The Laws of True Syllogistics 124
3.7 Rhetoric and Poetics: A Motley Collection of Rules 130
Conclusion: Innovations in the Medieval Humanities 139

4. Early Modern Era: The Unity of the Humanities 142
4.1 Philology: Queen of Early Modern Learning 143
4.2 Historiography: Philology Spreads and the Secularization of the World View 161
4.3 Linguistics and Logic: Under the Yoke of Humanism 184
4.4 Musicology: The Missing Link between Humanism and Natural Science 198
4.5 Art Theory: A Turning Point in the Representation of the Visible World 211
4.6 Rhetoric: The Science of Everything (or Nothing)? 229
4.7 Poetics: Classicism *in extremis* 233
Conclusion: Was there Progress in the Early Modern Humanities? 240

5. Modern Era: The Humanities Renewed 250
5.1 Historiography: The Historicization of the World 251
5.2 Philology: A Completed Discipline? 272
5.3 Linguistics and Logic: The Laws of Language and Meaning 281
5.4 Musicology: The Systematic Versus the Historical 301
5.5 Art History and Archaeology: Towards a Visual Philology 311
5.6 Literary and Theatre Studies: The Curious Disappearance of Rhetoric and Poetics 326
5.7 The Study of All Media and Culture: From Film Studies to New Media 339
Conclusion: Is there a Break in the Modern Humanities? 346

6. Conclusions: Insights from the Humanities that Changed the World 352

Appendix A: A Note about Method 364
Appendix B: The Most Important Chinese Dynasties 366
Index 367

List of Illustrations

1. *Doryphoros* by Polykleitos, Roman copy of the Greek original from *c.*450 BCE. 46
2. Detail of a wall painting in cave I, second half fifth century, Ajanta, India. 49
3. Gu Kaizhi, detail of the *Admonitions of the Instructress to the Court Ladies*, end of the fourth century. (Possibly a copy from the Tang Dynasty.) 51
4. Masaccio, *The Holy Trinity*, *c.*1425, Santa Maria Novella, Florence. 212
5. Example of a tiled floor with objects in perspective, after the description in *De Pictura*, Leon Battista Alberti, 1435. 213
6. Piero della Francesca, detail of *The Flagellation of Christ*, *c.*1450, Urbino. 214
7. Albrecht Dürer, artist with mechanical grid, in *Underweysung der Messung*, 1525. 217
8. Frans Hals, *Malle Babbe*, *c.*1630. 221
9. Adriaen Brouwer, *The Bitter Tonic*, *c.*1635. 221
10. Leon Battista Alberti, *Palazzo Rucellai*, 1446–51, Florence. 225
11. Leonardo da Vinci, *Vitruvian Man*, *c.*1487. 226
12. A stemma for six hypothetical texts A, B, C, D, E, and F. 277
13. Grimm's Sound Shift Law. 282
14. A context-free grammar for a small part of English. 291
15. Giovanni Morelli, study of the depiction of ears by eight different Renaissance artists, in *Kunstkritische Studien*. 313
16. Jan van Eijck, *Arnolfini Wedding*, 1434. 317

List of Table

1. Truth table for conjunction, disjunction, implication, and equivalence. 54

Preface to the Paperback Edition

Five years after the publication of the original Dutch book (*De Vergeten Wetenschappen*, 2010) and two years after its English translation (*A New History of the Humanities*, 2013) the academic landscape has changed. The "History of the Humanities" has developed from a non-existing field into a flourishing discipline with its own journal (*History of Humanities*), an annual conference (*The Making of the Humanities*), an academic society (*Society for the History of the Humanities*) and several monographs. An increasing number of universities across the globe are teaching the History of the Humanities on par with the History of Science, and the premier journal in the History of Science, *Isis*, has recently devoted a special Focus section on *The History of Humanities and the History of Science* (June 2015). It seems that the humanistic disciplines have been brought back to their rightful place in the family tree of knowledge.

Nevertheless, in terms of funding and student numbers the humanities continue to be under immense pressure. Few people realize that without the humanistic disciplines the entire academic system would collapse. A general history of the humanities and their relations to the sciences remains thus more urgent than ever. But there are also signs of hope: the idea of the unicity of humanities and science has hit a nerve among natural scientists. The June 2015 issue of *Scientific American* dedicated a column to *A New History of the Humanities* concluding that "Regardless of which university building scholars inhabit, we are all working toward the same goal of improving our understanding of the true nature of things, and that is the way of both the sciences and the humanities, a *scientia humanitatis*."

However intriguing these developments are, my greatest source of inspiration remains the animated discussions with my readers, especially with the students who use my work as a textbook. The book and its translations have found their way in courses in philosophy, history of science, cultural history, media studies, literary criticism, and in liberal arts programs. I am most grateful for the feedback and criticism I received from my readers. News and updates about the book and about the history of the humanities in general can be found on the weblogs devergetenwetenschappen.blogspot.com (for all translations) and historyofthehumanities.wordpress.com (for the English translation).

Preface

The humanities are under pressure all over the world. Once the pinnacle of education and intellectual development, today they suffer from a serious image problem. Disciplines like philology, art history, linguistics, literary studies, and musicology are seen as a luxury pastime which is of little use to society and even less to the economy. Arguments in favour of the humanities emphasize their importance for critical thinking,[1] cultural consciousness,[2] historical responsibility,[3] and for creating competent, democratic citizens.[4] While these arguments may all be true, humanistic scholars seem to overlook the possibility that the very assumption behind the image problem may be wrong. They appear to have taken for granted that their field does not contribute to the economy or society. They stress that the humanities do not solve concrete problems and that their value lies elsewhere.[5] Yet a quick glance at the history of the humanities shows a different picture. Not only did humanistic insights change the world, many of these insights dealt with concrete problems and resulted in applications in entirely unexpected fields. As if humanists have no clue about their own history, these insights and applications have even been credited to the sciences.

Take Panini's discovery from around 500 BCE that the Sanskrit language was based on very precise rules, a so-called 'grammar'. This not only changed our perspective on language, it also contributed to the development of the first programming languages many centuries later. Or take Lorenzo Valla's fifteenth-century demonstration that 'The Donation of Constantine' was a fake. Suddenly the papal claim to worldly power appeared to be based on fiction. When in the seventeenth century Joseph Scaliger discovered that the earliest Egyptian kings were older than the presumed age of the earth, this led to a secular world view where it was no longer theologians but citizens who had the last word. And when Leon Battista Alberti gave the first description of linear perspective, it not only *literally* changed our view of the world but it led to revolutionary architectural design techniques as well. A more recent humanistic breakthrough is the discovery that languages in Europe and Asia are related via sound shift laws. This pointed to a common linguistic origin known as Indo-European which completely changed our view of the relationships between peoples, for better and worse.

[1] See e.g. Stefan Collini, *What Are Universities For?*, Penguin Books, 2012.

[2] Edward Said, *Orientalism*, Vintage Books, 1978; John Carey, *What Good Are the Arts*, Oxford University Press, 2006.

[3] Jörg-Dieter Gauger and Günther Rüther (eds), *Warum die Geisteswissenschaften Zukunft haben!*, Herder, 2007.

[4] Martha Nussbaum, *Not for Profit: Why Democracy Needs the Humanities*, Princeton University Press, 2010.

[5] Jonathan Bate (ed.), *The Public Value of the Humanities*, Bloomsbury Academic, 2010.

These examples of humanistic insights and discoveries are just the tip of the iceberg. The list also includes findings by Chrysippus, Sibawayh, Chen Kui, Ibn Khaldun, De Laet, Bopp, Lachmann, Propp, Panofsky, Todorov, and many more. Few of these scholars are widely known today, yet their insights have changed the world.

Why then is there no overview of the history of the humanities, while there are dozens of overviews of the history of science? Such a history is badly needed, if only to highlight the feats and deeds of humanistic inquiry. This book wants to fill this gap and offers the first overarching history of the humanities from Antiquity to the present day. The study of literature, art, theatre, and music is not only important to keep alive the great works from the past. In their critical investigations, humanists have revealed patterns that led to entirely unforeseen applications and insights. Many of these are forgotten. The aim of this book is to bring them back to life.

My personal interest in the history of the humanities was triggered when I discovered the lack of it. This happened in 1980 when I read some classics on the history and philosophy of science at the age of fifteen. I was immediately fascinated by the topic and wanted to read similar books on the history of 'humanistic disciplines'—which in the non-Anglophone world are also referred to as 'sciences' (see chapter 1). To my disappointment I could not find any work that focused on the history of the humanities. In the meantime I chose a different academic path. But the interest in the general history of the humanities remained: how did the study of language, music, art, literature, theatre, and the past originate, and how did they develop? Why is their history treated so differently from the history of the sciences? Apart from studies on the history of individual humanistic disciplines, there was no overview. Yet it had to be possible to write such a general history of the humanities.

My first, very modest start was the organization of a symposium on the history of the humanities together with Jaap Maat at the Three Societies' Meeting organized in Oxford, 2008. The reactions were encouraging enough to organize a much larger conference, now also together with Thijs Weststeijn, on the history of the humanities in the early modern period. This conference, The Making of the Humanities (23–5 October 2008), brought together around eighty historians and other scholars (with keynote speakers: Floris Cohen, David Cram, Anthony Grafton, and Ingrid Rowland) who shared a common interest in the comparative history of the humanities.[6] Although the conference was organized at short notice, the response far exceeded our expectations. It was clear that the subject triggered interest. The day after the conference I started writing this book.

It was as if I had entered a new world. I 'discovered' the Chinese theories of rhetoric and poetics. I learned about Arab historiography with its precise *isnad*

[6] The conference has become an annual event, the next one being held in Baltimore in October 2016. Three edited volumes with papers from the conferences have been published: Rens Bod, Jaap Maat and Thijs Weststeijn (eds.), *The Making of the Humanities, Volume I: Early Modern Europe*, Amsterdam University Press, 2010; Rens Bod, Jaap Maat and Thijs Weststeijn (eds.), *The Making of the Humanities, Volume II: From Early Modern to Modern Disciplines*, Amsterdam University Press, 2012; Rens Bod, Jaap Maat and Thijs Weststeijn (eds.), *The Making of the Humanities, Volume III: The Modern Humanities*, Amsterdam University Press, 2014.

method. And I read about the fascinating Indian drama theory which was even more detailed than its Greek counterpart. And not to forget the African historiography with its specialists of the word, and the humanistic text-critical methods that continue to be used even today. How could it be that these fields of knowledge were never brought together? Again and again I fell in love with a period, a region, or a person. In the summer of 2010 the Dutch manuscript was finished and published in the same year.[7] The book was well received in the Netherlands and Belgium, and a year later the text was translated into English by Lynn Richards. Her translation forms the basis of the current, updated, and extended work.

I am indebted to a great many people. Floris Cohen was most encouraging and helpful while I was writing both the Dutch and the English versions of this book. His work on the comparative history of science is a continuous source of inspiration for my own work. Also Paul Voestermans's enthusiastic interest in the book's various stages and his many suggestions, also on the English version, have been of great importance. I am greatly indebted to Peter Engelfriet and Wim Raven for their excellent comments and suggestions on the Chinese and Arabic humanities respectively. They have been truly wonderful. Dirk van Miert is gratefully acknowledged for checking the Latin and for several other suggestions. OUP's editor Christopher Wheeler has been most helpful with his advice and with finding the best possible readers for the last versions of the book. It was also a great pleasure to work with Leston Buell on the copy-editing of this book. As always, all errors are my responsibility. I also wish to express my gratitude to Ruud Abma, Pieter Bakker, Wouter Beek, Johan van Benthem, Albert Blankert, Anne Blankert, Livio Bod, Bart van den Bosch, Jeroen Bouterse, David Cram, Marjolein Degenaar, Kees van Dijk, John van Eck, Els Elffers, Mark Hannay, Theo A. J. M. Janssen, Theo M. V. Janssen, Bart Karstens, Bram Kempers, Ellen Kempers, Wessel Krul, Charley Ladee, Michiel Leezenberg, Fenrong Liu, Jaap Maat, László Marácz, Marita Mathijsen, Daniela Merolla, Hajo G. Meyer, Wijnand Mijnhardt, Jan Noordegraaf, Peter van Ormondt, Henk Schultink, Floris Solleveld, Martin van Staveren, Siep Stuurman, Stan Verdult, Rienk Vermij, Gerlof Verwey, Thijs Weststeijn, Co Woudsma, Joost van Zoest, OUP's anonymous readers, and others I may have forgotten to mention.

Most important of all are Daniela and Livio. Without them this book could never have been written. Daniela's knowledge of the African humanities is unsurpassed and she saved me from falling into many of the Eurocentric pitfalls, while Livio gave most valuable feedback on the first version of this book. Finally, I was greatly inspired by two marvellous places during the last stages of this work. First, Guadaiona, in the middle of the frightening but magnificent *calanchi* of Bagnoregio where I lost my heart. And Colle San Giacomo near the tall and graceful Monte Velino with amazing vistas on the town of Tagliacozzo. Every view from the window was a blissful experience which kept me going.

[7] Rens Bod, *De Vergeten Wetenschappen: Een Geschiedenis van de Humaniora*, 1st edition 2010, 2nd and 3rd revised and updated editions, 2011, 4th updated edition, Prometheus, 2012, 5th updated edition, Prometheus 2015.

It is impossible to be a specialist in all fields of the history of the humanities. For a new undertaking like this, criticism is essential. I thus call on the reader: if you find on the basis of your own expertise any conspicuous omission, inaccuracy or plain error, I would be grateful if you would let me know, either via email to <rens.bod@gmail.com> or via the weblog of this book: <http://historyofthehumanities.wordpress.com>. Of course, I reserve the right to disagree with you, but I will gratefully acknowledge every serious suggestion in a revised edition of this book—were such ever to appear.

1

Introduction: The Quest for Principles and Patterns

This is the first overarching history of the humanities in the English language.[1] Unlike the sciences and the social sciences, the humanities lack a general history. This is puzzling if we realize that for many centuries there was no distinction between humanities and science. Whether one wanted to grasp the secrets of the human or the natural world, it was part of the same intellectual activity. Pythagoras investigated both music and mathematics, and al-Biruni was both a historian and an astronomer. Even the icons of the scientific revolution—Galileo, Kepler, and Newton—were engaged in philology and the study of the natural world. This raises the question as to what extent the distinction between the humanities and science is essential or artificial. Where do their research methods differ? When did they develop in different directions? Is the famous Two Cultures debate sparked off by C. P. Snow in 1959 just a phenomenon of the last fifty years or has it existed before?[2] And have insights and discoveries in the humanities ever led to 'scientific' breakthroughs? A historiography of both the humanities and the sciences is indispensable in answering such questions.

What are the humanities? It is like the notion of 'time' in St Augustine: if you don't ask, we know, but if you ask, we are left empty handed.[3] Since the nineteenth century the humanities have generally been defined as *the disciplines that investigate the expressions of the human mind*.[4] Such expressions include language, music, art, literature, theatre, and poetry. Thus, philology, linguistics, musicology, art history, literary studies, and theatre studies all belong to the realm of the humanities, unlike

[1] The term 'humanities' is ambiguous in the Anglophone world. While today's use of the term commonly refers to a branch of academic *disciplines* such as literary studies, historiography, musicology, art history, theatre studies, and the like, it can also be used to refer to the *subjects* studied by these disciplines, such as literature, music, art, theatre. And sometimes the two meanings are even conflated. In this book I will use the term humanities to refer to the disciplines, or better (as I will argue below) to the *studies* of literature, music, art, theatre, etc. This use of the term corresponds to the German *Geisteswissenschaften* ('sciences of the spirit'), the Italian *scienze umanistiche* ('humanistic sciences'), or the Dutch *alfawetenschappen* ('alpha sciences').

[2] Charles Percy Snow, *The Two Cultures and the Scientific Revolution*, Cambridge University Press, 1959.

[3] St Augustine, *Confessions*, Book XI, chapter XX.

[4] Wilhelm Dilthey, *Einleitung in die Geisteswissenschaften: Versuch einer Grundlegung für das Studium der Gesellschaft und der Geschichte*, 1883, reprinted by Teubner, 1959. For an English translation, see Wilhelm Dilthey, *Selected Works*, volume 1, translated and edited by Rudolf Makkreel and Frithjof Rodi, Princeton University Press, 1991.

the study of nature, which belongs to the domain of science (such as physics, astronomy, chemistry, and biology). Similarly, the study of humans in their social context is one of the social sciences (such as sociology, psychology, anthropology, and economics). But these definitions are unsatisfactory. Mathematics is to a large extent a product of the human mind, and yet it is not considered a humanistic discipline. A pragmatic stance may be more workable: the humanities are the disciplines that are taught and studied at humanities faculties. According to this pragmatic 'definition', the humanities currently include linguistics, musicology, philology, literary studies, theatre studies, historical disciplines (including art history and archaeology), as well as more recent fields such as film studies and media studies. In some countries theology and philosophy are also taught in humanities faculties, whereas in others they are faculties in their own right.

The humanities come in different forms. They have a memory function by keeping alive the works from the past and the present, often through collections. They have an educational function by teaching these works to new generations. They also have a critical function by interpreting these works for the public at large. In addition to all this, the humanities have a research function by asking questions and posing hypotheses regarding humanistic artefacts. While often intertwined, these functions have not been equally prominent in all historical periods. Yet, as we will see, the research function of the humanities is conspicuous in all eras. It is exactly this *empirical* dimension of the humanities that forms the main focus of the current book.

This raises an immediate conceptual problem—to what extent can expressions of the human mind, such as language, literature, music, and art, be called 'empirical' if they are created by people? Is it not the case that the humanities study primarily 'the world in the mind' rather than an external one? Indeed, the products of the humanities have been created by people, but when the products manifest themselves in the form of manuscripts, pieces of music, literary works, sculptures, grammar books, plays, poems, and paintings, they are obviously just as open as other objects to empirical research and the development of hypotheses. We will see that since Antiquity humanistic material has indeed been exposed to hypotheses and evaluation relating to assumed patterns and interpretations.

In this book I show how scholars, from the ancient world to today, have explored humanistic material—language, texts, music, literature, theatre, art, and the past—and what insights they gained from it. I want to stress, perhaps unnecessarily, that a history of the humanities is not about the history of music, art, or literature, but about the history of musicology, art theory, and literary theory. This history begins with the birth of the first humanistic activities in Antiquity. It is often assumed that the humanities did not form a separate field of study before the nineteenth century.[5] In part this is true—at least for musicology which was until the

[5] See e.g. Hans-Georg Gadamer, *Wahrheit und Methode: Grundzüge einer philosophischen Hermeneutik*, Mohr, 1960, translated into English as *Truth and Method*, by Joel Weinsheimer and Donald Marshall, Continuum, 1975, pp. 3ff. See also Albert Levi, *The Humanities Today*, Indiana University Press, 1970. And see also Jörg-Dieter Gauger and Günther Rüther (eds), *Warum die Geisteswissenschaften Zukunft haben!*, Herder, 2007.

eighteenth century (also) regarded as a mathematical activity in the so-called *quadrivium* of the *artes liberales* (see 2.4 and 4.4). But it should not be forgotten that already around 1700 the conceptual distinction between a science of the human and a science of the natural was worked out by Giambattista Vico (see 4.2). And as early as from the fourteenth century onwards, we find a branch of thriving disciplines known as the *studia humanitatis* from which the (early) modern humanistic disciplines emerged (see 4.1). We can even discern an unbroken tradition in the study of humanistic material that goes back to the Roman *artes liberales* and further to the Hellenistic curriculum known as *enkyklios paideia* (see 2.7). In writing a general history of the humanities, we thus need to start where we first find these studies—in Antiquity.

But what would be the reason to separate the history of the humanities from the history of other disciplines—be it the natural or social sciences or even from the general history of knowledge? The endeavour to write a history of *all* disciplines was attempted by George Sarton in the 1930s.[6] However, the result of his work, which was based on a highly positivistic concept of progress, did not go beyond the fourteenth century, and even within that period the humanities occupied an extremely marginal position in Sarton's history. Although he included linguistics and musicology to some extent, he left out other humanistic disciplines such as art history and literary theory. According to Sarton, unlike the study of music, the history of the visual arts (painting, architecture, and sculpture) only throws light upon scholarship from 'the outside' and does not contribute to academic 'progress'.[7] Sarton did not elaborate any further on this issue, but it seems that he was pointing to the history of art *itself* rather than art history as a discipline.

We will see that art history, like literary theory, is an essential component in the history of the humanities. From as early as the third century BCE, Alexandrian scholars tried to shed light on artists' quests for the 'correct' proportions when depicting reality. In the first century CE, Pliny described in detail how classical sculptors kept to exact proportions, for example between the sizes of the head and the body, and Vitruvius reported on the proportions in classical temples. Surprisingly enough, these ratios correspond with the proportions that were found in the study of musical harmony (by Pythagoras, Ptolemy, and others). Similar relationships were discovered in the study of Indian and Chinese art and music (for instance by Bharata Muni and Liu An). We should therefore include the study of music (musicology) and of art (art history) if we want a proper understanding of the historical development of the humanities. Of these two, however, Sarton only addressed musicology, and then primarily because of its importance to scientific progress. Sarton's work is thus not a general history of scholarly disciplines, let alone the humanities. The same can be said of Hans-Joachim Störig's overview

[6] George Sarton, *Introduction to the History of Science*, 3 volumes, Williams and Wilkins, 1931–1947.

[7] Sarton, *Introduction to the History of Science*, volume 1, p. 5.

of the history of science in 1953.[8] Although his work was not based on a positivistic belief in progress, Störig included only linguistics and historiography as humanistic disciplines.[9]

The general history of the humanities has thus remained underexposed in terms of both content and period. This is all the more striking because a large number of histories of science have been written from the nineteenth century onwards.[10] And, more recently, general histories of the social sciences have also been produced.[11] In other words, from a historiographical point of view, a general history of the humanities is conspicuous by its absence. How can this manifest gap in intellectual history be understood? One explanation, which will emerge from this book, is that the humanities have become increasingly fragmented over the last two centuries—unlike the sciences, where the opposite seems to have taken place. Current historiographies of science usually take physics as the central discipline, alongside which other sciences (chemistry, biology, and geology) are discussed and compared. Such an approach is much harder if not impossible to maintain for the history of the humanities. There is no central humanistic discipline on which all other disciplines can be modelled—although we will see in this book that there are common humanistic practices and methodologies. So far, the histories that have been written are almost exclusively of *single* humanistic disciplines, such as histories of linguistics,[12] histories of literary theory,[13] and histories of historiography.[14] Connections between methods and principles among different disciplines have rarely been made. This has led to peculiar situations. For instance, in seventeenth-century England, William Holder wrote both linguistic and musicological works that were interrelated, but he is usually treated as two different people. And during

[8] Hans-Joachim Störig, *Kleine Weltgeschichte der Wissenschaft*, Fischer, 1953.

[9] Additionally there are the great works by Michel Foucault (*The Order of Things*, 1966) and Georges Gusdorf (*Les Sciences humaines et la pensée occidentale*, 1967), but these are more of a philosophical nature, and focus on the social or 'human' sciences rather than on the humanities: linguistics and historiography are included, but no other humanistic disciplines are.

[10] Among the many histories of the natural sciences, one of the first is William Whewell, *History of the Inductive Sciences*, 3 volumes, Parker, 1837. Later classics are Stephen Mason, *A History of the Sciences*, Macmillan, 1962; William Dampier, *A History of Science and Its Relation to Philosophy and Religion*, Cambridge University Press, 1966. Some more recent ones include James McClellan and Harold Dorn, *Science and Technology in World History: An Introduction*, Johns Hopkins University Press, 1999; Frederick Gregory, *Natural Science in Western History*, Wadsworth Publishing, 2007; Patricia Fara, *Science: A Four Thousand Year History*, Oxford University Press, 2009; H. Floris Cohen, *How Modern Science Came into the World*, Amsterdam University Press, 2010.

[11] Examples of histories of the social or human sciences (not be confused with the humanities) are Roger Smith, *The Norton History of the Human Sciences*, W. W. Norton, 1997; Scott Gordon, *The History and Philosophy of Social Science: An Introduction*, Routledge, 1993; Theodore Porter and Dorothy Ross, *The Cambridge History of Science, Volume 7: The Modern Social Sciences*, Cambridge University Press, 2003.

[12] See e.g. R. H. Robins, *A Short History of Linguistics*, Longman, 1997; Esa Itkonen, *Universal History of Linguistics*, John Benjamins, 1991; Pieter Seuren, *Western Linguistics: An Historical Introduction*, Blackwell Publishers, 1998.

[13] See e.g. Richard Harland, *Literary Theory from Plato to Barthes*, Palgrave Macmillan, 1999; Harry Blamires, *A History of Literary Criticism*, Macmillan, 1991.

[14] See e.g. Ernst Breisach, *Historiography: Ancient, Medieval and Modern*, The University of Chicago Press, 2007; Markus Völkel, *Geschichtsschreibung: eine Einführung in globaler Perspektive*, Böhlau Verlag, 2006; Daniel Woolf, *A Global History of History*, Cambridge University Press, 2011.

the Chinese Han Dynasty, Sima Qian developed a narrative scheme that related to both historiography and poetics, yet he is only known as a historian.

This means that a *comparative*, *interdisciplinary* history of these fields is essential. And moreover, we cannot restrict ourselves to one region. It emerges that there is almost nowhere that the history of the humanities can be considered in isolation. Panini's Indian linguistics, for instance, first filtered through to China and Islamic civilization, and after that had profound effects on the study of language in Europe. As far as the discovery of patterns is concerned, historians in Greece (Herodotus and Thucydides), China (Sima Qian), and Africa (Ibn Khaldun) all 'discovered' the constantly recurring historical pattern of rise, peak, and decline. In spite of this, the historiographies of the separate humanistic disciplines are often confined to the Western tradition, with no attempt to unravel the fascinating interactions between the different areas.[15] This book endeavours to reveal this interplay at least to a degree, although a history of the humanities from a global perspective is difficult because many sources are not yet accessible or remain untranslated.[16] I realize that I devote a disproportionate amount of attention to the Western humanities in this book. But besides Europe and the USA I will also deal with the humanities in India, China, Islamic civilization, and Africa, with some excursions to Byzantium and the Ottoman Empire. Any future world history of the humanities should also encompass other regions—from pre-Columbian America to Japan (which I only briefly discuss in chapter 6).

How, though, can we compare the different humanities disciplines not only across periods but also across regions? The contexts of these disciplines, as well as their concepts, can differ endlessly. While it seems problematic to directly compare linguistics, poetics, art theory, musicology, and historiography across regions and periods, we may be able to compare the underlying *methods* used in these disciplines as well as the *patterns* found with these methods. Humanities scholars typically employed one or more *methodological principles* to investigate their humanistic material. And in using these principles they searched for some kinds of *patterns* in the material. These principles and patterns were sometimes literally mentioned while at other times they remained implicit but could often be extracted from the texts. While the contexts of these humanistic studies differ immensely across disciplines, regions, and periods, there appear to be deep commonalities at the level of principles used and patterns found. A comparison between humanistic practices across disciplines, regions, and periods thus seems to be possible in terms of these two concepts.

[15] Even the nine-volume *Cambridge History of Literary Criticism*, Cambridge University Press, 1989–2005, is restricted to Western literary criticism. Not all historical overviews suffer from Western limitation, for example Völkel, *Geschichtsschreibung: eine Einführung in globaler Perspektive*, Woolf, *A Global History of History*, and Itkonen, *Universal History of Linguistics*, aim at a worldwide coverage, albeit for one humanistic discipline only. In the history of the natural sciences, a worldwide perspective is also gaining ground, such as McClellan and Dorn, *Science and Technology in World History: An Introduction*, and Floris Cohen, *How Modern Science Came into the World*, Amsterdam University Press, 2010.

[16] See e.g. Khaled El-Rouayheb, 'Opening the gate of verification: the forgotten Arab-Islamic florescence of the 17th century', *International Journal of Middle East Studies*, 38, 2006, pp. 263–81.

But there is another reason, too, why it makes sense to compare principles and patterns across cultural contexts. On the way, it became crystal clear to me that many of the methods invented in very specific disciplines had been applied by humanists to new problems in other disciplines (often from different periods and regions) *without* taking into account their original religious or cultural contexts. For example, Panini's formal grammar method (see 2.1) originally served the Vedic ritual practice, but when it was (re)discovered in nineteenth-century Europe, his grammar was stripped of its ritual connotations and was used by 'modern' linguists for their own theories of language (see 5.3). A similar thing occurred in the Arab world where the eighth-century *isnad* method of reconstructing the words and deeds of the Prophet (*hadith*) was later used by historians such as Al-Dinawari, Al-Tabari, and Al-Masudi as a successful method for historical source reconstruction without religious connotations (3.2). Their method may even have influenced textual criticism in Renaissance Europe (see 4.1). Thus, the sophisticated source reconstruction that initially had a religious purpose could be applied to non-religious source reconstructions as well, and this was done by scholars themselves. Very specific methods that were developed for solving one particular problem in the humanities in a specific context could be cut loose and reinserted into a different context for solving other, new problems.

The focus on principles and patterns also allows us to discern new patterns *not* found by humanities scholars themselves. These I will call *metapatterns*. For example, it appeared that there was a process from descriptive to prescriptive approaches in all humanistic disciplines in Antiquity. The regularities in Greek tragedies found by Aristotle were quickly turned into prescriptive rules by later poeticists such as Horace (see 2.8 for details). And the mathematical proportions found in classical Greek art and architecture by Pliny and Vitruvius were taken as normative prescriptions by later art theorists (see 2.5, 4.5). The same can be observed in Chinese and Indian poetics and art theory. Surprising enough, this process was reversed at the end of the early modern period—that is, it went from prescriptive back to descriptive again, in Europe and China alike. Another meta-pattern that emerged, is that the time pattern in historical writings from a particular region corresponded with the time pattern used in the canonical texts of that region. This was found in China, Islamic civilization, Europe, Africa (Ethiopia), and India (see 3.2). Thus, by using the concepts of principles and patterns, it is possible to find novel metapatterns across disciplines and even regions. Next, these patterns can be interpreted again in the context of each specific region, and be understood by the cultural products themselves, e.g. the canonical texts of a civilization. But without the concepts of principles and patterns to begin with, it would be hard to find such metapatterns.[17]

[17] My way of working thus differs from Geoffrey Lloyd and Nathan Sivin, *The Way and the Word: Science and Medicine in Early China and Greece*, Yale University Press, 2003, and from Geoffrey Lloyd, *Disciplines in the Making*, Oxford University Press, 2009. These authors do not introduce additional concepts in their cross-cultural comparisons, and consequently find more divergences than common patterns. If we remain too specific, we will not discover commonalities. On the other hand, we should of course make sure that even our most general concepts still remain historically meaningful.

In this book I thus concentrate on the apparently unbroken strand in the humanities that can be identified as *the quest for patterns in humanistic material on the basis of methodical principles*. This strand has not been the only thread in the history of the humanities, but it can be found in all disciplines, periods, and regions. Moreover, it gives my historiography a degree of cohesion alongside which I can also find a place for other approaches that are not searching for patterns. One of the conclusions in chapter 6 will be that there is only a gradual differentiation between the humanities and the sciences, and that there is a continuum in the nature of the patterns and their possible 'exceptions'. The history of the humanities appears to be the missing link in the history of science.

My approach to the history of the humanities challenges a very dominant view in the *philosophy* of the humanities. This view, initiated by Wilhelm Dilthey, contends that the humanities (*Geisteswissenschaften*) are concerned primarily with *verstehen* (understanding), whereas science (*Naturwissenschaften*) is about *erklären* (explaining).[18] According to Dilthey, humanities scholars would be failing if they observed, counted, measured, or hunted for apparent regularities. What they should be doing is searching for the motives and intentions of important historical figures. Laying bare these *inner* mainsprings is more important than studying the *external* manifestations of the human mind. In this context one also uses the distinction introduced by Wilhelm Windelband between an *idiographic* approach to knowledge (which is the study of the unique, the special) and a *nomothetic* way of studying (which seeks to generalize).[19] Although this vision has been very influential in the philosophy of the humanities,[20] it proves to bear less relation to humanistic practice. Even when Dilthey's vision was gaining ground (in the late nineteenth and early twentieth century), there were both idiographic and nomothetic practices in every humanistic discipline, and the latter were often dominant. We have found nomothetic, pattern-seeking components not only in the linguistics of e.g. de Saussure and Jakobson but also in the philology of Lachmann, the musicology of Schenker, the literary theory of Propp, the art history of Wölfflin, and the historiography of the *Annales* school, just to name a few. In spite of Dilthey's and Windelband's constitutive recommendations, there was a boom in efforts to search for and find patterns in the humanities. The fact that the view of the humanities that Dilthey's and Windelband's works represented was nevertheless influential springs primarily from the powerful identity it gave the humanities, which enabled them to differentiate and emancipate themselves from the up-and-coming natural sciences (see chapter 5). This book will, however, show that *the quest for principles and patterns in the humanities is a continuous tradition*. Historiography thus appears to be ideally suited to the refutation of philosophical visions.

[18] Dilthey, *Einleitung in die Geisteswissenschaften*, pp. 29ff.

[19] Wilhelm Windelband, *Geschichte und Naturwissenschaft*, 3rd edition, Heitz, 1904.

[20] See e.g. Gadamer, *Truth and Method*, pp. 6, 56ff. See also the anthology of (abridged) texts in the philosophy of the humanities, in Gauger and Rüther, *Warum die Geisteswissenschaften Zukunft haben!*. And see Gunter Scholz, *Zwischen Wissenschaftsanspruch und Orientierungsbedürfnis: zu Grundlage und Wandel der Geisteswissenschaften*, Suhrkamp, 1991.

Our comparative approach calls for a few further decisions to be made. I have opted for a 'classical' division into periods, namely Antiquity, the Middle Ages, the early modern era, and the modern era. A classification like this is unsatisfactory when I come to describe the humanities in China, India, Islamic civilization, and Africa. I will therefore also refer regularly to periodization within a particular region, for example the dynasties in China. Obviously, any periodization falls short when we want to establish links between civilizations, whether we opt for Chinese dynasties, the Greek Olympiads, or the ages of al-Tabari. Working within the traditional periodization, I address the history of the humanities primarily chronologically and by discipline, but I try to make as many comparisons as possible between disciplines and regions. In so doing I have concentrated more on the internal development of the humanities and less on their external cultural context, although I have tried to integrate these two as much as possible. I have selected a chronological structure rather than a treatment based on themes since it appears that a sequential overview of the humanities is a requirement for recognizing themes that go more across history.[21] We will therefore only reveal the underlying themes as we go along and not specify them beforehand, with one major exception that we meet in all periods and regions—the ongoing search for methodical principles and empirical patterns in humanistic material.

Any intellectual history is faced with a terminological-conceptual problem—which designations can best be used to describe scholarly activities in the past? Can we refer to the study of music and the study of art in the ancient world by using contemporary terms like 'musicology' and 'art history' without lapsing into misleading anachronisms? If we squeeze historical intellectual activities into a straight-jacket of present-day expressions, we run the risk of descending into an undesirable kind of 'presentism', in which the past is interpreted in terms of current concepts and perspectives. The preferred starting point is to use contemporary terms for an intellectual activity, for example *poetics* for the study of poetry and theatre in ancient Greece and *grammar* for the study of language. But sometimes these expressions are ambiguous, as is the case with *musica*, which can mean the study of music or the music itself (and more besides). Specific terms are lacking in other cases; for instance in the absence of anything better, the study of art was put under mineralogy and the application of materials in Pliny's *Naturalis historia*. In order to tackle these problems, at least to some extent, in many cases I mention the contemporary or regional designation of the humanistic activity concerned, and then replace it with what I consider to be the most coherent term. On some occasions this is historical (*poetics* for instance) and on others it is current (such as *musicology*). I do not believe that every form of presentism can be avoided—and it does not even need to be always avoided. It emerges that there is greater continuity between the humanities of Antiquity, the Middle Ages, and the modern era than could originally be suspected, both with respect to questions asked, methods used, and patterns found. Such continuity was also remarked upon with regard

[21] See e.g. John Pickstone, *Ways of Knowing: A New History of Science, Technology and Medicine*, Manchester University Press, 2000.

to the development of the natural sciences from the fourteenth century onwards (by Pierre Duhem and others), but it goes even further back in the humanities (see the conclusions of chapters 4 and 5). It is not just that 'humanities' reads and sounds better than the 'study of the products of the human mind' or the like—there also appears to be a historical justification for generalizing the term to cover different periods.[22] Some conceptual anachronisms are not only useful, they are also justifiable.[23]

The various different terms used to describe humanistic activities in regions outside Europe, for example India and China, are another problem. It is precisely for this reason that I will not so much focus on *disciplines*—the latter being a Western concept originating from the medieval universities—but rather on the *study* (or *studies*) of language, literature, art, music, theatre, and the past, which are found in all regions independent of whether these studies were carried out privately or academically, in a religious or in a secular context. It is only for convenience that I often refer to this activity (i.e. the study of language, literature, music or art, etc.) as a 'discipline'.

Thus as a whole, this book is about the history of *the methodological principles that have been developed and the patterns that have been found in the study of humanistic material (texts, languages, literature, music, art, theatre, and the past) with these principles.* The patterns found can consist of a regularity (often with exceptions) but they can also consist of a system of rules such as a grammar, or a system of interpretations, and they may even be similar to 'laws' such as the sound shift laws in linguistics and the laws of harmony in music. My concept of 'patterns' is in fact an umbrella that covers everything that can be found between inexact regularities and exact laws. For the time being I will not make this concept more specific because in my quest I do not want to exclude any 'pattern' in advance. The concept of 'pattern' will gradually crystallize, and will be compared with similar concepts in other sciences and disciplines. (I have given a more detailed description of my working approach in Appendix A.)

At this point it may be important to briefly come back to what this book is *not* about. In my history of the quest for principles and patterns I have not included the social sciences, not even those social sciences that have humanistic aspects, such as (parts of) geography, anthropology, sociology, and psychology. The reason is that there are already excellent books on the general history of the social sciences (see the references in footnote 11). What is missing in the historiography of knowledge is a general history of the humanities, which is exactly what the current work is about. This is not to say that I treat the humanities as a fixed bundle of activities that have remained unchanged since Antiquity. In fact, the study of language, texts, art,

[22] A justification of a generalization of the term humanities (*Geisteswissenschaften*) to other periods can also be found in Helmut Reinalter and Peter Brenner (eds), *Lexicon der Geisteswissenschaften*, Böhlau Verlag, 2011, pp. 258ff. And also in Julie Thompson Klein, *Humanities, Culture and Interdisciplinarity: The Changing American Academy*, SUNY Press, 2005, pp. 13ff.

[23] Cf. David Hull, 'In defense of presentism', *History and Theory*, 18, 1979, pp. 1–15. Nicholas Jardine, 'Uses and abuses of anachronism in the history of the sciences', *History of Science*, 38, 2000, pp. 251–70.

music, literature, rhetoric, and the past has changed dramatically, also under the influence of the upcoming social sciences in the modern era. At various points I will therefore discuss the influences from sociology, anthropology, and psychology on the humanities—from Comte, Weber, Ehrenfels, and Lévi-Strauss to Geertz. But I will not go into the history of these disciplines themselves. A general history of all sciences, i.e. of all knowledge-making disciplines, will have to await its publication.

My decision to focus on principles and patterns will, however, often lead to surprising choices. Many a famous humanist, historian, or philologist will be mentioned only briefly—if at all—while other scholars are dealt with at length. More than once I will describe a well-known work with a single sentence, not because I consider it unimportant or not influential, but because it did not contribute much to the quest for principles and patterns. Of course, another focus would lead to a different history of the humanities. Some humanistic activities will even fall largely outside the scope of my story. We find empirical searches less frequently in philosophy and theology, part of which I therefore do not address. For example, I go into the linguistics of Panini and Apollonius Dyscolus, but the 'language philosophy' of Confucius and Plato gets no more than an honourable mention. The part of theology that is concerned with investigating textual sources will be discussed in some detail, whereas speculative theology will only be mentioned in passing. Having said this, I will often go into the immense impact of theology and philosophy on the humanities, but these disciplines will not receive separate chapters—they simply play a role (almost) everywhere.

My concentration on principles and patterns does not mean that I omit once-only, fortuitous discoveries. Who can leave out from the history of the humanities the archaeological discovery of Troy by Heinrich Schliemann? A more interesting question, however, is whether this discovery was indeed coincidental or whether it was based on methodical principles. Additionally, I will also consider scholars who on the contrary sought to *refute* the concept of patterns—from the Pergamon *anomalists* in the third century BCE to the European *deconstructivists* in the twentieth century. Yet I will argue that seeking and finding patterns is timeless and ubiquitous, not only when observing nature but also when examining texts, art, poetry, theatre, languages, and music. Just as in all other scholarship, it is about trying to make a meaningful distinction between fortuitous and non-fortuitous patterns. Of course the humanities are also concerned with acquiring insights into our culture and its values, and through this, into our own humanity. This book shows that there is also a centuries-old humanistic tradition that seeks principles and patterns while at the same time giving us an understanding of what makes us human. For a long time this tradition was neglected and almost exclusively attributed to science.

My way of approaching the source material is explained in some detail in Appendix A. For the moment it suffices to say that when reading into the history of a certain discipline (of a certain period, region, and civilization), I usually started out with secondary material and worked from there to compare primary sources—which I read in their original languages as far as I could, otherwise in translations. I was surprised how much Chinese, Sanskrit, and Arabic, but also Ge'ez, Russian, and Turkish material was available in English, French, or German. While many of these sources had already been translated more than a century ago, they had never been brought together, let alone compared. They seemed to have remained in the specialized academic communities. My way of referring to these sources is as follows: if a source—primary or secondary—was not originally in English, I also tried to find an English translation to which I refer in the footnotes.[24] At the same time, I refer to the original source as well but only if I could read it, that is, when it was written in German, Dutch, English, French, Italian, Latin, or Spanish. For texts written in other languages, such as Sanskrit, Arabic, Russian, Chinese, Ge'ez, Fulani, Greek, and Turkish, I had to rely on translations. To verify the reliability of these translations, I consulted Arabists, Indologists, Sinologists, Africanists and other scholars, who also helped me out on a variety of other issues—I acknowledge them on the way and I have gratefully mentioned their names in the Preface.

In sum, in writing an overarching history of the humanities we are confronted with at least four major challenges: the problems of demarcation, of comparativism, presentism, and source selection. There are no straightforward solutions to these problems—if any at all—but we can make motivated choices and see how far we can get. For this book my choices have been the following:

Demarcation. No hard distinction can be made between the humanities, social sciences, and natural sciences. Yet since exactly those disciplines that make up the humanities have been historiographically neglected, we have to investigate their joint history before we can write a general history of science or knowledge.

Comparativism. While it seems problematic to directly compare the study of language, art, literature, music, theatre, and texts—the more if they come from different regions or periods—we can compare them at the level of *methods* used and *patterns* found. We can also do this because humanists themselves often (re)used methods and patterns from different disciplines, periods, and regions in new contexts.

Presentism. Using the present-day meaning of 'humanities' in earlier periods is a conceptual anachronism. But given the continuity between the 'humanities' in Antiquity, Middle Ages, and modern era, this conceptual anachronism is useful rather than harmful.

Source selection. If we want to write an overarching history of the humanities, as well as of other disciplines, it is practically impossible for a scholar to consult all sources in their original languages. We have to work together with other scholars to check the sources and to verify the reliability of translations.

Finally, for anyone who is puzzled by the word 'New' in my book's title, I have used it to contrast my work with previous histories. As I explained above, these previous works focus either on a single humanistic discipline or just on a couple of disciplines. Instead, this book covers eight humanistic disciplines, and several more from the twentieth century onwards. These disciplinary histories are intertwined, but to a certain degree they can also be read independently of one another.

[24] I made an exception for texts by well-known Greek and Latin authors whose English translations can be easily found in the Loeb Classical Library. These texts are quoted without reference to their English translations.

Someone who is only interested in the history of the study of music for instance, can confine themselves to reading the sections on musicology—a field for which no overarching historical overview has so far been written. But anybody who wants to experience the whole adventure of the quest for principles and patterns in the humanities from Antiquity till today will have to read the book from cover to cover. In order to give the reader something to go on, I end every chapter with a comparative conclusion of the period covered. If, after reading this book, someone feels the call to write a different history of the humanities, my objective will have been achieved. As the old Vossius said, 'after me there will be others, and again others, who will do it better than me.'[25]

[25] Gerardus Vossius, *Poeticae institutiones*, Praefatio, in *Opera*, III, 1647 (without page numbers): 'Exsurgent post me alii, et alii, qui felicius conentur.'

2

Antiquity: The Dawn of the 'Humanities'

The humanities came about in a variety of ways—as part of a ritual, as a consequence of philosophy, and sometimes as a political instrument. I discuss the methodical principles that have been developed for each of the humanities and the results (patterns) that have been obtained with these principles. I will moreover address the approaches that, conversely, reject the quest for patterns. More than once we find that there is a surprising correspondence between the humanities in different parts of the world—from China to India to Greece—yet there appears to have been little or no sharing of knowledge.

2.1 LINGUISTICS: THE BIRTH OF GRAMMAR

It is often said that all learning and science began in Greece. Not, though, the study of language. The history of linguistics begins not with Plato or Aristotle, but with the Indian grammarian Panini.[1] Admittedly the first dictionaries go even further back—Mesopotamian clay tablets in the second millennium BCE—and Confucius was philosophizing about language in the sixth century BCE, but the first attempt to systematically describe a language as a whole was made in India by Panini.

Panini and the discovery of grammar. Although Panini is recognized as the father of linguistics, we know virtually nothing about his life, not even in which century he lived. All we know is that he was born in Ghandara, in former India (currently Afghanistan), and that it must have been between the seventh and fifth centuries BCE.[2] His insights did not become known in Europe for over two thousand years, but when they did they changed Western linguistics for ever (see 5.3). What is so special about Panini's work and how does he differ from his Greek contemporaries—who, after all, knew nothing of him?

In his *Ashtadhyayi* ('Eight Books') Panini characterizes a language—in this case Sanskrit—as a system containing a finite number of rules that can be used to describe a potentially *infinite* number of linguistic utterances (sentences).[3] A system like this is currently called a 'grammar'. Panini's Sanskrit grammar might be finite, but it is very big indeed. It comprises 3,959 grammatical rules. Panini's grammar is truly heroic: as far as we know he was the first person to undertake the task of creating a complete system of rules with which it was possible to predict with precision whether or not a sequence of sounds represented a correct linguistic utterance in Sanskrit.[4] Panini's grammar is moreover unsurpassed. After two and a half thousand years, the efficacy of this system of nearly four thousand complex interconnected rules remains undisputed.[5]

Panini was not just a brilliant linguist. The underlying formalism and the method he developed in the *Ashtadhyayi* are just as interesting. Most of us were probably brought up using normative school grammar textbooks that quote a number of hard cases but rarely try to be comprehensive. We probably learned, for example, that *less* should not be used with countable nouns, but that *fewer* should be used instead (i.e. not 'ten items or less' but 'ten items or fewer'). And we were doubtless also taught that in conditional sentences with *if*, the correct form is 'if I had known . . . ', not 'if I would have known . . . '. But interestingly we are never

[1] Panini is also commonly transcribed as Pāṇini, where the accent lies on the first syllable ('Pā').

[2] S. Shukla, 'Panini', *Encyclopedia of Language & Linguistics*, 2nd edition, Elsevier, 2006. See also Paul Kiparsky, 'Paninian Linguistics', *Encyclopedia of Language and Linguistics*, 1st edition, Elsevier, 1993.

[3] For an (abridged) English translation of the *Ashtadhyayi*, with examples and commentaries, see Panini, *The Ashtadhyayi*, translated into English by Srisa Chandra Vasu, Nabu Press 2011 (reprint of 1923). For a German translation, see Otto von Böhtlingk, *Panini's Grammatik*, Buske Helmut Verlag, 1998 (reprint of 1839–40).

[4] Paul Kiparsky, *Panini as a Variationist*, The MIT Press/Poona University Press, 1979. See also George Cardona, *Panini: His Work and its Traditions*, Motilal Banarsidass, 1988.

[5] Esa Itkonen, *Universal History of Linguistics*, John Benjamins, 1991, pp. 83–4.

told about the more obvious cases of English grammar. For example, in English it is sometimes possible to leave out words, a phenomenon known as *ellipsis.* But no school English grammar book tells us that in conjoined sentences we can leave out a verb *only* in the second part of the sentence. So while it is correct to say 'John ate an apple, and Peter a pear', the following is incorrect: 'John an apple, and Peter ate a pear'. Apparently this is so trivial that there is no need to mention it, let alone have a rule for it. But in languages like Japanese and Korean, it is precisely the opposite: verbs can be left out in the first part of a conjoined sentence. This may seem unnatural to native English speakers, but it shows that the phenomenon of ellipsis in conjoined sentences needs to be explicitly addressed. Otherwise a grammar is incomplete, and ungrammatical sentences might not be recognized as such.

In contrast, Panini's approach in the *Ashtadhyayi* was to make his grammar system explicit and comprehensive. He devised a set of rules that, using a combination of a finite number of lexical units (the *word stems*), could cover all correct Sanskrit utterances.[6] Panini invented an ordered system of rules in order to achieve this goal. His rules are applied in a certain order so as to arrive at a linguistic utterance. This corresponds to the concept of an algorithm: a procedure that generates a result in a finite number of sequential steps. Panini's rules are also *optional*,[7] which means there is always more than one possible choice (otherwise it would only be possible to cover one linguistic utterance). He introduced a metarule in order to make his system consistent: 'If two rules conflict, the last rule prevails.'[8] Panini organized his grammar so that this metarule is always valid.

One of the most interesting ideas in Panini's system of rules is that a grammar rule can invoke itself. This is known as *recursion.*[9] Recursion is also known as the *Droste effect* and occurs for instance in an English sentence like 'she was harassed by the individual who was caught by the policeman who was spotted by the photographer'. We can make this sentence longer, as long as we want, by repeatedly applying the same grammatical relative clause rule and (optionally) putting in other words (e.g. by extending the sentence above with 'who was chased by a dog'). In this way Panini could really cover an infinite number of sentences with a finite number of rules and a finite lexicon.

Although we will not go into Sanskrit in any detail, we would like to illustrate the formalism of Panini's grammar. The fundamental productive units of his grammar are the optional rules referred to above, which have the following form: $A \rightarrow B/C\_D$. In effect this means that A can be replaced by B in the context of C and D, where A, B, C, and D are linguistic units that can range from word stems and word categories to whole word groups. In modern linguistics this type of rule is called

[6] Vidyaniwas Misra, *The Descriptive Technique of Panini: An Introduction*, Mouton, 1966, pp. 43ff.

[7] Panini's rule 2.1.11 (*vibhasa*) in the *Ashtadhyayi*. See also Itkonen, *Universal History of Linguistics*, p. 62.

[8] Panini's rule 1.4.2 in the *Ashtadhyayi*. For a discussion about this metarule, see Frits Staal, *Universals: Studies in Indian Logic and Linguistics*, University of Chicago Press, 1988, p. 155. See also Hartmut Scharfe, *Panini's Metalanguage*, American Philosophical Society, 1971.

[9] John Kadvany, 'Positional value and linguistic recursion', *Journal of Indian Philosophy*, 35, 2007, pp. 587–20.

a *context-sensitive* rule (see 5.3) and proves, despite its simplicity, to be powerful enough to describe not just Sanskrit but other languages too. The phenomenon of recursion is achieved if we substitute A for B in the rule stated above, or in other words if A can be replaced by itself ($A \rightarrow A/C\_D$).

However, Panini's grammar does not focus solely on word form and structure (morphology), pronunciation (phonology), and sentence structure (syntax). It also covers meaning (semantics).[10] Like Latin, Sanskrit permits substantial freedom in word order, and so the most information about meaning is contained in the words and their linguistic context. Panini's grammar assigns semantic roles to words that are strung together into more complex parts and ultimately into a meaningful sentence. Panini's grammatical process is a 'mini-play' acted out by an Agent (the leading player) and a number of other roles, such as Goal, Recipient, Instrument, Location, and Source.[11] The finale is an interpretation of the whole sentence.

Since Panini specifies a clear procedure for his grammar, which he expresses as a system of rules, we will designate his method as the *procedural system of rules principle.*

Evaluation of Panini's system of rules. How successful is Panini's grammar? In other words how accurate is his procedural system of rules and can we assess this using the existing corpus of classical Sanskrit? Panini's grammar has indeed been evaluated in detail with the result that (after correction of a few inconsistencies by Panini's followers—see below) so far not one sentence in classical Sanskrit has been found that is not accepted by the grammar. It is moreover the case that similarly no ungrammatical sentences—in so far as these can be created with certainty—are accepted by Panini's grammar. It is more difficult to test whether Panini's grammar also assigns the correct meanings to Sanskrit's grammatical sentences because we cannot find out what these meanings are in all cases from the Sanskrit corpus that has come down to us. As grammar, however, his system of rules is still undisputed.[12] As a counterargument one could contend that classical Sanskrit only exists in the form of a finite corpus, and that consequently Panini's grammar cannot be evaluated in regard to language variation and change. To a degree this is correct. Panini's grammar can only be assessed using the finite surviving corpus of classical Sanskrit. Nevertheless Panini defined extra rules for spoken Sanskrit through which he tried to take language variation into account. Moreover, later Sanskritists extended Panini's system of rules by including new words and forms that were not known in classical Sanskrit (the great Indian grammarian Nagesa was still doing this in the eighteenth century).[13]

[10] Johannes Bronkhorst, 'Panini's view of meaning and its Western counterpart', in Maxim Stamenov (ed.), *Current Advances in Semantic Theory*, John Benjamins, 1992, pp. 455–64. See also S. D. Joshi, 'Sentence structure according to Panini', in G. V. Devasthali (ed.), *Glimpses of Veda and Vyakarana*, Popular Prakashan, 1985.

[11] Kiparsky, 'Paninian Linguistics', *Panini as a Variationist.*

[12] According to Kiparsky, 'Modern linguistics acknowledges it as the most complete generative grammar of any language yet written, and continues to adopt technical ideas from it.'—see Kiparksy, 'Paninian Linguistics', *Panini as a Variationist.* The term 'generative grammar' refers to a system of rules that can generate an unrestricted number of sentences (see 5.3).

[13] Itkonen, *Universal History of Linguistics*, p. 30.

What drove Panini to develop something so unbelievably complicated as a complete Sanskrit grammar? This question is difficult to answer because we know so little about Panini and his predecessors. The *Ashtadhyayi* mentions a number of earlier grammarians, but most of their works have never been found. We do know that there was a tradition of linguistic philosophy in India, with such philosophers as Sakatayana (eighth century BCE) and Yaska (fifth century BCE). For example, Sakatayana asserted that all nouns could be derived etymologically from verbs,[14] and Yaska argued that all complex meanings could be developed from the combination of the meanings of the smallest units (a principle that would later become known as the *principle of compositionality*—see 3.1 and 5.3).[15] The idea that all more complex utterances could be generated using a finite number of units was therefore already in the air, but there was as yet no grammar. However, there arose a need in Hindu India for a strict interpretation of religious Vedic texts (see 2.8), which were memorized and recited with the very greatest accuracy. It is possible that the intellectual origin of Panini's linguistics lies in this Vedic 'learning'.[16] But his grammar concentrates more on the (then) spoken language than the language of the Vedas. It is therefore sometimes asserted that his work was first of all intended to record how to speak Sanskrit 'correctly'. In fact, Panini's grammar was the cornerstone of Brahman education.[17] Pupils were trained from the age of six in memorizing Panini's grammar rules. This was followed by many years of study. But it does not explain what moved Panini to formulate his Sanskrit grammar as an extremely systematic and ordered system of interlinked, recursive rules.

Panini's work had a huge and enduring impact on both linguistics and logic in the Indian world. His grammar generated many schools of linguists which wrote commentaries on commentaries on his grammar, in which they also resolved some inconsistencies and gaps in Panini's system of rules: from Katyayana's explanation in the fourth century BCE and Patanjali's impressive 'Great Commentary' in the second century BCE to the Nyaya school, which based its system of logic on Panini's method (see 2.6).[18] Panini's grammar also produced the model for languages like Tamil and Tibetan, which demonstrated de facto that its rule formalism was not limited to Indo-European Sanskrit. It could also be used to describe other, non-Indo-European languages (see 5.3).

However, it was more than a thousand years before Panini became known outside the Indian world, initially in the seventh century CE by Chinese Buddhist pilgrims who visited India, and then in the eleventh century through the book *Indica* by the Persian scholar al-Biruni, albeit in a very abridged form (see 3.1). It was not until much later, with the advent of nineteenth-century comparative linguistics, that Panini's grammar was also 'discovered' in Europe (see 5.3). The

[14] Bimal Krishna Matilal, *The Word and the World: India's Contribution to the Study of Language*, Oxford University Press, 1990, p. 9.

[15] Eivind Kahrs, *On the Study of Yaska's Nirukta*, Bhandarkar Oriental Research Institute, 2005.

[16] Frits Staal, 'The origin and development of linguistics in India', *Hymes*, 1974, pp. 63–74.

[17] Hartmut Scharfe, *Education in Ancient India*, Brill, 2002, pp. 20ff.

[18] For an overview of these commentaries, see Frits Staal, (ed.), *A Reader on Sanskrit Grammarians*, Motilal Banarasidass, 1985.

classical and Hellenistic Greeks seem never to have known Panini's work—either through Alexander the Great or any other route. They would certainly have appreciated it given their quest for principles. But the Greek philosophers would probably have found Panini's nearly 4,000 rules somewhat on the high side for a compact, axiomatic description of language. However, we now know that no human language exists with fewer rules—there are likely to be more.

Dionysius Thrax's grammar textbook. Compared with Panini, the other linguistics from Antiquity appears to be from a different world. No other work from Chinese, Greek, or Roman literature comes close to Panini's grammar in terms of complexity or precision. However, there was a flourishing linguistic philosophy (which actually comes outside the scope of this history): in the sixth century BCE Confucius was philosophizing about the meaning of names and in the fourth century BCE Plato was contemplating the origin of words and their relationship with reality in the *Cratylos*. This was followed by a study of word categories and word forms by Aristotle at the end of the fourth century BCE, Chrysippus in the third century BCE (see also 2.3) and others. But there was no trace of a descriptive grammar in the form of a system of rules.

The first attempt handed down by the Greeks is a prescriptive grammar textbook by Dionysius Thrax, the *Téchne grammatiké* (first century BCE).[19] There was widespread interest among well-to-do Romans in learning Greek as a second language, and grammar textbooks satisfied this need. Dionysius defines linguistics as 'practical knowledge of the elements of a language as used by poets and prose writers'. This definition reveals that his goal was very different from Panini's. Dionysius Thrax and other Western grammarians had a practical, educational goal—to teach Greek on the basis of normative instructions. Panini, on the other hand, wanted to design a procedural grammar based on formal and complete rules, but it could not be used to learn Sanskrit as a foreign language.

Dionysius's slim school grammar, which runs scarcely thirty pages, was used for centuries as a textbook. Dionysius concentrates on correct pronunciation, stress, punctuation, the alphabet, syllables, nouns, verbs, articles, prepositions, adverbs, and conjunctions in order to arrive ultimately at an overview of the different metres. His grammatical terminology was used in all European grammars until the end of the eighteenth century,[20] but Dionysius's grammar barely goes beyond a description of the conjugations and declensions. Syntax or word order is almost completely ignored. However, others attempted to describe the 'natural word order' of Greek. They included Dionysius of Halicarnassus (end of the first century BCE), but later he refuted the eight rules he had proposed on the basis of examples from Homer (see 2.8, Poetics).

Apollonius Dyscolus's sentence structure and the Roman tradition. We see a profound interest in Greek *syntax* for the first time in the work of Apollonius

[19] J. Alan Kemp, 'The Tekhne Grammatike of Dionysius Thrax translated into English', *Historiographia Linguistica*, 13(2/3), 1986, pp. 343–63.

[20] On the influence of Dionysius's grammar, see Pieter Seuren, *Western Linguistics: An Historical Introduction*, Blackwell Publishers, 1998, p. 22.

Dyscolus (second century CE), although his analysis remains largely based on the Aristotelian concepts of subject and predicate. Apollonius's most important innovations were (1) an analysis of the argument structure of the verb, and (2) an extension towards the idea of syntactic agreement in terms of number and gender, as it exists between article and possessive pronoun in Greek.[21] Apollonius also describes the complex Greek case system, and he remarks—and expresses surprise—that the subject sometimes has the first (nominative) case but can also have the fourth (accusative) case if it is construed with an infinitive. However, Apollonius does not give the underlying rule for this phenomenon; he addresses it on the basis of a number of examples. His grammar is consequently partially *example based*: if no rules are found, the linguistic phenomena are discussed using examples without making generalizations about those examples. It is indeed not always possible to generalize about linguistic phenomena. The existence of idiosyncratic and idiomatic expressions (for example, 'by and large' and 'long time no see') represents the standard example for a lack of rules. But Apollonius also uses examples where there could well be an underlying rule, but he did not find it.

Apollonius's attempt aside, virtually no classical linguist tackled the challenge of using a rule-based grammar to describe the capricious syntax and semantics of the Greek language. Although Apollonius's son Aelius Herodian (*c.*180–250) developed a system of rules for the simpler problem of accentuation,[22] the Greeks soon appeared to realize that a system of rules for syntax or word order could not be created from a small number of axioms or principles. Incidentally, we will see in 2.5 that the musicologist Aristoxenus undertook the construction of a grammar of Greek melodies, but his results have had little impact outside musicology.

Grammars in the tradition of Dionysius Thrax and Apollonius Dyscolus were also developed for Latin, for example by Varro, Donatus, and Priscian.[23] Varro (first century BCE) put the emphasis primarily on formal (syntactic) categories and Donatus and Priscian on functional (semantic) categories. Donatus's *Ars minor* and *Ars maior* (fourth century CE) became the most widely used educational works in the first half of the Middle Ages, until Priscian's *Institutiones grammaticae* (sixth century CE) was rediscovered in the Carolingian Renaissance. Although Priscian introduced the idea of a rule (*regula*) for describing the declension of nouns, his grammar is largely concerned with word structure and similarly does not go into syntax and (sentence) semantics. Western linguistics continued to be dominated by the taxonomic study of words after Priscian, and this situation did not change until the later Middle Ages (see 3.1).

Humanities or science? Panini's grammar is 'one of the greatest monuments of human intelligence'.[24] The only work from Antiquity that is comparable in originality and scope is *The Elements*—the axiomatic analysis of geometry by the

[21] Fred Householder, *The Syntax of Apollonius Dyscolus*, John Benjamins, 1981, p. 2. See also David Blank, *Ancient Philosophy and Grammar: The Syntax of Apollonius Dyscolus*, Scholars Press, 1982.

[22] Augstus Lentz, *Herodiani technici reliquiae*, Olms, 1965 (reprint of 1867–70).

[23] Vivian Law, *The History of Linguistics in Europe*, Cambridge University Press, 2003, pp. 42ff.

[24] Leonard Bloomfield, *Language*, University of Chicago Press, 1984 (reprint of 1933), p. 11.

Hellenistic mathematician Euclid in the third century BCE. In both cases a finite number of rules are used to cover an infinite number of possible expressions, where Euclid's work is concerned with expressions in mathematical language and Panini's in human language. Yet Panini's grammar is language-specific: it addresses only Sanskrit. This gives rise to the question of how *general* Panini's formalism of context-sensitive grammar is. Can we also describe other languages with this kind of grammar? That is, is it universal or specific? As we have seen, Panini's formalism was also used to describe the Tamil and Tibetan languages, but we do not encounter any notion of 'Universal Grammar' until the work of Roger Bacon in the thirteenth century (3.1). This idea was worked out in greater depth in the seventeenth century, and a pinnacle was reached in the twentieth century by Noam Chomsky, who contended that a Universal Grammar is innate (5.3). Meanwhile we have learned that Panini's formalism of a context-sensitive grammar can indeed be used for describing many, and perhaps all, human languages. His formalism even served in the twentieth century as the basis for the first high-level programming languages, such as ALGOL60, which also work on the basis of a fully specified system of rules (see chapter 6).[25] Virtually all programming languages are written in a formalism that uses Panini's linguistic notion of a grammar. Such a grammar can determine whether a given sequence of statements forms a correct expression in the particular programming language or not.[26]

But if the nature of Panini's system of rules is so formal, can his grammar still be considered as a product of the humanities? Should it not rather be regarded as a scientific study of language? Putting it another way, is Panini's grammar representative of the humanities? Definitions of the humanities and science are of no help to us here. There was no differentiation between these forms of knowledge in the eras of Panini and Euclid. We will see that over the centuries, but primarily from the nineteenth century on, linguistics used an increasingly formal approach. But more importantly, its subject is the study of human language, which is a pre-eminent expression of the human mind. As such linguistics is taught in all humanities faculties worldwide. No matter how one looks at it, Panini is the father of linguistics, not mathematics.

[25] P. Z. Ingerman, 'Panini-Backus form suggested', *Communications of the ACM*, 10(3), 1967, p. 137.

[26] On the relation between Panini's linguistics and computer science, see also Rens Bod, 'Discoveries in the humanities that changed the world', *Annuario 53, 2011–2012*, Unione Internazionale degli Istituti di Archeologia, Storia e Storia dell'Arte in Roma, 2011, pp. 189–200.

2.2 HISTORIOGRAPHY: THE SOURCE PROBLEM AND THE FORM OF THE PAST

How do we understand our past? Historiography emerged independently in different places around the world, but when exactly did it begin? This question is a tough one to answer because myth and historiography are so often intertwined (myths can contain social truths but they do not amount to historical writing).[27] For example, in his *Theogony* and *Works and Days* the Greek Hesiod (*c.*700 BCE) describes a world history of gods, heroes, and mortals in five ages: the golden age, the silver age, the bronze age, the heroic age, and the iron age. It is a story of growing chaos and misery, but historiography it is not. The Chinese *Book of Documents* dating from the sixth century BCE may be the oldest historical source book that has been handed down, the compilation of which is attributed to Confucius. In fifty-eight chapters it tells of the words and deeds of illustrious rulers, starting with the legendary emperors Yao and Shun, followed by the emperors of the Xia, Shang, and Zhou dynasties.[28] It is beautiful prose, describing a succession of extravagant, benevolent, wise, and murderous rulers, but is it historiography? It makes sense to let the history of historical writing begin with Herodotus's *Histories* (440 BCE).

Herodotus and Thucydides: the cyclical pattern of rise, peak, and decline. As far as we know, Herodotus was the first historian to collect his material systematically and test its accuracy in some way. This practice was the beginning of historiography as a 'principle-based activity'. Can we talk about historiography at all without a test of the reliability or accuracy of sources? That would mean having to assume that every report and every source is true, even contradicting sources, which would be logically impossible. Thus Herodotus subjected his material to a rudimentary form of critical analysis. Rather than trusting (verbal) sources from his immediate environment, he travelled around the then known world in an attempt to get as close as possible to the 'source' relating to his main subject: the Persian Wars (490, 480–479 BCE). If Herodotus was not sure about a source or if sources contradicted one another, he analysed them all and then selected the one he considered to be the *most probable*. He was not afraid to voice his own opinion in this process. Because of the subjectivity of this method Herodotus was accused by later historians (including Claudius Aelianus in *Varia Historia, c.*200 CE) of embellishing or even distorting sources. As a consequence, Herodotus acquired the dubious reputation of being both the 'father of history' and the 'father of lies'.[29]

Current opinion about Herodotus is considerably kinder: we know that objectivity in historiography is an unattainable ideal, and that Herodotus's method, while it was far from perfect, did at least include a critical component. We will designate

[27] Percy Cohen, 'Theories of myth', *Man: Journal of the Royal Anthropological Institute*, 4(3), 1969, pp. 337–53.

[28] James Legge, *The Chinese Classics, Volume III: The Shoo King or the Book of Historical Documents*, Trubner, 1865.

[29] Detlev Fehling, *Herodotus and His 'Sources': Citation, Invention, and Narrative Art*, Arca Classical and Medieval Texts, Papers, and Monographs 21, Francis Cairns, 1989.

his principle as the *most probable source principle*. Although one could ask oneself whether such a subjective rule deserves the word 'principle', Herodotus had to rely much more on oral sources than later historians. Whereas we only have the *result* of the work of many other scholars (such as Panini's Sanskrit grammar), Herodotus also describes *how* he used his principle.[30] On Egypt, for instance, he said: 'So far the Egyptians themselves have been my authority, but in what follows I will describe what others are also inclined to accept about the history of this country, and I shall add something of what I myself have seen.'[31] And: 'With regard to this point I feel it necessary to give an opinion that I know will be contradicted by most people. Nevertheless, since I believe it to be correct I will not suppress it',[32] after which he selects the source he considers to be the most probable. This principle gave Herodotus a licence to choose the source that he thought of highly—but not until he had compared and evaluated the other sources. Moreover, he used his principle in a principled manner, for instance when he wrote: 'The third opinion is by far the most plausible, yet the most erroneous of all. It has no more truth in it than the others.'[33]

Why did Herodotus look for an explicit method based on a critical principle? We should keep in mind that Herodotus's historiography was competing with the 'many tales' of the Greeks,[34] in particular the Homeric epic and its illustrious heroic deeds, which people learned at their mothers' knees. Oral traditions like these enjoyed huge authority, and only a fundamental method that was also based on oral tradition could legitimize a form of written history. Herodotus's goal was therefore the same as that of the Homeric bard: keeping alive remembrance of the past, in which the standards of propriety were set down.

The greatest historian immediately after Herodotus was Thucydides, whose description of the Peloponnesian Wars (431–404 BCE) tolerated no second-hand or third-hand sources, but *eyewitness accounts* only. He distrusted every source that was not based on direct evidence.[35] Thucydides had a good point here of course, but unlike Herodotus he could describe a history of events whose eyewitnesses, including himself, were still alive. Thucydides had another point. Because he wanted to base his work on the *eyewitness account principle* he rejected the ethnographical and geographical descriptions that are abundant in Herodotus. The broad cultural approach was lost on Thucydides, and he stressed time and again that the historiographer should concentrate on the history of the *lives of people*—no more and also no less. This is exemplified to perfection in his famous funeral oration for Pericles. Although Thucydides has (had) a better reputation than Herodotus,[36] he

[30] On Herodotus's way of working, see also Ernst Breisach, *Historiography: Ancient, Medieval and Modern*, University of Chicago Press, 2007, p. 19.

[31] Herodotus, *Histories*, 2.147.

[32] Herodotus, *Histories*, 7.139.

[33] Herodotus, *Histories*, 2.21.

[34] On this issue, see D. H. Fowler, 'Herodotus and his contemporaries', *Journal of Hellenic Studies* 116, 1996, pp. 62–87. See also G. E. R. Lloyd, *Disciplines in the Making: Cross-Cultural Perspectives on Elites, Learning, and Innovation*, Oxford University Press, 2009, p. 67.

[35] Thucydides, *History of the Peloponnesian War*, 1.21.

[36] Clifford Orwin, *The Humanity of Thucydides*, Princeton University Press, 1994.

offered no solution to the problem of conflicting sources for which Herodotus had devised a heuristic, no matter how subjective it was. Thucydides's stories moreover lack a precise indication of time, but to a lesser extent than the chronological vacuum found in Herodotus.

Despite these shortcomings, both historians applied their historical principles with vigour, and this transformed historiography from uncritical storytelling into a critical activity. As well as writing a description of the Persian and Peloponnesian Wars, Herodotus and Thucydides also 'discovered' something remarkable. They both believed they had recognized a *cyclical pattern in history*. Herodotus's history, for instance, reflected a repeating pattern of rise, peak and decline. We see this pattern in his descriptions of both people and states, for example the tyrant Pisistratus and Athens, King Croesus and Lydia, and Darius and Persia: their fortunes rose and fell. Herodotus considered the cyclical pattern to be the basic structure of history: 'For many states that were once great have now become small, and in my lifetime those that are great used to be small.'[37]

Thucydides also contended that the rise and fall of Athens and its disintegration during the Peloponnesian Wars had parallels with other historical periods, and believed that the cyclical pattern was analogous to human nature and therefore could even serve as an 'aid for interpreting the future'.[38] This vision of the future, present, and past as an eternal cyclical pattern can of course also be found in the works of the Pythagoreans, Anaximander, Parmenides, and Plato (*Timaeus*), as well as in the Greek tragedies.[39] The new element in Herodotus and Thucydides is that they believed to have recognized these cyclical patterns on the basis of their methodical principles. Herodotus found the pattern in the lifes of people and states through source comparison, while Thucydides found the cyclical pattern through eyewitness accounts.

Alongside the cyclical model, Herodotus also found a cultural pattern. After he had compared the Persians, Greeks, and Callatians he felt he was able to conclude that *all peoples believe their own way of life is the best*: 'Assume that we let all people select the very best laws and customs from all over the world. After careful consideration they would all choose their own laws and customs, because they are all convinced that their own way of life is the best one.'[40] Thus Herodotus also conceived the concept of cultural relativism.[41]

Berossus, Manetho, Timaeus: chronology and the written source principle. Who came after Herodotus and Thucydides? After the Peloponnesian Wars the Greek world of the *polis* (city-state) produced no new historical methods or principles. Some historians wrote in the broad Herodotean manner, like Ephorus of Cyme who produced a compilation of all Greek history, by some referred to as

[37] Herodotus, *Histories*, 1.5.

[38] Thucydides, *History of the Peloponnesian War*, 1.22.

[39] For an overview, see David Bebbington, *Patterns in History*, Regent College Publishing, 1990.

[40] Herodotus, *Histories*, 3.38.

[41] Siep Stuurman, *De uitvinding van de mensheid: korte wereldgeschiedenis van het denken over gelijkheid en cultuurverschil*, Prometheus, 2009, p. 160.

the first universal history.[42] But most historians in the early fourth century BCE applied Thucydides's approach, such as Xenophon who wrote the impressive *Anabasis*, a record of the expedition against the Persians and the journey home, which was later used by Alexander the Great (356–323 BCE) as a field guide. Even in the time of Alexander there were no great historians of the stature of Herodotus or Thucydides. Yet a new genre of biography emerged and a long tradition of descriptions of India was initiated by Megasthenes's *Indica* (*c.*300 BCE). These too, though, were based on Thucydides's eyewitness account principle.

The most important historiographies during Hellenism were outside Greece, for example a history of Mesopotamia by Berossus and a history of Egypt by Manetho, both in the third century BCE. It is here that we see a *chronological organization* of the material. Whereas Berossus appears to use a random chronology in his *Babyloniaca*, Manetho's *Aegyptiaca* contains a meticulous arrangement of Egyptian royal dynasties based on lists of kings. Many such lists have survived, such as the *Annals of the Old Kingdom* and the *Turin Papyrus of Kings*. However, it is not known which lists of kings Manetho used.[43] He was, though, one of the first Hellenistic historians to work on the basis of written sources, so we can describe his method as the *written source principle.* Manetho's work was used later as a written source for the Jewish history of Flavius Josephus (first century CE) and in Christian chronicles, such as those by Eusebius (fourth century CE). We will see how Manetho's lists of kings led to the early modern crisis in the Christian world view, after their accurate reconstruction by Joseph Scaliger in the sixteenth century (see 4.2).

The preoccupation with chronological organization can also be seen in the historians of *Magna Graecia* in southern Italy. Take for example the description by Timaeus of Tauromenium ('Taormina') around 300 BCE of the creation of Rome and Carthage. He calculated that Carthage was established in 814–813 BCE. To do this Timaeus used the Olympiads as units of time and also a list of winners of the Pythian Games (which had already been drawn up by Aristotle). This made it possible to synchronize different historiographies. Timaeus's chronological time scale was generally adopted by other scholars, and even by the astronomer and geographer Eratosthenes.[44]

Early Roman historiography and Polybius's personal experience principle. Roman historiographers are viewed as rarely showing significant originality. Most historiographers contend that the Romans adopted Greek ideas about historical theory and combined them with Timaeus's chronological innovations.[45] And indeed the first histories of Rome were written in Greek both by Romans such as Fabius Pictor and by Greeks, for instance Polybius. Like Thucydides, Polybius assumed in his *Histories* (200 BCE) that the most reliable sources were accounts by *eyewitnesses*, or at least had to be based on *personal experience*. He had no access to

[42] Polybius, *Histories*, 33.2.

[43] Gerald Verbrugghe and John Wickersham, *Berossos and Manetho, Introduced and Translated: Native Traditions in Ancient Mesopotamia and Egypt*, University of Michigan Press, 1996.

[44] For the wonderful life of Timaeus, see Truesdell Brown, *Timaeus of Tauromenium*, University of California Press, 1958.

[45] See e.g. Stephen Usher, *Historians of Greece and Rome*, Duckworth Publishers, 2001.

Greek libraries for many years because he was a Roman hostage and was dependent on other people's contacts or his own experience for sources. For instance, he personally tracked Hannibal's journey through the Alps, and we will therefore designate his principle as the *personal experience principle.*

Polybius expressed great admiration for the way Rome succeeded where the Greeks had failed. Rome, he argued, was an exception to the cyclical pattern of rise, peak, and decline that had occurred in the history of Athens, i.e. a cycle of monarchy, aristocracy, oligarchy, democracy, and via tyranny back to monarchy again.[46] Unlike Athens, Rome was immune to this cycle—and therefore to decline—because of its *mixed* constitution. In Rome there was a monarchy (the consuls), an aristocracy (the senate), and a democracy (the people's assemblies) all at the same time.[47] According to Polybius this simultaneity broke the cyclical pattern, which turned the history of Rome into a universal history.[48] Polybius later changed his mind and came to believe that Rome would lose its mixed and simultaneous constitution sooner or later as a result of corruptive power and prosperity, and that it would subsequently become vulnerable to the eternal cycle.

Roman historians and the annals: written source principle. Roman historians nevertheless introduced a historiographical innovation: the *Annales* (annals), in which events were described in summary form and chronologically year after year. Initially the *Annales* were updated on a daily basis by the Roman high priest, the *pontifex maximus*, who recorded the names of magistrates and also all manner of details about miraculous portents, temple consecrations, and shortages. Starting in about 120 BCE the annals were replaced by the *Annales maximi*, which comprised short texts and a strict chronological structure. The *Annales maximi* had a substantial influence on Roman historians, who also used a chronicle-based approach, resulting in the many historical overviews starting with 'the founding of the city'.[49] These linear rather than cyclical descriptions, usually referred to as *Origines* but also as *Annales*, have a stricter chronological approach than the works by Manetho discussed above. The annalists wanted to describe every year without there being any gaps. The *Annales* produced by the high priests were of help to them, but so too were the works of previous historians and philologists' text reconstructions (as we will see in 2.3). For example Polybius used the researches of Fabius Pictor, whereas Livy (Titus Livius), in his large-scale *Ab urbe condita* (end of the first century BCE), included both Polybius and Fabius Pictor and also Varro's chronological reconstruction (first century BCE) in his historical overview.[50]

These historians' annals were therefore based first and foremost on textual rather than oral sources, and thus on the *written source principle.* These historiographies

[46] Polybius, *Histories*, 1.1–2.

[47] Because of his praise of a mixed constitution, Polybius is often credited with the invention of the separation of powers, later elaborated by Montesquieu (see 4.2).

[48] Polybius, *Histories*, 1.4.

[49] Bruce Frier, *Libri Annales Pontificum Maximorum: The Origins of the Annalistic Tradition*, University of Michigan Press, 1999.

[50] On the intertwining of Roman historiography and philology, see also Andreas Mehl, *Roman Historiography*, Blackwell Publishers, pp. 66ff.

were moreover linear without cyclical patterns. Yet the annalistic tradition did not displace other forms of historiography. Both Sallustius and Julius Caesar concentrated primarily on relatively short periods (such as the late republican era and the conquest of Gaul). Consequently there were now two sorts of historiography: the *Historiae*, which recorded the events observed by the author himself, and the *Annales*, which described events from the past. These two traditions came together in the works of Publius Cornelius Tacitus (55–120 CE). In his best known work, which is called *Annales* because of its structure, Tacitus begins with a summary of the historical highlights. He then continues from where Livy left off—the death of Augustus.

Biography as a genre also received a substantial boost from *Parallel lives* by Plutarch (*c.*46–120 CE) and the famous biographies of the emperors by Suetonius (*c.*70–140). After Suetonius the volume of historical writing dropped sharply, and there were essentially no new principles or patterns. Cassius Dio (*c.*155–229) wrote a historical overview of Rome, but he was no longer interested in the republican era. In his *Res Gestae* the last major classical historian, Ammianus Marcellinus (*c.*330–400), extended Tacitus's work up to the year 378. During Marcellinus's time the annals continued to consist of a linear history of Rome without any degeneration. Even though the decline of Rome in the fourth century could not be denied, Marcellinus interpreted the later Roman Empire as a form of 'maturity'. To him an approaching end was inconceivable. Like so many classical historians, Marcellinus attributed the problems of the City not to structural changes but only to failing individuals.

The annal form, with its linear pattern, lived on in the Christian chronicles. They began not with the founding of the 'city' but with Abraham, Adam, or even the biblical creation of the world. A very influential work was the *Chronicon* of Eusebius (265–340), in which a linear time scale was given from Abraham to Emperor Constantine. The *Chronicon* integrated biblical and classical history and is one of the greatest chronological overviews from the classical world. It brought together the old works of Berossus and Manetho, the time scales of the Roman annals and even the chronology of the Libyan historian Sextus Africanus (221 CE). It initiated a tradition that is known as *Universal History* (see chapter 3).

Sima Qian: the most extensive historiography in Antiquity. Outside the Graeco-Roman world, China is the most important region with a historiographical tradition. Chinese historiographers used a strictly annalistic approach, as had historians in the Roman Empire. The Book of Documents referred to earlier describes a chronological succession of rulers. But it is especially the *Bamboo Annals* and the *Spring and Autumn Annals* that record events in an extremely concise way and in sequence.[51]

The first great Chinese historian, Sima Qian[52] (*c.*145–86 BCE), initially followed the existing annalistic tradition when he laid the foundations for many later

[51] There are also commentaries on the Spring and Autumn Annals dating from the classical period, such as the *Zuo Zhuan* written by Mister Zuo in the fourth century BCE.

[52] This book I use the *pinyin* transcription, except when titles of books and articles or literal citations use the *Wade-Giles* transcription.

Chinese historians during the Han Dynasty (see Appendix B for an overview of Chinese dynasties). As court historian he described the entire history of all the peoples and regions he knew of inside and outside China in the 130 volumes of *Shiji* ('Records of the Grand Scribe').[53] The time span addressed by Sima Qian extended from the era of the mythical Yellow Emperor in the twenty-seventh century BCE up to the first century of the Han Dynasty (second century BCE). Never before had a historian covered such a huge period. The *Shiji* is the most extensive historiography from the ancient world.

His father Sima Tan was head of the imperial library and had acquired a collection of historical material to which Sima Qian added sources from the imperial archives and private libraries (the 'written source principle'). In his *Shiji* Sima Qian also used oral sources that he had collected during his many journeys around the country.[54] Unfortunately the selection principles in the *Shiji* and the way Sima Qian dealt with contradictory sources are a matter of conjecture. However, he did complain about the lack of sources as a result of the book burning in 213 BCE at the time of the strict legalistic regime of the Qin Dynasty. During this regime the dissemination of Confucian ideas was considered to be subversive. The prime minister Li Si decided in 213 BCE to put an end to the Confucian 'agitators' once and for all and ordered that all philosophical books had to be burned by their owners. At the same time, 460 Confucian scholars were said to have been buried alive, but there is no evidence that this execution took place. However, the damage to Chinese cultural heritage was considerable.

In his Records of the Grand Historian, Sima Qian used the *written* sources (Book of Documents, Book of Songs, Spring and Autumn Annals) that were handed down and the many *oral* sources he had collected. The annalistic approach alone was no longer adequate for dealing with so many different sources. Sima Qian therefore defined five historical genres, each with its own form and style. These defined Chinese court historiography for more than 1,800 years and also had an immense impact on Chinese poetics:

(1) Annals: imperial biographies in strict annalistic form.
(2) Tables: tabular overview of the governments, with the most important events.
(3) Treatises: descriptions of the different areas of state involvement, ranging from rites, music, and astronomy to rivers and canals.
(4) Hereditary lineages: descriptions of states and people in chronicle form.
(5) Illustrative traditions: biographies of important people.

Sima Qian almost failed to finish his life's work. After he had completed about half of the *Shiji*, he dared to speak out against the emperor in support of a general who

[53] The *Shiji* is partly translated into English by Burton Watson, *Records of the Grand Historian*, Columbia University Press, 1993. For a complete English translation of the *Shiji*, see William Nienhauser (ed.), *The Grand Scribe's Records*, 9 volumes, Indiana University Press, 1994–2012.

[54] Burton Watson, *Ssu-ma Chien: Grand Historian of China*, Columbia University Press, 1958; William Nienhauser (ed.), *The Grand Scribe's Records*, volume 1, Indiana University Press, 1994.

had fallen out of favour. Sima Qian was sentenced to castration, the second most severe punishment. Instead of committing suicide, which was normal in such situations, he endured his punishment so that he could complete his history, as he had promised to his father.[55]

In his historiography Sima Qian concluded that he recognized a *cyclical pattern* in the sequence of dynasties. Every dynasty began with a virtuous ruler chosen by Heaven (*tianming* or Heaven's Mandate), after which each successive ruler was increasingly less sound until Heaven lost patience and withdrew the mandate of the last, worthless ruler.[56] After that everything began again, including the calendar. This pattern of the rise and fall of dynasties resembles the cyclical pattern found by Herodotus and Thucydides, but in Sima Qian's case it is rooted in Taoism. According to the Taoist world view, the cosmos consists of opposite phases with the same 'energy' that continually exert influence on each other (*yin* and *yang*). Everything in existence was subjected to this fundamental principle. This became the path or way (*Tao* or *Dao*) which became visible in the changes that were caused by the progress of the present. People believed they could discover patterns in these changes, and it was the task of the historian to find these patterns, which could serve as an explanation of the present. In Sima Qian's view, the present could be described as a repetition of an identical situation in the past.

Ban Gu and Ban Zhao: basic Confucian virtues principle. Ban Gu (32–92 CE) more or less continued where Sima Qian had stopped. In his *Hanshu* ('Book of the Han'), Ban Gu advocates the most objective possible account of history, in his case the history of the Western Han dynasty. However, he too was subject to the reality that objective recording could only be the result of a subjective selection process. The criteria that Ban Gu used in his choices for the reliable sources arose from Confucianism that was accepted during the Han Dynasty based on the six basic virtues, that may be loosely translated as: humanity, obedience, justice, decency, loyalty, and reciprocity. Ban Gu restructured Confucian political ideas into a new notion of Heaven's Mandate. Unlike Sima Qian, who claimed that Heaven assigned or removed its sanction based upon moral merits, Ban Gu held that the ruling dynastic family's Mandate was permanently bestowed, and thus irrevocable, regardless of the ruler's good or bad behaviour.[57]

Ban Gu's younger sister Ban Zhao (45–116) completed her brother's work after he was executed in 92 CE because of his involvement with the Empress Dowager Dou. Possibly the first female historian, Ban Zhao was also the author of *Lessons for Women*, in which she recommends submissive obedience to men by women,

[55] It has also been noted that Sima Qian sacrificed his honour and reputation because he hoped that his work would clear his name—see John Minford and Joseph Lau (eds), *Classical Chinese Literature, from Antiquity to the Tang Dynasty: An Anthology of Translations*, Columbia University Press, 2002, p. 330.

[56] The concept of Heaven's Mandate is already found in the *Book of Documents* and is further elaborated on by Sima Qian.

[57] Anthony Clark, *Ban Gu's History of Early China*, Cambria Press, 2008.

although she also advocates good education for women.[58] Ban Gu's and Ban Zhao's work was continued by Sima Biao who wrote a history of the Eastern Han, the *Houhanshu*, which is considered to be highly accurate historiography.[59]

The fall of the Han dynasty, in 220 CE, marked the end of an era in China. The period of four centuries between the Han and the Tang dynasties (from 220 to 618) shows some similarities to the Middle Ages in Europe.[60] There are many expressions of nostalgia, as well as efforts to construct a continuity with the Han. Over 1,100 historical works appeared during these four centuries. It is clear that there is a goldmine here where a substantial quantity of methodological research remains to be done.

India: a civilization without historiography? We can be relatively brief about Indian historiography in Antiquity. It was primarily mythical and not based on sources. These myths could contain truths, social or religious, but they do not correspond to an evaluation of material from sources. In India they exist in the form of *Puranas*, which give an overview of the cosmos from creation to annihilation. The Puranas cover vast periods of millions of years, which in turn are subdivided into *Yugas*. Although the Puranas reveal a clear preoccupation with the history of the world—with legendary dynastic lists and royal genealogies—the Indians appear to have deemed the human world unworthy of historical investigation. The absence of historical studies in Indian Antiquity has given numerous historiographers concerns.[61] How can it be that a culture that has distinguished itself in so many areas in the humanities—from linguistics, logic, rhetoric, and poetics to art theory and musicology—has apparently no interest in investigating its own history? Apart from inscriptions, almost all historical knowledge about India up to the twelfth century comes from external sources. This is such a remarkable phenomenon that some historiographers simply do not accept it and assume that the Indian historical works written at some time or another were all lost.[62] Others explain this phenomenon by referring to the awesome epochs, lasting millions of years, in the mythical Puranas that nullify any human history. Whatever the truth may be, already in the eleventh century al-Biruni was surprised while staying in India that there did not appear to be any interest in the historical sequence of things: 'Unfortunately the Hindus do not pay much attention to the historical order of things, they are careless in relating the chronological succession of their kings, and when they are pressed for information and are at a loss, not knowing what to say, they invariably take to

[58] Barbara Bennet Peterson (ed.), *Notable Women of China: Shang Dynasty to the Early Twentieth Century*, M. E. Sharpe Inc., 2000.

[59] B. J. Mansvelt Beck, *The Treatises of Later Han*, Brill, 1990.

[60] On-cho Ng and Q. Edward Wang, *Mirroring the Past: The Writing and Use of History in Imperial China*, University of Hawai'i Press, 2005, pp. 80ff.

[61] For a critical discussion on this issue, and a plea for treating the Sanskrit epics and inscriptions as history writing as well, see Romila Thapar, 'Historical traditions in early India: *c.*1000 BC to *c.*AD 600', and 'Inscriptions as historical writing in early India: third century BC to sixth century AD', in Andrew Feldherr and Grant Hardy (eds), *The Oxford History of Historical Writing, Volume 1: Beginnings to AD 600*, Oxford University Press, 2011, pp. 553–600.

[62] R. C. Mujandar, 'Ideas of history in Sanskrit literature', in Cyril Philips (ed.), *Historians of India, Pakistan and Ceylon*, Oxford University Press, 1961, p. 25.

story-telling.'[63] Arguably the oldest known Indian historiography, Kalhana's *Rajatarangini*, which describes the local history of Kashmir, appeared in the twelfth century (see 3.2).

The principle-based method in historiography. In conclusion we can say that classical historiography in the Graeco-Roman world and China employed principle-based methods. These methods could be based on Herodotus's most probable source principle, the eyewitness account principle as employed by Thucydides, Polybius's personal experience principle, the written source principle as used by Manetho, Livy, and others, a combination of oral and written sources as in the work of Sima Qian, or finally Ban Gu's basic virtues principle. No matter how subjective and questionable the historical principles may be, they appear to have been fruitful enough to establish (or should we say impose?) patterns in historical material. And these patterns can be falsified. For example, the cyclical pattern of rise and decline can be empirically refuted (as seemed to be the case for some time with the linear development of Rome) or empirically supported by new sources or facts (as ultimately emerged with the fall of Rome). There are, of course, also micro- and mesohistories that concentrate on a shorter period or even one 'event', such as Sallust's accounts of the Catiline conspiracy and the Jugurthine war. But these single-event histories are also prone to falsification. For instance, Timaeus and generations after him focused on the precise dating of a single event, the foundation of Rome, which eventually, like a sort of pendulum coming to rest, ended up as 753 BCE in Varro. This date was accepted for centuries until Joseph Scaliger demonstrated in 1606 that no source material existed that permitted establishment of Rome's foundation date with such accuracy.

Like linguistics with its rule-based grammar, historiography is an empirical discipline. At a very abstract level there is only little difference between Panini's linguistic method and the historical methods of Herodotus, Thucydides, and others. Both try to cover a maximum number of phenomena with a minimum number of principles. There is, of course, a much bigger difference at a more practical level. Classical historiography is not 'replicable'. It remains a personal, albeit critical, choice as to whether or not a source is identified as 'probable'. The way that Panini's procedural grammar works, on the other hand, can be replicated exactly. Everyone who applies the rules gets the same result.

[63] Edward Sachau, *Alberuni's India*, volume 2, Trübner & Co., 1888, p. 349.

2.3 PHILOLOGY: THE PROBLEM OF TEXT RECONSTRUCTION

There was a time when texts represented an empirical world without equal. Scholars developed hypotheses about thousands of extant manuscripts which could be confirmed or refuted by the findings of new manuscripts. The goal of philology, or textual criticism, was the reconstruction of a supposed original source on the basis of surviving manuscripts. This made philology one of the most interdisciplinary activities, where knowledge of grammar, rhetoric, history, and poetics came together. Philology is older than many people think. If we believe tradition, the Athenian tyrant Pisistratus gave instructions as far back as the sixth century BCE to establish the 'official' text of Homer for the Panathenaic festival.[64] Homer was recited and studied continuously, from the poet Solon in the sixth century BCE to Aristotle at the end of the fourth century BCE.

The analogists of Alexandria. The systematic philological study of literary texts did not really start until the establishment of the library of Alexandria by Ptolemy II in about 300 BCE. The gathering of hundreds of thousands of manuscripts[65] from all parts of the Hellenistic world resulted in one of the greatest problems in the history of learning and science. Among the often dozens or even hundreds of copies of the same text, no two were alike. In some cases the differences were modest and had come about because of copying errors, but the discrepancies could also be substantial, consisting of whole sentences that appeared to be deliberate changes, additions, or omissions. And there were also texts that had only survived in the form of incomplete fragments. How could the original text—the archetype—be deduced from all this material?

The first person to systematically tackle this problem was Zenodotus of Ephesus (*c.*333–*c.*260 BCE), who was also the first librarian of the Alexandrian library. Zenodotus compiled a dictionary using typically Homeric words, with which he hoped to be able to formulate the 'perfect' text from the many corrupt remnants of manuscripts.[66] Unfortunately there was no theory underlying Zenodotus's attempt and his criteria appear to have been based on aesthetic preferences and guesswork.

His successors, Aristophanes of Byzantium (*c.*257–180 BCE) and Aristarchus of Samothrace (*c.*216–*c.*144 BCE) tried to provide such a theory so as to keep philology as free as possible from subjective elements. The problem of corrupted words represented one of the biggest challenges. How could an unknown word form be identified as an archaic word or an error? Aristophanes approached this problem on

[64] Norbert Ehrhardt, *Athen im 6. Jh. v. Chr. Quellenlage, Methodenprobleme und Fakten*, in *Euphronios und seine Zeit*, Antikensammlung Staatliche Museen zu Berlin, Landshut-Ergolding, 1992, pp. 12–23.

[65] According to most estimates the Alexandrian library grew from around 200,000 manuscripts in the third century BCE to over 700,000 manuscripts in 50 BCE—see Luciano Canfora, *The Vanished Library: A Wonder of the Ancient World*, University of California Press, 1990.

[66] Klaus Nickau, *Untersuchungen zur textkritischen Methode des Zenodotos von Ephesos*, De Gruyter, 1977.

the basis of a concept of *analogy*.[67] If he could establish that *an unknown word was formed and conjugated or declined in the same way as a known word*, he believed that he could reconstruct the original form with a certain degree of reliability. Aristophanes defined five criteria that word forms had to comply with among themselves in order to be described as 'analogous'. The word forms had to correspond in regard to (1) *gender*, (2) *case*, (3) *ending*, (4) *number of syllables*, and (5) *stress* (or sound). His pupil Aristarchus of Samothrace added a sixth criterion—when comparing two word forms, both had to be *compound* (complex) or *non-compound* (simplex).[68] Historical philology actually started with Aristophanes. We will refer to his method as the *analogy principle*. The Alexandrian philologists used the same designation: *analogía*.

Aristophanes's philological method appears to be a kind of metagrammar that discovered regularities between words on the basis of a few analogy criteria. Yet his method is very different from Panini's grammatical system of rules in 2.1. The Alexandrians had no precise rules for establishing corresponding word forms. Searching for analogies was a complex undertaking in which the meaning of a word, style, metre, and aesthetic aspects often had to be taken into account. Consequently, the discovery process remained associative and subjective to a significant degree. However, the Alexandrian method has withstood the test of time as a critical approach to text reconstruction. We owe a debt of gratitude to the insights of the Alexandrians for the editions of Homer, Hesiod, Pindarus, Archilochus, and Anacreon, and of the tragedians and historians, which have been handed down to us. Moreover, the analogical methods of comparing words set the standard for the first Greek grammar by Dionysius Thrax (see 2.1).

The anomalists of Pergamon. At the time of the analogists in Alexandria, there was also a competing, Stoic school that was established in Pergamon by Zeno of Citium (334–262 BCE). The philologists of this school searched for exceptions rather than regularities. They worked on the basis of the *anomaly principle*, which was introduced by Chrysippus of Soli (*c*.280–*c*.207 BCE) using the term *anomalía*.[69] Instead of looking for analogies between word forms, Chrysippus stressed the importance of seeking the *differences* and *exceptions* between word forms. According to the anomalists it was impossible to deduce the original form of a text on the grounds of analogies. The most fervent supporter of the anomalistic approach was Crates of Mallos (died *c*.150 BCE). According to Crates all the efforts expended by the analogists were vain and superficial. The only way to arrive at the original text was to select the surviving document that came as close as possible to the author's intentions and, once this selection had been made, the chosen

[67] For an in-depth study on Aristophanes, see Christopher Callanan, *Die Sprachbeschreibung bei Aristophanes von Byzanz*, Vandenhoeck und Ruprecht, 1987. See also Rudolf Pfeiffer, *History of Classical Scholarship: From the Beginnings to the End of the Hellenistic Age*, Clarendon Press, 1968, pp. 202–8.

[68] Francesca Schironi, *I frammenti di Aristarco di Samotracia negli etimologici bizantini: Etymologicum genuinum magnum, Symeonis, Megalē grammatikē, Zonarae lexicon*, Vandenhoeck und Ruprecht, 2004.

[69] Pfeiffer, *History of Classical Scholarship: From the Beginnings to the End of the Hellenistic Age*, p. 203.

document had to be adhered to as closely as possible.[70] To a degree this approach appears very similar to Herodotus's historical most probable source principle. Just as Herodotus dealt with inconsistent sources, so the anomalists compared inconsistent texts and selected the text that to the best of their knowledge was the 'most probable'. According to the anomalists all words, sentences, and interpretations were exceptional, and no deeper system of regularities existed. This method was therefore wilder than the method based on analogies and often resulted in highly fanciful and allegorical text interpretations.[71]

Nevertheless, the anomalistic approach produced a number of extraordinarily original works. The anomalists—unlike the analogists, whose work was mostly formal—produced erudite commentaries. For example Demetrius of Scepsis wrote a series of thirty books about the Trojan forces, which were addressed in fewer than sixty-two lines in the entire *Iliad.* Every point of view was dissected by the author, using a vast quantity of literature, local and oral traditions, history, mythology, geography, poetry, and observations by travellers—in other words he called upon the entirety of classical knowledge to contribute to the interpretation of the text.[72] This detailed and particularistic approach was developed in Pergamon and since then has never disappeared from the humanities. We will come across anomalism in many different guises. It survived both the Middle Ages and the early modern age, was regenerated in nineteenth-century hermeneutics, and reached a provisional pinnacle in twentieth-century poststructuralism (see chapter 5).

The resolution of the analogy–anomaly controversy in Rome. In a way both the analogists and the anomalists were right. The analogists had an important point as regards the regularities between words, but they underestimated the semantic and pragmatic purport of a text and ignored the intention of the author, whereas the anomalists addressed in depth the (possible) interpretations and the assumed intentions of a text but were wrong to dismiss any form of order between words. The two schools appear to have opposed each other from about 250–50 BCE, after which the controversy seems to have been settled or in any event to have gone away. Roman philologists in the first century BCE, such as Marcus Terentius Varro and Verrius Flaccus, would appear to have been influenced by both schools, given the attention they devote to both the regular and the extraordinary in their text analyses.[73] In 55 BCE even Julius Caesar wrote a book about the controversy in which he sought a balance between the two extremes. But Rome had no Homer to promote philological text reconstructions as the Alexandrians had had. It was not until the Palatine Library was set up in Rome in 28 BCE that philologists like Julius Hyginus (who was also the first Roman librarian) started to write commentaries and Valerius Probus began to reconstruct Latin texts using surviving manuscripts.

[70] Maria Broggiato (ed.), *Cratete di Mallo: I frammenti*, Agorà Edizioni, 2001. See also David Greetham, *Textual Scholarship: An Introduction*, Garland Publishing, 1994, p. 300.

[71] See the discussion in John Edwin Sandys, *A Short History of Classical Scholarship from the Sixth Century B.C. to Present Day*, Cambridge University Press, 1915, p. 49.

[72] For fragments of Demetrius of Scepsis, see <http://www.digitalclassicist.org>.

[73] Thomas Baier, *Werk und Wirkung Varros im Spiegel seiner Zeitgenossen von Cicero bis Ovid*, Franz Steiner Verlag, 1997.

Roman philology does not appear to have generated new principles or methods. It was primarily eclectic, as was virtually all Roman learning, science, and philosophy. However, Roman philology did pay more attention to the chronology and dating of texts. This is consistent with the large-scale production of chronologies in Roman historiography, which was in turn the consequence of the annalist tradition of the high priests (see 2.2). Conversely, philology also provided productive feedback for historiography, and Roman historians could make good use of texts conserved by philologists. We can therefore rightly talk about fruitful cross-fertilization here. For example, the philologist and linguist Varro drove chronological historiography forward with his reconstruction of *annals* texts.

Can we also say that there was cross-fertilization between philology and linguistics? As we have seen, the Alexandrian philologists had a significant impact on the development of the first Greek grammar, but we should realize that there was barely any differentiation between these two disciplines in the Hellenistic world. It was not until the end of Antiquity that grammars such as Dionysius Thrax's *Téchne grammatiké* were considered to be linguistic works which were not concerned with the philological goal of reconstructing the source. It would therefore appear that there was a process in which linguistics became independent from philology rather than a process of cross-fertilization.

Later Roman philology was primarily encyclopedic. As early as the first century BCE Varro wrote a general work, the *Disciplinarum libri novem,* about the liberal arts, the *artes liberales*, which covered the knowledge that every free man was expected to master. Varro's work has not survived, but we know from other texts that the liberal arts were classified as (1) grammar, (2) logic, (3) rhetoric, (4) geometry, (5) arithmetic, (6) astronomy, (7) music, (8) medicine, and (9) architecture. More or less the same classification was used as a general 'complete' education in the Hellenistic world from Sicily to the Punjab (see 2.7, Rhetoric). The first seven subjects of this series correspond to the *artes liberales* of St Augustine, which were elaborated by Martianus Capella in his allegory *Nuptiae Philologiae et Mercurii* (fifth century CE). The disciplines of grammar, logic, and rhetoric formed the *trivium,* and the disciplines of geometry, arithmetic, astronomy, and music made up the *quadrivium*. Martianus's work became one of the most important textbooks in the Christian Middle Ages and for centuries represented the basic knowledge in medieval monastery and cathedral schools.[74] But his work has nothing to do with source reconstruction. This is to be found in late Antiquity, but usually in the form of text exegesis such as Macrobius's influential commentary on Cicero's *Somnium Scipionis* ('Dream of Scipio', about the rewards in the hereafter for a righteous life) in the fifth century.

Evaluation of the two schools. How reliable were the methods of the Alexandrian philologists? A reconstruction of a source can be evaluated if the source (or a text older than the texts that have been handed down) is subsequently found. However, it is seldom if ever the case that a source can be reconstructed *exactly* from

[74] Reinhold Glei (ed.), *Die sieben freien Künste in Antike und Gegenwart,* Bochumer Altertumswissenschaftliches Colloquium 72, 2006.

surviving copies (see also 4.1 and 5.2), and we can therefore only talk about the degree of accuracy of a reconstruction, if at all. In the case of the Alexandrian method, unfortunately none of the analogistic text reconstructions has survived, let alone original source texts (the Alexandrian reconstructions only survive in the form of medieval copies). The only 'surviving' classical library, the *Villa dei Papiri* in Herculaneum, was buried under three metres of mud and lava during the eruption of Vesuvius in 79 CE. The carbonized scrolls have been partially reconstructed by computer tomography and it emerges that on the whole they contain sources of the then influential Epicurean scholar Philodemus of Gadara.[75] Sadly no later copies of Philodemus's work have survived, so in this case we have the reverse situation of a source text without any copies.

The anomalistic method of the Stoic philologists is a very different story. The erudite, detailed text exegeses from Pergamon are perhaps the finest that the classical humanities generated. But is it scholarship or literary art? Whatever the case may be, the results of the anomalists cannot be verified empirically, let alone replicated. But then again, that was far from their intention.

Philology in China and India. The philological study of literary texts also flourished in ancient China. While there were many commentaries, text compilations, and text reconstructions,[76] it is hard to find any precise principles of textual criticism. But since there was a flourishing reconstruction practice, it should be possible to distil the underlying principles of Chinese text reconstruction. Yet, much work needs still to be done to understand philological method in early China.

The oldest known Chinese text reconstruction has been attributed by Mencius (372–289 BCE) to Confucius, who was believed to have compiled the first overview of the Chinese classics—the so-called Confucian classics.[77] But we do not know on which principles the reconstruction of these classics was based. There is a mythical story that after the book burning under Li Si (213 BCE during the Qin Dynasty) the Confucian classics could be reproduced virtually effortlessly thanks to the exceptional memories of the Confucians. According to the *Shiji*—the grand historiographical overview by Sima Qian, see 2.2—we have Fu Sheng to thank for some of the sources that have been handed down. He hid works by Confucius in a wall in his home. The construction of the Bamboo Annals, which describe Chinese history from 2000 BCE to about 300 BCE, is another example. After their discovery in 297 CE the bamboo slips, which had been left loose, had to be restored to their original sequence. However, the anomalous divergent script of the characters used was not comprehended and the reconstruction by Shu Xi (*c.*261–*c.*302 CE) during the Jin Dynasty was so controversial that for centuries the Bamboo Annals were thought to be a forgery.[78] A firm philological methodology, as we know it from the

[75] Marcello Gigante, *Philodemus in Italy: The Books from Herculaneum*, 2nd edition, University of Michigan Press, 2002.

[76] For an overview, see Michael Loewe (ed.), *Early Chinese Texts: A Bibliographical Guide*, Society for the Study of Early China, 1993.

[77] We now know that the classics were not compiled by Confucius but by later scholars—see Chin Annping, *The Authentic Confucius: A Life of Thought and Politics*, Scribner, 2007.

[78] Edward Shaughnessy, *Rewriting Early Chinese Texts*, SUNY Press, 2006, pp. 191ff.

Alexandrians,[79] seems to have developed in China only in the Ming period (1368–1644) when explicit principles of textual criticism were formulated (see 4.1)..

The Indian Vedas were handed down with the utmost precision, and they are known as the oldest continuous oral tradition in the world.[80] The goal was the complete and perfect memorization of the holy Vedic texts on the basis of songs (*pathas*). Every two adjoining words in the text were learned (and presented) first in their original sequence, after which they were repeated in reverse order, and finally in their original sequence again. The reverse had no implications for the meaning because of the free word order in Sanskrit. Special verse structures and rhythms ensured that the texts remained intact down to each syllable. This method proved to be so effective that a collection of the oldest Indian Vedas, the *Rigveda* (*c.*1500 BCE), is claimed to exist as one single text without variants.[81] It is of course true that in India, as in other countries, not all texts were memorized.[82] Even so it appears that the problem of inconsistency was limited and did not trigger a scholarly practice of text reconstruction.

It therefore seems that philology as the search for principles of reconstructing the original source was created where there were large quantities of inconsistent manuscripts. This occurred first of all in the Hellenistic world, where hundreds of thousands of manuscripts were collected and which represented an empirical world on their own.

[79] Geoffrey Lloyd and Nathan Sivin, *The Way and the Word: Science and Medicine in Early China and Greece*, Yale University Press, 2002, p. 137.

[80] Gavin Flood (ed.), *The Blackwell Companion to Hinduism*, Blackwell, 2003; Arthur MacDonell, *A History of Sanskrit Literature*, Kessinger Publishing, 2004.

[81] Friedrich Max Müller, *The Hymns of the Rigveda, with Sayana's Commentary*, 4 volumes, 2nd edition, Oxford University Press, 1890–2. See also Karl Friedrich Geldner, *Der Rig-Veda: aus dem Sanskrit ins Deutsche Übersetzt*, Harvard Oriental Studies, volumes 33–7, 1951–7.

[82] Radha Kumud Mookerji, *Ancient Indian Education*, 2nd edition, Motilal Banarsidass Publications, 1998, pp. 535ff.

2.4 MUSICOLOGY: THE LAWS OF HARMONY AND MELODY

The oldest law in 'science' is a law of music.[83] It is about harmonic (or consonant) intervals and is attributed to Pythagoras (sixth century BCE). Consonant intervals are harmonies in which the separate notes dissolve into one another, so to speak. Pythagoras's law of music states that consonant intervals correspond to the ratios between the first four whole numbers (1, 2, 3, and 4), while the dissonances reflect more complicated ratios.[84] The ratios can be established by plucking a string, and then plucking a half, two-thirds, or three-quarters. You then hear an octave, a fifth, and a fourth, whereas non-harmonic intervals, such as the second, relate to ratios like 8/9.[85] Although the Pythagorean law of music appears to predict the 'simplest' harmonic intervals well, we run into problems with such intervals as the third and the sixth. The thirds are generally perceived to be consonant, but in the Pythagorean system they correspond to a more complex ratio (64/81) than the dissonant second (8/9). Pythagoras rejected the thirds, probably for 'numerological' reasons. Likewise Pythagoras's law of music does not explain why a relatively simple ratio, 8/9 (the second), sounds horrendously dissonant while 3/4 (the fourth) sounds beautifully consonant.

Pythagoras's harmony theory. The central question in *harmony theory* is *whether the consonant intervals are based on an underlying system.* Pythagoras believed they were, and made the consonances coincide with simple arithmetical proportions of the first four whole numbers, the sum of which is equal to ten—the holy Pythagorean number. This result was considered to be very important by the Greeks, who generalized it to the world view that is known as the *harmonia mundi.* According to this world view the cosmos, like music, can be described using simple mathematical proportions. In the *Timaeus,* Plato even contended that a scale could be derived from the cosmos in which the ideal distances between the seven heavenly bodies 'resound'. The relative distances between the then seven known 'planets' (Sun, Moon, Venus, Mercury, Mars, Jupiter, and Saturn) were thought to correspond to the ratios of the musical intervals. This world view, known as the Music of the Spheres, survived for over 2,000 years.[86] It is attributed to Pythagoras, who included music in the other mathematical disciplines (arithmetic, geometry, and astronomy) under the name *Mathēmata.* A century later the sophists added the *Mathēmata* to the three 'linguistic' disciplines (grammar, logic, and rhetoric), which jointly comprised the liberal arts (*artes liberales*) (see also 2.3). Within this

[83] H. Floris Cohen, *Quantifying Music: The Science of Music at the First Stage of the Scientific Revolution, 1580–1650,* Reidel Publishing Company, p. 2.

[84] Although the law has been attributed to Pythagoras, it was already known centuries earlier by the Babylonians—see M. L. West, 'The Babylonian musical notation and the Hurrian melodic texts', *Music & Letters,* 75(2), 1994, pp. 161–79.

[85] The non-musical reader is advised to consult <http://en.wikipedia.org/wiki/List_of_musical_intervals> for (links to) sound fragments—where it should be noted that interval fractions such as 1/2, 2/3, 3/4, etc. are also represented by their reciprocals 2/1, 3/2, 4/3, etc.

[86] Even in the 17th century Johannes Kepler spared no effort to let the relative distances between the planets correspond with ideal mathematical proportions.

curriculum the study of music was considered to be a mathematical discipline that focused on the relationships between the consonant intervals or 'good' harmonies.

But when does a musical harmony actually sound 'good'? Can a clear orderly structure really be found here? The Pythagoreans' results suggest an orderly structure for the simplest intervals, but it proved not to be possible to generalize this orderliness in the Pythagorean system to all 'harmonious' intervals, such as the sixth or the third. This soon gave rise to controversies, for example in the work of Aristoxenus of Tarentum (fourth century BCE). Aristoxenus, a pupil of Aristotle, is considered to have been the first 'musicologist' of Antiquity.[87] He was bitterly disappointed when Theophrastus was chosen instead of himself to succeed Aristotle as the head of the Peripatetic School. Nevertheless, Aristoxenus became one of the most significant music theorists in the history of musicology. He asserted that intervals, and therefore harmonies, did not have to be assessed on the basis of their simplest arithmetical ratios but on the grounds of *hearing*. He argued that because the Music of the Spheres could not be observed, it would be better to sing and play what sounds good to us, without its being governed by the simplicity of mathematical proportions. In other words Aristoxenus contended that empirical findings (based on observation) took precedence over theoretical calculations. It is important to stress that Aristoxenus did *not* reject a mathematical approach. On the contrary, he was fiercely opposed to the work of his non-mathematical contemporaries, referred to as 'harmonists', who did not want to apply any formal principles at all to the study of music and abandoned all laws of music. Aristoxenus's approach remained formal, but empiricism had the last word.[88]

Aristoxenus's laws of melody. One of Aristoxenus's most important innovations in his *Harmonic Elements* and *Rhythmic Elements* is the insight that music consists of laws that are empirical.[89] According to Aristoxenus these laws can be discovered by studying the pieces of music themselves instead of using preconceived mathematical or philosophical ideas. Here he departs from his teacher Aristotle, who assigned no independent value to music and whose position was still close to the Pythagoreans. Another, possibly even more significant innovation by Aristoxenus is the extension of musicology to the study of *melody*. *Which sequences of tones form orderly melodies and can we devise a system of rules for them?* This question seems similar to the one that Panini asked about language. Aristoxenus was the first to study specific pieces of music, an activitiy which we will designate as musicology.

Aristoxenus unearthed regularities, which he called 'natural laws of melody' and which specify possible successions of intervals for a 'correct' melody.[90] Firstly he

[87] Sophie Gibson, *Aristoxenus of Tarentum and the Birth of Musicology*, Routledge, 2005.

[88] Cf. Andrew Barker, 'Music and perception: A study in Aristoxenus', *Journal of Hellenic Studies* 98, 1978, pp. 9–16.

[89] For a translation of the extant parts of Aristoxenus's harmonic and melodic work, see Andrew Barker, *Greek Musical Writings*, volume 2, Cambridge University Press, 1989, pp. 119–89; and Lionel Pearson, *Aristoxenus: Elementa Rhythmica, the Fragment of Book II and the Additional Evidence for Aristoxenean Rhythmic Theory*, Oxford University Press, 1990.

[90] Edward Lippman, *Musical Thought in Ancient Greece*, Columbia University Press, 1964, pp. 22ff.

observed that not every random series of intervals between notes produces a melody. When singing, the voice appears to rank the intervals and notes on the basis of principles of *continuity* and *succession.* As is normal in Greek music, Aristoxenus establishes a close link here between melody and harmony. He describes in detail the possible intervals in the rise and fall of a melody, and to this end he introduces the ideas of *tetrachord* and *pyknon.* A tetrachord is half a scale; a series of four tones in a scale. A pyknon is a tone unit; its size depends on the scale and consists of two equal dieses—where one diesis is the basic unit of an interval, which in Aristoxenus corresponds roughly to a quarter tone. Aristoxenus then created an axiomatic system—once again in the tradition of Aristotle and Euclid—in which the laws (theorems) that he had found empirically were 'proven' deductively on the basis of axioms or first principles that he defined. Aristoxenus's axioms were particularly detailed, for example:[91]

Axiom I

Let it be accepted that in every genus, as the melodic sequence progresses through successive notes both up and down from any given note, it must make with the fourth successive note the concord (consonant interval) of a fourth or with the fifth successive note the concord of a fifth.

Axiom III

On the magnitudes of intervals, those of the concords appear to have either no range of variation at all, being determined to a single magnitude, or else a range which is quite indiscernible, whereas those of the discords (dissonant intervals) possess this quality to a much smaller degree.

These axioms are primarily harmonic but according to Aristoxenus they also form the basis for the melodic succession of notes. To this end, Aristoxenus introduces theorems that describe the restrictions on melodies and that are deduced from the axioms. Aristoxenus calls these theorems 'natural melodic laws', as the following two rules illustrate:

- No pyknon, neither a complete pyknon nor a part of one, is melodically adjacent to another pyknon.
- Starting from a tone, there is a progression in two possible directions: a third down or a pyknon up.

Aristoxenus gives a total of twenty-five rules or 'natural melodic laws'. Needless to say these rules do not represent universal laws but regularities of the Greek musical tradition at the time of Aristoxenus. Greek music consisted primarily of improvisations that were rarely recorded in musical notation. The singer or instrumentalist was at the same time the composer and soloist. This is not to say that the musical execution was spontaneous and unprepared. The performer had to obey generally accepted but unwritten rules covering melodic form and style. It was probably these unwritten conscious or unconscious rules that Aristoxenus tried to formalize. This is why we can define his 'laws' as a *restriction-based grammar* of the class of all the possible Greek music within which the composer-singer could operate. The

[91] Gibson, *Aristoxenus of Tarentum and the Birth of Musicology*, pp. 66–8.

twenty-five laws still give the singer tremendous freedom in terms of melodic options. They only describe the boundaries of the possible melodies and not the melodies themselves. Initially these laws were descriptive, but they were soon considered to be prescriptive, particularly by the later Aristoxenean school (see below).

Is there a procedural system of rules in Aristoxenus's theory of music? While there is a collection of rules, they are not procedural in the way that Panini's grammar is. This is because Aristoxenus's rules do not result in new melodies. They only describe or 'declare' the boundary conditions with which melodies must comply without giving an algorithmic procedure. We will therefore refer to Aristoxenus's working principle as the *declarative system of rules principle*. There is a system of declarative rules underlying the foundations of Greek melodies and this is what Aristoxenus wanted to expose.

It is striking to see how far Aristoxenus's method departs from the methods of Greek linguists like Dionysius Thrax and Apollonius Dyscolus (see 2.1). In the study of language it is likewise possible to state which words and groups of words can be combined, or not, to form larger units or sentences, together with their meaning (as Panini had done for Sanskrit). But no Greek linguist appears to have looked into this. They were primarily interested in the word classes and their subdivisions (see 2.1). A syntactic tradition in musicology apparently does not guarantee a syntactic tradition in linguistics or philology (nor is the converse true, as we will see with regard to Indian musicology). The extent to which Aristoxenus's melodic laws were accurate and empirically adequate has yet to be addressed. Unfortunately this cannot be established directly because the oldest surviving Greek melodies were composed approximately two centuries *after* Aristoxenus.[92] We will see how musicologists in the early modern age once again tackled the challenge of defining the rules underlying musical melodies, and we will be able to evaluate Aristoxenus's method indirectly (see 4.4).

The influence of Pythagoras and Aristoxenus. Although Pythagoras and Aristoxenus appear to be diametrically opposed as regards method and subject, both sought to find precise musical laws. Pythagoras used arithmetical proportions, even when they did not correspond to empiricism, while Aristoxenus worked on the basis of empirical findings that he traced back to first principles in the best Aristotelian tradition.[93] Pythagoras and Aristoxenus stood shoulder to shoulder in their opposition to the 'harmonists', from whom nothing has been handed down, but about whom Aristoxenus wrote that they ignored all laws. To a degree this opposition is comparable to the contrast that we find in philology (see 2.3) between the Alexandrian analogists, who pursued a system of rules, and the Pergamon anomalists, who believed that rules did not exist. As in the interpretation

92 There are only very few surviving Greek melodies from Antiquity: two Delphic hymns from *c.* 150 BCE, a drinking song from the same period, and a few hymns by Mesomedes of Crete from the second century BCE—see M. L. West, *Ancient Greek Music*, Oxford University Press, 1994.

93 See the discussion in Carl Huffman (ed.), *Aristoxenus of Tarentum: Texts and Discussions*, Transactions Publications, 2012.

of texts, there were two traditions in musicology: one formal (Pythagoreans and Aristoxeneans) and one informal (harmonists), and within the formal tradition there were theoretical (Pythagorean) and empirical (Aristoxenean) schools.

Aristoxenus's influence is less far-reaching than one would think on the grounds of his work. Although he gathered a following including such disciples as Cleionides, Aristides Quintilianus (not to be confused with the rhetorician Marcus Quintilianus or Quintilian—see 2.7), Bakcheios, and Psellos, his adherents ignored all axioms, proofs, and even the experimental-empirical foundation of Aristoxenus's work. This resulted in a considerable misunderstanding in the reception of Aristoxenus. Euclid (third century BCE) and Ptolemy (83–168 CE) incorrectly labelled his work as anti-mathematical. However, Euclid did prove that not all of Aristoxenus's empirically established theorems could be deduced from the axioms or first principles he established—but this did not detract from the theorems themselves, which in fact were not deductive but on the contrary empirical and inductive. After the death of Aristoxenus, Pythagorean musicology became dominant once again albeit after the introduction of a few empirical adjustments. In his *Canon* Euclid describes a Pythagorean version of the Aristoxenean theory of harmony, and in the work of Eratosthenes and Ptolemy we see a compromise between the two approaches to the theory of harmony. Based on Aristoxenus's criticism, Ptolemy conducted very precise experiments with the monochord, and described an earlier experiment by Aristoxenus as being too inaccurate. However, none of these changes affected the theory of melody, which Aristoxenus considered to be the most 'musicological'. The two harmonic approaches were discussed in *De institutione musica* by Boethius (480–525), who had an obvious Pythagorean stance. A particularly influential aspect of this work, which kept the controversy going into the Middle Ages (see 3.4), was Boethius's tripartite classification of music based on Plato: *musica mundana* (the inaudible music of the cosmos, or the harmony of the spheres), *musica humana* (the music of humans, or the harmony between body and mind), and *musica instrumentalis* (instrumental music, in which he also included vocal music). The harmony theory in Martianus Capella's textbook *Nuptiae Philologiae et Mercurii* (fifth century CE, see 2.3), in which the seven *artes liberales* are addressed, is primarily Quintilianist and Aristoxenean in nature. But it was to be the Pythagorean theory of harmony, with its strict mathematical ratios, that had the greatest influence in the Middle Ages and the early modern age. There was not a real Aristoxenus revival until the sixteenth century, thanks to humanists like Vincenzo Galilei, the father of Galileo Galilei (see 4.4).

Bharata Muni's theory of harmony and melody. The oldest musical tradition in India is linked to the ritual memorization of the Vedas on the basis of hymns (*pathas*), as explained in 2.3. There was a very precise yet prescriptive system of rules that is described in the third of the four Vedas, the *Samaveda* (which essentially means 'melodic knowledge'). The Samaveda gives specific rules for singing the hymns, but there is no musicological analysis of the songs whatsoever.[94]

[94] Ralph Griffith, *The Sāmaveda Saṃhitā*, translated by Ralph Griffith, 1893, revised and enlarged edition, by Nag Sharan Singh and Surendra Pratap, Nag Publishers, 1991.

The first musicological studies in India were conducted by the mythical figure Bharata Muni (*c.* first century BCE–first century CE), who in his *Natya Shastra* ('Treatise on the Performing Arts') laid the foundations for both Indian musicology and Indian dance and drama (see 2.8).[95] Until the thirteenth century his work was considered to be the 'definitive' study of Indian classical music. Bharata Muni primarily described musical instruments, but he also discussed the underlying system of rules that he found in (or prescribed for) Indian music. Bharata presented the system of rules in the form of principles, which seem to have been derived empirically from the musical material:

1. The principle of *shadja*: this is the first tone of a scale; if this tone has been specified, the other tones of the scale are also fixed.
2. The principle of *consonance*, which consists of two sub-principles:
   a. There is a fundamental note in the scale that is always present and is unalterable. (This is the tonic in a scale, which divides an octave into twelve semitones with seven basic tones: *sa re ga ma pa dha ni*, which in sequence correspond to *do re mi fa sol la ti*.)
   b. There is a natural consonance between tones, the best being between *sa* (the *shadja*) and the following *sa* (*tar shadja*), and the second best between *sa* and *pa*. (These two consonances correspond to the octave and fifth in Greek harmony theory, respectively.)
3. The principle of *emotion*: there are emotions for every tone, on the grounds of which musical modi or scales are defined, through which musical forms can be produced. (This resembles the melodic idioms that are used in Indian shrutis and *ragas*.)

The principles given above are very far removed from the formal melodic rules we have seen in the work of Aristoxenus. Yet Bharata Muni devotes so much attention to musical forms under the third principle that they could be considered as guidelines for 'possible' Indian melodies,[96] which would loosely correspond to a declarative system of rules. But Bharata's musical principles never come anywhere close to the great Indian linguistic tradition by Panini who sought an explicit and complete system of rules for possible Sanskrit utterances (see 2.1). With respect to harmonic theory, however, there is a closer link between Indian and Greek musicology, particularly with regard to the consonant tone harmonies with which a piece of music must comply. Like Pythagoras, Bharata identifies the octave and the fifth as the most consonant, although he does not go into the arithmetic proportions between these harmonies.

This resemblance between Pythagoras and Bharata may not be coincidental. It is very well possible that aspects of Greek musical knowledge were known in India at the time of Bharata Muni. Indians had already been exposed to Babylonian

[95] Anupa Pande, *A Historical and Cultural Study of the Natyasastra of Bharata*, Kusumanjali Prakashan, 1991.

[96] Pradip Kumar Sengupta, *Foundations of Indian Musicology*, Abhinav Publications, 1991, pp. 104ff.

knowledge due to the invasion of the Persians in the sixth century BCE, and later they encountered Greek learning after the invasion by Alexander the Great in 327–326 BCE. While it is usually assumed that knowledge transmitted was primarily about astronomy,[97] the possibility that musical knowledge, which was considered to be of cosmological importance, was covered too, is more than likely.[98]

Musicology in China: Liu An. The first analysis of Chinese music is found in the Book of Rites, the *Liji*, which was attributed to Confucius and in which a chapter is devoted to musical practice. It sets forth the relationship between music and reality on the basis of the pentatonic Zhou scale (named after the Zhou Dynasty). 'The basic note *kung* represents the sovereign; *shang* (one note above *kung*) represents the minister; *chiao* (one third above *kung*) represents the people; *chih* (one fifth above *kung*) represents national affairs and *yu* (one sixth above *kung*) represents things. If these five notes are not in disarray, harmony will prevail in the country.'[99] There are a few references to the 'rules of melody' in this chapter, but what they are is not discussed. In addition, according to Sima Qian (see 2.2), Confucius compiled 305 songs in the *Shijing* ('Book of Songs'), which still has a significant influence on contemporary Chinese literature and music. Whether Confucius really compiled the *Shijing* is highly doubtful.

The most important Chinese musicological work in Antiquity is credited to Liu An (179–122 BCE). This colourful character was King of Huainan during the Han Dynasty and also the legendary founder of the martial art of tai chi. *Huainanzi* ('The Masters of Huainan'), which is attributed to him, is one of the cornerstones of Taoist philosophy, including the famous theory of yin and yang (see also 2.2).[100] The work contains a collection of essays that were said to have arisen from the literary and philosophical debates between Liu An and his guests at court. But the *Huainanzi* also contains one of the most detailed musicological studies of Chinese tone temperaments and it gives a complete analysis of the *Pythagorean comma.*[101] Until the discovery in 1978 of the inscriptions on the famous bells of Zenghouyi (433 BCE), the *Huainanzi* was the oldest known analysis of the Chinese twelve-tone temperament, with approximations accurate to six figures and two decimals.[102] However, Liu An's approach is primarily mathematical and does not aim to describe pieces of music, let alone the class of possible melodic sequences.

[97] Nataraja Sarma, 'Diffusion of astronomy in the ancient world', *Endeavour*, 24, 2000, pp. 157–64.

[98] For example, in the ancient Indian astrological text Yavanajataka (which is a versification of an earlier translation into Sanskrit of an Alexandrian text), there are many references to Greek astrological and other knowledge. See David Pingree, *The Yavanajataka of Sphujidhvaja*, Harvard University Press, 1978.

[99] Walter Kaufmann, *Musical References in the Chinese Classics*, Detroit Monographs in Musicology No. 5, 1976, p. 37.

[100] For an English translation of the *Huainanzi*, see John Major, Sarah Queen, Andrew Meyer, and Harold Roth, 2010, *The Huainanzi: A Guide to the Theory and Practice of Government in Early Han China, by Liu An, King of Huainan*, Columbia University Press, 2010.

[101] The Pythagorean comma is the tiny (proportional) difference between twelve fifths and seven octaves. This interval has huge consequences for tuning musical instruments because in most music twelve fifths and seven octaves are viewed as the same interval.

[102] Ernest McClain and Ming Shui Hung, 'Chinese cyclic tunings in late Antiquity', *Ethnomusicology*, 23(2), 1979, pp. 205–24.

2.5 ART THEORY: THE VISUAL REPRODUCTION OF THE WORLD

How can we depict visual reality as faithfully as possible? This question is at one and the same time philosophical, technical, and artistic, and it was addressed from different points of view in Graeco-Roman Antiquity.[103] That art imitated nature was a truism, but this *mimesis* was also problematic. Plato, for example, thought that while painting and sculpture sought to reproduce reality, such reproductions did not represent the essence of things. They remained an outward semblance. Plato's vision of art is a result of his doctrine of ideas, in which the essence is created by ideas and not phenomena.[104] Aristotle is significantly more positive about art as an imitation or visual rendition of reality. According to him imitation is an element of the learning process: a step-by-step process from experience to understanding (see 2.8, Poetics).

Illusionism according to Pliny. The oldest surviving art history (and theory) is not Greek but was written by the Roman author Pliny the Elder (23–79 CE) in his *Naturalis historia* or Natural History (*c.*77 CE).[105] Pliny made use of earlier art historians, such as Xenocrates of Sicyon and Antigonus of Carystus (both in the third century BCE), but their works have not survived.[106] Pliny considered the golden age of Athens (fifth century BCE) as the beginning of the great age of painting, which reached a pinnacle in the fourth century BCE. Although his history begins with a complaint about the decline in painting in his own time, Pliny found a continuous line in art that can be described as an attempt to represent the world as *realistically* as possible, which is also known as *illusionism.* He explained that illusionist painting began with the drawing of a line around the shadow of a person, which led by way of a monochrome painting to a polychrome rendition of reality.[107] Pliny recounted the efforts made to achieve illusionism in the fifth century BCE with an anecdote about the rivalry between the painters Zeuxis and Parrhasius. Zeuxis had painted a bunch of grapes that was so lifelike that birds flew to it. Assured of his victory, he asked Parrhasius to unveil his painting, whereupon Zeuxis discovered that what he thought was a veil was in fact the painting itself. At this point Zeuxis acknowledged defeat because while he had been able to delude birds, Parrhasius had succeeded in making a fool of an artist.[108]

The rendition of emotions was also part of the quest for illusion. The sculptors Phidias and Praxiteles became famous because they had instilled the passions of the soul in stone. The painter Timanthes, on the other hand, was celebrated for the art of omission. In his painting of the sacrifice of Agamemnon's daughter Iphigenia,

[103] J. Pollitt, *The Ancient View of Greek Art*, Yale University Press, 1974.

[104] Plato, *Meno*, 71–81, 85–6; *Symposium*, 210–11.

[105] Silvio Ferri, *Plinio il Vecchio: storia delle arti antiche*, Fratelli Palombi, 1946.

[106] On Xenocrates of Sicyon (or of Athens), see Bernard Schweizer, 'Xenokrates von Athen: Beiträge zur Geschichte der antiken Kunstforschung und Kunstanschauung', *Schriften der Konigsberger Gelehrten Gesellschaft, Geistwissenschaftliche Klasse*, 9(1), 1932, pp. 1–52.

[107] Pliny, *Naturalis historia*, 35.15–16.

[108] Pliny, *Naturalis historia*, 35.65.

the king's face is covered by a veil, stimulating the viewer's imagination. The representation of a personality's deeper characteristics was even more important than the depiction of emotion. Apelles of Kos was highly praised for his portrayal of the physical likeness of Alexander the Great but also the striking reproduction of his heroic character, spirit, and temperament. According to Pliny, Apelles surpassed all the painters born before him or since.[109] Unfortunately, no single work by Apelles has survived. Pliny described illusionism as lifelike representation and the portrayal of emotions and character, but he also saw it as the capacity to depict *abstract* ideas. For example Zeuxis painted a picture of Penelope, Odysseus's wife, which according to tradition was the personification of virtue.

Can the pictorial representation of reality be formalized on the basis of rules? Although Pliny dealt with a number of the technical aspects of the depiction process, he emphasized coincidence and inspiration in his anecdotes. Pliny told the story of the painter Protogenes, who tried to portray a dog foaming at the mouth. After many failed attempts, Protogenes threw his sponge at the panel in a fury, and this produced precisely the visual effect he wanted. 'And so chance gave the painting naturalness!' wrote Pliny.[110] As regards inspiration, he recorded Phidias's refusal to use a model for his gigantic statue of Zeus in Olympia (one of the Seven Wonders of the Ancient World): Phidias took Homer's text as his sole inspiration.

We can of course take these anecdotes with a pinch of salt, but the message is clear. According to Pliny there is no general system of rules that can cover the entire (observable and non-observable) world. A new solution had to be found for a new pictorial problem, and a fortuitous discovery or a literary source of inspiration could be a golden opportunity. If there was a principle underlying the working methods of Protogenes and Phidias it was the *anomaly principle*, namely that new situations in the artistic process cannot be approached by rules but are exceptional.

The canon and the mathematical proportions principle. While there may not be a general system of rules, Pliny wrote about basic principles that artists draw on for their art—a *canon*. The concept of a canon was known before Pliny and originated in a lost treatise by the sculptor Polykleitos (fifth century BCE). A canon serves as a model or standard for other works. Pliny described the way Polykleitos created a canon with his sculpture the *Doryphoros* (spear-bearer) through which 'he created the art itself by means of a work of art'.[111] Sculpture is summarized in Polykleitos's statue itself: the correct pose of the spear-bearer (the contrapposto stance), the depiction of the different parts of the body, the refined detail, but above all the way in which the definable parts of the statue are related to each other in accordance with simple mathematical proportions based on Pythagoras's whole number ratios—see Figure 1.[112]

[109] Pliny, *Naturalis historia*, 35.79–97.
[110] Pliny, *Naturalis historia*, 35.104.
[111] Pliny, *Naturalis historia*, 34.55.
[112] Warren Moon (ed.), *Polykleitos, the Doryphoros and Tradition*, University of Wisconsin Press, 1995.

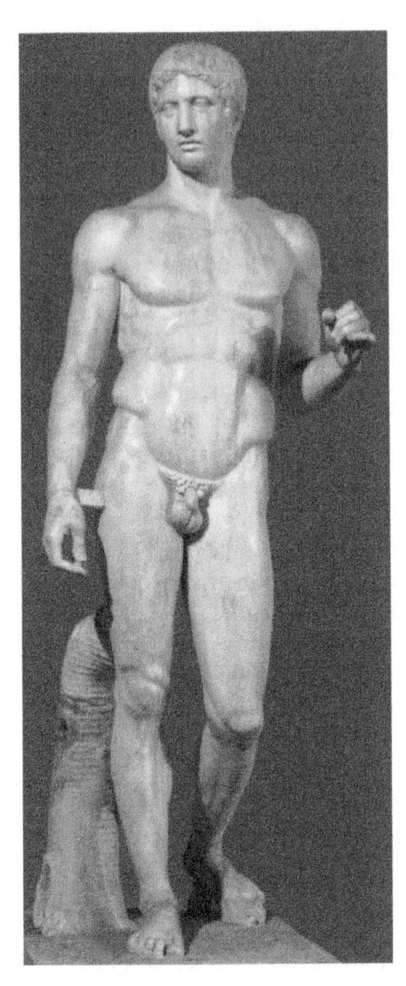

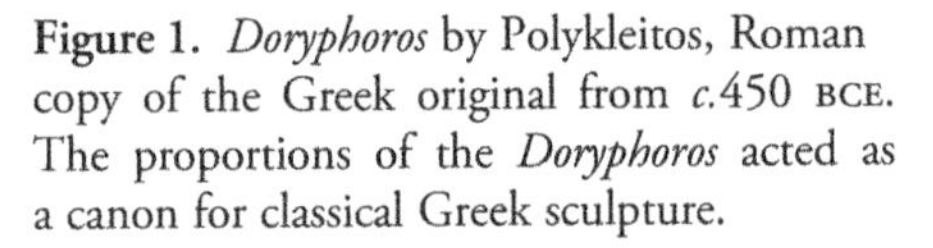

Figure 1. *Doryphoros* by Polykleitos, Roman copy of the Greek original from *c.*450 BCE. The proportions of the *Doryphoros* acted as a canon for classical Greek sculpture.

Polykleitos's canon comprises exact relationships between the dimension of the head and the height, between the width of the head and the shoulders, between the palm of the hand and the fingers, etc. When conceived as a canon, a system of proportions like this specifies the boundary conditions that other statues must meet. It does not prescribe what other statues should look like. It just states the rules within which a sculptor has the freedom to move. This appears to correspond to the *declarative system of rules principle* found in Aristoxenus's music grammar (see 2.4). However, given that these rules consist of mathematical ratios, we can also refer to it as the *mathematical proportions principle,* as in Pythagoras's theory of harmony (see 2.4).

The ratios in Polykleitos's canon seldom correspond to the exact anatomical body proportions of a specific person. But provided that the 'pure' relationships are given due regard, harmony and equilibrium can be achieved in the representation of the world. Illusionism in itself was therefore not enough to attain harmony. No matter how lifelike a sculptor might mold the world if he did not obey the rules of pure proportions his efforts would be in vain. Polykleitos's canon was warmly embraced by his contemporaries, as the hundreds of copies of the Doryphoros (of which five Roman copies still exist) testify. Polykleitos's canon and pure arithmetical ratios in sculpture and architecture are also referred to in approving terms in various other classical works, such as in the surviving text fragments of Philo Mechanicus and Galen. His sculpture was considered to be a masterpiece that referred to art itself. This self-reflexive aspect of art recurs in other periods, but we find the oldest (surviving) reference in the work of Pliny.

Proportions in architecture: Vitruvius. Pliny indicated the importance of mathematical proportions not only in sculpture but also in his overview of architecture. However, these architectural rules had already been treated by Marcus Vitruvius in *De architectura* (*c.*15 BCE).[113] Vitruvius's treatise is the only work on classical architectural theory to have survived. It was not until 1452 that Leon Battista Alberti published a new book about this subject in Europe (see 4.5).

The concept of *symmetria* is key in *De architectura*. According to Vitruvius, *symmetria* is the harmonious ranking of the parts of the whole according to the correct proportions. *Ordinatio*, that is the proper application of *symmetria*, results in *venustas*, or beauty, which is the value added by an architect to a building. According to Vitruvius, the qualities of good architecture are firmness (*firmitas*), commodity (*utilitas*), and beauty (*venustas*). These three qualities have dominated architectural theory to the current day.

Vitruvius described how the Greeks designed architectural styles: the Doric, Ionic, and Corinthian orders. It gave them a feel for proportions that led to an understanding of the ideal relationships between the parts of a building. These ideal ratios were explained by Vitruvius using the numbers ten and six. The number ten is a 'natural' perfect number because it is the number of fingers that nature gave to man, and six is a mathematically perfect number because it is equal to the sum of its factors (one, two, and three). All ratios of the parts to the whole, in both the human body and classical temples, can be defined in terms of ten and six and combinations of their factors. Everything was expressed in exact proportions, from the smallest anatomical details to the whole body, and from the minutest details of a temple (such as the triglyphs) to the building as a whole. The ratio between chin and head, between head and height, or between the diameter and height of a column, between the length and width of a temple, and even the positions of the doorways: it was all described in terms of numerical ratios.

According to Vitruvius (book III) even the different human positions correspond to the relationships between the two fundamental figures of the cosmic order—the square and the circle—which was depicted fifteen centuries later in a unparalleled

[113] For an English translation, see Vitrivius, *On Architecture*, translated by Richard Schofield, with an introduction by Robert Tavernor, Penguin Classics, 2009. See also Ingrid Rowland and Thomas Howe (eds), *Vitruvius: 'Ten Books on Architecture'*, Cambridge University Press, 2001.

way in Leonardo da Vinci's *Vitruvian Man*: a man who is standing in different poses in a square and a circle at the same time (see 4.5).

Overview of principles and patterns in Roman art historiography. Pliny's 'art-historical' method was based on several principles: he used the anomaly principle to demonstrate that good art cannot be spelled out in rules. At the same time he used the mathematical proportions principle—that had already been used by Pythagoras in the study of music—to formalize the regularities in the relationships in sculptures and buildings, which were employed as boundary conditions in a (declarative) system of rules. These ratios had already been described in the work of Vitruvius, and probably also in the treatises of earlier painters, sculptors, and art historians that have not survived. As a historian Pliny once again utilized the written source principle (and not the eyewitness account principle for instance). In addition to written sources, Pliny also seems to have made use of direct visual sources—a number of the works of art he described were in Rome. The patterns that Pliny identified were *progress in the faithful reproduction of reality* (illusionism), the *prototype work of art that establishes art itself* (the canon), and *harmonic proportions* (the link to Pythagoras). As we have seen, it is improbable that these patterns were discovered by Pliny himself. He based his writings on the results of earlier art theoreticians who were moreover artists in their own right. However, we have to rely almost entirely on Pliny as the only source of classical art history, apart from Vitruvius on architecture, that has been handed down.

The Sadanga and the Tala proportions. Rule systems with precise proportions for pictorial representation were also developed in India and China. The oldest known text about art in India is a theoretical treatise about Buddhist painting, the *Sadanga* or Six Limbs. According to tradition it was written around the first century BCE, but in the form that has been handed down it refers to a document by Vatsyayana from the third century CE.[114] The Six Limbs describe the principles of painting, which are:

(1) Knowledge of the appearances,
(2) Correct observation, measure, and structure,
(3) Action, feeling, form,
(4) Grace,
(5) Similitude,
(6) Use of brush and colour.

The Six Limbs seem to be largely prescriptive, but it is possible that the rules were derived from earlier Indian frescoes of the second century BCE. In terms of dimensions, the famous *Ajanta Cave* wall paintings in Maharashtra comply more or less with the second principle of the Six Limbs—correct observation, measure, and structure—which describes the rules of proportion, anatomy, and foreshortening (see Figure 2).

[114] Prithvi Agrawala, *On the Sadanga Canons of Painting*, Prithivi Prakashan, 1981.

Figure 2. Detail of a wall painting in cave I, second half fifth century, Ajanta, India.

The *Vishnudharmottara* (*c.*400 CE), a later treatise on Hindu art, contains similar rules that specify the constraints within which the artist is free to move.[115] The dimensions of limbs are spelled out down to the smallest detail in accordance with the *Tala* proportions, as are the forms and dimensions of the eyes, the different poses, foreshortening methods, the effect of shadows, the representation of kings, courtiers, warriors, animals, rivers, etc. In Indian art, as in Greek art, a declarative system of rules was developed that might initially have been derived empirically, but subsequently had a prescriptive role as a canon.

The Six Principles of Xie He, and early references to oblique perspective. Chinese art theory, like the Indian *Sadanga,* also specifies Six Principles. They were developed around 500 CE by Xie He, who lived during the Liu Song Dynasty and the Southern Qi Dynasty.[116] His critical text *Gu huapin lu* ('Classification of

[115] C. Sivaramamurti, *Indian Painting*, National Book Trust India, 1970, pp. 20ff.

[116] Osvald Sirén, *The Chinese on the Art of Painting: Texts by the Painter-Critics, from the Han through the Ch'ing Dynasties*, Dover Publications, 1936, reprinted in 2005, p. 219.

Painters') appears to have been inspired by the Six Limbs, which might have been brought to India by Chinese Buddhist monks. Xie He described the Six Principles as follows:

1. Spiritual resonance or vitality: this is the energy transmitted by the artist in the work (according to Xie He there was no point in looking further at a work of art without spiritual resonance),
2. 'Bone method': brushstrokes that express self-confidence, power, and elasticity (like bones),
3. Resemblance to the subject,
4. Suitability: the appropriate use of colour and tone,
5. Division: making a composition, space, and depth,
6. Transmission by copying: the pictorial representation of models as imitations of earlier works.

Xie He's Six Principles started a long tradition in Chinese art criticism (see 3.5 and 4.5). He also produced a list of the names of twenty-seven painters with a brief description of their works and the way they were—or were not—able to combine the Six Principles. Xie He used a point system to rank painters into six classes. The painter Wei Xi came out top of the ranking because of the fine details in his brushwork and his integration of the Six Principles, in which 'he surpassed all the great classical masters'.

Only a few works by the painters classified by Xie He have survived. The best known of the painters he identified is Gu Kaizhi, who was active in the fourth century.[117] His *Admonitions of the Instructress to the Court Ladies* is among the most important works of art in Chinese Antiquity (see Figure 3). Nevertheless, Xie He only granted Gu Kaizhi a place in the third class when he wrote, 'His style was fine and subtle, his brush without a flaw, yet his workmanship was inferior to his ideas, and (in his case) fame surpassed reality.'[118] Xie He's work is the earliest Chinese art history.

Chinese painting also displayed knowledge of perspective. This type of perspective is often referred to as 'oblique perspective' and represents the receding lines as parallel obliques, as also found in Graeco-Roman painting.[119] But just as with Graeco-Roman painting, it is hard to find any contemporary analysis of this kind of perspective, let alone a theoretical underpinning. The closest is a text attributed to the aforementioned Gu Kaizhi, *On Painting*, which has not been preserved except in corrupt fragments, and which mentions the use of ruled lines as a painting

[117] Michael Sullivan, *The Arts of China*, 4th edition, 1999, University of California Press, p. 97. See also Richard Barnhart, Yang Xin, Nie Chongzheng, James Cahill, Lang Shaojun, Hung Wu, and Wu Hung, *Three Thousand Years of Chinese Painting*, Yale University Press, 2002.

[118] Sirén, *The Chinese on the Art of Painting: Texts by the Painter-Critics, from the Han through the Ch'ing Dynasties*, p. 220.

[119] On perspective in Graeco-Roman art, see Erwin Panofsky, *Renaissances and Renascences in Western Art*, Harper and Row, 1969, pp. 128ff.

Figure 3. Gu Kaizhi, detail of the *Admonitions of the Instructress to the Court Ladies*, end of the fourth century. (Possibly a copy from the Tang Dynasty.)

technique.[120] This technique became later known as 'ruled-line painting', in which a ruler was used as a tool to guide the brush.[121] The first treatise that dealt with the principles of the well-known 'linear perspective' was not published before the fifteenth century by Leon Battista Alberti (see 4.5).

[120] Susan Bush and Hsio-yen Shih (eds), *Early Chinese Texts on Painting*, Harvard University Press, p. 20.

[121] Bush and Shih, *Early Chinese Texts on Painting*, pp. 248ff. See also Sullivan, *The Arts of China*, p. 185.

2.6 LOGIC: THE RULES OF REASONING

What is valid reasoning, and when can a statement be correctly considered as true or untrue? These are the key questions in logic, also known as dialectics. Logic probably developed as the art of disputation in legal disputes where the cogency of an argument could make or break a case. Together with rhetoric and grammar, logic made up the *trivium* in the *artes liberales*, which formed the classical curriculum in Europe for many centuries. But classical logic had not one but three major traditions: Greek, Indian, and Chinese, all of which flourished the most in the fourth century BCE.

Zeno, Plato, and the deductive logic of Aristotle. Zeno of Elea developed the first Greek principle of reasoning in the fifth century BCE, known as the *reductio ad absurdum*: an evidently untrue or absurd conclusion deduced from an assumption (premise) is used to demonstrate that the assumption itself is untrue. For example, Plato (428–347 BCE) describes in his *Parmenides* how Zeno is said to have proved monism on the basis of the absurd consequence of assuming a multiplicity of gods.[122] Plato himself was more interested in the philosophical aspects of logic, for instance, what is the nature of the connection between an argument's assumptions and the conclusion, and what is the nature of a definition? These questions come to the fore in his dialogues, for example *Theaetetus*, the *Republic*, and the *Sophist*.

Aristotle was the first person who tried to expose the system of rules between premises and conclusion. The composition and structure of reasoning was analysed in a collection of six works called the *Organon*. Aristotle's logic, and in particular his theory of *syllogism*, had an unprecedentedly significant impact. His system of logic was used in virtually all intellectual activities in Europe until well into the nineteenth century.[123] A syllogism is a logical argument in which a proposition (the conclusion) is deduced from two other propositions (the premises).[124] Aristotle considered syllogism as the core of *deductive* reasoning in which facts were determined by combining statements, contrary to *inductive* reasoning in which facts are determined by repeated observations. A syllogism consists of three parts: a major premise, a minor premise, and a conclusion. A well-known example is the following valid reasoning:

(1) *Major premise*: All men are mortal.
(2) *Minor premise*: All Greeks are men.
(3) *Conclusion*: All Greeks are mortal.

This syllogism is of the type:

[122] Plato, *Parmenides*, 126a–128e.

[123] For a history of Western logic, see William Kneale and Martha Kneale, *The Development of Logic*, Clarendon Press Oxford, 1978.

[124] Or as Aristotle put it in his *Prior Analytics*, a syllogism is 'a discourse in which, certain things having been supposed, something different from the things supposed results of necessity because these things are so' (24b18–20).

(1) *Major premise*: All A are B.
(2) *Minor premise*: All C are A.
(3) *Conclusion*: All C are B.

Instead of the term 'all', syllogisms can also contain the terms 'some', 'no', and 'not'. For example, the following reasoning includes the term 'all', 'some', and 'not':

(1) *Major premise*: All informative items are useful.
(2) *Minor premise*: Some books are not useful.
(3) *Conclusion*: Some books are not informative.

This syllogism is of the type 'All A are B. Some C are not B. Some C are not A.' To put it more precisely, a syllogism comprises three propositions, of which the first two, the premises, have exactly one term in common, and of which the third, the conclusion, contains the two non-common terms. Although there is an infinite number of possible syllogisms, a total of 256 different types using the four predicates 'all', 'some', 'no', and 'not' can be identified; but only a maximum of sixteen are valid. For example, the following syllogism is *not* valid reasoning:

(1) *Major premise*: All men are mortal.
(2) *Minor premise*: Some Greeks are mortal.
(3) *Conclusion*: Some Greeks are men.

The conclusion may perhaps be correct, but the reasoning is invalid. It does not necessarily follow from the premises that all people are mortal and some Greeks are also mortal that some Greeks are therefore men. One can easily see this if 'Greeks' is substituted by e.g. 'dogs'. In other words, the syllogism of the type 'All A are B. Some C are B. Some C are A.' is invalid.

The maximum of sixteen valid types of syllogism could be used repeatedly to construct new valid reasoning, which according to Aristotle established the underlying basic reasoning system, or dialectics. Nevertheless, Aristotle's logic was far from perfect. It could not even be used to describe the most elementary reasoning in Euclid's *Elements*. However, logic could be used to a certain extent to describe 'everyday' reasoning, as used in speeches (see 2.7).

In his *Metaphysics* (Book IV), Aristotle proposed some even more basic principles of human reasoning in addition to these syllogisms. These principles are the *law of non-contradiction* (an assertion and its negation can never be true at the same time) and the *law of excluded middle* (every assertion is either true or untrue). These laws can be considered as the criteria with which all valid reasoning and correct argumentation must comply.

Stoic propositional logic. A rather different type of logic developed in the century after Aristotle: *propositional logic*, which arose from the Megaric and Stoic schools.[125] In this extremely original branch of logic the truth or untruth of combinations of (or operations on) propositions is deduced from the truth or

[125] Susanne Bobzien, 'Stoic logic', in K. Algra, J. Barnes, J. Mansfeld, and M. Schofield (eds), *The Cambridge History of Hellenistic Philosophy*, Cambridge University Press, 1999, pp. 92–157.

Table 1. Truth table for conjunction, disjunction, implication, and equivalence.

A	B	A and B	A or B	A → B	A ↔B
untrue	untrue	untrue	untrue	true	true
untrue	true	untrue	true	true	untrue
true	untrue	untrue	true	untrue	untrue
true	true	true	true	true	true

untruth of the propositions themselves. Operations on propositions take place on the basis of *connectives*: the *negation* ('not'), the *conjunction* ('and'), the *disjunction* ('or'), the *implication* ('if..then..'), and the *equivalence.* For example, the conjunction used in the proposition 'John is clever and Peter is stupid' can only be true if both propositions 'John is clever' and 'Peter is stupid' are true. If one of the two conjugated propositions is untrue, the whole proposition is untrue. The situation is different with disjunction. The proposition 'John is clever or Peter is stupid' is only untrue if both 'John is clever' and 'Peter is stupid' are untrue. And by implication the proposition 'If John is clever, then Peter is stupid' is only untrue if 'Peter is stupid' is untrue and 'John is clever' is true.

In this way it is possible to construct a truth table that states the truth value of a complex statement on the basis of the possible combinations of truth values of the individual statements. In the following truth table (Table 1) A and B represent propositions, and the arrow → represents the implication 'if A then B', and the double arrow ↔ represents 'A is equivalent to B'. Negation is not considered in this table because it only connects to one proposition and not two: if the truth value of A is equal to 'true', then the truth value of its negation is equal to 'untrue', and vice versa. The truth values of the combined propositions are given for every combination of the truth values of A and B (in the first two columns). For example, the first row in the table shows, among other things, that if the truth values of both A and B are equal to 'untrue', the truth value of the conjugated proposition 'A and B' is also equal to 'untrue'.

The power of a truth table is that the truth values of more complex propositions, such as 'If A and B then A or B' can be calculated by dividing them up into the smallest sub-propositions. In one of the surviving fragments of Stoic logic,[126] Philo of Megara (*c.*300 BCE) does indeed give a truth table of the implication 'A → B', the oldest in the world.[127] A generation later, under Chrysippus of Soli, propositional logic was given an axiomatic foundation and was forged into a coherent, systematic whole.[128] This was an unprecedented result in logic. Yet surprisingly Chrysippus was also the founder of the *anomalistic* school of philology, where all patterns and regularities were rejected (see 2.3). At the end of the book we will return to this fascinating scholarly disunity that we find among some thinkers (see the conclusion

[126] 'Stoics', in K. Hülser (ed.), *Die Fragmente zur Dialektik der Stoiker*, 4 volumes, Frommann-Holzboog, 1987–88.

[127] W. V. O. Quine, *Mathematical Logic*, W. W. Norton, 1940, p. 15.

[128] 'Stoics', in K. Hülser (ed.), *Die Fragmente zur Dialektik der Stoiker.*

to chapter 5). In any event the Stoics believed that rule systems did not apply to areas beyond dialectics, for example philology or linguistics.

Despite the precise axiomatization of propositional logic, it was no match for Aristotle's logic. Syllogistic logic was simply much more practical and was adopted by the Neoplatonists and others in the third century CE. Roman logic, from Cicero to Boethius, produced no new insights and Aristotle also remained the basis for logic in the Middle Ages and thereafter.[129] Until the nineteenth century Stoic logic was looked upon as a formalistic system that was not essentially different from the Aristotelian system. It was not until the development of modern propositional calculus that the value of the Stoics' achievement was recognized and became part of *predicate logic* (see 5.3).

Classical logic is probably the only Greek discipline that adequately compares with Panini's grammar (see 2.1). A formal system was devised for both Aristotelian reasoning and Paninian ideas of grammar that covered all possible lines of reasoning or linguistic utterances on the grounds of a finite rule-based procedure. Did this make logic, like linguistics, an empirical subject? In other words, can we test rules or patterns of reasoning for correctness in specific, practical situations? Yes, we can, as will we see when we discuss rhetoric in 2.7.

The Rigveda and inductive Nyaya logic. In India there was also a flourishing tradition in the study of logic that went back even further than the Greek one. The oldest speculations about logic are to be found in the *Rigveda*[130] (*c.*1500 BCE, see 2.3) in which various logical distinctions were made, such as 'A', 'not A', 'A and not A' and 'not A and not not A'.[131] The first school of logic appears to have been founded by Medhatithi Gautama in the sixth century BCE.[132] However, it was Punarvasu Atreya (*c.*550 BCE) who is known as the author of the first logical-rhetorical work, the *Charaka-Samhita*, which we will consider in 2.7 (Rhetoric). Panini's grammar, as described in the *Ashtadhyayi*, is also sometimes considered to be a logical treatise. As we commented in 2.1, his grammar is based on an underlying system of rules that has the nature of an algorithm, in other words a procedure of ordered rules that produces a result on the basis of a finite number of basic operations.

The most important Indian school of logic is *Nyaya* (Sanskrit for 'recursion' or 'inference'). This school was based on texts by Aksapada Gautama around 200 CE, known as the *Nyaya Sutras*.[133] Nyaya followers saw knowledge and logic as a way to

[129] R. W. Sharples, *Stoics, Epicureans and Sceptics: An Introduction to Hellenistic Philosophy*, Routledge, 1996, pp. 24–6.

[130] Arguably the most accurate translation of the Rigveda is Karl Friedrich Geldner, *Der Rig-Veda: aus dem Sanskrit ins Deutsche Übersetzt*, Harvard Oriental Studies, volumes 33–7, 1951–7. Existing English translations of the Rigveda are based on older, nineteenth-century translations, such as Ravi Prakash Arya and K. L. Joshi, *Ṛgveda Saṃhitā*, Parimal Publications, 2001.

[131] RV 10.129 (129th hymn of the 10th Mandala of the Rigveda).

[132] Satis Chandra Vidyabhusana, *A History of Indian Logic: Ancient, Mediaeval, and Modern Schools*, Motilal Banarsidas Publishers, 1971; see also Radha Kumud Mookerji, *Ancient Indian Education*, Motilal Banarsidass Publications, 2nd edition, 1998, pp. 319–23.

[133] *Nyaya-Sutras of Gautama*, 4 volumes, translated by Ganganatha Jha, 2nd edition, Motilal Banarsidass Publications, 1999.

escape from suffering, and therefore made an effort to attain the correct forms of reasoning. Four sources of knowledge were identified: *observation, inference* (or *deduction*), *comparison*, and *evidence*. Inference, or *Anumana*, is of primary importance to logic. It is considered to be a combination of inductive and deductive reasoning and one of the most significant Nyaya insights.[134] Nyaya inference comprises five steps, as in the following examples:

(1) There is fire on the hill. (*pratijna*, what has to be proved)
(2) Because there is smoke. (*hetu*, the reason or cause)
(3) Where there is smoke, there is fire, as in an oven. (*udaharana*, the example)
(4) Just as on the hill. (*upanaya*, applying the example to the case)
(5) Therefore there is fire on the hill. (*nigamana*, the conclusion)

The characteristic property of the Nyaya inference is the emphasis on the example and its application in a new situation. Contrary to Aristotelian logic, this form of reasoning is not deductive and therefore less 'strict', but consequently more broadly applicable in practice. It was used in ancient India for making medical diagnoses by trying to establish the most probable clinical description on the basis of symptoms. The Nyaya inference is also easy to use in rhetoric or oratory, where inductive argumentation was usually dominant, whereas Aristotelian syllogistics had to follow a tortuous path, such as making premises reasonable instead of absolute (see 2.7). However, with the exception of medicine, the Nyaya logicians restricted themselves to philosophy and theology. Despite the many commentaries on their logic, their theory of inference remained virtually unchanged until the tenth century.

Mohism and analogical logic. Logical themes emerged in China as early as the 'Book of Changes' (the *Yijing*) and among the Chinese sophists, such as Gongsun Long's paradox 'white horses are not horses'.[135] But the most important school of logic, the *Mohist* school, was founded by Mozi (also called Mo Tzu) in the fifth century BCE. Mozi was the first to dispute the Confucianist school by putting logic at the centre of his studies. This resulted in the *Mohist canons*, which used logic as the basis for all other disciplines from economics to optics—although Mozi wanted nothing to do with music, which he thought useless and elitist.[136] The Mohists concentrated on *analogical* reasoning rather than formal logic or explicit ideas of logical inference. This was done using four different basic techniques:

Illustrating (*pi*): propounding similar situations
Parallelizing (*mou*): placing statements side by side and accepting them all
Adducing (*yuan*): putting forward a precedent
Inferring (*tui*): an analogical extension of what the opponent asserts

[134] Sue Hamilton, *Indian Philosophy: A Very Short Introduction*, Oxford University Press, 2001, pp. 4ff.

[135] Gongsun Longzi, 'On the white horse', in Philip Ivanhoe and Bryan Van Norden, *Readings in Classical Chinese Philosophy*, Hacket Publishing Company, 2005, pp. 363–5.

[136] For an in-depth study of the Mohist canons, see A. C. Graham, *Later Mohist Logic, Ethics and Science*, The Chinese University Press, 1978.

None of these techniques was described by using formal, step-by-step inferences as in Greek and Indian logic. Although this reasoning method is less reliable than deductive and inductive reasoning, Mohist logic is probably closer to everyday human analysis, which according to (present-day) cognitive scientists is based more on making analogies than applying formal rules.[137] But it was not the case that Mohist logic permitted every analogical argumentation. For example, the Mohists formulated *basic principles* to which all reasoning and argumentation had to be subjected. The first of these principles says that of two contradictory statements one must be false: they cannot be true at the same time. This corresponds to the *law of non-contradiction* as put forward by Aristotle (see above). The second principle says that two contradictory statements cannot be both false, one of them must be true. This corresponds the *law of the exluded middle.*[138] It appears that the Mohists were the first to formulate these two basic logical laws (see 2.7).[139]

Yet things did not turn out well for Mohist logic or for logic in general in ancient China.[140] During the Qin Dynasty (221–206 BCE) Mohism was banned by the strict legalistic regime, and it was not until the seventh century CE that the practice of logic reappeared under the influence of Buddhist philosophy.

Parallels between Greece, India, and China. It is fascinating to observe that the development of logic in Greece, India, and China was parallel in several respects. Deductive, inductive, and analogical logic were created independently of each other before Hellenism (prior to Alexander the Great's conquests). The same applies to the basic principles of logical reasoning, such as the laws of non-contradiction and excluded middle discovered by the Mohists and Aristotle, who were unaware of each other's existence. We found a similar parallel in historiography, for example the discovery (or recognition) of the cyclical pattern of rise, peak, and decline, and to a certain degree in the theory of musical harmony. There is still no explanation for these remarkable parallel developments, particularly not where they are pre-Hellenistic.

[137] Dedre Gentner, Keith Holyoak, and Boicho Kokinov (eds), *The Analogical Mind: Perspectives from Cognitive Science*, The MIT Press, 2001.

[138] These two principles correspond respectively to A74 and A75 in Ian Johnston, *The Mozi: A Complete Translation*, The Chinese University Press, Hong Kong, 2010.

[139] Jialong Zhang and Fenrong Liu, 'Some thoughts on Mohist logic', in Johan van Benthem, Shier Ju, and Frank Veltman (eds), *A Meeting of the Minds: Proceedings of the Workshop on Logic, Rationality and Interaction*, College Publications, 2007, pp. 85–102.

[140] Chad Hansen, *Language and Logic in Ancient China*, University of Michigan, 1983.

2.7 RHETORIC: ORATORY AS A DISCIPLINE

Nowadays rhetoric no longer exists as an independent discipline. As the art of persuasion it is sometimes part of applied linguistics or literary studies, and as the practice of oratory it has a dubious reputation at best.[141] How different this is from the situation in Antiquity and for many centuries thereafter. Rhetoric was the crowning glory on the foundations of grammar and logic. As the 'art of speaking well' it represented the completed education of every cultured young person and was directly applicable in law and politics. From Aristotle to Caesar and Hermogenes, philosophers, statesmen, and philologists spent time analysing speeches. What is the structure of a successful speech? What sort of arguments does an audience find convincing? Where do people get their arguments from? And how should a speech be delivered and using which gestures? These questions were of obvious importance in an oral culture that was (more or less) democratic. You had to persuade your listeners that you were right, and more importantly you had to be proved right.

Sophists versus Plato. What made rhetoric in Antiquity an empirical discipline, was the realization that there was a system underlying oratory and that this system could be applied in practical situations such as politics and the administration of justice. And indeed the first Western rhetoricians were teachers—the *Sophists* in the fifth century BCE such as Gorgias, Protagoras, and Isocrates.[142] Their starting point was that absolute truth did not exist and that there were arguments for and against every point of view. Rhetoric consisted of training in verbal techniques and searching for persuasive arguments. It is a commonly held view that Western rhetoric was indeed developed in fifth-century Athens thanks to democratic structures in which a substantial proportion of the population participated. This generated a demand for education in verbal techniques for convincing people, which resulted in Sophist schools and analytical studies of methods of persuasion.[143]

But Plato was extremely critical of the Sophists. According to him, claims that the Sophists could teach their pupils were unmerited. They contributed no specific relevant knowledge and their rhetoric had no subject. In contrast, Plato propagated the dialectic method (of Socrates). In his *Gorgias* he rejected the practice of a debate with a winner and a loser, believing on the contrary that a dialogue between two people with different perspectives was a way of achieving an understanding of both points of view. But Plato did not provide any formal teaching in presenting arguments.

Aristotle's rhetorical argumentation: the *enthymeme*. Aristotle was the first to attempt to expose the system underlying rhetoric. Although Aristotle is considered first and foremost as a philosopher and logician, he was also an empirical scholar, in the first instance as a 'biologist', but also as a historiographer (as evidenced by his

[141] See e.g. the discussion in Brian Vickers, *In Defence of Rhetoric*, Oxford University Press, 1989.

[142] Gilbert Dherbey, *Les Sophistes*, Presses Universitaires de France, 1995.

[143] Susan Jarratt, *Rereading the Sophists: Classical Rhetoric Refigured*, Southern Illinois University Press, 1991.

history of Greek theatre and the Pythian Games) and as a philologist and student of literature (for example his analysis of Homer and his *Poetics*—see 2.8). In his discourse *Rhetorica* Aristotle positioned himself between philosophy on the one hand and learning and science on the other. It was the first complete and systematic treatment of the discipline from the perspective of the speaker, argumentation, and audience.[144] According to Aristotle, rhetoric—like dialectics—is within everyone's competence. Some were more adept than others, but the tools of rhetoric that someone successfully uses can be systematized and therefore represent a *téchne*, an *ars*: an *ordered set of rules*.[145] We see the same ambition in Aristotle that we see in Panini (see 2.1): to expose the procedures or system of rules that underlies a competence that everyone has, be it grammar or rhetoric. Aristotle's working principle therefore seems to be the same as Panini's: the procedural system of rules principle.

In order to make his argument as specific as possible, in his discourse *Rhetorica* Aristotle began by explaining how he characterized the discipline: rather than the art of persuasion, he defined rhetoric as 'the skill to establish, for any subject, which means of convincing people are available'.[146] In so doing Aristotle showed that he wanted to put all the emphasis on *inventio*, the discovery of arguments, rather than on a rhetoric of success as practised by the Sophists.[147]

Aristotle's most important insight was that rhetoric is based on the same sort of argumentation pattern as dialectics or logic. In other words the rhetorical presentation of arguments is not essentially different in terms of structure from logical deduction (see 2.6). According to him the only relevant distinction was in the nature of the premise: in rhetoric the premise was not necessarily a *true* statement but one that was *plausible* to a particular audience. This was expressed through the concept of an *enthymeme*. An enthymeme is an abbreviated syllogism with the same type of premise–argument structure as a deductive proof, but one where the premise is only deemed 'acceptable' by a particular target group. Everything else in a rhetorical presentation of arguments is only an addition to the process of persuasion. Using historical examples, Aristotle showed that a target group is easily convinced if it believes that something has been proved. Formally speaking, the basic idea of a rhetorical proof is as follows: in order to get a particular target group to believe that statement q is the case, the speaker must first select a collection of statements P that are already accepted by the target group. The speaker must then demonstrate that q can be deduced from P where the statements in P are taken to be premises.[148]

How can we find premises or arguments that the different audiences consider acceptable? According to Aristotle this is done on the basis of 'examples', which fall into two categories: *historical* examples or precedents, and *fictional* examples such as

[144] Eugene Garver, *Aristotle's Rhetoric: An Art of Character*, University of Chicago Press, 1994.

[145] Aristotle, *Rhetorica*, I.1.

[146] Aristotle, *Rhetorica*, I.2.

[147] A. Braet (ed.), *Taalbeheersing als nieuwe retorica: een historisch, programmatisch en bibliografisch overzicht*, Wolters-Noordhoff, 1980, pp. 9ff.

[148] Braet (ed.), *Taalbeheersing als nieuwe retorica.*

parables. Aristotle explained that in a normal speech, that is to say based mainly on enthymemes, only one or at most a few examples are necessary. On the other hand, in order to present convincing inductive argumentation one would need to have a whole series of examples, which is rather long-winded in a speech.

Aristotle referred to another book, the *Topica*, as somewhere to find examples. In it he demonstrates how arguments can be found using abstract search strategies. These search strategies can be fairly general, for example a cause–effect, a whole–part, a difference, an agreement, or an analogy. But the sources can also be very concrete, such as eyewitness accounts, contracts, oaths, and confessions. For instance, let us assume that an orator wishes to persuade his audience of the idea that a certain Dionysius, who has requested a bodyguard, is trying to establish a tyranny. Based on Aristotle's heuristics the orator can set to work as follows: first of all he thinks about all previous tyrants (the 'example'), and then he names all the people known to the audience who also requested a bodyguard before they became tyrants (the plausible statement, a 'premise'), after which he can use an inductive argument to contend that Dionysius is also bent on tyranny. Ultimately this results in a rhetorical proof (the enthymeme) which, while it may not hold water, is nevertheless convincing to a particular target group.

Evaluation of Aristotelian rhetoric. To a degree Aristotle's approach is comparable to the way that Panini tackled language: he proposed a general procedure that, if correctly used, results in a rhetorical presentation of proof (enthymeme). Aristotle gave rules for (1) finding a precedent or example, (2) a statement that is plausible to a target group, and (3) the final rhetorical argumentation in the form of an abbreviated syllogism. We will therefore refer to Aristotle's method as the *procedural system of rules principle* (see 2.1). But his rhetorical procedure is not in the same class as Panini's. Although Aristotle said that he was aiming at a systematic set of rules (and therefore used the same principle as Panini), his rhetorical theory did not produce an explicit system of rules that guaranteed a valid rhetorical proof. Aristotle's system is also not a compilation of declarative rules that specifies the general constraints for possible enthymemes, which is what Aristoxenus's musical grammar does for the class of possible pieces of music (see 2.4). It would be better to describe Aristotle's intricate system as a collection of *heuristics*, in the sense that they are solution strategies that might well produce a result but not necessarily a correct one. Heuristics generate guidelines about possible solutions and save a lot of time and effort by limiting the solutions to those that have the best chance of being applicable. In this regard it may be better to compare Aristotle's rhetorical method with the analogical method of the Alexandrians (who were active a generation after Aristotle—see 2.3). The analogists did not have an unambiguous procedure for word reconstruction either, but only a number of heuristics with which correspondence between word forms could be established in order to find a possible solution.

In his *Rhetorica* Aristotle also addresses putting confidence in character (*ethos*), the emotions of the audience (*pathos*), and the style and structure of speech itself (*logos*, which was also used to denote the rhetorical argumentation). These extra elements in the *Rhetorica* are primarily an 'addition' to the process of persuasion.

No matter how important these supplements may be to the manipulation of the audience, they are essentially separate from the rhetorical argumentation.

There is no answer to the question of whether Aristotle's rhetorical method was put to use in his time to construct an effective oratory because the Aristotelian corpus was not published until two centuries after his death. According to Plutarch, Andronicus of Rhodes published an edition of the works of Aristotle which were brought to Rome by the dictator Sulla in 84 BCE.[149] In the interim it was not Aristotelian rhetoric that flourished but the more prescriptive Hellenistic form, which was used in all parts of the Roman Empire. Aristotle was not considered to be the absolute authority in all areas of learning and science until the Middle Ages, when his works were disseminated in Europe by Islamic civilization.

Hellenistic and Roman rhetoric: *stasis* and *bene dicendi*. It was during Aristotle's lifetime that Alexander the Great died at the age of thirty-three (323 BCE), leaving behind a huge empire that extended from Sicily to the Punjab. After twenty years of internal strife this empire split into three parts: west Asia and Asia Minor under Antigonus, Egypt, and the Aegean region under Ptolemy, and the areas from Syria to India under Seleucus. Despite this fragmentation and the enormous diversity of these areas, Greek culture was so dominant that a general educational programme emerged that prepared young men for the Greek way of life. This standard curriculum was known by the name *enkyklios paideia*,[150] the 'general education', and broadly speaking included the same disciplines as the later *artes liberales*: grammar, logic, rhetoric, arithmetic, geometry, astronomy, and music. It was also during this period that the classic subdivision of rhetoric developed. It is normally referred to in Latin:

(1) *inventio* (discovering arguments),
(2) *dispositio* (ranking the arguments and the speech),
(3) *elocutio* (the style),
(4) *memoria* (memorizing the speech),
(5) *actio* (the delivery).

This five-part classification was still in use in the eighteenth century.[151] The most important innovation in rhetoric during Hellenism was the *stasis* theory as part of the *inventio*. This theory was attributed to Hermagoras of Temnos (150 BCE), but it has only survived in summary form in *De inventione* by Marcus Tullius Cicero (106–43 BCE). The theory bears most resemblance to Aristotle's sources theory (*Topica*), but then in a simplified form by asking such questions as *who, what, how, where, for what, why, when* and *with what?* In the context of judicial argumentation, answers to these questions can generate the right points to make in a speech.

For a long time it was incorrectly believed that Cicero was the originator of the stasis theory. But Cicero did not emerge as an original thinker or scholar in

[149] Lucius Mestrius Plutarch, *Sulla*, XXVI.

[150] Teresa Morgan, *Literate Education in the Hellenistic and Roman Worlds*, Cambridge University Press, 1998, pp. 50ff.

[151] Thomas Conley, *Rhetoric in the European Tradition*, University of Chicago Press, 1990.

practically any field at all. The fact that his *De oratore* (55 BCE) became one of the most influential rhetorical works in the European tradition is primarily due to the eclectic overview of the discipline that Cicero wrote. However, *De oratore* is more than just a work of rhetoric: it combines philosophy, oratory, statesmanship, and ethics in the form of a normative compilation of procedures describing 'how it should be done'.[152] The traditional imitation of the classical examples then became the most important, and there was no empirical slant in Cicero. Similarly there was no empirical revival in *Institutio oratoria*, written by the greatest rhetorician after Cicero, Marcus Fabius Quintilian (35–100 CE)—even though his influence was very substantial in the early modern period, particularly after an intact manuscript of the *Institutio* was rediscovered in the fifteenth century (see 4.1). Rhetoric transformed from a largely critical discipline under Aristotle into a normative manual under Quintilian. This proved to be extremely valuable for teaching oratory (*bene dicendi*), but virtually useless for rhetoric as an empirical subject.

The situation was not essentially different in the Greek part of the Roman Empire, although this is where the prodigy Hermogenes of Tarsus (155–225) brought about a revival and the further development of the stasis theory. Hermogenes was a teacher in rhetoric as early as at the age of fifteen, attracting the interest of the emperor himself. In his *Progymnasmata* Hermogenes defined a new organization of rhetoric, which varied somewhat from Cicero's and Quintilian's five-part classification and was soon turned into a prescriptive scheme.[153] His work was still being used in the Byzantine Empire in the tenth century.

Oratory in India and China. How did rhetoric fare outside the Graeco-Roman world? The distinction between dialectics and rhetoric was primarily a Greek and in particular an Aristotelian phenomenon. Rhetoric as oratory formed part of the logic of argumentation and debating in both (pre-Hellenistic) India and China.[154] It should be remembered that Indian Nyaya logic was partially inductive (as was the Aristotelian enthymeme) and that Chinese Mohist logic had a primarily analogical slant (resembling Aristotle's source theory).

The 'methods for debating' were explained in the Indian text *Charaka-Samhita* (550 BCE), which is attributed to Punarvasu Atreya, but it may actually have been written much later. These methods were adopted by the Nyaya philosophers (see 2.6) and were subdivided into the following elements:[155]

(1) a definition of the subject of the debate,
(2) a proposition,
(3) a counter-proposition,
(4) the speech, or the source of knowledge,

[152] Elaine Fantham, *The Roman World of Cicero's De Oratore*, Oxford University Press, 2007.
[153] Malcolm Heath, *Hermogenes On issues: Strategies of Argument in Later Greek Rhetoric*, Oxford University Press, 1995.
[154] Robert Oliver, *Communication and Culture in Ancient India and China*, Syracuse University Press, 1971.
[155] Radha Kumud Mookerji, *Ancient Indian Education*, 2nd edition, Motilal Banarsidass Publications, 1998, p. 320.

(5) the application (this corresponds to the Nyaya logical induction),
(6) the conclusion,
(7) the response,
(8) the example,
(9) the truth established by experts or the proof by deduction,
(10) doubt or uncertainty that is accepted by both parties.

The emphasis on the debate in Indian rhetoric corresponds with the later logical works of Aksapada Gautama (see 2.6). Points (9) and (10), in which the two debating parties achieve a joint result, are moreover reminiscent of Plato's dialectics, in which new insights can emerge from different points of view and do not necessarily lead to a divergence of opinion. Indian rhetoric is therefore very far removed from Aristotelian rhetoric, where the focus is not on the debate but very much on the argumentation. There is however a similarity with Aristotle's approach in the use of plausible statements which are also inductive in Aristotelian rhetoric (such as the generalization about all previous tyrants who had requested a bodyguard before they seized power). But where Aristotle had to bend over backwards to insert inductive reasoning into the argumentation, Indian rhetoric warmly embraces induction as an essential ingredient of both reasoning and debate.

The Chinese study of rhetoric began as far back as Confucius's doctrine of virtues. In the *Lunyu* (the Analects) it is explained which virtues had particular powers of persuasion, such as integrity, cordiality, benevolence, and respect. Rhetorical elements also appeared in the writings of Mencius (372–289 BCE). But it was with the Mohists that rhetoric was investigated from an empirical perspective, where it reached a pinnacle in the form of disputations.[156] The Mohists criticized one another and tried to convince their king with their proposals. They were interested in both the practicalities of debating and in its metatheory. Under the Mohist approach the practical aspects of rhetoric coincided with the basic techniques of logical argumentation, which—as discussed in 2.6—is analogical in nature. This has some similarity to the Aristotelian *Topica*, or source theory, where analogies were also deemed to be acceptable as 'plausible statements'. However, it is in the field of the theory of disputation (*bian*) that the Mohists made the most significant discoveries—the principles of reasoning, which were discussed when we considered Chinese logic. A 'debate' was defined as *a disagreement about assertions that are in contradiction*, in which the first debating principle is that one of two contradictory assertions must be untrue. Moreover, the second principle states that it is not possible for both such assertions to be untrue; one of the two must be true. It was thus in the context of 'debate' that the Mohists defined their two famous principles of argumentation.[157] There is a degree of consensus that the Mohists were among the first to formulate these

[156] A. C. Graham, *Later Mohist Logic, Ethics and Science,* The Chinese University Press, 1978. See also Joseph Needham and Christoph Harbsmeier, *Science and Civilization in China: Volume 7, Part I: Language and Logic,* Cambridge University Press, 1998.

[157] Respectively A74 and A75 in Johnston, *The Mozi: A Complete Translation.*

fundamental laws of logic, that is, the *law of non-contradiction* and the *law of excluded middle*.[158]

There are many other studies on persuasion in China, but few if none search for underlying rule systems.[159] In later rhetorical studies, such as the impressive *Wenxin diaolong* ('The Literary Mind and the Carving of Dragons') by Liu Xie, the rhetorical investigations move towards an analysis of the writing process, the *inventio*. But since these investigations are closely connected to Poetics, they will be discussed in 2.8.

158 Zhang and Liu, *'Some thoughts on Mohist logic'*. But see also the earlier discussion in Donald Leslie, *Argument by Contradiction in Pre-Buddhist Chinese Reasoning*, Centre of Oriental Studies, Australian National University, Canberra, 1964.

159 Geoffrey Lloyd and Nathan Sivin, *The Way and the Word: Science and Medicine in Early China and Greece*, Yale University Press, 2002, pp. 249–50.

2.8 POETICS: THE STUDY OF LITERATURE AND THEATRE

Literature, poetry, and theatre appear to have originated as parts of celebrations: epic heroic poems were presented at ceremonies, and plays lent style to official festivals and other celebrations. However, literary and theatrical products themselves soon became subjects of study, which was designated as *poetics*.

Poetics as *mimesis*: Plato's criticism. Although some Greeks considered poetics and rhetoric to be the same subject (Theophrastus asserted that 'poetics and rhetoric are concerned with informing and guiding the audience'), at the same time the two disciplines differed from each other. This was in particular the case with regard to what Plato called *mimesis* or imitation: poetics (drama, epic, lyric) sought to reconstruct a complete experience and tried to let the audience live through events as if they had been there.[160] Rhetorical tools might well have been employed, but the essence was the mimetic experience. Plato thus immediately made poetics the subject of debate.[161] He felt that imitation only represented observable phenomena and not the essence of things. Rather than getting closer to the 'truth', which in Plato's view was abstract (the 'idea'), poetics moved further away from it. Moreover, there was a moral danger lurking in literature and theatre because they could have a surreptitious influence on the audience, particularly young people, through uncontrollable emotions. In Plato's opinion this criticism also applied to imitation in other arts, such as painting and sculpture (see 2.5).

***Catharsis*: narrative structure according to Aristotle.** Aristotle took a somewhat different position to that of Plato. Although he saw eye to eye with his teacher on the definition of *mimesis*, Aristotle considered the human need for imitation to be a natural and healthy impulse. 'Art imitates nature' was his dictum,[162] and the challenge was to reveal the concealed characteristics of things. Through this Aristotle ascribed a higher order of truth to poetics—the deeper universal truth is the essence rather than reproduction of the actual details, and the poet needed to understand its principles. Aristotle therefore went in search of rules underlying a good 'story'. His implicit question was: *is there a system of rules that provides the basis for creating a good play, poem, or story?* Aristotle believed he had found a number of such rules and described them in his *Poetica*.[163] The rules were presented as 'requirements' that one can set for a 'good' tragedy. Aristotle illustrated the rules using existing tragedies (by Aeschylus, Sophocles, and Euripides) and epic poetry (particularly by Homer). Therefore, his system of rules applied at first to classical Greek stories and not to narratives in general.

[160] D. A. Russell and M. Winterbottom (eds), *Ancient Literary Criticism: The Principal Texts in New Translations*, Oxford University Press, 1988, p. 71.

[161] Plato, *Republic*, II, III, and X.

[162] Aristotle, *Physics*, II 2, 194a21 f.

[163] For an English translation, see e.g. Aristotle, *Poetics: A Translation and Commentary for Students of Literature*, translated by L. Golden, commentary by O. B. Hardison, Jr, Prentice-Hall, 1968. See also Russell and Winterbottom, *Ancient Literary Criticism: The Principal Texts in New Translations*.

Aristotle began by listing the six elements of tragedy: plot, character, diction, thought, spectacle, and song. He then spelled out rules for these elements, such as:

- A story needs a plot that consists of a beginning, followed by a middle and finishing with an ending. These three stages of beginning, middle, and ending have to be clearly recognizable.
- The highest level of suspense in a story should coincide with the actual middle of the story.
- The story has to have a hero or protagonist who represents an important person in the city-state, because these characters are crucial to the existence of the city-state.
- The suspense in the story comes from a conflict that is contained in the character of the opponent or antagonist. The plot has to concentrate on resolving this conflict.
- The story has to evoke feelings of pity and fear in members of the audience, who will identify themselves with the hero and who, by working with the hero to resolve the conflict, will ultimately have a sensation of *catharsis*: a sort of mental form of 'purification' or 'new understanding'. The aim of the story as a whole is to generate *catharsis* in the members of the audience.
- The poet should prefer probable impossibilities to improbable possibilities.[164]

These rules are not as exact as Aristoxenus's rules of melody or Pliny's proportion rules, let alone the logical rules of reasoning or the Paninian grammar rules. The rules of poetics provide the necessary scope for ambiguity, and in this respect they resemble the historiographical rules for working with sources developed by Herodotus. Aristotle's rules were unambiguous as far as the *structure* of a tragedy or epic was concerned, but they left a lot open as regards the *content* (except for the obligatory elements of hero, antagonist, and conflict). Aristotle's system of rules can therefore best be compared to a 'grammar of constraints' (declarative system of rules principle) within which a drama or epic had to be played out. As a generalization about classical tragedies, Aristotle's poetics may be descriptively adequate, but as a *general* system of rules for theatre and epics it is at best prescriptive.

It is important to remember what Aristotle was trying to achieve with his system of rules. His work was not aimed at poets or dramatists but at analysing the requirements for a 'good' story.[165] As he usually did, to this end Aristotle described which resources could be used to achieve a goal. In his view the objective of poetics was to bring about *catharsis* in members of the audience, and the resources needed to do this were obtained from *mimesis*—the imitation of an action such that the people in the audience were, so to speak, eyewitnesses of an event.

From *mimesis* to *imitatio*: Horace and Dionysius of Halicarnassus. The concepts of mimesis and catharsis were the two most significant characteristics of

[164] Aristotle, *Poetica*, XXIV, 60a16.
[165] Richard Harland, *Literary Theory from Plato to Barthes*, Palgrave Macmillan, 1999, pp. 14–15.

classical poetics. They lived on in the poetics of the Hellenistic and Roman world, where Horace (Quintus Horatius Flaccus) (65–68 BCE) was one of the primary exponents. Horace's treatise *Ars Poetica* was based almost entirely on Aristotle, whose works had been brought from Athens to Rome by Sulla (if we can believe tradition). However, Horace's system of rules became normative and imitative. He wrote, as if he were the author of a strict school textbook, 'I shall impart to the poet his obligations; I shall tell him where he can find his sources . . . what he may and may not . . . '.[166] Whereas we can still give Aristotle the benefit of the doubt, Horace's instructions are so imperative that they have nothing to do with the empirical humanities: searching for observable patterns in epic, lyric, and drama.

The 'imitation of other authors' rather than the 'imitation of nature' became a goal in itself and was worked out as a full-fledged methodological principle by Dionysius of Halicarnassus (*c.*60 BCE–*c.*7 CE). Although Dionysius is often seen as a rhetorician, his fragmentarily extant work *Peri Mimeseos*, or *On Imitation*, was of primary importance for poetics.[167] By *imitatio*, Dionysius meant the practice of emulating, adapting, reworking, and enriching a source text by an earlier author. He described the best models in literature that ought to be imitated. Dionysius's principle of *imitatio* marked a departure from Aristotle's principle of *mimesis*.[168] It initiated a long tradition which dominated Western poetics well into the eighteenth century. Yet Dionysius was not an imitating scholar himself: among other things he initiated a surprisingly original *empirical* approach to poetics, as we will see below.

Longinus and the sublime: *ecstasy* versus *catharsis*. A rather different view on poetics was laid down in the book *Peri Hupsous*, commonly translated as *On the Sublime*, which has been attributed to Longinus (first century CE, although this attribution is in all probability incorrect).[169] This influential work is a compendium of literary examples taken from over fifty authors and a thousand years of literary history. The examples are very diverse, including excerpts from Homer, Sappho, Plato, Cicero, and even a passage from Genesis, which was exceptional for the first century CE. Longinus praised and vilified literary works as examples of good or bad style. He recommended an elevated style of writing that at the same time expressed the essence of simplicity. Longinus also introduced the concept of 'the sublime'. 'The first and most important source of the sublime is the power to create great conceptions.'[170] In Longinus's opinion the 'sublime' referred to a style of writing that rises 'above the ordinary'. Longinus gave five sources of sublimity: great thoughts, strong emotions, figures of speech, noble diction, and dignified usage. He believed that the sublime does not result in persuasion or catharsis among the audience (as was the case with Aristotle) but in *ecstasy*: 'for what is wonderful always goes together with a sense of dismay, and prevails over what is only

[166] Horace, *Ars Poetica*, 306–8.

[167] Malcolm Heath, 'Dionysius of Halicarnassus on imitation', *Hermes*, 117, 1989, pp. 370–3.

[168] David West and Tony Goodman, *Creative Imitation and Latin Literature*, Cambridge University Press, 2007.

[169] For an English translation of *Peri Hupsous*, see James Arieti and John Crossett, *Longinus: On the Sublime*, Mellen, 1985.

[170] Longinus, *Peri Hupsous*, I, 3.

convincing or delightful, since persuasion, as a rule, is within everyone's grasp: whereas, the sublime, giving to speech an invincible power and strength, rises above every listener.' Longinus's description of the sublime is so arresting that in the view of later aestheticians, such as Edmund Burke in the eighteenth century, he had identified the essence of poetic perception (see 4.5). Obviously, the wish to discover an underlying system of rules for the *creation* of the sublime or the beautiful was not part of Longinus's approach.

Dionysius of Halicarnassus and empirical poetics. Where Longinus did not search for an underlying system of beauty, Dionysius of Halicarnassus (see also above) did. Drawing an analogy with the rules for accomplishing harmony in the visual arts, Dionysius tried in his *De compositione verborum* to find rules for writing poetic sentences according to a concept of 'natural' word order (which brought his poetics close to rhetoric). Dionysius contended as a hypothesis, which he adopted from the Stoics, that word order based on natural principles led to a beautiful and poetic composition.[171] As natural principles he adduced rules such as (1) nouns go before verbs, (2) verbs go before adverbs, and (3) earlier events are recounted before later ones. Dionysius explained the naturalness of these rules on the basis of philosophical considerations. One such consideration was that nouns must come before verbs because the former give the substance and the latter the 'accident', and the natural mode is for the substance to precede its accidents.

Dionysius's work was special in that he also actually tried to verify his rules using a text that no one doubted was beautiful and poetic, namely Homer. But when he demonstrated that Homer did not comply with the three natural rules mentioned above, Dionysius also rejected the other rules that he (and the Stoics) had proposed. The experiment did not produce the result that Dionysius had hoped for. It had emerged that the beauty of Homeric poetry was not founded on natural rules. This negative result may have led Dionysius to turn to his *imitatio* principle we discussed above, i.e. that beauty can be better achieved by imitating the great authors rather than by applying formal rules. In any case, Dionysius did not attempt to come up with a modified or improved rule system for beautiful sentences; instead he discarded the theory that beauty could be created by a 'natural' system of rules.[172] Here we have a clear example of a falsification in the humanities.[173]

Two approaches thus appeared to have existed in Greek poetics: the pursuit of 'good' narratives, resulting in catharsis, and the aspiration to create 'beautiful' or sublime narratives, leading to ecstasy. A system of rules could be constructed for the former, but not for the latter—if it was not by imitating great works from the past (the *imitatio* principle). There were likewise rules for good music and art (the exact proportions principle), but not for sublime music and art (the anomaly principle)—see 2.4 and 2.5.

[171] Casper de Jonge, *Between Grammar and Rhetoric: Dionysius of Halicarnassus on Language, Linguistics, and Literature,* PhD thesis, Universiteit Leiden, 2006, pp. 248ff. (Also published by Brill, 2008.)

[172] De Jonge, *Between Grammar and Rhetoric: Dionysius of Halicarnassus on Language, Linguistics, and Literature*, p. 278.

[173] See section 2.2 for another falsification in the humanities: the refutation of Polybius's thesis that the history of Rome follows a linear pattern without decay.

Systems of rules in Indian poetics: rasas and Mimamsa. There was a quest for systems of rules in Indian poetics too. As in Indian musicology, the most important work in this field was the *Natya Shastra* ('Treatise on the Performing Arts') attributed to Bharata Muni (see 2.4).[174] For over fifteen centuries this work provided the basis for all literary and dramatic production in Sanskrit. The status of the *Natya Shastra* was so exalted that it was often considered to be a separate Veda.[175] The *Natya Shastra* gave a detailed overview of the different 'rhetorical sentiments', which were also called *rasas*. There were eight types of rasa: erotic, comical, pathetic, furious, heroic, terrible, odious, and marvellous. There was then an explanation of how the emotional condition could be represented in these rasas. The emotional condition of paralysis, for instance, was expressed by sweating, a broken voice, shivering, shuddering, change of colour, tears, loss of consciousness, or fainting. The accompanying instruments and singing were also specified. After these rules, the *Natya Shastra* went into the structure of drama, describing no fewer than twenty-five types, ranging from one act to fifteen acts. There were meticulously detailed accounts ranging from make-up, sets, and costume to bodily dynamics in terms of the smallest movements of the lips, eyebrows, and even the ears. There were also elaborate descriptions of the different types of people in the audience and their ability or inability to grasp certain emotions.

No Greek work on drama came close to the precision of the *Natya Shastra*. In terms of 'formalization', however, the *Natya Shastra*'s system of rules was closer to Aristotle's *Poetica* than to Panini's grammar or Aristoxenus's theory of melody. For example, 'shivering' was a reflex action that did not appear to need further specification. Otherwise the *Natya Shastra* seemed to be a comprehensive traditional manual for performing plays, dance, and music. The intricate details of the work suggest that it was created in a descriptive way. The *Natya Shastra* could have come about as a record of an existing theatrical tradition, from which—possibly after one or more additions—there were few if any further departures, after which it became prescriptive.

Systems of rules were not devised just for the process of creation. They were also developed for the interpretation process or, in other words, text exegesis. For example, the *Mimamsa* school set out to develop a rule-based exegesis of the Vedas. Such an exegesis became topical when Vedic rituals became more and more marginalized around the second century CE by 'new' Indian philosophies like Buddhism. As a counterweight, the Hindu manuscript scholars wanted to demonstrate the validity of the Vedas on the basis of fully specified interpretation instructions so that everyone could understand them. Jaimini's *Purva Mimamsa Sutras*, dating from around the second century BCE, represented the most important work in this field.[176] It will come as no surprise to read that Jaimini's interpretation

[174] Anupa Pande, *A Historical and Cultural Study of the Natyasastra of Bharata*, Kusumanjali Prakashan, 1991.

[175] G. N. Devy (ed.), *Indian Literary Criticism: Theory and Interpretation*, Sangam Books, 2002, p. 3.

[176] R. A. Ramaswami Shastri, *A Short History Of The Purva Mimamsa Shastra*, Annamalai University Sanskrit Series No. 3, 1936.

system was strongly prescriptive and therefore less interesting from our point of view. But Jaimini's attempt seemed to have been successful because a long period of stagnation in Indian Buddhism occurred.

Liu Xie's 'Literary Mind': composition methods and literary history. In China, Confucius discussed the didactic and allegorical goal of literature as far back as the sixth century BCE, but the oldest surviving work about poetics is an essay *On Literature*, which is a mere six hundred words long, by Cao Pi (187–226 CE).[177] Cao Pi, son of the legendary general Cao Cao, rose to become the first emperor of the Wei Dynasty, but he was also a poet and critic. His most significant assertion was that 'the Vital Breath is of the greatest importance in literature' and in his opinion this varied from 'singer to singer when performing the same song'.

The oldest surviving *systematic* work about Chinese literature is the *Wenxin diaolong* ('The Literary Mind and the Carving of Dragons') by Liu Xie (465–521 CE).[178] It consisted of fifty chapters and began with an explanation of the thirty-two genres that were known in Liu Xie's time, ranging from the most aesthetic to the most practical. For instance, he gave an overview of myths and sagas, the classics (in particular Confucius) and their imitations, the different styles of poetry (*yuefu*) and poetic prose (*fu*), hymns and eulogies, prayers and vows, laments, historical works (including an extensive stylistic description of Sima Qian's writings—see 2.2), philosophical works, offerings to heaven and earth, exam papers, reports and memoranda, and even declarations of war. In short, the first part of The Literary Mind is as complete an outline as possible of the entirety of Chinese poetics and rhetoric.

In the second part Liu Xie addressed the writing process itself, such as devising a literary text (the *inventio*), the first rough draft of a text, revision, and the necessary adaptation to a situational context. Compositional methods were also discussed, and there was an explanation of the structural elements, such as words, sentences, and paragraphs. He also dealt with rhetorical figures of speech, the rendition of emotions, vitality, musicality, parallelism, and so on. Finally, Liu Xie considered external factors such as the physical environment, the role of the critic, and individual talent and personal aspirations.

Liu Xie's Literary Mind is one of the most impressive overviews of literature and its history in Antiquity from the points of view of both analysis and production.[179] His work is comparable to Longinus's history of a thousand years of classical literary history, but far surpasses it in its systematics. It is difficult to categorize Liu Xie's work as poetics or rhetoric. Like the works of Dionysius of Halicarnassus, to a degree it belongs to both.[180]

[177] John Minford and Joseph Lau (eds), *Classical Chinese Literature, from Antiquity to the Tang Dynasty: An Anthology of Translations*, Columbia University Press, 2002, p. 425.

[178] Yang Guobin, *Dragon-Carving and the Literary Mind: An Annotated English Translation and Critical Study of* Wenxin Diaolong *by Liu Xie*, Foreign Language Teaching and Research Press, 2003.

[179] Zong-qi Cai (ed.), *A Chinese Literary Mind: Culture, Creativity and Rhetoric in Wenxin Diaolong*, Stanford University Press, 2001.

[180] H. Zhao, 'Rhetorical invention in "Wen Xin Diao Long"', *Rhetoric Society Quarterly*, 24(3/4), 1994, pp. 1–15.

CONCLUSION: COMMON PATTERNS IN THE HUMANITIES OF THE ANCIENT WORLD

In pursuit of a system of rules. If we try to consider the humanities of the ancient world as a whole, we can state that while various methods and principles were employed, everywhere *efforts were made to formulate a system of rules (derived empirically or otherwise)*: grammars in linguistics, rules for working with sources in historiography, heuristics for word analogies in philology, harmonic proportions and declarative grammars in musicology and art history, procedural grammars in logic, a heuristic system of rules in rhetoric, and a narrative system of rules in poetics. We find this quest for precise systems of rules in all disciplines and in all regions, from China and India to the Greek-Hellenistic world. There is one notable exception. In philology, alongside an approach based on rules, there is also a tradition that rejects rules and contends that only exceptional cases exist. This anomalistic tradition has remained rare and has not created a dominant, let alone unique, approach anywhere (we also find anomalism in art theory, but it goes hand in hand with the canon of mathematical proportions). Although there is no general method, all classical humanities have an empirical component: people searched the observable linguistic, historical, musical, literary, logical, or artistic material for underlying rules, heuristics, analogies, or proportions.

Parallel discoveries of patterns. Together with the general pursuit of systems of rules, there were remarkable parallel discoveries to be found in the humanities. The cyclical pattern in history was established in Greece (Herodotus) and China (Sima Qian). In musicology the basic harmonic principles of tonic, octave, and fifth were discovered in all three regions: Greece (Pythagoras), India (Bharata Muni), and China (Liu An). In logic, the laws of the excluded middle and non-contradiction were established in China (the Mohists) and in Greece (Aristotle). And in art theory the concept of numerical proportions for visual harmony were described by Pliny and in the Sadanga. Parallel discoveries are not unique to the humanities. They have also occurred in science.[181] Moreover, there are many patterns that are dissimilar: the rhetorical, grammatical, and narrative systems of rules and patterns are different in all regions.

From descriptive to prescriptive. It has not been possible in all cases to determine the degree to which observed regularities and patterns are descriptive or prescriptive. The initially empirical regularities found in musicology, art history, and poetics appear to have been used very soon as obligatory instructions. This *process of changing from descriptive to prescriptive* seems to be a recurring characteristic in the humanities: regularities and patterns in language, art, and music were transformed into normative rules. This may differentiate the humanities from the sciences. Regularities in natural phenomena are never prescriptive. Nature does not take the slightest notice of normative instructions. It is nevertheless the case that

[181] See e.g. H. Floris Cohen, *How Modern Science Came into The World: Four Civilizations, One 17th-Century Breakthrough*, Amsterdam University Press, 2010.

once regularities have been found—for example in the movements of the planets or falling bodies—they have a drastic effect on their human observation, as in the humanities. An example can be found in the observed order in planetary movements, which was interpreted by Plato, Eudoxus, and others as a system of circles. This assumed order muddied our observation of how planets move for almost 2,000 years. Planets had to and would move in circles—if necessary in a complex combination thereof—until Johann Kepler believed he could reduce planetary orbits to ellipses on the basis of new observations by Tycho Brahe. There is a similar history to be found in the apparent order of falling bodies. Whether we like it or not, regularities found in previous observations impose themselves on new observations. In this regard there is no essential difference between the study of scientific phenomena and humanities phenomena.

Seldom deductive. There is perhaps another characteristic difference between the humanities and the sciences in Antiquity. Compared with the classical sciences, the deductive method (in which assertions are derived from necessary first principles) in the humanities is marginal. We only find a deductive style in Aristotelian logic and in Aristoxenian musicology, but then Aristoxenus was a pupil of Aristotle. However, Aristoxenus's derivations come across as laboured and in some cases are demonstrably incorrect. Moreover, deductive Aristotelian syllogistics is an exception among the classical logical systems. Indian Nyaya logic is inductive (assertions are 'derived' from repeated observations rather than first principles), and Chinese Mohist logic is analogical (contentions are 'derived' through analogy with similar situations). We could interpret Panini's grammar as a deductive system in view of the fact that linguistic utterances are derived deductively by combining grammatical rules. However, it is difficult to consider 4,000 rules as 'necessary first principles'. The use of the deductive method in the humanities appears to have been limited to the Aristotelian tradition.

Falsification, and rules for the 'good' versus the 'beautiful'. Prevailing opinion notwithstanding, it is striking that the humanities of the ancient world have often emerged as falsifiable and often replicable. It emerged, for example, that the hypotheses of Polybius (about the acyclic pattern in Roman history) and Dionysius of Halicarnassus (regarding the coincidence of natural and poetic word order) could be *refuted* in their own day and age on the basis of empirical facts (historical and poetic materials, respectively). On the other hand it proved possible to *substantiate* other hypotheses and theories, such as Panini's Sanskrit grammar and the Ptolemaic theory of the Pythagorean harmonic intervals. And some theories turn out to be neither verifiable nor falsifiable, not because of epistemological shortcomings (in fact this only applies to anomalistic philology) but as a result of the loss of humanities material: Aristoxenus's theory of melody and Aristophanes's theory of philology can no longer be tested using the material for which they were devised. Finally, we can establish a metapattern in the validity of the patterns of rules that has been discovered. In the classical humanities there do *not* appear to be valid systems of rules for the beautiful or the sublime, but there *are* for the correct or the good. There are therefore no rules for poetic language, sublime rhetoric, and beautiful art (see Dionysius, Longinus, and Pliny), but there are rules for grammatical language, valid rhetoric, and harmonious art (see Panini, Aristotle, and Vitruvius).

Unanswered questions. We do not have all the answers. Why did Greek linguistics lag so strikingly behind the other Greek humanities? Why, for example, was no system of rules developed for linguistic syntax whereas one was for musical syntax? Why is there no Indian historiography while the Indians contributed so much in all other fields of the humanities? And we should ask ourselves why Roman learning and science delivered nothing and remained eclectic. Although Roman learning and science, including the humanities, included Greek learning and science de facto from about the second century BCE, virtually all the scholars continued to be of Greek origin. The current explanation is that the Romans were interested first and foremost in a political career, in which learning and science that was not immediately applicable was considered irrelevant. In some areas the classical Chinese humanities also appear to have remained 'backward' compared with India and Greece, in particular in linguistics and philology. According to tradition, it was the book burning in 213 BCE during the Qin Dynasty that led to the destruction of many Chinese texts. While the impact of the book burning may have been exaggerated by later historians, the brilliant works that were created immediately after the Qin Dynasty, such as Liu An's musicological writings and Sima Qian's historical reports, fill us with sadness about all that might have been lost and will never be known.

3

The Middle Ages: The Universal and the Particular

It is virtually impossible to tell when the humanities of the ancient world became the medieval humanities. For example, one could say that medieval Christian historiography began with Eusebius in the fourth century, whereas European musicology may well not have started earlier than the eighth century. Usually the 'end' of Graeco-Roman humanities is marked by the flight to Persia of the last followers of Plato's academy when it was closed by Emperor Justinian in 529. The expansion of Islam established a flourishing culture of learning and science that extended from Persia to Africa, and reached the European world via Sicily and Spain. The humanities were at their most vigorous in this Islamic civilization. Although China and India seem to have gone their own way, they exerted noticeable influence on Islamic culture, with al-Biruni acting as a pivot between India and the Arab world, and Buddhist monks fulfilling the same role between India and China. There was also interaction between China and Islamic civilization in Samarkand that reached Europe by way of Byzantium.

3.1 LINGUISTICS: FROM RULES TO EXAMPLES

During the Middle Ages there were three great linguistic traditions that operated largely independently of one another. They were the continuation of Panini in India, the example-based linguistics of Sibawayh in Islamic culture, and speculative grammar in Europe. There were also a number of identifiable interactions. Indian linguistics percolated into China and Islamic civilization, Arabic linguistics was based on that of the Greeks, and European linguistics—although derived from the Roman tradition—did not really come to life until after the Islamic influence from southern Spain and Sicily.

India: Panini's legacy. Panini's accurate system of rules for Sanskrit (see 2.1) must have made such an overwhelming impression that for nearly twenty-two centuries Indian linguists worked almost exclusively on writing commentaries on, interpretations of, and a few changes to the work of the great master. The distinction between (late) Antiquity, the Middle Ages, and the early modern era is almost meaningless in Indian linguistics. Until far into the eighteenth century Panini's grammar was looked upon in India as an almost 'complete' system that was essentially not open to improvement.[1] Although Panini's grammatical method was applied to other languages, for example Tamil and Tibetan, this was to the credit of linguists outside India.

Yet there is also a non-Paninian tradition in India. For instance, Bhartrhari (*c.* sixth century) studied, among other things, linguistic communication in his *Vakyapadiya* ('Treatise on Words and Sentences').[2] Bhartrhari divided linguistic communication into three stages: (1) conceptualization by the speaker, (2) production of language by the speaker, and (3) comprehension of language by the listener. This three-fold classification was a surprising forerunner of many contemporary theories of linguistic communication (see 5.3). Bhartrhari was also the founder of the Sphota school, which was concerned with the question of how the human mind can organize linguistic units into a coherent whole, such as a conversation or a discourse. As regards semantics, the Sphota school propagated *semantic holism*, which means that the significance of the whole cannot be derived from the meanings of the parts.[3] On this issue the Sphota school was diametrically opposed to the logical Nyaya school, which—like the language philosopher Yaska (see 2.1)—defended the concept of *compositionality*, according to which the significance of the whole, a sentence for instance, can indeed be deduced from the import of the components (the words).

During the course of the seventh century travelling monks from China made the Indian linguistics tradition available in Chinese. Through his travels and translations, the famous Buddhist monk Xuanzang (*c.*600–64) contributed more than

[1] Itkonen, *Universal History of Linguistics*, pp. 70–8, p. 335.

[2] There seems to be no English translation of the *Vakyapadiya*. An accessible German translation is Wilhelm Rau, *Bhartṛharis Vākyapadīya I & II*, Steiner, 1977–91.

[3] Harold Coward, *The Sphota Theory of Language: A Philosophical Analysis*, Motilal Banarsidass, 1980.

anybody to the exchange between India and China.[4] In the late seventh and early eighth centuries I Ching and Fazang also translated many texts from Sanskrit into Chinese. Apart from these translations, however, Indian linguistic tradition appeared to have had virtually no impact on Chinese linguistics, which either remained philosophical, following the practice of Confucius, or was concerned with compiling dictionaries.

Islamic civilization: Sibawayh and example-based grammar. When by 750 Islam extended from Persia to North Africa and from Arabia to Spain, the Islamic humanities flourished like nowhere else in the world. Yet, their history has yet to be written.[5] The period of 750 till 1258 is usually referred to as the Islamic Golden Age. Under the Abbasids, Baghdad became the most important centre of knowledge in the world. The pinnacle was reached in the ninth century with the establishment of the House of Wisdom (*Bayt al-Hikma*). During the al-Ma'mun caliphate a huge collection of Greek, Indian and Persian documents, and manuscripts from the library of Constantinople were translated into Arabic en masse. Libraries were also established in Islamic Spain and Sicily. They played a significant part in the transfer of knowledge to Christian Europe during what is called the 'twelfth-century Renaissance'. However, Baghdad was to remain the most important centre for four centuries, until the city was destroyed by the Mongols in 1258.

One could say that Arabic linguistics began with Ibn Abi Ishaq (died in 736, not to be confused with the translator Hunayn Ibn Ishaq—see 3.3). Ibn Abi Ishaq based his normative grammar on the language of the Bedouins, which he believed to be the purest of all languages. However, his grammar was such a long way from a descriptive system that it cannot be considered a scholarly account of a natural language.

Scholarly Arabic linguistics began with the Persian linguist Sibawayh (*c*.760–793), who worked in Baghdad. A non-Arab, Sibawayh wrote the first Arabic grammar in his *Al-kitab fi al-nahw* ('The Book of Grammar'), called *Kitab* for short.[6] Sibawayh's *Kitab* was intended to enable non-Arab Muslims to understand the Koran, in the same way that Dionysius Thrax's grammar was meant to help speakers of other languages learn Greek. But the *Kitab* went into much greater detail than Dionysius's slim grammar textbook, which had barely thirty pages. In over 900 pages Sibawayh addresses essentially all facets of Arabic. Even so, his basic linguistic concepts seem to have come directly from the Greek grammatical tradition, such as the ideas of word form, declension, and the identification of two genders and three verb forms. Although most Greek works were not translated into Arabic until the eighth century, it is assumed that Dionysius Thrax's linguistics was known to Sibawayh. This was because Dionysius's work was translated at an early stage into Syriac, which was

[4] Sally Wriggins, *Xuanzang: A Buddhist Pilgrim on the Silk Road*, Westview Press, 2003.

[5] The *Encyclopedia of the History of Arabic Science* (Routledge, 1996) aims to provide an overview of the Arab sciences, thereby leaving out the humanistic disciplines.

[6] Surprisingly, Sibawayh's *Kitab* has not been translated into English. There are various French and German translations, for example Gustav Jahn, *Sîbawaihis Buch über die Grammatik übersetzt und erklärt*, Berlin, 1895–1900, reprinted Hildesheim, 1969.

understood and read in large parts of the Persian Empire and subsequently the Arab Empire.[7]

Although the elementary categories in Sibawayh's grammar are Greek through and through, in his *Kitab* he took a decisive step in the direction of an *example-based description* of a language. Such a description already existed in a rudimentary form in Apollonius Dyscolus (see 2.2). It was based on the following idea. Where rules could be found, they were stated, such as for conjugations and declensions. If not, the phenomena were described as best one could on the basis of an enumeration of examples. For instance, Apollonius Dyscolus found that when taken as a whole, word order—or syntax—was too complex to cover with rules. Sibawayh solved this by using a very large number of specific cases to show how Arabic works. While such an enumeration may be helpful for the learner, it is not possible to produce or understand *new* sentences purely on the basis of examples (as we explained in 2.1). Arguably for this reason Sibawayh introduced two original linguistic concepts: *analogical substitution* and *lexical dependence*. He used the first concept to show how words or combinations of words can substitute for each other provided that they are in similar, analogous contexts. For example, in the English sentences 'John likes Mary' and 'Peter hates Suzan' the words 'likes' and 'hates' appear in similar contexts, namely between singular proper nouns. This means that 'likes' and 'hates' may be substituted for each other, resulting in sentences that are again grammatical (i.e. 'John hates Mary' and 'Peter likes Suzan'). Additionally, Sibawayh employed the idea of lexical dependence to show how the form of a word depends on the form of another word. In many languages, like English, the form of the verb depends on the subject, for example 'I like it' but 'she likes it'.[8]

Sibawayh's linguistics seems to consist first of all of an enumeration of examples or, as the Arabist Kees Versteegh puts it: 'Sibawayh's *Kitab* is a collection of all the peculiarities and exceptions in the Arabic language.'[9] But thanks to the concept of analogical substitution, Sibawayh can in principle construct an infinite number of new sentences by substituting words and series of words in the examples, as Panini did with rules for Sanskrit (where Panini included any exceptions in the rules themselves). However, the comparison with Panini does not really hold water. Firstly, the idea of constructing new sentences from earlier sentences is not literally found in Sibawayh's work. One has to distil it indirectly from his examples. Secondly, Sibawayh does not give an exact definition of analogical substitution, but only illustrates it, once again, with examples. Consequently, his system is in fact not verifiable. It is better to compare Sibawayh's system to those of the Alexandrian analogist philologists in the third century BCE or the example-based grammar of Apollonius Dyscolus. Sibawayh's intent seemed to have been that the language speaker (or the student of Arabic) could generalize about the examples provided,

[7] On the influence of Greek linguistics on Arabic linguistics, see Kees Versteegh, *Greek Elements in Arabic Linguistic Thinking*, Brill, 1977.

[8] According to some historians of linguistics this made Sibawayh a dependency grammarian before the term was coined—see Jonathan Owens, *Early Arabic Grammatical Theory*, John Benjamins, 1990, pp. 13ff.

[9] Versteegh, *Greek Elements in Arabic Linguistic Thinking*, p. 11.

and to this end he gave them a tool in the form of analogical substitution, which people could interpret as a metarule. Sibawayh and his descriptive, example-based grammar represent the beginning of a long tradition, which is still very much alive in the modern age (see 5.3).

Sibawayh is also sometimes compared with Panini because of his treatment of phonetics and phonology, which like the Paninian grammar went into staggering detail. The pronunciation of the verses in the Koran was crucial to Muslims because in this case it was about the enunciation of the language of the Creator. Later Islamic linguists could have known about Panini's Indian linguistics, but no more than superficially. In 1030, in his anthropological description of India, the *Kitab al-Hind* or *Indica* (see 3.2), al-Biruni devoted an entire chapter to Indian linguistics and he addressed the phonological aspects of Panini's grammar. But he remarked straight away that 'we Muslims cannot learn anything of it, since it is a branch coming from a root which is not within our grasp—I mean the language itself'.[10] According to al-Biruni, Sanskrit is so different from Arabic that a strict rule-based method of describing it is not applicable to Arabic.

Al-Biruni's *Indica* did not appear until centuries after the death of Sibawayh, who probably did not know Panini's work. But this did him no harm. For centuries Sibawayh was considered by Baghdad and the rest of the Islamic world as the greatest linguist of Arabic, and generations after him were inspired by his approach. Farra' (761–822), who specialized in the properties of syntactic gender, al-Akhfash (*c.*835) and Mubarrid (825–98) were among the most prominent linguists. However, from the tenth century onwards Arabic linguistics became progressively more rigid, and linguists appeared to have been interested in classifying and codifying all kinds of minor details about Arabic that had already been formulated by linguists in the eighth and ninth centuries. There are nevertheless very many Arabic linguists who are of interest to linguistics historians, such as Ibn Jinni in the tenth century, al-Jurjani in the eleventh century (whose grammar was translated in the seventeenth century in Leiden by Thomas van Erpe), and al-Suyuti in the fifteenth century (who was also a historian—see 3.2). But there is little or nothing of interest to us in our quest for underlying methodical principles or empirical patterns.

We will designate Sibawayh's method as the *(analogical) example-based description principle*. This principle is related to the analogy principle of the Alexandrian philologists but differs from it in that the Alexandrians tried to unearth *rules* on the basis of analogies whereas Sibawayh was satisfied with a collection of *examples* and his concept of—analogical—substitution.

Europe: Modists, universal grammar, and hierarchical sentence structure. In the centuries following the fall of the Western Roman Empire in 476 CE, we can perhaps talk for the first time about 'European' linguistics. The centres of learning and science were no longer the cities, such as Alexandria or Rome, which through their impact covered a large part of the former known world, but monasteries,

[10] Edward Sachau, *Alberuni's India*, volume 1, Trübner & Co., 1888, p. 135.

whose influence was limited to the educated part of Christian Europe or even less. And that part of Europe was in a bad way. It suffered terribly from the fifth to the tenth centuries from attacks and raids by Vandals, Goths, Huns, and Norsemen. This in turn had an impact on the humanities. Until the ninth century there was little theoretical or empirical development whatsoever in linguistics. The main exception may have been a descriptive grammar of Celtic in the late-Roman tradition, the *Auraicept na n-Éces* ('The Scholars' Primer'), which could date from the seventh century and was written in relatively tranquil Ireland.[11] We have philologists like Cassiodorus (*c.*485–*c.*585) to thank for the classical linguistics that has survived. Out of his love for Graeco-Roman culture, during his long life as a Roman statesman he had as many classical works as possible copied, as a result of which the survival of many such texts hung by this one thread, while the city libraries met with disaster.

It was not until the era of Charlemagne (742–814) that education was restored to a degree by the scholar and educator Alcuin of York (*c.*735–804), but by then knowledge of Greek had largely disappeared from Europe and the linguistic treatises that were still studied were confined to Latin. During this *Carolingian Renaissance*, monastery schools used Donatus's *Ars minor* as well as Martianus Capella's general textbook (see 2.3) for linguistic education until the linguistic treatise *Institutiones grammaticae* by Priscian was rediscovered. However, this rediscovery did not lead to any new developments in linguistics. The situation did not change until the European world was opened by Islamic civilization arriving from Sicily and above all from Spain. Finally one could again study the writings of Aristotle, who was largely unknown in Christian Europe except for two works of logic from the *Organon* (see 3.6). Many Greek and Arabic texts were translated into Latin in harmonious collaboration between Jewish, Christian, and Islamic scholars in the course of what is called the *twelfth-century Renaissance*. Hundreds of translations were generated on the border between the Islamic and Christian cultures, especially in Toledo. For example, the Italian monk-translator Gerard of Cremona spent no less than forty years in Spain, where he translated eighty-seven works from Arabic. However, the translators proved to be primarily interested in scientific and philosophical texts, and they ignored linguistic, musicological, and historiographical writings.[12] The growth of medieval universities in the twelfth century helped in the propagation of the translated texts and started a new infrastructure needed for learned communities.[13] The European universities put many of the translated texts at the centre of their curriculum. This reacquaintance with Greek and above all Aristotelian writings radically changed both the sciences and the humanities in Europe.

[11] James Acken, *Structure and Interpretation in the Auraicept na nÉces*, VDM Verlag, 2008.

[12] For a list of medieval translations from Arabic and Greek into Latin, see Edward Grant, *A Source Book in Medieval Science*, Harvard University Press, 1974, pp. 35–41.

[13] It should be kept in mind that the schools of Bologna, Oxford, and Paris, which grew very considerably in the twelfth century, are not usually described as universities until the early thirteenth century, when they acquired constitutions and legal identity.

Initially Aristotle's impact on European linguistics came not from his works on logic, but from his metaphysics.[14] According to Aristotle, knowledge can be divided into practical know-how and theoretical understanding, and only the latter leads to truth.[15] In Aristotle's view only three disciplines were truly theoretical—physics, mathematics, and theology. The thirteenth-century linguists in Europe then asked themselves whether language could also be studied in a theoretical manner, alongside the established practical, descriptive method. These thirteenth-century linguists included outstanding scholars such as Roger Bacon, Boetius of Dacia, and Thomas of Erfurt, all of them from northern Europe.[16] Their linguistic movement has been termed *speculative grammar*, and it reached its pinnacle between 1270 and 1320. The word 'speculative' should be interpreted as 'theoretical'. These speculative grammarians concentrated intently on a quest for 'universal' aspects of language and the relationship with reality. Obviously words could not be universal because they differed from one language to another, but according to these scholars the grammatical categories were. The term they used for grammatical categories was *modi* (modes). And although almost every speculative grammarian had his own collection of *modi*, there was some agreement about the basic classification. The *modi* were subdivided into (1) categories of 'being' (*modi essendi*), (2) categories of 'understanding' (*modi intelligendi*), and (3) categories of 'signifying' (*modi significandi*). In fact, all categories were designated as *modi*: word classes, cases, genders, conjugations, and so on.[17]

According to speculative grammarians, or Modists as they were also called, grammatical categories epitomized reality on the grounds of these modes. Every verb, for example, could be traced back to a mode independent of the specific meaning of that verb. This led the Modists to postulate that *every verb can be reduced to a copula and an adjective*. The sentence 'John grows' can for instance be paraphrased as 'John gets taller'. Using this reduction, all sentences could be converted into 'simpler' sentences with only a conjugation of a copula as a verb (that is, a conjugation of *to be, to become, to get, to feel*, or *to seem*). Boetius of Dacia contended that the universal rules of language were to be found on the basis of continuing linguistic reduction like this.[18] And Roger Bacon thought through linguistic reductionism to the ultimate implication and deduced that a Universal Grammar that embodied all languages did indeed exist.[19]

[14] On Aristotle's influence on European medieval linguistics, see Vivian Law, *The History of Linguistics in Europe*, Cambridge University Press, 2003, p. 171.

[15] Aristotle, *Metaphysics*, VI, 1025b27.

[16] G. L. Bursill-Hall, *Speculative Grammars of the Middle Ages: The Doctrine of the Partes Orationis of the Modistae*, Mouton, 1971.

[17] Pieter Seuren, *Western Linguistics: An Historical Introduction*, Oxford University Press, 1998, p. 32.

[18] Boetius of Dacia, *De modis significandi*, *c.*1270. See also Umberto Eco, *The Search for the Perfect Language*, Wiley-Blackwell, 1995, p. 44.

[19] Roger Bacon, *Opus maius*, part III, 1267. Bacon's famous quote on Universal Grammar is: 'Grammatica una et eadem est secundum substantiam in omnibus linguis, licet accidentaliter varietur.' ('In substance grammar is the same in all languages, though it may vary in accidental ways.'). See G. Wallerand, *Les Œuvres de Siger de Courtrai*, Louvain, Institut supérieur de philosophie de l'Université, 1913, p. 43.

Although the Modists did not produce any practical grammars for specific languages, their theoretical considerations did lead to concrete hypotheses. Some of them, for example the Universal Grammar hypothesis, were difficult to test. But for others it was relatively easy, such as the assertion that a complex sentence could be reduced to simpler sentences resulting in a smaller number of concepts (such as the reduction we described earlier of a verb to a copula and an adjective). The Modists assumed that an infinite number of linguistic phenomena could be covered by using a finite number of concepts. In so doing they embraced the procedural system of rules principle without using it in a concrete way.

As well as coming up with the idea of a Universal Grammar, the Modists are also seen as the originators of the extremely influential notion of *hierarchical sentence structure* (see 5.3), in which a sentence is broken down into parts (parsed), and these parts are further parsed into individual words.[20] The Modist Thomas of Erfurt illustrated this concept with the sentence *homo albus currit bene* (the white man runs well), which he first split into two parts—the subject *homo albus* and the predicate *currit bene*—and he then revealed the dependence relationships between the words, where *albus* depends on *homo* and *bene* on *currit*.[21] We encountered the idea of lexical dependence earlier in Sibawayh's grammar. Although we have no indications that the Modists knew Sibawayh, it is possible that they saw his work or that of his followers. Many Christian scholars had access to the libraries in Islamic Spain, particularly after the conquest of parts of al-Andalus during the *Reconquista*. But there are also numerous differences between the Modists and Sibawayh. The Modists had a primarily theoretical approach whereas Sibawayh designed a practical grammar for Arabic, and while the Modists searched for universal rules, Sibawayh worked on the basis of specific examples.

Modism disappeared as quickly as it had arisen. This was probably linked to the triumph of *nominalism* at the expense of *realism* in the fourteenth century. These two movements dominated the philosophical landscape of the European Middle Ages for a long time, which culminated in the dispute known as the 'Problem of Universals'. In short, the realist school defended the view that universal concepts really existed (depending on individual things or otherwise) whereas the nominalist school believed that only individual things had an actual existence and that the universals represented only a spiritual concept. The logician William of Ockham stated that individual things could have something in common, but that these commonalities were only mental ways to refer to individual things. The nominalist movement gained support thanks to the resounding success of Ockham's syllogistics (see 3.6), and gradually all logic and linguistics that smacked of 'universalism' were rejected. Nowhere in his standard work, *Summa logicae*, did Ockham attack speculative grammar, but this was certainly done by others in the fourteenth

[20] Michael Covington, *Syntactic Theory in the High Middle Ages: Modistic Models of Sentence Structure*, PhD thesis, Yale University, 1982.

[21] Thomas of Erfurt, *Grammatica speculativa*, *c.*1310, translated by G. L. Bursill-Hall, 1972.

century.[22] However, since the nominalists offered no replacement linguistic theory, the less theoretical and more grammatical ideas of the Modists endured. The notion of a Universal Grammar also remained an extremely attractive one. It returned in the sixteenth and seventeenth centuries in the work of Dalgarno, Wilkins, and Leibniz (4.3), and was revived in twentieth-century linguistics by Noam Chomsky (5.3).

Parallel worlds: vernacular grammar. At this point the impression may have been created that European linguists in the Middle Ages worked primarily in a hypothetical and speculative way, and did not dirty their hands writing real grammars. Such an impression is incorrect. One remarkable phenomenon in European humanities is that during the Middle Ages there was both an upper echelon of theoretical scholars and a lower echelon of practical workers. This was the case in linguistics, historiography, art theory, and poetics. There were two parallel worlds that interpreted the humanities in different ways: speculative–hypothetical versus descriptive–hands-on. It was not always the case that the upper echelon was populated by highly educated people and the lower by the less well educated. Scholarly monks also wrote practical grammars. But the two traditions—theoretical and practical—operated side by side and sometimes even independently for many years.

We see this in linguistics in the many pedagogical grammars that were written, not just for Latin but also for colloquial language or the vernacular. We have already mentioned the seventh-century Celtic grammar. In the ninth century the Bulgarian monk Chernorizets Hrabar wrote the first grammar for Old Bulgarian or Old Church Slavonic.[23] In the tenth century the Anglo-Saxon monk Aelfric of Eynsham wrote a Latin grammar, albeit *in* Old English.[24] He apologized repeatedly for this in his introduction, but he explained that his use of the vernacular was indispensable to avoiding illiteracy among young people. Aelfric made all sorts of comparisons between Old English and Latin and in his introduction he wrote that his grammar can be used to learn both Latin and Old English. However, in reality he made exclusive use of Latin linguistic categories, which are seldom adequate for Old English (for example Old English uses articles, which do not exist in Latin as a word class).

No new methods or ideas are to be found in these vernacular grammars. They are largely modelled on Latin grammars with eight word classes, and they are also partially based on examples, as in Sibawayh's *Kitab*. At the same time, they are different from Sibawayh's method. None of the medieval grammars used the concept of analogical substitution that Sibawayh knew how to use so effectively. The medieval vernacular grammars are therefore closer to the tradition of the hybrid Graeco-Roman linguistics. Because of the use of both rules and examples,

[22] Earline Ashworth, *The Tradition of Medieval Logic and Speculative Grammar*, Pontifical Institute of Mediaeval Studies, 1977.

[23] David Huntley, 'Old Church Slavonic', in Bernard Comrie and Greville Corbett (eds), *The Slavonic Languages*, Routledge, 1993, pp. 125–87.

[24] Julius Zupitza (ed.), *Aelfrics Grammatik und Glossar*, Weidmann, 1880, repr. 1966.

we shall designate the approach of these vernacular grammars as the *combination of the example-based description principle and the procedural system of rules principle.* The procedural system of rules in these grammars is limited to the morphology (word forms) and phonology (pronunciation), while the syntax (word order) is generally example based. We also see a similar combination of rules and examples in later grammars for the vernacular, for example the fourteenth-century *Leys d'amors* ('Laws of Love') for Occitan (see 3.7)—but here too the influence of the Latin grammatical model is immense.

In another medieval linguistic manuscript, however—a twelfth-century grammar of Old Icelandic (or Old Norse)—we find a surprising new method. This anonymous manuscript is known as the First Grammatical Treatise, not because it would be the first European linguistic theory but because the discourse was the first of four grammatical works in the Icelandic manuscript Codex Wormianus.[25] In the treatise, the *technique of minimal pairs* was developed, which was to be rediscovered in the twentieth century in structural linguistics (see 5.3). Minimal pairs are pairs of words that differ by only a *single* sound and have a *different* meaning. These pairs are used to demonstrate that the two sounds are two separate 'phonemes' of a language (that is to say: sounds that distinguish words from each other). For example, in English /v/ and /b/ are phonemes because the word pair /vet/ and /bet/ has two different meanings, whereas in Spanish the sounds /v/ and /b/ are interchangeable and therefore refer to the same phoneme. If a language's phonemes have been identified, one obtains the smallest building blocks of the language. Panini was already using phonemes in the fifth century BCE as was Sibawayh in the eighth century CE. However, by means of the principle of minimal pairs, the First Grammatical Treatise additionally provides a 'discovery procedure' for finding these phonemes. However, the use of minimal pairs alone does not result in a grammar of a language. The technique focuses on deriving the phonological units.

Principles and patterns in medieval linguistics. Most of the grammars described above use a combination of *example-based* and *rule-based description.* Rules were given where they could be found. If not, examples were discussed in order to cover a particular phenomenon. Sibawayh provided a further tool for generalization about examples, but usually the users of European grammars had to manage without this help. There was also the tradition of speculative grammar, where there was a quest for deeper, general rules, resulting in the concept of a Universal Grammar. But this tradition did not lead to practical grammars. It is only in India that we find a completely formal system of rules without examples, and then only as a continuation of an older tradition.

[25] Einar Haugen (ed.), *First Grammatical Treatise: The Earliest Germanic Phonology*, 2nd edition, Prentice Hall Press, 1972.

3.2 HISTORIOGRAPHY: UNIVERSAL HISTORY AND FORMAL TRANSMISSION THEORY

Medieval historiography was a hotchpotch compared with the reasonable degree of methodological unity in the historiography of Antiquity. There was great divergence between the principles employed, from the highly informal principle that we will call biblical coherence to the formal principle based on precise transmission chains, the *isnad.* But almost all over the world there were chronicles or Universal Histories, for example in Roman Africa, Europe, the Byzantine Empire, the Arab world, and Ethiopia. The interest in Universal History appeared to have been a consequence of a deeply religious society that tried to reconcile its history with its holy book. These forms of historiography were often far from critical, particularly in Christian Europe and Ethiopia. In China on the other hand there was detailed historical criticism, and in Islamic civilization a formal transmission theory emerged for the first time. In India there was essentially no historiography, with the exception of a chronicle of Kashmir.

Christian historiography in Roman Africa: biblical coherence principle. The first Christian histories appeared in Africa and continued the custom of the former Roman annalistic tradition, where a historical overview was produced from when the city was founded (*Ab urbe condita*). The Christian chroniclers went one step further than the pagan annalists and constructed an overview of history from creation to their present day or even to the end of time. As we saw in 2.2, Sextus Julius Africanus and Eusebius initiated chronicles like these, which were also the start of the *Universal History* (or *Salvation History*). These histories focused on the person of Jesus and were heavily preoccupied with periodization. For example, in his *De Civitate Dei* the Berber church father St Augustine (354–430)—who was working in Hippo Regius (present-day Algeria)—proposed a periodization in six ages.[26] This subdivision, which had already been made in the third century by Sextus Africanus, is analogous to the six days of the creation and contained the following periods: (1) from Adam to the Flood, (2) from the Flood to Abraham, (3) from Abraham to David, (4) from David to the Babylonian exile, (5) from the Babylonian exile to the birth of Christ, and (6) from the birth of Christ to the end of the world. St Augustine compared these six ages with the six stages of people's lives: from childhood, puberty, adolescence, maturity, and serenity to old age. The reference to the human-life ages is somewhat surprising as it was essentially pagan, reaching back to Cicero, Livy, and others.

Through St Augustine's periodization, the history of the world was equated de facto with biblical Judeo-Christian history, and only a very modest place remained for contemporary history. The same applies to alternative periodizations, for instance those of St Jerome (*c.*347–420) and Orosius (*c.*375–418), in which world history was divided up into four empires: Babylonian, Macedonian, African

[26] St Augustine, *The City of God*, translated by Henry Bettenson, Penguin Books, 1972, book XVff. See also Karla Pollman, *Saint Augustine the Algerian*, Edition Ruprecht, 2003.

(Carthaginian), and Roman. The end of time would happen during the Roman Empire.

Although Christian historiography remained officially based on the ancient principles of written sources, eyewitness accounts and/or personal experience, the sources were hardly ever verified for reliability or accuracy. What counted was the authority of the source and above all the extent to which the source agreed with biblical history. There was even a belief that developed among more fanatical Christians that all other classical sources and texts should have been rejected. It is to St Augustine's great credit that he convinced these radical Christians that pagan views should not be mistrusted but should in fact be appropriated. In *De doctrina christiana*, St Augustine explicitly stated that 'all good and true Christians should understand that truth, wherever they may find it, belongs to their Lord.'[27] The study of the classics during the Middle Ages could be legitimized time after time thanks to St Augustine's authority.

World history was reinterpreted using what we shall call the *biblical coherence principle*. According to Justin Martyr, Moses had a decisive influence on Homer.[28] According to Melito of Sardis the Roman Empire was created in order to facilitate the spread of Christianity.[29] And according to Clement of Alexandria, Plato and Aristotle were not altogether wrong but were not fully informed.[30] The interpretation of the last age, from the birth of Christ to the end of the world, was a problem for all Christian historians. While there appeared to have been a clear pattern in history before Christ (at least in St Augustine's periodization), it seemed that there was little divine structure to be found in the post-Christian period up to the end of time. Miracles and prophecies were therefore considered to be very important because they bore witness to the omnipresence of God.

There was a departure from the pattern of cyclicity. The classical Roman historians had already exchanged this pattern for their linear historiography, but it was the Christian historians who gave acyclicity a theological foundation. The Universal World History followed a *linear pattern* from a *unique* beginning (the creation) to an *ultimate* goal (the last judgment). With a little goodwill this linearity could also be observed in the history of Rome itself. Despite the raids and the pillaging of the city, Rome continued to exist. It was unthinkable to St Augustine that the Roman Empire would not reach the end of time. Previously he had talked about Old Rome and New Rome. The latter would attain the perfection that the first had not achieved. Even after the 'fall' of the Western Roman Empire, this New Rome remained the centre of papal power.

Christian historiography in Europe: Gregory of Tours, miracles, and prophecies. While North Africa was the intellectual centre of the early Christian humanities until the fifth century—Sextus Africanus, Tertullian, Origen, Orosius, and St Augustine

[27] St Augustine, *De doctrina christiana*, 18.28. See also St Augustine, *Teaching Christianity: De doctrina Christiana*, translated by Edmund Hill, New City Press, 1996, p. 144.

[28] E. J. Goodspeed (ed.), *Die ältesten Apologaten*, Vandenhoeck & Ruprecht, 1915.

[29] Bernhard Lohse, *Die Passa-Homilie des Bischofs Meliton von Sardes*, Brill, 1958.

[30] Eric Osborn, *Clement of Alexandria*, Cambridge University Press, 2008.

all worked in Africa—it soon moved to Europe. One of the causes was without doubt the invasion of North Africa by the Vandals in the fifth century, during which St Augustine died. The conquest of North Africa by the Arabs at the end of the seventh century was the final blow to the Christian humanities in this region.

Gregory of Tours (*c.*539–94) was the first major historian north of the Alps.[31] Gregory was working in Frankish Gaul, which converted to Christianity en masse in around 500 following the example of their sovereign Clovis. Gregory furthered the integration of the Roman annalistic tradition and the Christian chronicles. In his *Decem libri historiae*, better known as the *Historia Francorum* ('History of the Franks'), he began by giving a historical overview from the creation to the death of his predecessor Martin of Tours, after which he concentrated on the history of Gaul.[32] For some of his information Gregory was able to draw on earlier works, thus making use of the written source principle. However, Gregory could chronicle the contemporary history of the Franks at first hand. As Bishop of Tours he was in a position to observe politics at close quarters, either on the basis of personal experience or from direct reports from witnesses. Gregory gave an exhaustive overview of the sixth-century Frankish sub-kingdoms—royal marriages, local politics, relations with the government, and the many intrigues in other parts of the empire. However, where there were no sources, Gregory referred to miracles, omens, and other events that were not based on either sources or eyewitness accounts. Moreover, Gregory wanted to explain historical events and did so on the basis of prophecies and allegorical Bible interpretations. For example, he described the plundering entourage of a Frankish princess and argued that the biblical text Joel 1:4 prophesied the event. On top of all this Gregory also used ad hoc prophecies (to explain the death of a king or a political conflict, for instance) without distinguishing between biblical and post-biblical prophecies, let alone addressing the problem of using a prophecy as an explanation of a historical fact. If we wanted to characterize his work using a principle, it would be at worst an allegory principle and at best the biblical coherence principle referred to above. The historiography had to agree with the religious doctrine or with an allegorical interpretation of the Bible as a written source. But in practice any rumour, miracle, omen, or prediction could be acceptable provided that it did not conflict with ecclesiastical doctrine. This meant that Gregory had a free hand in explaining historical facts. Despite these epistemological shortcomings, Gregory's historiography is the only available source about the Frankish Empires after the death of Clovis.

The Venerable Bede and historical time reckoning. The Venerable Bede (*c.*673–735), who worked in the Anglo-Saxon kingdom of Northumbria, set in motion the dissemination of the influential *Anno Domini* (AD) dating method. Apart from this innovation, there are no new principles to be found in Bede's historiography. This does not of course detract from the fact that his history of the Anglo-Saxon people from Caesar to 731 (*Historia ecclesiastica gentis Anglorum*) is of

[31] Martin Heinzelmann, *Gregory of Tours: History and Society in the Sixth Century*, Cambridge University Press, 2001.

[32] Gregory of Tours, *The History of the Franks*, translated by L. Thorp, Penguin 1974.

inestimable value because of its uniqueness.[33] Like Gregory, he described the miracles of his time primarily based on hearsay. Although his reports sound improbable to us, according to medieval historians they pointed to a divine plan, and as such they formed an essential component of Universal History.[34]

Bede's two manuals *Liber de temporibus* and *De temporum ratione* were of the utmost importance to historiography. These works provided a foundation for the discipline that is known as time reckoning or chronology, and also gave a decisive impulse to the *Anno Domini* dating system. This system was not invented by Bede himself but by the Scythian monk Dionysius Exiguus (*c.*470–*c.*544). Dionysius wanted to extend the Easter dates that had been calculated by the Alexandrian church on the basis of the *Anno Diocletiani* dating system in the Julian calendar. This *Anno Diocletiani* was started during the last great persecution of Christians under Diocletian on 29 August 284 CE. But now Dionysius did not want to keep alive the memory of the tyrant Diocletian in his calculations of the Easter dates, and so he introduced—still in the Julian calendar—a new time reckoning in which he departed from the tradition of naming calendar years after consuls. Dionysius simply stated that the consulate of Flavius Probus took place '525 years after the incarnation of our Lord Jesus Christ', which meant that he de facto introduced the *Anno Domini* dating system.[35] It is not known exactly how Dionysius arrived at the number 525. However, it was not Dionysius but Bede who used *Anno Domini* to date historical events for the first time in his *Historia*. And Bede was also the first to use the Latin equivalent of *before Christ*, and, moreover, the custom among historians of not indicating a year 0 (the number zero was not known in Christian Europe). In continental Europe *Anno Domini* was adopted as the preferred dating system by Alcuin of York during the Carolingian Renaissance, which resulted in its definitive dissemination in Western Europe. It took many centuries before the dating system became fashionable in Eastern Europe too, and even later further afield as Europe expanded.

Bede also used the *Anno Domini* system when dating the Easter cycle, the dates for which he calculated as far ahead as 1595.[36] Easter Sunday was calculated as being the first Sunday after the first full moon in spring. The Easter cycle therefore consists of both lunar cycles (of nineteen years) and solar cycles (of twenty-eight years), and consequently has a period of 532 years. Through his Easter cycle Bede demonstrated that he had mastered all the knowledge available during his time, from astronomy and mathematics to history and theology. Thanks to Bede, time reckoning and therefore also astronomy became auxiliary disciplines for historiography, and in particular Universal History.

[33] For an English translation of Bede's *Historia ecclesiastica* see A. M. Sellar, *Bede's Ecclesiastical History of England*, Christian Classics Ethereal Library, 1907.

[34] N. J. Higham, *(Re-)Reading Bede: The Historia Ecclesiastica in Context*, Routledge, 2006.

[35] Bonnie Blackburn and Leofranc Holford-Strevens, 'Calendars and Chronology', *The Oxford Companion to the Year: An Exploration of Calendar Customs and Time-Reckoning*, Oxford University Press, 1999, pp. 659–937.

[36] John Heilbron, *The Sun in the Church*, Harvard University Press, 1999, p. 35.

It is debatable whether we should define Bede's *Anno Domini* time reckoning as a new methodical principle. Firstly, because Bede's dating system came within the existing Julian time reckoning, and only the starting year differed from earlier datings, and secondly because the new time reckoning was not a working principle with which patterns in historical material could be discovered—it was nothing more than the establishment of a new starting year. However, it is beyond doubt that the dissemination of the new time reckoning was hugely important. Although many other calendar systems remain in use (even the *Anno Diocletiani* dating system is still employed in Africa today, for example in the Coptic and Ethiopian calendars), the *Anno Domini* system of dates—currently also called the *Common Era*—is the only worldwide time reckoning.

We will see how, centuries later, Bede's time reckoning led to a crisis in chronological historiography. Aligning the profusion of resurrected historical facts with both biblical chronology and Bede's time reckoning became increasingly problematic. There was simply not enough time to cram world history from the first Egyptian kings to the birth of Jesus into the biblical chronology. However, it was not until the seventeenth century that this problem caused a revolution. For the time being everything in the garden was rosy.

National histories. During the Carolingian Renaissance and the reign of Alfred the Great, King of Wessex, there was a blossoming of annals, chronicles, and hagiographies. Although this vigour did not generate any new principles, it is striking that most annals, such as the royal Frankish annals, Carolingian chronicles and the annals of Fulda, were written wholly or in part by anonymous historians. However, it appears that these authors often worked under the supervision of famous historians, such as Einhard (Charlemagne's biographer), Rudolf of Fulda, or Meinhard. As we have already seen in European linguistics, in historiography there seems to have been a scholarly upper echelon and a more practical lower one. A few of the more learned or imaginative wrote the great chronicles and the theoretical works on time reckoning while for the most part historiography took the simpler form of the usually anonymous annals.

New genres also appeared—the *gesta* and the *national history*. The gesta are a record of the deeds of bishops or rulers, for example *Gesta Friderici imperatoris* ('Deeds of Emperor Frederick') by Otto of Freising (*c.*1113–1158), who also wrote the *Chronica*, one of the most extensive medieval chronicles.[37] The first national history dates from the seventh century—Paul the Deacon's *History of the Lombards*.[38] This was followed by others, for example a *History of the Britons*, including the basis of the Arthurian epic, which was largely written in the ninth century by the Celtic monk Nennius.[39] Histories of virtually every people in Europe appeared after the

[37] Otto of Freising, *Die Taten Friedrichs oder richtiger Cronica*, translated by Adolf Schmidt, Wissenschaftliche Buchgesellschaft, 1986.

[38] Paul the Deacon, *History of the Lombards*, translated by William Foulke, University of Pennsylvania, 1907.

[39] Nennius, *The History of the Britons*, translated by J. A. Giles, Henry G. Bohn, 1848.

Carolingian Renaissance—Widukind wrote a *Saxon History* in the tenth century,[40] Helmold composed a *Chronicle of the Slavs* in the twelfth century[41] and chronicles were likewise produced about Latvia, Estonia, Bohemia, Poland, Denmark, and West Francia. A new history of England and Normandy (after the Battle of Hastings) was written. These chronicles have little or nothing to do with Universal History. They are primarily a legitimization of new states or dynasties by means of a collective genealogy, which was usually imposed. The historians took carte blanche in working with sources and explanations, and they did not use a clear systematic principle.

City chronicles and encyclopaedic history: Ranulf Higden. Are there other methods or empirical patterns to be found in medieval Christian historiography? First of all it should be pointed out that many late medieval chronicles are still lying unread in repositories.[42] But the general picture that emerges from the chronicles that have been studied is that the monasteries lost influence in around 1300, as the universities and cities became more influential. The evidence for this includes the many city chronicles of London, Cologne, Florence, and elsewhere.

The fourteenth century saw the appearance of encyclopaedic world histories, which tried to summarize all knowledge from the past, including all known religions. Over forty of these *summae* or *breviaria* have survived. The fourteenth-century *Polychronicon* by the Anglo-Saxon Benedictine monk Ranulf Higden (1280–1363) was the most systematic. In his introduction Higden gave a methodical overview of eight sources of knowledge that he considered necessary for a complete understanding of world history:[43]

(1) Topography of historical events
(2) Two *status rerum* ('the state of things')
(3) The three ages (before the law, under the law, and under God's glory)
(4) The four world empires (Assyria, Persia, Greece, Rome)
(5) The five world religions (nature worship, idolatry, Judaism, Christianity, and Islam)
(6) The six ages of the world
(7) The seven types of history-makers: ruler, soldier, judge, cleric, politician, merchant, and monk
(8) The eight calendar systems: three Jewish (starting in January, February, or March), three Greek (Troy, Olympiads, Alexander), one Roman (foundation of the city), and one Christian (incarnation of God).

This quasi-scientific framework did not stop Higden from recording his world chronicle from a Christian moralist perspective. He also obtained his information primarily from earlier chronicles. The *Polychronicon* is therefore not much more than an eclectic compendium based on the written source principle. Nevertheless,

[40] Albert Bauer and Reinhold Rau (eds), *Die Sachsengeschichte des Widukind von Korvei*, Wissenschaftliche Buchgesellschaft, 2002.
[41] Bernhard Schmeidler, *Helmolds Slavenchronik*, Monumenta Germaniae Historica, 1937.
[42] Ernst Breisach, *Historiography: Ancient, Medieval and Modern*, University of Chicago Press, 2007, p. 144.
[43] Ranulf Higden, *Polychronicon*, translated by John Trevisa, Longman, 1876.

Higden's approach is impressive. His work was one of the first medieval Christian overviews of the different world religions, history-makers, and calendar calculations. It is striking that while Islam was discussed under world religions (where forms of nature worship were also addressed), it was not mentioned under calendar systems. Higden's *Polychronicon* enjoyed unprecedented popularity. An English translation appeared as early as 1387 and less than a century later his compendium was already printed (by William Caxton in 1482).[44] The *Polychronicon* was the standard work on world history in the fifteenth-century Christian world. If one pattern emerges from Higden's work, it is a pattern of pluriformity. No matter how moralistic his compendium may be, the picture of a multiform world without a preconceived plan seems inescapable.

Byzantine historiography: continuation of the Greek tradition. In European historiography the biblical coherence principle represented a break with earlier practice. Byzantine historiography, on the other hand, retained many features of classical history writing. This is not surprising. In 330 Constantine moved the capital of the Roman Empire to Byzantium, which would subsequently be named after him: Constantinople. The Eastern Roman Empire continued to exist after the fall of the Western Roman Empire in 476. The Byzantines saw themselves as inhabitants of the Roman Empire and their rulers as an uninterrupted continuation of the Roman emperors. However, knowledge of Latin deteriorated rapidly, and the empire's culture and language became predominantly Greek. Byzantine historiographers continued the systematic principles of their Greek predecessors.

For example, in his *History of the Wars*, Procopius (*c*.500–*c*.565) applied Polybius's personal experience principle (see 2.2). In 527 Procopius became private secretary to General Belisarius under Emperor Justinian and accompanied him during his campaigns against the Persians, Vandals, and Ostrogoths. He used his personal experiences as the basis for his first masterpiece, which comes across as remarkably objective and impartial. The nature of his later work, the *Anékdota* ('Unpublished Writings')—which did not appear until the tenth century—was completely different. Still using his personal experience, he portrayed Belisarius as a puppet of his domineering wife Antonina and gossiped quite freely about Justinian's corrupt wife Theodora.[45]

Agathias (*c*.536–582/594) collected his historical information on the basis of the eyewitness account principle following the tradition of Thucydides, whose style he also imitated. His *Histories* are the only source for the story of the closing of Plato's Academy by Justinian in 529 and the subsequent flight of the philosophers to Persia.[46]

It is not entirely surprising that military men who wrote history, like Bryennius (eleventh century) and Cinnamus (twelfth century), took Xenophon as their

[44] William Kuskin, *William Caxton and the English Canon: Print Production and Ideological Transformation in the Late 15th Century*, University of Wisconsin–Madison, 1997.

[45] Procopius, *The Secret History*, translated by Anthony Kaldellis, Hackett Publishing, 2010.

[46] Warren Treadgold, *The Early Byzantine Historians*, Palgrave Macmillan, 2007, pp. 279–90.

example (see 2.2). The theologically inspired historians, such as Joannes Malalas (sixth century), Theophanes Confessor (seventh and eighth centuries) and Joannes Zonaras (twelfth century), wrote chronicles and Universal Histories in the tradition of Sextus Africanus and Eusebius, chiefly employing the written source principle.

There were also hagiographies and biographies, for instance the *Alexiade* by Anna Komnene (1083–1153). She was the oldest daughter of Emperor Alexios Komnenos and from the seclusion of a convent she wrote an extremely biased biography of her father in which she also exposed the many intrigues at the Byzantine court. Her contempt for the crusaders, who were more interested in Constantinople's riches than 'liberating' Jerusalem, is a fascinating combination of the personal experience and eyewitness account principles.[47]

Even after the fall of Constantinople in 1453, Byzantine historians—for example Laonicus Chalcondyles (1423–1490) and Critobulus (*c.*1410–*c.*1470)—continued in the existing tradition, although from a Turkish perspective. There were no new principles or patterns in Byzantine historiography, which does not detract in the slightest from the historical importance and the fascination of these works.

Islamic historiography: transmission theory and the *isnad* principle. Universal Histories were also produced in Islamic civilization. However, Islamic historiography begins with the reconstruction of Muhammad's life, the *sira*, which authors tried to base on the most accurate possible foundations. All historical information had to be traced back to the Prophet himself as closely and precisely as possible. This 'transmission theory' was further developed in the study of *hadith* ('that which is told'). The *hadith* consists of thousands of fragments about what Muhammad had said or done. There were many fragments in circulation so it was important to establish which sources were more reliable than others. Different methods were developed, of which the *isnad* or 'chain of transmission' was the most common. Every *hadith* had its *isnad*, such as: 'A heard it from B, who heard it from C, who heard it from a companion of Muhammad.' *Isnads* were investigated in depth in order to verify that the transmission chain was feasible, for example by ensuring that all the links had indeed existed and had lived in the same area at the time of the transmission. During the first three centuries after the death of Muhammad in 632, Islamic scholars and theologians debated the question of which of the traditions that were handed down were authentic and which had been invented after the fact.[48]

Nonetheless, the traditional reconstruction of the early history of Islam has remained problematic.[49] To start with, there were very few primary sources dating from this period. The oldest biography of Muhammad, written by Ibn Ishaq (704–*c.*767), appeared over a century after the Prophet's death.[50] And then,

[47] Anna Komnene, *The Alexiad*, translated by E. R. A. Sewter, Penguin, 2009.

[48] For a critical analysis of the *hadith* and *isnad* (in Sunni Islam), see G. H. A. Juynboll, *Encyclopedia of Canonical Hadith*, Brill, 2007.

[49] Wim Raven, 'Sīra and the Qur'ān', in Jane Dammen McAuliffe (ed.), *Encyclopaedia of the Qur'ān*, volume 5, Brill, 2005, pp. 29–51.

[50] Alfred Guillaume, *The Life of Muhammad: A Translation of Isḥaq's 'Sirat Rasul Allah'*, Oxford University Press, 1955, reprinted in 2004.

the reconstruction of the transmitted information was executed by Islamic scholars in a highly politicized context during the Abbasid caliphate just after the overthrow of the Umayyad dynasty, while the groups from which the Sunnis and Shiites would later arise published rival versions of the history of Islam. Some scholars think that these factors made the problem of reconstructing the sayings and deeds of Muhammad so great that, despite the precisely formulated transmission theory, no single *hadith* can be considered reliable. However, it should be pointed out that there are major differences of opinion on this point between (religious) Muslim scholars and (secular) Islamic Studies scholars.[51]

Unification of the most probable source and eyewitness account principles. Although the principle of *isnad* has resulted in controversies in the reconstruction of the sayings and deeds of the Prophet, it became a success when it was applied outside of theology, such as in profane history and literature. Soon no text was credible if it was not accompanied by at least a transmission chain. Once the *isnad* methodology was established between 700 and 770, it provided a tool for determining the degree to which a source was reliable. One of the most important elements of a transmission chain are the informants, or *transmitters*. These were evaluated using the following points:

- Could the transmitters have met each other, given where they were and at which times?
- Are there any records of their meeting or collaboration or common interests?
- Are the transmitters of sound moral fibre and not politically motivated?
- Is the information that has been transmitted logically consistent? Is it rational?
- Is the transmission chain free of hidden defects?

Depending on the answers to these questions, every transmitted source was put into one of four categories:

- *sahih*, 'very reliable'
- *hasan*, 'good, but less reliable'
- *da'if*, 'dubious'
- *mawdu'*, 'invented'

These categories assigned a degree of probability to each source and can therefore be effectively considered as a formalization of Herodotus's most probable source principle (see 2.2). At the same time the *isnad* methodology also provided a solution for Thucydides's problem of second-hand or third-hand eyewitness accounts (see 2.2), provided that the transmission could be demonstrated to be feasible, consistent, and defect free. In other words, the *isnad* methodology effectively formalized and unified Herodotus's most probable source principle and Thucydides's eyewitness account principle.[52] Herodotus appeared to use his

[51] Fred Donner, *Narratives of Islamic Origins: The Beginnings of Islamic Historical Writing*, Darwin Press, 1998.

[52] Herodotus and Thucydides were at best only indirectly known by Islamic historians. Classical historiography was transmitted through Christian intermediaries, often via Syriac historians or

principle haphazardly without any method, whereas *isnad*-based historiography is founded on a well-defined system of rules. Thucydides foresaw problems with second-hand or third-hand eyewitness accounts, but the *isnad*-based historiography provides a method for incorporating such eyewitness reports. The formal integration of these two earlier principles was of overriding significance to the often exceptionally fine historical precision in Arabic historiography. We will argue in 4.2 that it may also have been of overriding significance to later European historiography and philology.[53]

A special property of the critical transmission theory of the *isnad* is that it arose out of religious endeavours. This was very different from practices in European Christian historiography, where miracles, omens, and prophecies were often recorded without criticism by Gregory of Tours and Bede as proof of God's omnipresence. An explanation might lie in the fact that the medieval Christian tradition had four Gospels that could serve as a history of the deeds and sayings of Jesus, whereas the Islamic tradition had to reconstruct the life of Muhammad almost completely from oral information and wanted to approach the task as critically and accurately as possible.

The *isnad*'s empirically motivated and logically refined tradition may be considered to be a forerunner of the blossoming of science during the Golden Age of Islamic civilization. However, the extent to which *isnad*-based history actually stimulated the development of Islamic science is not known. But in view of the fact that many Islamic scholars, from al-Dinawari to al-Biruni, were actively involved in both history and physics (and even medicine), it can be no coincidence that Islamic history and science *both* followed a strongly empirical pattern.

Universal History based on the *isnad*: al-Tabari. The Persian al-Tabari (838–923) produced one of the first examples of a Universal History employing the new historical principle. In his *Tarikh al-rusul wal-muluk* ('History of the Prophets and Kings') he wrote a salvation history from the creation to 915.[54] This magnum opus consisted of forty books, the first four of which correspond roughly to the Hebrew Bible. The *isnad* was unnecessary and even impossible for these volumes and the written source principle was employed. After one volume on pre-Islamic empires, in which he quoted the Byzantine historian Agathias (see above), al-Tabari devoted the remaining thirty-six volumes to a very detailed account of three centuries of Islamic history. He described the many caliphates,

Christians converted to Islam. Herodotus and Thucydides can be traced in al-Sijistani's list of pre-Islamic scholars, the *Siwan al-Hikma* ('Vessel of Wisdom') from the 10th century—see Franz Rosenthal, *The Classical Heritage in Islam*, Routlegde, 1975, pp. 36–7. Few Greek and Roman historians were quoted by Islamic historians, among which Eusebius and Orosius—see Franz Rosenthal, *A History of Muslim Historiography*, Brill, 1968, pp. 78ff. Despite this limited reference to Greek and Roman historiography, the problem of source selection was well known among Islamic historians (see also Chase Robinson, *Islamic Historiography*, Cambridge University Press, 2003, p. 49).

[53] Cf. George Makdisi, *The Rise of Humanism in Classical Islam and the Christian West*, Edinburgh University Press, 1990. While Makdisi gives maximal effort to show that Islam influenced Western humanism, he does not argue that the *isnad* may have had a major influence on early modern European historiography and philology.

[54] Ehsan Yar-Shater (ed.), *The History of al-Ṭabarī*, State University of New York Press, 1985–2007.

conquests, crises, downfalls, periods of recovery, uprisings, and of course the establishment of the Abbasids in Baghdad. His work is considered by many to be the one of the most accurate histories of early Islam. In order to enhance the legitimacy of his historiography, in the best *isnad* tradition al-Tabari referred to a chain of 'guaranteed informants'. This had already been done in less ambitious works by the ninth-century historians al-Ya 'qubi and al-Dinawari, who also wrote the first history of the Kurds. In addition to Arab history, al-Tabari's work also recounted the past of other peoples in the Middle East, including those of his Persian fatherland. At first sight al-Tabari's history appears to be neutral because he employed the *isnad* conventions, but it soon becomes clear that he was an adherent of the Abbasid ideology, in which everything always leads to one person: the Abbasid caliph.[55] Notwithstanding, his Universal History contains extremely valuable information about early Islamic civilization.

The apex of the Universal History: al-Masudi. The new style of historiography reached a pinnacle in the Universal History by the Arab al-Masudi (896–956), the *Muruj al-dhahab wa ma'adin al-jawahir* ('Meadows of Gold and Mines of Gems').[56] As Herodotus had done before him, al-Masudi combined historiography with geographical descriptions. Also his knowledge was not limited to the Islamic world alone. He included the history of Europe (he gave an overview of all kings starting with Clovis) and even China. After the Battle of Talas near Samarkand in 751, Islamic civilization came into contact with Chinese culture and technology, which is how the Arab world—and later Europe—acquired the knowledge to produce paper. In terms of breadth and depth, every medieval Christian historian was in al-Masudi's shadow. Scholarly culture under the Abbasids resulted in much greater knowledge about world history than that of Bede or Higden. Thanks to the introduction of Chinese paper mills, books were relatively cheap and—as well as city libraries—private collections containing thousands of volumes were no exception. Bede's famous monastery library held only a couple of hundred manuscripts.

Al-Masudi owed much of his knowledge to the journeys he made, again like Herodotus, in the Persian provinces, Armenia, India, eastern Africa, Sri Lanka, China, and probably Russia. His writings on the Tang Dynasty, the Khazars and the Russians were unique, and his approach is strikingly original, albeit hybrid. On the one hand Meadows of Gold and Mines of Gems uses historically documented facts that were derived using the *isnad* method from a chain of verifiable sources and stories.[57] On the other hand al-Masudi alternates these facts with less reliable anecdotes, poems, and even jokes without any investigation of how they were transmitted. In so doing al-Masudi in fact turns his back on the *isnad* methodology to an extent and uses the (first) eyewitness account or personal experience principle. The result is a literary gem in which al-Masudi devoted as much attention to social,

[55] For a narratological analysis of al-Tabari's work, see Johan Weststeijn, *A Handful of Red Earth: Dreams of Rulers in Tabari's History of Prophets and Kings*, PhD thesis, University of Amsterdam.

[56] Mas'udi, *The Meadows of Gold, The Abbasids*, translated by Paul Lunde and Caroline Stone, Kegan Paul, 1989.

[57] Tarif Khalidi, *Islamic Historiography: Histories of Mas'udi*, State University of New York Press, 1975, pp. 3ff.

economic, and cultural details as he did to political history. It is one of the most compelling historiographies to emanate from Islamic civilization, and it earned him the title of the 'Herodotus of the Arabs'.

Al-Biruni and the 'anthropological' method. The Persian al-Biruni (973–1048) was more of a universal scientist—from astronomer to mathematician—than a historian, even though by the age of twenty-six he had already written a chronology, which has not survived. George Sarton called al-Biruni 'one of the very greatest scientists of Islam, and, all considered, one of the greatest of all times'.[58] Al-Biruni is of interest to the humanities primarily as an Indologist. His exhaustive description of India, the *Kitab al-Hind*, is part of a long tradition of historiographies about India (which began with the historian Megasthenes during the era of Alexander the Great—see 2.2) and is often referred to as *Indica*. Al-Biruni started this work with a foreword about historical methodology that still appeals to the imagination. 'No one will deny that in questions of historic authenticity *hearsay* does not equal *eye-witness*.'[59] He then analysed the pros and cons of eyewitness accounts and information that has been transmitted, and went on to construct a subtle criticism of the *isnad*. 'The tradition regarding an event which in itself does not contradict either logical or physical laws will invariably depend for its character as true or false upon the character of the reporters.'[60] In other words, a lie that is transmitted with the very greatest possible accuracy and without defects remains a lie. This inherent but not previously expressed problem with the *isnad* method led al-Biruni to let Indians speak for themselves—primarily by means of quotations from Sanskrit literature but also by recording what Indians had to say—rather than a principle based on transmission. Al-Biruni is therefore sometimes referred to as the first anthropologist, although it should be emphasized that there was no independent, let alone participative, observation. Since in his description of India, al-Biruni also addressed Indian astronomy (and its exceptional heliocentric theories) at length, one would have expected his work to have generated some interest among Christian scholars during the extensive burst of translations in twelfth-century Europe. However, it appears that they knew nothing about him. It was not until the eighteenth century that a few chapters were translated into French.[61] It was even later that an integral translation of the *Indica* was produced.[62]

Al-Biruni was the author of a further historical work in the year 1000 with the beautifully poetic title *The Remaining Signs of Past Centuries* (*Kitab al-athar al-baqiya 'an al-qurun al-khaliya*).[63] This book contains a comparative study of the calendar systems of different civilizations punctuated with historical and

[58] George Sarton, *Introduction to the History of Science*, volume 1, Williams and Wilkins, 1931, p. 707.

[59] Edward Sachau, *Alberuni's India*, volume 1, Trübner & Co., 1888, p. 3. (Italics are as in Sachau's translation.)

[60] Sachau, *Alberuni's India*, p. 3.

[61] B. Boncompagni, *Intorno all'opera d'Albiruni sull'India*, Tipografie delle Scienze Matematiche e Fisiche, 1869.

[62] Sachau, *Alberuni's India*.

[63] This book is also known as *Chronology of Ancient Nations*, after the translation by Edward Sachau, W. H. Allen & Co., 1879.

astronomical information and descriptions of the customs and religions of the different peoples (from Manichaeism and Buddhism to Christianity). All traces of the *isnad* method were absent from this work too. For that matter, the *isnad* would not even have been applicable to *The Remaining Signs* because al-Biruni used existing historical writings, and therefore applied the written source principle. It is not entirely clear what purpose he had in writing this work. He criticized everything and everyone, from copying errors in extant texts to people's ignorance: 'Many people attribute to God's wisdom all they do not know of physical sciences.' Al-Biruni's goal may have been to show, on the grounds of a total description, that science can serve as an alternative to human superstition or credulity. We have seen that in fourteenth-century Europe Ranulf Higden also wrote an overview of all the calendar systems, religions, and peoples known to him. We can sense that al-Biruni's scepticism and relativism in view of such cultural diversity without clear unity is even greater than Higden's.

Ibn Khaldun: historical critique and the 'sociological' method. Ibn Khaldun (1332–1406) may possibly have been the most important exponent of historiography in Islamic civilization. He was born in Tunisia, so we could just as easily discuss Ibn Khaldun under the heading of African history.[64] Ibn Khaldun's principal work was the *Muqaddimah* dating from 1377.[65] It is also known as the *Prolegomenon* (the first part of his Universal History). In it he describes the history of North Africa, but he begins the *Muqaddimah* with a detailed and refined historical criticism of the errors made by fellow historians, and the problems of historiography in general. Ibn Khaldun warns that all historical reports are by their very nature prone to mistakes. He discusses among other things the partisanship of a historian towards a creed or opinion, the historian's over-confidence in his sources, the failure to understand what is meant, the inability to place an event in its real context, the common desire to gain the favour of those of high rank, and the ignorance of the laws governing human society.

With regard to the last of these, Ibn Khaldun has been called the forerunner or even the founder of sociology.[66] It is striking how critical and sceptical he is about his own discipline. He also attacks 'idle superstition' and the 'uncritical acceptance of historical data'. It was on these grounds that Ibn Khaldun introduced a scientific method for historiography in which he referred to the 'new science'. He then presented a few important definitions and methodologies:

- History is composed of the news about days, states, and previous centuries. It is a theory, an analysis and justification of creatures and their principles, and a science of how events happened and their causes.
- Myths have nothing to do with history and should be refuted.

[64] Joseph Ki-Zerbo (ed.), *General History of Africa I: Methodology and African Prehistory*, Heinemann, 1981, pp. 26–7.

[65] For a full translation of the *Muqaddimah*, see Franz Rosenthal, *The Muqaddimah: An Introduction to History*, Princeton University Press, 1958, reprinted in 1989.

[66] Syed Hussein Alatas, 'The autonomous, the universal and the future of sociology', *Current Sociology*, 54, 2006, pp. 7–23.

- To build strong historical records, the historian should rely on the necessary rules for the truth comparison.

Ibn Khaldun did not explain in detail precisely what the necessary rules referred to in the last point were, but it is clear that his historical method is encapsulated in knowledge of how society functions and its laws. Although it is tricky to formulate a principle to cover his method, it is practical to refer to Ibn Khaldun's 'new scientific' approach as the *sociologically analysed source principle*, by which we mean that every source has to be critically compared in regard to the social context and the laws that control the society. Ibn Khaldun's method is not a fully formalized principle in the way that the *isnad* method is.

The most impressive application of Ibn Khaldun's approach is his historical and sociological elaboration of the cyclical pattern of rise, peak, and decline. If a society becomes a leading civilization or even the dominant culture in a region, according to Ibn Khaldun the peak of this civilization is always followed by a period of decline. This means that the next cohesive group that conquers this civilization is a gang of barbarians by comparison. Once they have established their control over the conquered civilization, these barbarians are attracted by its more refined aspects, such as literature, art, and science, which are subsequently assimilated or appropriated by the oppressors. The upshot is that the next group of barbarians repeats this process, as a result of which the pattern of peak and decline actually leads to an accumulation of knowledge and culture. And we can indeed find a pattern like this with the Romans who conquered Greece and the Arabs who overran the Persian Empire.

While we found Ibn Khaldun's pattern earlier in the work of Herodotus, Thucydides, and Sima Qian, his analysis and the explanations he gives for this pattern go much further than Thucydides's simple analogy with human peak and decline, or the mandates granted and then withdrawn by heaven in the work of Sima Qian. Ibn Khaldun developed a historical–sociological *mechanism* that enables us to obtain a deeper understanding of the operation of the cyclical pattern, in which a new group is attracted to the knowledge of a previous dominant group, and as a result knowledge accumulates. As far as we know, Ibn Khaldun was the only Islamic historian who described a cyclic historical pattern rather than a purely linear one. However, he was active during the latter days of the great Islamic civilization and appears to have had little or no influence on other historians at the time. Ranulf Higden, as a Christian contemporary of Ibn Khaldun, represented a completely different world that was not really commensurate at all. Through his criticism Ibn Khaldun stood above his historical material and designed a new vision of world history. Higden could do or did nothing else than give summaries, and it is not clear to us how he critically evaluated his material before he accepted it.

Other Islamic historians. In my selection of Islamic historians I have been guided, as I have throughout this book, by their importance to the quest for methodical principles and the patterns found using them. Consequently, I do not do justice to the many other fascinating Islamic writers of history. For example, I have not discussed al-Saghani, who worked in the tenth century and was possibly the first historian of science.[67] He was also known as al-Asturlabi or the astrolabe maker. Similarly, I have not described the first history of the crusades to be written from an Islamic perspective by Ibn al-Qalanisi[68] in the twelfth century, or the enormous world history of Rashid al-Din,[69] which was written in the early years of the fourteenth century (based completely on written sources), and that of al-Suyuti, consisting of nearly 500 volumes dating from the fifteenth century (which addressed many other topics besides history, including linguistics).[70] I have not addressed their work here because I did not find new methodological approaches or patterns. However, I do not pretend in any way that there are not systematic principles or methods in the huge Arabic corpus other than those that I have found. But I do venture to contend that Islamic historians rarely if ever based their writings completely on the strict *isnad* method. This method was always used in combination with the freer personal experience, (first) eyewitness account, or the written source principle. This would appear to indicate that an overly formal historical principle is unlikely to endure, possibly because such a principle damages the historical record. We will return to this problem, particularly with regard to the twentieth century with the notorious principles of neo-positivist historiography (see 5.1).

Things did not go well with Islamic historiography, or any of the other Islamic humanities and sciences. The decline of Islamic learning actually started in the eleventh century, although most historians maintain that the plundering of Baghdad in 1258 marked the end of the Golden Age. However, parts of the Muslim world continued to flourish. But if we consider the crusades, the internal conflicts and the Spanish *Reconquista* of al-Andalus, we see that from the thirteenth and fourteenth centuries Islamic civilization became more and more inward looking and only rarely produced major scholars or scientists, with Ibn Khaldun as the great exception. We know of another revival of Islamic historians in seventeenth-century India during the Mughal Empire (see 4.2). This seems to have been followed by a period of historiographical silence. However, this apparent silence may be misleading because only a fraction of the large volume of Arabic humanities material after 1400 has so far been studied.[71]

China: history as a government matter. Sima Qian's *Shiji* represented the most important historical tradition in China (see 2.2). After a slow start, his historiographical classification into annals, tables, treatises, hereditary lineages, and illustrative traditions—together with the notion of unity of Heaven and humanity (*tianren heyi*)—served for centuries as the model for the dynastic chronicles that were

[67] Franz Rosenthal, 'Al-Asturlabi and as-Samaw'al on scientific progress', *Osiris*, 9, 1950, pp. 555–64.

[68] H. A. R. Gibb, *The Damascus Chronicle of the Crusades, Extracted and Translated from the Chronicle of Ibn al-Qalanisi*, Dover Publications, 1932, reprinted in 2002.

[69] Basil Gray, *The World History of Rashid al-Din: A Study of the Royal Asiatic Society Manuscript*, Faber & Faber, 1978.

[70] E. M. Sartain, *Jalal al-din al-Suyuti: Biography and Background*, volume 1, Cambridge University Press, 1975.

[71] See the discussion by Khaled El-Rouayheb, 'Opening the gate of verification: the forgotten Arab-Islamic florescence of the 17th century', *International Journal of Middle East Studies*, 38, 2006, pp. 263–81.

written from the Tang period (618–907) onwards.[72] The Tang Dynasty is often considered to be the pinnacle of Chinese civilization, with the art of printing as its most significant discovery. The oldest known printed book in the world, the *Diamond Sutra,* dates from 868 (587 years before the *Gutenberg Bible*) and the Confucian classics appeared in print as early as 932.

Writing the official court history was the job of civil servants who recorded all sayings and deeds of the emperor. This record keeping was highly formalized: after every audience notes were converted into formal records, usually in the form of a chronicle.[73] The practice of keeping records guaranteed that the ruler would obtain a kind of immortality: after the emperor's death the records could be immediately turned into an established narrative format following the annals-biography scheme.[74] The court history made Chinese historical production a state matter and it was given its own 'historiographical bureau' (*Shi guan*). This bureau was also charged with writing histories of the previous dynasties based on the records produced by court officials. It became a central task of a new dynasty to compile the history of the previous dynasty, and to legitimize the assumption of power of the new rulers. History functioned as an explanation of the relevance of the past to the present.[75] This started a historiographic tradition that would last till the fall of the Qing dynasty in 1912.

The court chronicles, which are known jointly as the Twenty-Four Histories, follow the example of the *Shiji* in terms of both form and content. These Twenty-Four Histories, which also include the *Shiji,* cover a period of no less than 1,832 years. They fill 3,212 volumes and contain about forty million words. Nearly two thousand years of history were described on the basis of essentially one method resulting in the same cyclical pattern of rise and fall of dynasties.

Ultimately the dynastic histories appeared to be a type of source publication, which was very thorough but dry as dust. Important exceptions were the *Shiji* by Sima Qian and the *New History of the Five Dynasties,* thanks to the anomalous but extremely lively style of court historian Ouyang Xiu (1007–1072).[76] Regional histories and biographies of dignitaries and the venerable were also written, but so far we have not found new principles or patterns in them.

Liu Zhiji's historical criticism: all possible sources principle. Despite—or perhaps because of—the leaden historiographical bureaucracy in China, there was historical criticism, in particular from Liu Zhiji (661–721), who objected to the mechanical character of the dynastic histories. In 710 he wrote *Generality of Historiography* (*Shitong*), the first Chinese work that was completely devoted to historiography.[77] After a critical analysis of the historical writings before the Tang

[72] Denis Twitchett, *The Writing of Official History Under the T'ang,* Cambridge University Press, 1992.

[73] Twitchett, *The Writing of Official History Under the T'ang,* pp. 35ff.

[74] On-cho Ng and Q. Edward Wang, *Mirroring the Past: The Writing and Use of History in Imperial China,* University of Hawai'i Press, 2005, p. 113.

[75] Ng and Wang, *Mirroring the Past,* p. 259.

[76] Ouyang Xiu, *Historical Records of the Five Dynasties,* translated by Richard Davis, Columbia University Press, 2004.

[77] Ng and Wang, *Mirroring the Past,* pp. 121ff. See also Liu Zhiji, *Shitong tongshi,* annotated by Pu Qilong, 2 volumes, Shanghai guji chubanshe, 1978.

Dynasty, Liu Zhiji addressed the different historiographical methods, for example (1) the style to be used, (2) the problems with documentation, (3) linguistic and writing skill, and (4) the use of criticism in the investigation. According to Liu Zhiji a historian must first and foremost remain as objective as possible. He must not base assessments on moral positions or other value judgments. He should moreover be sceptical about every theory. Proof is the only thing that counts, and when describing an event, the historian must give an overall picture that is obtained from *all possible sources*. In Liu Zhiji's opinion all factors—cultural, social, and economic—have to be taken into account and their presentation needs to be detached and unprejudiced.

Liu Zhiji's criticism corresponds to a degree with that of Ibn Khaldun, but his method was different. It was not based on sociological analysis as propagated by Ibn Khaldun. Liu Zhiji believed that historiography should employ as many sources and factors as possible. We shall therefore refer to his method as the *all possible sources principle*. Although we find Liu Zhiji's emphasis on objectivity and social-economic factors nearly seven centuries later in the work of Ibn Khaldun, it seems impossible that the latter knew about the work of the former.

Sima Guang, Zhu Xi, and the return to the basic virtue principle. During the Song Dynasty (960–1279) a total of some 1,300 historical works were composed, a number that was only exceeded during the Qing Dynasty with its staggering 5,478 works of history—see 4.2 and 5.1. In the context of this book it is not feasible to investigate all these works for methodical principles. It is nevertheless possible to sketch the following picture based on the works of a few major historians from that period, for instance Xue Juzheng (912–81), Ouyang Xiu (1007–72), and Sima Guang (1019–86). To begin with, all these historians emphasized the practical usefulness of history and the moral lessons that could be derived from it. Historiography should show all the facets of human behaviour in relation to the natural environment and social changes. As a result history can be used as a guide for the literate public. Historiography is a basic virtue for every intellectual whose greatest goal is to serve the state. We also came across this basic virtue principle—albeit in a more critical form—in the historiography of Ban Gu (see 2.2).

By far the most authoritative is the work of Sima Guang, the *Comprehensive Mirror for Aid in Government* (*Zizhi tongjian*).[78] The book chronologically narrates the history of China from 403 BCE until 959 CE. It consists of 294 volumes, sweeping through 11 Chinese historical periods, resulting in the first general history written in the chronicle or annalistic style. The *Zizhi tongjian* is probably the largest historical *magnum opus* in history. Particularly important was Sima's fascicle on the 'Examination of Discrepancies'. It discusses the inconsistencies among sources and the process from sources to narratives. According to Sima Guang, the historical narratives should be constructed on the basis of direct evidence. Where such evidence was not available, critical judgment had to be used to arrive at the best account closest to the truth.

[78] Sections of the *Zizhi tongjian* have been translated into English by Joseph Yap, *Wars With The Xiongnu: A Translation from Zizhi tongjian*, AuthorHouse, 2009.

A century later the philosopher Zhu Xi (1130–1200) wrote a synopsis of Sima Guang's work, known as *Zizhi tongjian gangmu* ('Outline and Details of the Comprehensive Mirror'), where Sima's work is reinterpreted in terms of good and bad, praising the good and condemning the bad. It is a typical example of a historical work based on the Neo-Confucianism that came into fashion during the Song Dynasty and combined Confucianism, Buddhism, and Taoism.[79] Given his prioritization of the classics over history, Zhu Xi was in scholarly terms at the opposite end of the spectrum from the work of historians such as Sima Guang. The historical criticism of Sima gave way to a moralizing appraisal of written events and people. Zhu Xi's philosophical works had a huge effect on Chinese culture: they became a standard during the Ming and Qing Dynasties when his explanation of Confucianism became the orthodox doctrine (see 4.2 and 5.1).

India and the chronicle of Kashmir. As far as we know, the first historical work in India—the *Rajatarangini* or the *River of Kings*—was written by Kalhana in the middle of the twelfth century.[80] This treatise gives a historical overview of the kings of Kashmir from the beginning of time. Prior to the twelfth century it is hard to find a work in India that could be described as historiography (unlike the profusion to be found in China, Byzantium, the Arab world, Europe, and Ethiopia), and it seems that the *River of Kings* was the only historiography in Sanskrit.[81] We have already discussed this remarkable hiatus in Indian humanities in 2.2. The list of kings in Kalhana's work goes back to 1900 BCE, and although some of these kings can be identified on the basis of inscriptions in archaeological finds, Kalhana's seemingly exact chronology is highly questionable. Because of the essentially complete absence of earlier sources, the *River of Kings* is considered primarily to be an account that is 'accurate' in the way in which contemporaries understood the knowledge of their past.

Ethiopia and Universal History. Ethiopia was probably home to the oldest human habitation in the world. In the earliest records the country was designated as *Punt*, but the Greeks called it *Aithiopìa*, and Herodotus described it in his *Histories*. According to tradition the first Ethiopian rulers were descendants of the biblical King Solomon, who had an affair with Makeda, the legendary Queen of Sheba from the Yemen. A son, Menelik, was born as a result of this relationship, and he became the ruler of the Aksumite Empire. Ethiopia was Christianized in the fourth century CE by the Greek-Syrian monk Frumentius, and from the fifth century onwards biographies of saints were written in Ethiopic (Ge'ez).

However, it was not until the fourteenth century that the first (surviving) Ethiopian chronicle appeared on the scene—*Kebra Nagast* or *Glory of Kings*—written

[79] Daniel Gardner, *Learning To Be a Sage*, University of California Press, 1990.

[80] Mark Aurel Stein, *Kalhana's Rajatarangini: A Chronicle of the Kings of Kashmir*, Saujanya Books, 2007.

[81] There have been older historical poems written in Pali on the history of Sri Lanka, such as the *Dipavamsa* ('Chronicle of the Island') and the *Mahavamsa* ('Great Chronicle')—see Wilhelm Geiger, *Mahavamsa: Great Chronicle of Ceylon*, Asian Educational Services, 1912; K. M. de Silva, *History of Sri Lanka*, Penguin, 1995. For a plea for treating epics as history writing, see Thapar, 'Historical traditions in early India: *c.*1000 BC to *c.*AD 600', in Feldherr and Hardy (eds), *The Oxford History of Historical Writing, Volume 1: Beginnings to AD 600*.

during the Golden Age of Ethiopian literature.[82] This anonymous work combines historiography with allegory and symbolism. The *Glory of Kings* begins with Adam and Eve and continues to about the fourth century CE with the Christianization of the Ethiopians, who were converted from the heathen worship of the sun, moon, and stars to the Christian worship of the 'Lord God of Israel'. In glorifying terms it describes the arrival of the Ark of the Covenant in Ethiopia, the abdication of Makeda (the Queen of Sheba) in favour of Menelik and the Christianization of the Aksumite Empire.

Although no clear methodical principle can be distilled from the text, it does seem that the biblical coherence principle applies. All 'events' are interpreted in biblical terms and put into a form that is as consistent as possible with Holy Scripture. The correspondence between the structure of the *Kebra Nagast* (from Adam to the then present day) and that of the Christian Universal Histories in North Africa and Europe is striking. Perhaps the Greek-Syrian monks who Christianized Ethiopia had influence during the writing of the *Kebra Nagast*. Earlier assumptions about the Coptic origin of the work cannot be substantiated.[83]

Metapattern in medieval historiography. Chronicles were written in all historiographical cultures—Africa, Europe, Byzantium, the Arab world, China, and India. Although the methodical principles differ substantially (for example the subjective principle of biblical coherence, the formal principle of the *isnad*, or the eclectic principle of all possible sources) we can nevertheless identify a metapattern in medieval historiography: *in a historiographical culture the time structure of the historical narrative tends to correspond with the time structure of the canonical texts.*[84] Hence Christian and Islamic historiographies follow the linear pattern with a unique beginning and an ultimate goal as represented in the Bible and the Koran, whereas Chinese historiography is characterized by an absence of any absolute beginning and end in accordance with the canonical Confucian and Taoist texts, often resulting in a cyclical rather than linear historical pattern. There are also exceptions, which are explicable up to a point. Contrary to the flow of the cycles of the holy Puranas, the Indian history of Kashmir had a linear pattern, but in fact the millions of years covered by the Puranas are too long for any human cycle. And contrary to the linear models of other Islamic historians, Ibn Khaldun offered some hope during the disintegration of the Arab Empire on the basis of his cyclical model of cultural enrichment. But his was a voice in the wilderness in the latter days of Islamic civilization.

[82] *The Kebra Nagast*, translated by E. A. Wallis Budge, London, 1932.

[83] David Allen Hubbard, *The Literary Sources of the Kebra Nagast*, PhD thesis, University of St Andrews, 1954, p. 370. See also Robert Beylot, *La Gloire des rois ou l'histoire de Salomon et de la reine de Saba*, Brepols, 2008.

[84] In no way do I want to reduce the narrative wealth of these texts to a linear or cyclical pattern. The noted commonality merely shows that despite the enormous variety there is also some unity.

3.3 PHILOLOGY: COPYISTS, ENCYCLOPAEDISTS, AND TRANSLATORS

Compared with the philology of the ancient world, medieval philology seems to have consisted primarily of copying, compiling, and translating activities. There were also text reconstruction activities in Europe, the Islamic civilization, and China, but we are left guessing as to precisely which procedures and methodical principles were employed.

Christianity and Judaism: biblical compilations, copies, and encyclopaedias. The first philological activity in both the Christian and Jewish traditions was the compilation of the text of the Bible. The existing aesthetic—and therefore secular—principles of Zenodotus (see 2.3) were not sufficient for this purpose. The text of the Jewish Bible (the Old Testament in Christianity) was undertaken by the Masoretes—'tradition keepers'—who worked as biblical scholars between the seventh and eleventh centuries in the Arab world, primarily in Tiberias, Jerusalem, and Babylon. They provided the Bible with vowel and pronunciation symbols and other symbols: the *nikud* and *teamim*. The Masoretes produced very authoritative biblical texts, but their compilation was made at the expense of the actual textual history.[85]

However, Christian philologists had even greater problems with the New Testament. The many Latin, Syriac, and Greek versions from the Early Middle Ages made every philological reconstruction a precarious business. What we see initially are compilations and translations—from both the Old and New Testament—without any attempt to reconstruct the 'original' on the basis of extant copies. The *Hexapla* of Origen from the third century CE or the influential *Vulgate* of St Jerome (*c.*347–420), commissioned by Pope Damasus I, are examples. St Jerome based his translation of the Old Testament on the original Hebrew text, and not just the Greek version (the *Septuagint*), which provoked fierce criticism from St Augustine. However, the more than 8,000 copies of the Vulgate are evidence of St Jerome's success at making his compilation the accepted version of the Bible. But the Vulgate was not based on any demonstrable methodical principle.[86]

Philologists also worked on producing encyclopaedic overviews in the tradition of late Roman philology, for instance the *Etymologiae* of the doctor of the church St Isidore of Seville (*c.*560–636).[87] Isidore's classical arrangement was very much in the pagan tradition. God did not appear on the scene until after an overview of the *artes liberales*. It earned Isidore the title of 'the last scholar of the ancient world'. During this period, knowledge of Latin decreased rapidly. Gregory of Tours (see 3.2) apologized for his poor Latin, and in the eighth century St Boniface heard a baptism being conducted with the words *In nomine patria et filia et spiritus sancti* (which 'translates' roughly as: in the name of the fatherland, the daughter and the

[85] Emanuel Tov, *Textual Criticism of the Hebrew Bible*, Brill, 2005.

[86] Bonifatius Fischer, *Beiträge zur Geschichte der lateinischen Bibeltexte*, Herder, 1986.

[87] *The Etymologies of Isidore of Seville*, translated by Stephen Barney, W. J. Lewis, J. A. Beach and Oliver Berghof, Cambridge University Press, 2006.

Holy Spirit).[88] There was no organized educational programme until Charlemagne brought the scholar Alcuin of York to the court. Knowledge of Latin and the classics then was restored to some degree and an extensive network of monastery and cathedral schools developed.

The urge to copy flourished during the Carolingian Renaissance and much of the still surviving Latin literature was secured. Meanwhile the Vulgate translation of the Bible had been copied so many times that countless errors had crept in. Virtually every church and monastery had its own accepted translation. Charlemagne's aim of uniformity resulted in an attempt to reconstruct the original Vulgate text by collecting and comparing manuscripts from all regions. In 801 Alcuin produced the first reconstructed Vulgate text, but it has been lost. Theodulf of Orleans (*c.*750–821) also created a revised text, which has survived in the *Codex Memmianus*. Unfortunately we do not know the methodical principles underlying these philological exercises. There was similarly little apparent method in later reconstructions of the Vulgate either. Roger Bacon, who was also a linguist (see 3.1), was one of the few scholars who devised principles for the Vulgate reconstruction. According to Bacon the old Latin manuscripts of the church fathers were the first authority. It was only if these old Latin manuscripts did not correspond with each other that it was necessary to refer to the original texts. In so doing Bacon was completely in line with the thinking in terms of authorities for which the Middle Ages, with their biblical coherence principle, are well known.

Knowledge of Latin also decreased at a dramatic rate in the Byzantine Empire where it disappeared completely from public life after the seventh century. On the other hand knowledge of Greek remained intact in the form of the many literary, historical, and scientific works written in the language. The most important 'philological' activities in Byzantium took place in the ninth century. Thanks to paper manufacturing know-how (which had also come to Byzantium from China via the Arab world—see 3.2), there was a more accessible medium for storing text. We therefore have the Byzantine copyists of that era to thank for most of the Greek literature that has survived, and above all those in the famous Stoudios Monastery in Constantinople.

The greatest Byzantine philologist by far was Photius (*c.*810–*c.*893), who had an unprecedentedly fast-track career as a patriarch and diplomat in the Byzantine Empire.[89] Shortly before he went on a dangerous mission to the Arab government, as consolation for his brother he wrote a summary of 280 books that he had read (from which he left out the standard texts that his brother already knew). This work, which is known as the *Bibliotheca*, is one of the most fascinating products of the Byzantine civilization.[90] It established Photius's name as the first book reviewer in history. If for no other reason, the work is of inestimable value because most of

[88] L. D. Reynolds and N. G. Wilson, *Scribes and Scholars: A Guide through the Transmission of Greek and Latin Literature*, 3rd edition, 1991, p. 92.

[89] Andrew Louth, *Greek East and Latin West: The Church ad 681–1071*, St Vladimir's Seminary Press, 2007.

[90] Nigel Guy Wilson, *Photius: The Bibliotheca*, Gerald Duckworth & Co., 1994.

the books that Photius discussed no longer exist. For example, not one work by twenty of the thirty-three historians whom Photius reviewed has survived.

Sadly we know of very few text reconstructions from this period. They certainly existed, as can be deduced from a few of Photius's letters, but there is no trace of scholarly text reconstruction. There was continuous study of the classics after Photius too, but methodical principles are few and far between. The impressive Byzantine encyclopaedia that was initiated in the tenth century under Emperor John I Tzimiskes is also worth mentioning. The *Suda* had 30,000 entries and left earlier compilations far behind.[91] Moreover, the *Suda* was the first encyclopaedic overview to be organized as a dictionary. Although the entries are not always equally reliable, it is an extremely important source.

Lupus of Ferrières's attempt to revive scholarly philology. We come across 'secular' philological text reconstruction, or what remains of it, briefly in the late Carolingian Renaissance. Lupus of Ferrières (*c.*805–862), who was working in Fulda under Rabanus Maurus ('the teacher of Germany') played a prominent role.[92] Using his contacts all over Europe he had manuscripts sent from Tours, York, Rome, and elsewhere. Lupus was not the only manuscript hunter in ninth-century Europe, but what made him unique was that he had manuscripts sent to him that his library already contained. Like the Alexandrian analogists before him, Lupus wanted to reconstruct the putative original text from surviving copies (see 2.3). In so doing he tried to mark the corruptions and variations in the manuscripts as accurately as possible. He annotated textual *lacunae* using spaces (rather than risk erroneous emendations). Lupus's greatest interest was in texts by Cicero, Livy, Macrobius, and Gellius. However, his own critical contributions are so modest that some people consider the use of the term philology to be inappropriate in describing Lupus's activities. Yet compared with the carelessness of most other 'classical' philologists, with the Christianization of the names of all classical authors by Hadoard in his *Collectaneum* as the nadir,[93] Lupus's text analysis is a model of meticulousness.

We would not be doing justice to the Carolingian Renaissance if we only judged it on the merits of its textual criticism. The continued existence of many texts was threatened by numerous hazards including that of the physical deterioration of the papyri on which they were preserved, and therefore depended heavily on copies made at this time. We have the diligence of these Carolingian copyists to thank for most of the extant Latin texts. For example, the best surviving manuscripts of Lucretius and Vitruvius were written in Charlemagne's palace scriptorium in 800. They are currently in Leiden as part of the *Codices Vossiani Latini*. We also have these copyists to thank for the reform of handwriting. To this day Caroline minuscule enjoys unprecedented popularity, as the font of the book you are currently reading demonstrates.

[91] *Suda On Line*: <http://www.stoa.org/sol/>.
[92] Robert Graipey, *Lupus of Ferrieres and the Classics*, Monographic Press, 1967.
[93] Charles Beeson, *The Collectaneum of Hadoard*, Classical Philology, 40(4), 1945, pp. 201–22.

During the turbulent centuries after Charlemagne's empire the production of books continued steadily in monasteries like Bamberg, Paderborn, and above all Montecassino. As we said earlier (see 3.1), the burst of translations of the many philosophical, mathematical, and scientific works from Arabic and Greek in Toledo was the zenith of the twelfth century. However, a revival of critical text reconstruction did not arrive until the era of humanism in the fourteenth and fifteenth centuries.

Islamic civilization: compilation of the Koran. The most important philological activity during early Islamic civilization was compiling the text of the Koran. According to tradition, in 634—two years after the death of Muhammad—the Caliph Abu Bakr ordered the collection of all the Koran verses that had been revealed to the Prophet. There were differing textual versions because Muhammad's words had been noted down by different writers. There were also many Koran verses that existed only in oral form. Abu Bakr assigned Zaid ibn Thabit the task of systematically compiling the Koran. Zaid was known as one of the few who had memorized virtually all the verses and had also noted down the most passages of text recording the words of Muhammad. The official codification of the collected texts began under the third caliph, Uthman ibn Affan (644–656), on the basis of diacritic and vowel symbols. According to tradition, six copies were sent to Mecca, Damascus, Basra, and Kufa, and Uthman kept one copy for himself. Two of these first Korans have survived and are kept in Tashkent and Istanbul. Virtually all current printed Koran texts are based on Uthman's compilation. However, this history of the Koran's development is disputed by many scholars.[94] As there is with regard to the *hadith* (see 3.2), there is a considerable difference of opinion here between Muslim scholars and Islamic Studies scholars.[95] It is in any event not clear which principles were used to construct the Koran, although the traditional vision is that it was primarily the eyewitness account principle: Zaid noted down Muhammad's revelations as literally as possible.

Ibn Ishaq's translation school in the House of Wisdom. If we can believe al-Biruni, the problem with corrupted texts in Islamic civilization was just as great as it was in Christian Europe. But little is known about philological textual criticism and subsequent text reconstruction. The Syrian Nestorian Hunayn Ibn Ishaq (808–873), who was a philosopher, physician, and translator in Baghdad, came closest to practising critical philology. Ibn Ishaq had a command of Arabic, Persian, Greek, and Syriac, and is well known mainly for establishing the illustrious translation school in the House of Wisdom.[96] He was a manuscript hunter and travelled all

[94] In fact, there are still several variants of the Koran—see Adel Khoury, *Der Koran: arabisch-deutsch*, 12 volumes, Gütersloh, 1990–2001. While most variants have been destroyed from the ninth century onwards, the extant versions still await further philological investigation. See Frederik Leemhuis, 'Readings of the Qur'an', in Jane Dammen McAuliffe (ed.), *Encyclopaedia of the Qur'an*, volume 4, Brill, 2004, pp. 353–63. A classic is Arthur Jeffery, *Materials for the History of the Text of the Qur'ān: The Old Codices*, Brill, 1937.

[95] This ambiguity is beautifully described in Thomas Bauer, *Die Kultur der Ambiguität: eine andere Geschichte des Islams*, Verlag der Weltreligionen, 2011. See also John Esposito, *What Everyone Needs to Know about Islam*, Oxford University Press, 2002.

[96] De Lacy O'Leary, *How Greek Science Passed to the Arabs*, Routledge, 1948, chapter 12.

over the world in search (initially) of medical manuscripts by Galen, Aristotle, Hippocrates, and Dioscorides, which he found in Greek monastery communities. In an extant letter to a friend who had asked him about the content of Greek medical works, Ibn Ishaq complained about the incompetent translators prior to him who had worked on the basis of damaged or illegible manuscripts and had not taken the trouble to track down an intact or less damaged version. Ibn Ishaq then explained his own way of working in which he compared a surviving translation with as many Greek manuscripts as possible in order to produce a better translation. Although he does not describe the method on which he based his comparisons, he may have adopted them from Galen, whom he quoted frequently as a physician and who had similar problems when establishing the Hippocratic corpus. Ibn Ishaq's systematic principles remain conjecture, but the results speak for themselves. The Arabic translations produced by Ibn Ishaq's school are among the best manuscripts that we have of the many lost Greek works.

There must have been much more textual criticism in Islamic civilization. For example Ibn al-Sikkit (*c.*802–857) and al-Suli (*c.*880–946) edited collections of poetry of the classical, partly pre-Islamic Arab poets. These 'philologists' had to determine the texts on the basis of handwritten manuscripts. But their philological methods remain unclear. This hiatus suggests that there is still a large amount of scholarly work to be done on the history of the (Islamic) humanities.

Text reconstruction in China: Zeng Gong and Su Song. Textual criticism flourished during the Tang and especially during the Song dynasty. Pre-Qin texts were meticulously reconstructed, and a full-fledged philological methodology seemed to emerge, but unfortunately we do not have enough information to distil this methodology. For example, Zeng Gong (1019–1083) was active as an editor of ancient documents in the Historical Commission between 1060 and 1067. In his reconstruction of a missing chapter from the renowned ancient historical work *Zhan guo ce* ('Strategies of the Warring States'),[97] Zeng searched for all copies that were available at the time in private collections. As such, he had to deal with many contrasting manuscripts. His new compilation became the authoritative version, on which all later editions of the *Zhan guo ce* were based, but we do not get insight in his actual method.[98]

In the work of the polymath Su Song (1020–1101), who also invented the famous water-driven astronomical clock tower in Kaifeng, we get a glimpse of insight into the nature of philological method. In his reconstruction of the *Fengsu tongyi* ('Common Meanings in Customs'), written by the social critic Ying Shao (*c.*140–204 CE),[99] Su Song compared many copies of the text by different authors from different periods. He renumbered ten extant chapters and succeeded in assigning titles to twenty chapters that had been lost before. In doing so, he discovered that

[97] *Chan-kuo Ts'e*, translated and annotated with an introduction by J. I. Crump, University of Michigan Center for Chinese Studies, 1996.

[98] Michael Loewe (ed.), *Early Chinese Texts: A Bibliographical Guide*, Society for the Study of Early China, 1993, p. 6.

[99] Michael Nylan, *Ying Shao's Feng su t'ung yi: An Exploration of Problems in Han-Dynasty Political, Philosophical, and Social Unity*, PhD thesis, University of California, Berkeley, 1982.

one particular chapter was missing. Thanks to Su Song's argument, it became clear that a loss of one chapter had taken place during the compilations in the Sui and Tang histories.[100] Although we do not know Su Song's exact method, we get enough insight to determine that he worked in a highly consistent manner. Besides textual criticism, the Song period also witnessed a groundswell of commentaries and textual analyses (see 3.7).

Thus, with respect to text reconstruction, there is some similarity between China, Europe, and Islam in the medieval period: despite the many philological activities, there is virtually no quest for the underlying rules of text reconstruction. The search for philological principles was not developed before the age of Humanism in Europe or the Ming period in China (see 4.1).

[100] Loewe, *Early Chinese Texts: A Bibliographical Guide*, pp. 106–7.

3.4 MUSICOLOGY: THE FORMALIZATION OF MUSICAL PRACTICE

Two works from Antiquity define early medieval musicological knowledge in Europe: Martianus Capella's *Nuptiae Philologiae et Mercurii*, in which the theory of harmony is addressed from an Aristoxenian perspective (see 2.4), and Boethius's *De institutione musica*, which was primarily a Pythagorean view of the theory of harmony.[101] There appears to have been no trace of any developments in the study of music in Europe until far into the ninth century. However, this does not mean that there were none. Much musicological understanding proved to have been handed down by word of mouth before it was written down. This can be derived from a few ninth-century manuscripts that refer to practices that were assumed to be known (see below). After the ninth century all at once new insights and discoveries followed hard upon one another's heels. In order to have a good understanding of this trend it is important to realize that medieval music *production* represented a break with the music of ancient times. Roman heathen music was associated with persecution of Christians in arenas, and—starting in about the fourth century—a new musical idiom was created that is known as liturgical Gregorian chant. In the ninth century the harmonic doubling of the Gregorian chants resulted in the scientific study of polyphony, which culminated in a system of rules for polyphonic singing. The emergence of the new musical (Gregorian) idiom may also explain why music theory was the only discipline within the *artes liberales* that underwent substantial changes in the Early Middle Ages, unlike the other *artes* disciplines in Europe—grammar, logic, and rhetoric—which long continued to build on the classical tradition.

The ongoing problem of musical notation: from Hucbald to Guido. The Flemish monk Hucbald (*c.*840–930) held the honour of having written the first work in European music theory, *De harmonica institutione.*[102] In his treatise Hucbald guided singers through the theory of music on the basis of well-known hymns in order to make them aware of the distances between the different tones and the harmonic way in which the tone system is ordered. Besides being one of the oldest musical anthologies, the most important contribution of Hucbald's work was the proposal to improve and extend the existing musical notation for Gregorian chants, known as *neume notation*, to include indications of the exact pitch. A neume was an element that specifies the shape of the melody, or melodic contour, in a sung syllable. Neumes did not signify pitch or interval. They were intended as a reminder at a time when music was handed down primarily by word of mouth. Hucbald was also the author of one of the most remarkable medieval poems—an ode to baldness—dedicated to Charles the Bald, king of West Francia.

[101] Besides these two works, also Macrobius's *Commentarius in Somnium Scipionis* (see 2.3) and Cassiodorus's *Institutiones divinarum et humanarum litterarum* (see 3.1) contain sources of musical knowledge.

[102] Yves Chartier, *L'Oeuvre musicale d'Hucbald de Saint-Amand: les compositions et le Traité de Musique*, Éditions Bellarmin, 1995.

It was long assumed that a document entitled *De alia musica*, which appeared in the early tenth century, was also written by Hucbald, but currently it is attributed to one Pseudo-Hucbald.[103] This treatise proposed a new notation for music with eighteen different pitches and where the syllables of the sung text were put on horizontal lines. For the first time this gave singers a clear representation of the rises and falls in pitch in whole and half tones. The first seven letters of the Latin alphabet were used to specify the tone.

Pseudo-Hucbald's innovation was the first step towards the famous *solmization* of Guido d'Arezzo (991–1033).[104] In his *Micrologus* Guido proposed a notation system in which a second could be signified with precision (see 2.4).[105] Using the hymn *Ut queant laxis*, which was very well known at the time, Guido had every line begin one second higher, and he used the first syllables *ut-re-mi-fa-sol-la* to define the tones, the hexachord. The only semitone in this series is between the *mi* and the *fa*. It was not until the nineteenth century that the initial letters of *Sancte Ioannes* were also added to define a seventh tone *si*, right at the end of Guido's hexachord series. The *ut* was not easy to sing, so it was replaced by *do*—from *Domine*—thus producing the well-known series *do re mi fa sol la si* (where *si* is often replaced by *ti* so that each note begins with a different letter).

***Musica enchiriadis*: a system of rules for polyphonic compositions.** The works discussed above are mainly concerned with musical notation. But then, at the end of the ninth century, one of the most remarkable works in the whole of medieval musicology appeared: the *Musica enchiriadis* ('Music of Many Parts'), attributed to either Pseudo-Hucbald or Odo of Cluny.[106] This work, which was followed by the commentary *Scolica enchiriadis*, was the first surviving attempt to establish a system of rules for polyphonic composition, the organum. Under the heading *symphoniae* the *Musica enchiriadis* defined two different ways to describe polyphony, which are known as the 'parallel organum' and 'modified parallel organum', and the rules for them can be summarized as follows.

Parallel organum:

(1) Take a Gregorian melody as *vox principalis.*
(2) Double the *vox principalis* with a fifth or a fourth by a second voice, the *vox organalis.*
(3) The *vox principalis* and the *vox organalis* can, if one wishes, be doubled further in a higher or lower octave, which creates a three-part or four-part parallel organum.

[103] Rembert Weakland, 'Hucbald as Musician and Theorist', *Musical Quarterly*, 42(1), 1956, pp. 66–84.

[104] See e.g. Angelo Rusconi (ed.), *Guido d'Arezzo, monaco pomposiano*, Olschki, 2000.

[105] *Micrologus Guidonis de disciplina artis musicae: Kurze Abhandlung Guidos über die Regeln der musikalischen Kunst*, translated by Michael Hermesdorff, Bayerische Akademie der Wissenschaften, Lexicon musicum Latinum, 1876.

[106] Dieter Torkewitz, *Das älteste Dokument zur Entstehung der abendländischen Mehrstimmigkiet*, Beihefte zum Archiv für Musikwissenschaft, volume 44, Steiner Verlag, 1999. See also Raymond Erickson, 'Musica enchiriadis, Scholia enchiriadis', *The New Grove Dictionary of Music and Musicians*, Macmillan, 2001.

Modified parallel organum:

(1) Take a Gregorian melody as *vox principalis*.

(2) The second voice, the *vox organalis*, continues the initial tone of the *vox principalis*.

(3) It is only when the *vox principalis* reaches an interval of a fourth or a fifth with the *vox organalis* that both continue in parallel fourths or fifths.

(4) At the end (the cadence) of the organum the *vox principalis* and the *vox organalis* come together again by reaching the same tone in the reverse order.

(5) The *vox principalis* and the *vox organalis* can, if one wishes, be doubled further in a higher or lower octave, which creates a three-part or four-part modified parallel organum.

These rules completely formalized the construction of parallel and modified parallel organa. It is important to mention that the text of the *Musica enchiriadis* refers to an assumed known musical practice. The *Musica enchiriadis* therefore does not appear to be a precept for making organa in general, but a descriptive system of rules for the *existing* musical practice of parallel and modified parallel organa.[107]

How can the system of rules discussed above be positioned among the systems of rules we have studied so far? Do the rules of the organa represent a declarative system of rules that only states the limits of the possible pieces (as developed by Aristoxenus) or is it a procedural system of rules that generates the complete class of possibilities (as defined by Panini)? Surprisingly enough it emerges that the latter is the case. Starting with a particular Gregorian melody, the rules of the *Musica enchiriadis* generate all possible parallel and modified parallel organa for that melody in a particular style (the early *Ars Antiqua*). The *Musica enchiriadis* therefore provides a procedural system of rules like Panini's description of language (whereas, as the reader will remember, Aristoxenus's melody grammar rules only give the 'restrictions'). Obviously *Musica enchiriadis* does not compare at all with Panini in terms of the number of rules, but it does provide an exact procedural system that, if complied with, produces an organum in a particular style. Such an approach had not been seen in the humanities since Greek logic.

Was this system of rules, which was initially intended to be descriptive, regarded with the passage of time as prescriptive—something we have often seen in the humanities of Antiquity (see the conclusion to chapter 2)? We cannot be certain, but if the system of rules was ever deemed prescriptive, it did not in any event last for long. In no time at all organa were being composed that did not comply with the system of rules in the *Musica enchiriadis*. Extant pieces of music from the tenth and eleventh centuries demonstrate how the two voices became progressively more independent melodically: besides parallel movements, alternating *sideways* and *contrary movements* were now also created, which are known as a melismatic or florid organum and a free organum. At first sight the polyphonic melodic lines do

[107] Thomas Christensen (ed.), *The Cambridge History of Western Music Theory*, Cambridge University Press, 2002, p. 480.

not appear to comply with rules, except the rule of consonance (and even this was not always adhered to). Upon closer inspection, however, it emerges that these organa were nonetheless based on very precise rules that were less 'free' than the term 'free organum' suggests. However, the rules were more complicated than in the *Musica enchiriadis*, and they became even more complex when, at the end of the eleventh century, polyphony developed to a point where composers combined two melodically independent lines on the basis of apparently all possible inclined movements. Nevertheless, this new organum style was described anew using rules in the treatise *De musica* by Johannes Cotto (John Cotton or Johannes Afflighemensis)[108] and the anonymous *Ad organum faciendum* ('Making an Organum').[109] These discourses from around 1100 are a snapshot of the tradition of organum construction, because after the system of rules for this more complex type of organum, other new and even more complex organa appeared in which the rhythm was also released. To a degree these treatises are like a description of a living 'language' without taking into account the fact that the 'language' changes from one generation to another.

European medieval music was therefore not constrained in any way by rule-based descriptions (which is precisely what happened with the descriptions of languages in linguistics, whose objective was to set aside Sanskrit and Arabic as 'classical' for instance). The wealth of musical forms that came out of the Notre Dame School in the twelfth century under Leoninus (*c.*1150–*c.*1200) and Perotinus (*c.*1160–*c.*1230) was almost too much to keep up with through rules, which were largely overtaken by other developments. The theoretical works that were still to be developed applied to a limited sub-style or period. Moreover, the rules that were devised were often so specific that there almost appeared to be an *example-based description* based on specific organa, as we also found in medieval linguistics (see 3.1). Apparently, polyphony became so complex that there was *a shift from rules to examples*. But this shift did not continue in musical theory. We will see that the music theory (*musica speculativa*) of the organa slowly but surely fused with the production of music (*musica activa*). But until the thirteenth century virtually all music scholars were composers and vice versa. This was an exception to our earlier observation of a theoretical upper echelon and a more practical lower rank, as we saw in European linguistics and historiography (see 3.1 and 3.2).

Rhythmical musical notation and the awareness of musical style. The development of polyphony was accelerated by improved musical notation and vice versa. The polyphonic organa could be recorded and handed down using Hucbald's or Guido's system of notes, but only so long as the note length was not important. This changed when organa were composed for *rhythmically* independent melodic lines, as was the case with Leoninus and Perotinus. Note length now became an essential component in the organum of the *Ars Antiqua*. Franco of Cologne

[108] Charles Atkinson, 'Johannes Afflighemensis as a Historian of Mode', Dobszay Festschrift 1995, pp. 1–10.

[109] *Ad organum faciendum*, translated by Jay Huff, Musical Theorists in Translation, volume 8, Institute of Mediaeval Music,? Brooklyn, 1963.

(thirteenth century) developed an extension of musical notation to include note length that seemed to be based on earlier work by Johannes de Garlandia (also thirteenth century). The central element of Franco's discourse, the *Ars cantus mensurabilis*, was a proposal to let the notes represent their length themselves.[110] Previously rhythm was based on the musical context where notes that were depicted in a similar way were given a corresponding note length. Franco's method made it possible to specify the length of a note exactly using new forms of the notes themselves (comparable to the notation we use today).

The treatise *Ars Nova* dating from around 1318, which was attributed to Philippe de Vitry (1291–1361), appears to have marked the end of the rule-based tendency in organum composition.[111] Philippe was an ardent advocate of a 'freer' way of composing, and he opposed what he considered to be the rigid style forms of the *Ars Antiqua*.[112] Countless innovations were contained in *Ars Nova*, and the fourteenth-century composer Guillaume de Machaut was the most important exponent. For example, jumping from one hexachord to another was made easier; in the *Ars Antiqua* this had been based on Guido's tone system in which it was only possible to change hexachord at the semitones. The fascinating feature of Vitry's discourse *Ars Nova* is that there was a discernible *process of growing awareness* about the concept of 'musical style'. We also see this in a treatise by Johannes de Muris (1290–1351), the *Ars novae musicae* (or *Notitia artis musicae*), in which the new art was propagated with great fervour.[113] Diametrically opposed to this was a work written by Jacques de Liège (*c.*1260–1330), *Speculum musicae*, in which he defended the old thirteenth-century art of music against the 'breakaway' new art.[114] We therefore see the concept of style as a set of shared aesthetic values or rules, together with the dialectical response to an earlier style, for the first time in European musicology in the High Middle Ages and not, as is often assumed, in Renaissance art theory (see 4.5).

Nicole Oresme and the discovery of overtones. The greatest musicologist in the Late European Middle Ages was Nicole Oresme (1323–1382), who was also one of the most universal medieval thinkers. He made significant contributions to mathematics, philosophy, and psychology as well as to musicology.[115] Oresme was active in nearly all musicological fields. His works include discourses about acoustics, musical aesthetics, the physiology of hearing, formal harmony theory, musical practice, and music philosophy.

[110] *Ars cantus mensurabilis*, translated by G. Reaney and A. Gilles, Corpus scriptorum de musica, volume 18, American Institute of Musicology, 1974.

[111] Daniel Leech-Wilkinson, 'The Emergence of Ars Nova', *Journal of Musicology* 13, 1995, pp. 285–317.

[112] Richard Hoppin, *Medieval Music*, W. W. Norton, 1978.

[113] *Notitia artis musicae et Compendium musicae practicae; Petrus de Sancto Dionysio Tractatus de musica*, edited by Ulrich Michels, Corpus scriptorum de musica, volume 17, Rome, 1972.

[114] *Speculum musicae*, edited by Roger Bragard, Corpus scriptorum de musica, volume 3/6, Rome, 1973.

[115] Ulrich Taschow, *Nicole Oresme und der Frühling der Moderne: die Ursprünge unserer modernen quantitativ-metrischen Weltaneignungsstrategien und neuzeitlichen Bewusstseins- und Wissenschaftskultur*, Avox Medien Verlag, 2 volumes, 2003.

Oresme's most significant achievement was the discovery of a new musical phenomenon: the overtone.[116] It is a sound component with a frequency that is higher than that of the fundamental tone of that sound as sensed by the ear. Oresme contended that the overtones play a major part in the concept of sound quality or timbre. Two different instruments, for example a lute and an organ, can play exactly the same note, yet they sound very different. Some overtones are harmonic, which means that their frequencies are whole number multiples of that of the fundamental tone. However, in less 'ideal' instruments, a glockenspiel for instance, the overtones are not harmonic. Overtones can also be sung, for example in throat singing (which, incidentally, was unknown to Oresme). Oresme's discovery represented the first real breakthrough in harmony theory since Pythagoras, and he also tried to use it to explain the relationship between consonances and dissonances. It was not until the seventeenth century that his findings were taken further by Marin Mersenne (see 4.4).

Transformation of Greek music theory by al-Kindi, al-Farabi, and Safi al-Din. The classical musicological heritage in Christian Europe was defined by Latin texts, but in the Islamic world it was Greek works that dominated. From the beginning of Islamic civilization, Arab scholars compared their musical practice with the musical theories they were exposed to. The theoretical and experimental development of Pythagorean music theory by Ptolemy made the greatest impression (see 2.4).[117]

One of the biggest challenges in early Islamic musicology was to provide a scientific basis for the empirical intervals in the Arabic twenty-four-part tone system. Al-Kindi (*c.*801–873) may have been the first person to apply mathematical Greek music theory to the Arabic idiom. In addition to a model for intervals and scales, al-Kindi also compiled an overview of the different rhythmic cycles that are so characteristic of Arabic music.

The most important Islamic musicologist was al-Farabi (*c.*872–*c.*950), who was also active in many other academic fields (see 3.6). In his three-part *Kitab al-musiqa al-kabir* ('Book of Great Music'), al-Farabi addressed music theory, instrument theory, and the melodic and rhythmic typology of Arabic music.[118] After an introduction in which he speculated about the origins of music and the nature of musical talent, in the first volume al-Farabi gave an overview of Pythagorean harmony theory in terms of four-tone series or tetrachords (see 2.4). Since many of these tetrachords were not being used, this work suggests that there was a renewed interest in the study of music as a purely mathematical science. His discussion of rhythm was also largely theoretical. Al-Farabi constructed a mathematical framework in which all possible rhythmic cycles were defined. In the second volume, al-Farabi used his theoretical approach to tackle a number of specific problems, in particular the technical peculiarities of musical instruments,

[116] Marshall Clagett (ed. and trans.), *Nicole Oresme and the Medieval Geometry of Qualities and Motions: A Treatise on the Uniformity and Difformity of Intensities Known as Tractatus de configurationibus qualitatum et motuum*, University of Wisconsin Press, 1968.

[117] Jean-Claude Chabrier, 'Musical Science', in Roshdi Rashed (ed.), *Encyclopedia of the History of Arabic Science*, volume 2, Routledge, p. 581.

[118] Touma Habib, *The Music of the Arabs*, Amadeus Press, 2003, pp. 17ff.

for example the correct placement of the frets on a lute so that acoustically consonant intervals are obtained, and the different possible consonances for two or three strings of the *tunbur*, or long-necked fretted lute. The third volume has the promising title *Musical Composition*, but in it al-Farabi was much less explicit. He described how the human voice can express literary and poetic forms, and how these can arouse feelings and stimulate the soul. Al-Farabi considered vocal music to be superior to instrumental music. He was at his most specific when he summarized different possible melodies in a typological scheme. However, this scheme consisted primarily of *abstract* tone series rather than *specific* (melodies from) pieces of music. Like his analysis of possible rhythmic cycles, al-Farabi's melody theory is therefore predominantly theoretical. It is more a description of *possible* music than of *actual* music. But even within this class of possible music it is not clear which musical idiom al-Farabi was describing. Most of all, his description seems to be a generalization of Arabic music after a musical idiom that al-Farabi presumably thought had enabled him to cover all music, but which was in fact still Arabic through and through. His system is not a rule-based grammar that specifies the boundary conditions (which we have called a declarative system of rules) but a sort of schematic classification of rhythms and melodies. To an extent al-Farabi's classification comes under the method that we designated earlier as the *example-based description principle*, but its examples had to be largely constructed. However, it should also be stressed that his discussion of Pythagorean harmony theory—as usual—accords with the *mathematical proportions principle*.

As well as the impressive research into music theory, encyclopaedic collections of Arabic music were also produced, of which the most important was the monumental *Kitab al-aghani* ('Book of Songs') by al-Isfahani (897–967).[119] In over twenty volumes containing 10,000 pages, al-Isfahani gave an overview of the versatile Arabic singing and poetry in the eighth and ninth centuries. According to the author it took fifty years to write. The songs were accompanied by a description of the associated rhythmic cycles and sometimes the melodic mode. But we do not find musicological definitions in his work, let alone a quest for underlying melodic or harmonic rules.

After al-Farabi and al-Isfahani there was not really a musicologist of any importance, although most scholars were involved in music as a sideline, including Avicenna (Ibn Sina). The Kurdish music theorist (and musician) Safi al-Din, who died in 1294, was a notable exception.[120] Spared by the Mongols after the fall of Baghdad in 1258, he restored Pythagorean harmony theory to a high point. Safi al-Din perfected both the circle of fifths and the calculation of the comma (see 2.4). But he is best known for developing the widely used seventeen-tone scale which was later expanded to the Arabic scale of twenty-four quarter tones (and which would lead to the quarter tone controversy—see 5.4). That seems to have been the last contribution from medieval Islamic musicology.

[119] Hilary Kilpatrick, *Making the Great Book of Songs: Compilation and the Author's Craft in Abu l-Faraj al-Isbahani's Kitab al-aghani*, Routledge, 2002.

[120] Owen Wright, 'A preliminary version of the *kitāb al-adwār*', *Bulletin of the School of Oriental and African Studies, University of London*, 58(3), 1995, pp. 455–78.

Arabic musicology appears to have been essentially unknown in Christian Europe, and vice versa. There was exchange in regard to instruments (the lute and the rebec are both Arabic in origin), but no Arabic musicological treatise seems to have been translated into Latin during the twelfth-century Renaissance.[121] Yet the increasing amount of example-based descriptions of both music and language (see 3.1) indicate a *common trend from rules to examples* in Christian and Islamic musicology and linguistics.

Continuation of Natya Shastra and the Tala. Until the thirteenth century, Bharata Muni's *Natya Shastra* represented the dominant musicological tradition in the Indian Sanskrit culture (see 2.4). A few new treatises also appeared, for example Narada Muni's discourse *Sangita Makarandha* in around 1100, but the rules remained largely based on Bharata Muni's scheme.[122]

The most important Indian musicological text from the Late Middle Ages was the *Sangita Ratnakara* ('Ocean of Music and Dance') written by Sarngadeva in the thirteenth century.[123] This treatise came to be seen as the 'definitive' text for both Hindustani and Carnatic music.[124] The basic elements of the *Sangita Ratnakara* are the *sruti* (the relative tone), the *swara* (the musical 'sound' of a single tone), the *raga* (the mode or melodic formula)—which we have already discussed—and the *tala* (the rhythmic cycle). These elements were developed further on the grounds of a very precisely formulated system of rules. There are seven *tala* families, each of which is subdivided into specific rhythmic ratios (resembling the *tala* ratios in the pictorial representation—see 2.5). In one way the work seems to be descriptive, but for centuries—up to the present day—it has been used as a basic book for improvization and composition. The form of the *Sangita Ratnakara* is rule-based and declarative. The boundary conditions have been defined but there is no procedure that generates new compositions.

Music as a government matter and refutation of musical cyclicity. There were major musical developments in China during the Tang Dynasty (618–907). The creation of Chinese opera and the foundation of the first conservatory under Emperor Xuanzong speak volumes. Musical history was part of court chronicles in the historiographical style of Sima Qian (see 2.2), and official reports of music output were made. However, there is little evidence of a quest for underlying musical principles and patterns, although historical research into this is still in full swing.[125] Major music historical works, such as the first music encyclopaedia with all classical texts (the *Yueshu* by Chen Yang in 1104), appeared during the Song

[121] Gerard of Cremona translated al-Farabi's *Classification of the Sciences* (*Kitab ihsa' al-'ulum*), but this work only digressed briefly on music.

[122] M. Vijay Lakshmi, *A Critical Study of Sangita Makaranda of Narada*, Gyan Publishing House, 1996.

[123] R. K. Shringy, *Sangita Ratnakara of Sarngadeva: text and English translation*, 2 volumes, Munshirm Manoharlal Pub. Pvt. Ltd., 2007.

[124] Daniel Bertrand, *La Musique carnatique*, Éditions du Makar, 2001.

[125] See e.g. Laurence Picken and Noël Nickson (eds), *Music from the Tang Court*, Cambridge University Press, 2000.

Dynasty (960–1279). We will find this interest in encyclopaedic overviews and criticisms in Chinese art historiography too (see 3.5).

The most important musicological study during the Song era was the work of Cai Yuanding (1135–1198). In his *Lülü xinshu* ('New Treatise of Music Theory') he described how the tones in the traditional circle of fifths contradicted the widespread cosmological interpretation of the twelve standard tones.[126] According to this interpretation, the twelve tones would be equidistant and cyclical and form complete octaves. Nevertheless, Cai demonstrated that this interpretation is incorrect if scales are transposed, that is if they are produced at another, higher or lower, tone. In so doing Cai refuted the cosmological concept of cyclicity (which also played a role in historiography) on purely musicological grounds. He proposed the use of six extra notes, but they were not adopted in Chinese music. However, Cai's work does demonstrate that mathematical harmony theory was studied in all its glory during the Song Dynasty.

No new principles, but new patterns by combining old principles. All in all no new principles were developed in medieval musicology, but new patterns were discovered. The mathematical proportions principle (which defines consonant intervals) was widely used in Arabic and European musicology, for example in the analysis of scales and polyphony. We can also see the declarative system of rules principle in Indian musicology. Christian musicologists in the Middle Ages were a long way behind their Islamic, Chinese, and Indian contemporaries in the mathematical–theoretical field. This is particularly striking in the continuation of Pythagorean harmony theory, which was propelled to new heights in Islamic and Chinese musicology. But Christian musicologists did produce innovations in the study of specific pieces of music. Surprisingly enough this was *not* done on the basis of a new methodical principle but by employing a combination of existing principles—the mathematical proportions principle (known from Pythagorean music theory) and the procedural system of rules (known from linguistics, such as Donatus's grammar). This amalgamation resulted in two surprising patterns. The first was the oldest known procedural system of rules for pieces of music, the *Musica enchiriadis*, and the second was the concept of musical style as a set of shared rules. We also came across a trend from rule-based to example-based descriptions of musical structures, but only when there no longer appeared to be any stable rules. Just as important were the more individual musicological discoveries of overtones by Oresme and the acyclicity of transpositions by Cai Yuanding.

[126] Gene Cho, *The Discovery of Musical Equal Temperament in China and Europe in the Sixteenth Century*, Edwin Mellen Press, 2003, pp. 172–5.

3.5 ART THEORY: FOLLOWING AND BREAKING RULES

An art historiography like that in Pliny's *Naturalis historia* is hard to find in the Christian Middle Ages. For more than a millennium no single art historical work appeared in Europe. This can be explained to a degree by the low status of the visual arts. Unlike music, which had been considered of cosmological importance since Pythagoras, painting and sculpture never belonged to the *artes liberales* that every free man was expected to master. The *artes liberales* remained the cornerstone of Christian education well after Antiquity, and in medieval times artists were perceived to be humble artisans and craftsmen. This was still the general view even during the era of Dante in the fourteenth century. For example, Benvenuto da Imola, one of the first commentators of Dante's *Divine Comedy* and a university professor in Bologna,[127] wrote that many were surprised that Dante had immortalized 'men with unknown names and a lowly occupation', such as Cimabue and Giotto.[128] Apart from a few anecdotes about these painters, however, Dante did not give a broader overview of the artists working during his time in the way that Pliny did.

Instruction and anagogy: from Procopius and Gregory to Suger. Procopius's *Peri ktismaton* ('On Buildings') dating from 560 CE is a possible exception to the absence of medieval art historiographies.[129] In his work Procopius described the churches, forts, aqueducts, and other public buildings that were constructed during the era of Emperor Justinian, with the Aya Sofia as the pinnacle. Architecture had enjoyed a status superior to painting and sculpture since early Roman Antiquity. In the first century BCE Varro even proposed including architecture in the *artes liberales* (see 2.3). However, Procopius's overview of Byzantine building is an enumeration rather than an art historiography.

A few handbooks with technical descriptions of the applied arts have also survived, for example *De diversibus artibus* ('On Various Arts') by Theophilus Presbyter, which had to have been written between 1100 and 1120.[130] In the eighteenth century Gotthold Lessing found this manuscript by chance in the huge Herzog August Bibliothek in Wolfenbüttel, where it can still be seen. The work describes the production of painting and drawing materials, stained-glass window techniques, and the working of precious metals. It also makes one of the earliest references to oil paint and even includes an introduction to organ building. Of similar importance was the famous thirteenth-century sketchbook of Villard de Honnecourt, which served as a model-book for architecture and sculpture.

Although a general medieval art historiography is conspicuous by its absence, there were writings about art theory or what we could take for it. The second

[127] Franco Quartieri, *Benvenuto da Imola: un moderno antico commentatore di Dante*, Longo, 2001.

[128] Rudolf and Margot Wittkower, *Born under Saturn*, W. W. Norton, 1969, p. 8.

[129] Procopius, *On Buildings*, translated by B. H. Dewing and G. Downey, Harvard University press, 1940.

[130] Erhard Brepohl, *Theophilus Presbyter und das mittelalterliche Kunsthandwerk*, Böhlau, 1999. For a translation see J. G. Hawthorne and C. S. Smith, *Theophilus: On Divers Arts*, University of Chicago Press, 1963.

biblical commandment was one of the problems for Christian art because it forbade images not just of God but of other heavenly or worldly things. 'You shall not make for yourself an idol, or any likeness of what is in heaven above or on the earth beneath or in the water under the earth. You shall not worship them or serve them.'[131] There is some ambiguity in these sentences. Was this a ban on images in general, as the first sentence implies, or a prohibition on making likenesses in order to worship them, as specified in the second sentence? It was hugely important to European art that Pope Gregory I believed the second interpretation (moreover, there was also a view that *every* biblical text should be construed allegorically—see 3.7). For example, in his letter to Bishop Serenus of Marseilles in 599, which became famous, he wrote that pictures can be useful when teaching people who cannot read. After expressions of excessive devotion, Serenus had had likenesses of saints destroyed. Gregory rapped him on the knuckles and wrote,

> For a picture is provided in churches for the reason that those who are illiterate may at least read by looking at the walls what they cannot read in books. Therefore, your Fraternity should have preserved them and should have prohibited the people from their adoration, so that both the illiterate might have a way of acquiring a knowledge of history, and the people would not have been sinning at all in their adoration of a picture.[132]

Gregory's 'art as instruction' principle gave visual reproduction legitimacy and a specific goal—the instruction of the illiterate. This goal had implications for what could and could not be represented. According to Gregory, images serve to instruct the minds of the illiterate. If the depiction did not contribute to instruction, it was consequently superfluous and even culpable. It was not the classical illusionism and the harmonic proportions that were important, but the essence and transfer of the biblical message. The level of the images had to correspond to the level of the illiterates. This vision of art from the highest ecclesiastical authority legitimized the visual rendition of biblical stories, but not much more than that.

Was there a collection of rules that could be used to convert biblical writings into picture writing for illiterates? Gregory's words were too vague to contain any sort of system of rules for pictorial representation. Artists were still free to choose *how* to depict, but in practice they soon lost interest in classical proportions, anatomy, and realistic depictions. Just as in historiography, the only thing that counted was the coherence of the image to church teaching, the hagiographies, and the Bible. We can therefore effortlessly generalize the earlier *biblical coherence principle* to medieval art theory. Even when icon worship was combatted with a general ban on images in the Byzantine Empire under Emperor Leo III (roughly between 730 and 843), a biblical coherence principle was used, but it was one that involved a *literal* interpretation of the second commandment.

[131] Exodus 20: 1–5.

[132] John Martyn, *The Letters of Gregory the Great*, Pontifical Institute of Mediaeval Studies, 2004, volume 2, p. 674. For a detailed description of the episode, see Helmut Feld, *Der Ikonoklasmus des Westens*, Brill, 1990, pp. 11–13.

As well as Gregory's 'art as instruction', the Augustinian distinction between literal and metaphorical meaning also had a substantial impact on art theory in the Middle Ages.[133] In his *De doctrina christiana* (397/426) St Augustine explained that every symbol—from biblical passage to visual representation—can have both a literal and an abstract import.[134] In the case of a literal sense, the depiction referred to a historical event. In an abstract or spiritual meaning the temporal and physical fell away and the image represented the true, heavenly state of things at the end of days. The fifth-century Byzantine mystic Pseudo-Dionysius the Areopagite developed this further in his *De hierarchia coelesti* ('On the Heavenly Hierarchy').[135] According to Dionysius the whole Bible should be read metaphorically and all symbols refer to the way things are in heaven, a mode that he calls anagogic (uplifting). As we shall see, Thomas Aquinas in the thirteenth century also adhered to this tradition with his textual interpretation theory (see 3.7).

We moreover find this metaphorical and anagogic vision of the world in the work of Abbot Suger when he described in exuberant language the Gothic style that was put into practice in the expansion of his church in St Denis in 1144. The opulence of the stained glass windows and, above all, the costly objects made of gold and precious stones should be viewed, according to the abbot, as anagogic, referring to celestial beauty. Suger's use of the costliest materials provoked a bitter polemic with Bernard of Clairvaux, who rejected Suger's luxurious style and sought the sober and essentialist, as found in the early Christian art at the time of Gregory I.[136]

Roughly speaking there were two Christian 'art theories': the 'art as instruction' theory of Gregory and the 'art as anagogy' theory of Dionysius. The art resulting from the first was sober, and from the second it was opulent. Faithfulness to nature and classic proportions were not important. What counted were the abstract, heavenly forms. Both art theories were based on the biblical coherence principle, but it would be a gross exaggeration to contend that medieval art was determined by these two theories. There were, for example, many regional styles and also upsurges of the classical idiom, such as during the Carolingian and the twelfth-century Renaissances (see 3.1). In fact, the classical style never disappeared altogether from the medieval idiom. But as the art historian Erwin Panofsky has shown, it was always associated with temporary revivals that consisted of imitations and not with a systematic resurgence like the fifteenth-century Renaissance (see also 5.5).[137]

In addition to his concept of anagogic, uplifting references, Dionysius's hierarchy of heavenly and earthly beings had great influence on the medieval world view.

[133] Robert Williams, *Art Theory: An Historical Introduction*, Blackwell, 2004, pp. 47ff.

[134] St Augustine, *De doctrina christiana*, book II.

[135] Gunter Heil and Adolf Martin Ritter (eds), *Corpus Dionysiacum: Pseudo-Dionysius Areopagita, De Coelesti Hierarchia, De Ecclesiastica Hierarchia, De Mystica Theologia, Epistulae*, Walter De Gruyter, 1991.

[136] Lindy Grant, *Abbot Suger of St Denis: Church and State in Early Twelfth-Century France*, Longman, 1998.

[137] Erwin Panofsky, *Renaissances and Renascences in Western Art*, Harper and Row, 1969.

Dionysius reinterpreted Aristotle's theory of celestial spheres from a Christian point of view. Every Aristotelian sphere (with stars, planet, moon, or sun) was henceforth propelled by a class of angels. The only exception was the outermost sphere of the *Primum Mobile*, which needed no mover. Dionysius divided heaven and earth into nine and eight ranks respectively. The celestial beings ranged from seraphim and cherubim to archangels and angels, whereas the population on earth had bishops and priests but also sinners and those possessed by demons. His work became known in Western Europe thanks to the Byzantine emperor Michael, who sent a copy to Louis the Pious in 827 (and who had it translated from the Greek by the Irish philosopher Johannes Scotus Eriugena). Dionysius's integration of the Aristotelian and Christian world views was eagerly embraced by theologians like Thomas Aquinas. The hierarchy proposed by Dionysius surfaced in all arts—from painting to literature (including Dante's *Divine Comedy*). Aristotle's theory of celestial spheres became inextricably bound up with the Christian–European world view.

Classification according to Xie He and 'rule-free' art. In contrast to the break in European art historiography, we find a high degree of continuity in China.[138] The Six Principles and the point system that Xie He described in his classification of painters (see 2.5) was the model for later art histories, for example the *Xu huapin* ('Continuation of the Classification of Painters') by Yao Zui in around 550.[139] Whereas Xie He used a points system, Yao Zui did not, allowing him to put forward his own imperial patron as the best painter without any hesitation!

During the Tang Dynasty (618–907), when historiography became an affair of state, the critical painter classifications became more and more like a 'standard' art history.[140] A brief biography of each painter and a description of his painting style were written in the historiographical fashion of Sima Qian. One of the most important texts emanating from the Tang era was the *Lidai minghua ji* ('Record of Famous Painters of Successive Dynasties') by Zhang Yanyuan, which contained an overview of all well-known painters up to 847.[141] His work was modelled on the Chinese court chronicles and contained an introduction to the origins of the art of painting, after which he listed the different stylistic traditions and made recommendations about art connoisseurship. By contrast, his contemporary Zhu Jingxuan (*c.*840) went in his *Tangchao minghua lu* ('Famous Tang Paintings') back to the painter classifications by introducing three classes: *shen* (inspired), *xiu* (excellent), and *neng* (competent).[142] He also added a separate class for unorthodox artists, *yi* (unlimited). This classification system was continued by Huang Xiufu's work *Yizhou minghua lu* ('Famous Paintings of Yi State') completed in around

[138] The continuity in (ancient) China may be explained by the inclination of Chinese scholars to seek general agreement on basic issues; for details, see Geoffrey Lloyd and Nathan Sivin, *The Way and the Word: Science and Medicine in Early China and Greece*, Yale University Press, 2002, p. 238.

[139] William Acker, *Some T'ang and Pre-T'ang Texts on Chinese Painting*, Brill, 1954, pp. 33–58.

[140] For an overview of these standard art histories, see Susan Bush, 'China: painting theory and criticism', *Grove Art Online*, Oxford University Press, 2010.

[141] Nicole Vandier-Nicolas, *Peinture chinoise et tradition lettrée*, Éditions du Seuil, 1983, pp. 8ff.

[142] Mary Fong, 'Tang Tomb Murals Reviewed in the Light of Tang Texts on Painting', *Artibus Asiae*, 45(1), 1984, pp. 35–72.

1006.[143] For the first time in art historiography the unlimited *yi* class was referred to as the highest achievable—it was for the genius who worked outside the scope of all existing rules.

The traditional classification system fell into disuse during the Song period (960–1279), and artists were assessed on the basis of their depictions of tangible subjects like figures, landscapes, and animals.[144] Even individual brushstrokes for representing rocks were discussed, for instance in the *Tuhua jianwen zhi* ('Overview of Painting') by Guo Ruoxu in 1075.[145] Historical overviews continued to appear, now usually organized by subject, and once again it emerged how influential Xie He's Six Principles and Sima Qian's schematic historiography were. However, the long uninterrupted chain of art criticism texts in the tradition of Xie He ended with the work of Xia Wenyan in 1365, in which he produced an encyclopaedic overview of all known artists, their biographies, and Xia Wenyan's own opinion, which he compared with the assessments of earlier art historians.[146]

We could designate the continuation of Xie He as the declarative system of rules principle, but in fact the Six Principles and the associated classification system are too vague and too subjective for this. However, the Six Principles are a sort of guideline, as we find in the classical canon. There is also the anomaly principle in Chinese art theory, such as Huang Xiufu's unlimited, rule-free class (*yi*) for the artistic genius. Pliny also employed the anomaly principle in addition to the mathematical proportions principle. Moreover, we found it in classical poetics (2.8). There was therefore observable convergence between 'east' and (classical) 'west', to wit that *good art complies with the rules, but art of genius is rule-free or rule-breaking.*

Continuation of Sadanga and rules for foreshortening. In India the Six Limbs theory (or *Sadanga*) and the Tala proportions (see 2.5) remained the traditional art theory texts throughout the Middle Ages. These texts were extended and clarified by encyclopaedic handbooks like the *Samarangana Sutradhara* in the eleventh century and the *Manasollasa* in the twelfth century.[147] Visual art is only one of the forms of art that were discussed in these works, in which the close ties to all types of art were emphasized, from architecture, literature, music, and drama to rhetorical art and poetry. These manuals gave a detailed description of painting practice and techniques, from paintings on walls and wood to canvas. As in the Tala theory, great weight was given to the proportions of the different elements in the subjects portrayed. We also find one of the oldest known procedures for the

[143] Robert Ames, Thomas Kasulis, and Wimal Dissanayake, *Self as Image in Asian Theory and Practice*, SUNY Press, 1998. p. 64–6.

[144] Bush, 'China: painting theory and criticism', *Grove Art Online*, 2010.

[145] *Kuo Jo-Hsu's Experiences in Painting: Tuhua jianwen zhi: An Eleventh Century History of Chinese Painting Together with the Chinese Text in Facsimile*, American Council of Learned Societies, 1951.

[146] Richard Edwards, *The Heart of Ma Yuan: The Search for a Southern Song Aesthetic*, Hong Kong University Press, 2011, p. 47.

[147] Pushpendra Kumar, *Bhoja's Samarangana-Sutradhara: Vastushastra*, 2 volumes, New Bharatiya Book Corporation, 2004.

representation of *foreshortening*. This refers to the effect of perspective on a part of the body or another object that is pointing to or away from the painter—it becomes significantly shorter. Foreshortening should not be confused with linear perspective, which uses one or more vanishing points (see 4.5). The technique of foreshortening was known as long ago as Greek Antiquity, but the oldest surviving description in Europe dates from the fifteenth century. The discussion of foreshortening in eleventh- and twelfth-century Indian handbooks consisted of a guideline for the artist without going into the mathematical aspects of the pictorial phenomenon.

The historiographical hiatus in Islamic art history. The absence (or unavailability) of contemporaneous sources in Islamic art theory and art historiography is one of the riddles in the history of the humanities—as is the centuries long absence of an Indian historiography (see 2.2). The Islamic arts flourished on all fronts. Take the famous geometric decorative art and the figurative miniatures in the Persian and Ottoman tradition. And we should not forget the impressive architecture in all regions of the Islamic world. Despite all this, there is not one work on Islamic visual art to be found until the twentieth century.[148] One could argue, as one does in regard to Christian medieval art historiography, that the visual arts were not a part of the official curriculum, as a result of which these arts were not practised from a scholarly point of view. But, as we contended at the beginning of this section, this explanation is unsatisfactory. An all-rounder like al-Biruni wrote about everything that interested him (see 3.2), and he did not allow his choice to be dictated by the academic curriculum. Yet he did not write a single word about the visual arts, not even in his exhaustive account of India. It has also been suggested that there was such a strong taboo in the Islamic world on the depiction of reality that there could not have been art theory or art history. But this explanation is not entirely convincing either, because the 'ban' on figurative representations did not exist in large parts of the Islamic world, and it is also not discussed in the Koran.[149]

However, there are technical handbooks that have survived, for example a treatise by a fourteenth-century potter Abu'l Qasim[150] and a geometry manual with implications for visual art written in 1427 by the mathematician al-Kashi.[151] There were moreover descriptions of buildings, for instance in the travel writings of Ibn Battuta (1304–*c.*1368). Yet there are no texts to be found that address the history of visual art, but this does not detract at all from Islamic art.

[148] See the discussion in Oleg Grabar, 'Islamic Art', in *Grove's Dictionary of Art*, Oxford University Press, 1996.

[149] The ban on figurative representations is mentioned in the *hadith*. But this appears to have been a lightweight problem in Islam: when centuries later photography was invented, the ban on images was quickly abolished—everyone wanted a snapshot.

[150] James Allan, *Abu'l Qasim's Treatise on Ceramics*, Ashmolean, 1975.

[151] Oleg Grabar, *The Mediation of Ornament*, Princeton University Press, 1992.

3.6 LOGIC: THE LAWS OF TRUE SYLLOGISTICS

As in musicology, there do not appear to have been any new developments in (Christian) European logic during the centuries immediately after Antiquity. But, unlike in musicology, where logic was concerned, this impression was the reality. Medieval logic is traditionally divided into old logic (*Logica vetus*) between 500 and 1200 and new logic (*Logica nova*) from 1200 to the Renaissance.[152] Aristotle was the absolute authority in both periods, and the division reflects the availability of his texts. Until Peter Abelard in the twelfth century, Europeans were only familiar with Boethius's translations of two books from Aristotle's *Organon* (*Categoriae* and *De interpretatione*) as well as the *Isagoge* by Porphyry of Tyre, which served as an introduction to Aristotle. Boethius (480–525) had planned to translate all the works of Plato and Aristotle into Latin, but things went differently. As a consul in the service of the king of the Ostrogoths, Theodoric the Great, Boethius was suspected of conspiracy with the Byzantine Empire, locked up, and ultimately executed—but not before he had completed his masterpiece, *Consolation of Philosophy*, in prison.

Peter Abelard and the *consequentiae*. Abelard (1079–1142), who was not familiar with the ins and outs of Aristotelian syllogistics, created an exceptionally original and independent work on logic, the *Dialectica*.[153] In it he developed the basic principles (criteria) for logical conclusions or *consequentiae*. Two of the criteria that Abelard proposed were: *If an affirmative statement is true, its negation is untrue* and *If a negation is true, its affirmation is untrue*. He also gave criteria for the logical implication A→B, where A is called the *antecedent* and B the *consequent*: *If the antecedent is true, the consequent is also true* and *If the consequent is untrue, the antecedent is also untrue*. (Note that these criteria are not trivial, because the following is not valid: 'If the consequent is true, the antecedent is also true.') These two criteria for logical implication became known over the years as the *modus ponens* and the *modus tollens* respectively. These 'rules' were already implicitly known in the propositional logic of Antiquity in the form of a truth table (see 2.6), but they were given the status of basic principles by Abelard. Abelard's most important contribution was the attempt to develop a procedural system of rules for valid conclusions using criteria or principles. Although the system is not perfect (it is not clear in which sequence and where the criteria should be applied in order to deliver valid conclusions), Abelard's system was to exert significant influence on later medieval logicians.[154]

The new logic. When Aristotle's remaining works on logic, in particular the important *Analytica priora*, became available in Europe during the burst of translations in the twelfth century (see 3.1), extensive commentaries on syllogistics began

[152] This division goes back to the period itself, see Henrik Lagerlund, 'The assimilation of Aristotelian and Arabic logic up to the later thirteenth century', in Dov Gabbay and John Woods (eds), *Handbook of the History of Logic, Volume 2: Mediaeval and Renaissance Logic*, Elsevier, 2008, p. 282. See also William Kneale and Martha Kneale, *The Development of Logic*, Clarendon Press Oxford, 1978.

[153] Peter Abelard, *Dialectica: First Complete Edition of the Parisian Manuscript by L. M. de Rijk*, Van Gorcum, 1956.

[154] Ian Wilks, 'Peter Abelard and his Contemporaries', in Gabbay and Woods (eds), *Handbook of the History of Logic, Volume 2: Mediaeval and Renaissance Logic*, pp. 83ff.

to appear. Textbooks that dealt with both the old and new logic were written, for instance the thirteenth-century *Tractatus* by Peter of Spain.[155] Translations were also made from Arabic of works by Islamic logicians, including al-Farabi, Avicenna (Ibn Sina), and Averroes (Ibn Rushd). Avicenna had a significant influence on Christian logicians, but the dominant Aristotelian tradition of the Late Middle Ages is due primarily to the commentaries by Averroes.

In Spain the influence of Islam was perceived as being so strong that an occasional person wanted to employ logic to the conversion of Muslims. For example the Majorcan writer and logician Ramon Llull (1235–1315) developed in his *Ars generalis ultima* ('The Ultimate General Art') a system for the mechanical combination of concepts so that all alternatives could be investigated.[156] Different combinations were produced by using groups of rotatable circles with signs for concepts on them. However, Llull's system did not testify to great logical insights and did not produce a single result. His system did, though, have an influence on the development of philosophical languages in the seventeenth century, and in particular Leibniz's *Ars combinatoria* (see 4.3). Llull seems to have been convinced that his invention, by systematically doing calculations on all concepts, could rebut Islamic believers and could spread Christian truth. He died during a proselytizing mission in Africa.[157]

European logic reached a pinnacle in the fourteenth century thanks to William of Ockham and Jean Buridan. This period, which was part of the scholastic era, had a dubious reputation for a long time. Scholastic logicians would have accomplished little else except split hairs and write commentaries on commentaries on Aristotle's work. Even so, these logicians made some impressive discoveries.

William of Ockham: terministic logic. Ockham (*c.*1287–1347) in particular is of prime importance to our quest for principles and patterns. He discovered two laws of propositional logic that were later to be called the De Morgan laws (named after the logician who rediscovered them in the nineteenth century).[158] Expressed in words, these two laws of logic describe the following equivalences: (1) the negation of a conjunction of two propositions is equivalent to the disjunction of both negations and, analogous to this, (2) the negation of a disjunction of two propositions is equivalent to the conjunction of both negations. In a somewhat more abstract form, using the *connectives* of propositional logic in 2.6, we can also express the two laws as follows:

(1) **not** (A **and** B) ↔ (**not** A) **or** (**not** B)
(2) **not** (A **or** B) ↔ (**not** A) **and** (**not** B)

We can illustrate these by employing the example we used in 2.6, where A stands for the proposition 'John is clever' and B represents the proposition 'Peter is stupid'.

[155] Peter of Spain, *Tractatus called afterwards Summule logicales: First Critical Edition from the Manuscripts with an Introduction by L.M. de Rijk*, Van Gorcum, 1972.

[156] Anthony Bonner, *The Art and Logic of Ramon Llull: A User's Guide*, Brill, 2007, pp. 121ff.

[157] Erhard-Wolfram Platzeck, *Raimund Lull: Sein Leben—seine Werke—die Grundlagen seines Denkens*, 2 volumes, Schwann, 1962–4.

[158] *Ockham's Theory of Propositions*, Part II of the *Summa Logicae*, translated by Alfred Freddoso and Henry Schuurman, University of Notre Dame Press, 1980, sections 32 and 33.

We can then express the first law as follows. The complex proposition 'It is not true that John is clever and Peter is stupid' is equivalent to the complex proposition 'John is not clever or Peter is not stupid'. The second law can be illustrated the same way. These laws are universal. It does not matter which propositions one chooses for A and B, it always yields a logical equivalence. Although observations on these laws had already been made by Aristotle, it was Ockham who described them in full for the first time (but in words and not as a formula).

As far as Ockham was concerned, the Aristotelian syllogisms formed the most important logical inferences. In his *Summa logicae* ('Compendium of Logic'), Ockham designed a version of Aristotelian logic that is called *terministic* because he used the separate terms or words in propositions as a starting point.[159] Words could refer to things in reality or to other words. If words were used in a sentence, they replaced the actual thing, which was called *suppositio*. Ockham made a distinction between different suppositions, for example:

- *Personal supposition*: the term was used to designate an object (for example 'rose' to designate a particular rose).
- *Simple supposition*: the term was considered in its own right (for example if one is talking about words or concepts: 'rose is a noun').
- *Material supposition*: the term was considered as a word (for example 'rose' has four letters).

Ockham's supposition theory soon became widely accepted among scholastic logicians. His theory could be used successfully to establish the validity of syllogistic reasoning. If one wanted to know if argumentation was valid, one had to define with precision what the words referred to. The general acceptance of Ockham's supposition theory resulted in *nominalistic* dominance during the Later Middle Ages (see 3.1) and contributed to the decline of the more *realistic* trends in some disciplines, for example Modist linguistics.

Ockham also worked on the development of trivalent logic. In so doing he took the revolutionary step of departing from Aristotle's law of excluded middle (see 2.5). Trivalent logic is useful when there are also propositions that cannot be said to be true or untrue, for instance because they are undefined. Ockham's idea of trivalent logic was not adopted until the modern age.

Jean Buridan: overarching logic. Buridan (*c.*1295–1358) was probably the most important logician of the Late Middle Ages. In his hands syllogistics was transformed into something that had never previously been shown. In the *Tractatus de consequentiis* ('Treatise on Inferences') Buridan considers the syllogism as a special case of an overarching theory of inferences or *consequentiae*.[160] Building on the tradition established by Abelard, he developed a system of *inference rules* with which he demonstrated, on the basis of a proof of completeness, that it included the whole of syllogistics.

[159] *Ockham's Theory of Terms*, Part I of the *Summa Logicae*, translated by Michael Loux, University of Notre Dame Press, 1974.

[160] Peter King, *Jean Buridan's Logic: The Treatise on Supposition and the Treatise on Consequences*, Reidel, 1985.

Buridan also worked on paradoxes. Besides his proposals for resolving the liar paradox (A man says he is lying. Is what he says true or untrue?), Buridan's name is linked to the renowned case of Buridan's ass. The ass dies of hunger because it cannot choose between two identical bales of hay, which demonstrated a paradox in rationality. This paradox is not (literally) found in Buridan's work, however, whereas we do find a variant of it in Aristotle's *De caelo*.[161]

Buridan's greatest fame was based on his *Summulae de dialectica* ('Compendium of Logic').[162] In this textbook Buridan explained the basis for *Logica moderna*, modern logic, which aimed at replacing the tradition of Aristotelian logic by the terminist logic of 'moderns' such as William of Ockham. Buridan's textbook was highly influential and went on to be used at all newly established universities in Europe—from Heidelberg and Prague to Vienna—such that the new logic displaced the older logic that had been handed down via Boethius (the *Logica vetus*). Until far into the Renaissance the *via Buridani* pointed the way through the linguistic trivium to the mathematical quadrivium, which for many was followed by the study of theology.[163] In present-day terms, medieval students thus started with the humanities, moved on to science, and finally arrived at God.

The logical heritage of Islamic civilization. Greek logic texts were being translated via Syriac into Arabic as early as the eighth century. But the Arabic translation movement accelerated in the ninth century when a start was made on the integral translation of the books of Aristotle's *Organon*, once again usually via Syriac translations dating from before Arab domination.[164]

Al-Farabi (*c*.872–*c*.950) was the first major logician in the Aristotelian tradition. As we have seen, he was also the greatest musicologist in Islamic civilization (see 3.4) and his *Kitab ihsa' al-'ulum* ('Classification of the Sciences') became one of the most widely read works in both the Islamic and Christian worlds.[165] Sadly most of al-Farabi's commentaries on Aristotle's *Organon* have been lost, but thanks to Avicenna (Ibn Sina) we know that al-Farabi's influence was huge. When Avicenna described his own version of syllogistics, he identified all the points in regard to which he differed from al-Farabi.

Avicenna's inductive and temporal modal syllogistics. While the translation of the *Organon* was still in full swing, from the east—Persian Khorasan—Avicenna (980–1037) tried to make it superfluous again. Avicenna was the first Islamic scholar with his own independent logic (*mantiq* in Arabic), which is mostly contained in his *Al-Isharat wa-'l-tanbihat* ('Remarks and Admonitions').[166] Avicenna produced two significant innovations. Firstly, he developed a *temporal modal*

161 Aristotle, *De Caelo*, translated by W. K. C. Guthrie, Heinemann, 1938, p. 237.

162 John Buridan, *Summulae de dialectica*, translated by Gyula Klima, Yale University Press, 2001.

163 Bernd Michael, *Johannes Buridan: Studien zu seinem Leben, seinen Werken und zu Rezeption seiner Theorien im Europa des späten Mittelalters*, PhD thesis, University of Berlin, 1985, pp. 128, 168.

164 Tony Street, 'Logic', in Peter Adamson and Richard Taylor (eds), *The Cambridge Companion to Arabic Philosophy*, Cambridge University Press, 2005, pp. 247–65.

165 Al-Farabi's work was translated twice into Latin, most famously by Gerard of Cremona.

166 *Al-Isharat wa-'l-tanbihat* (*Remarks and Admonitions*), edited by S. Dunya, Organisation Générale des Imprimeries Gouvernementales, 1960. Parts of Avicenna's work are translated by S. C. Inati, *Remarks and Admonitions, Part 1: Logic*, Pontifical Institute for Mediaeval Studies, 1984.

syllogism, in which the premises also contained predicates like 'always', 'usually', or 'sometimes'. The second innovation was more radical. As an alternative to Aristotelian deductive logic, Avicenna constructed inductive logic, which he considered to be applicable primarily in medicine—the field in which his fame was greatest (his *Canon of Medicine* was used as a textbook in Europe until the seventeenth century). Like the Nyaya logicians in India (see 2.6), Avicenna produced inductive reasoning schemes that could be used to make—on the basis of symptoms—a medical diagnosis that was not absolute but 'possible'. Avicenna's inductive logic placed him in an old Islamic tradition. In the seventh century, analogical and inductive reasoning was used in jurisprudence (the *fiqh*). Here a Koran text was compared and contrasted with a *hadith* (see 3.2) in order to create an analogy with a judicial verdict. Although Avicenna and his followers assumed that they had developed a completely new logic, their modal and inductive logic was practised mainly within syllogistics.

Al-Ghazali, Averroes, and Ibn al-Nafis. Although al-Ghazali (1058–1111) has gone down in history as one of the biggest sceptics, he considered logic as the only non-normative discipline without metaphysical presuppositions.[167] Later on in his life he asserted that logic is even indispensable to attaining true knowledge. He succeeded in elevating logic to a basic discipline in Islamic education. However, al-Ghazali's logic was too impoverished to still be recognizable as Farabian or Avicennian.[168]

Averroes (1126–1198), who worked in al-Andalus (Spain), was another Aristotelian commentator who appeared on the scene. In his extensive writings he fervently defended the work of the Greek master against the attacks of Avicenna.[169] Moreover, Averroes refined al-Farabi's analysis of Aristotelian logic by giving a new interpretation of modal syllogistics. Yet we find no new procedures, principles, or 'laws' in Averroes's work. But his explanations and commentaries had an unprecedented effect on medieval logic in Europe.

After Averroes, Arab logicians appear to have split into two camps: Avicennian or Aristotelian-Averroist. Nevertheless, new principles or patterns remained very rare or in any event difficult to find. The work of Ibn al-Nafis (1213–1288) is still the most worthy of mention because of his formalization of the concept of 'reliability' in the *isnad* methodology of source transmission. Using Aristotelian categories, he developed his influential classification system for establishing the reliability or unreliability of a historical transmission chain (see 3.2).

Islamic logic did not stop after the fourteenth century. Between roughly 1350 and 1800, thousands of pages were written by Arab logicians, but unfortunately only a fraction of them are accessible. The historiography of

See also Dimitri Gutas, *Avicenna and the Aristotelian Tradition*, Brill, 1988, as well as Lenn Goodman, *Avicenna*, Routledge, 1992.

[167] Charles Lohr, 'Logica Algazelis: introduction and critical text,' *Traditio*, 21, 1965, pp. 223–90.

[168] Tony Street, 'Arabic and Islamic philosophy of language and logic', *Stanford Encyclopedia of Philosophy*, 2008.

[169] Charles Butterworth, *Averroes' Middle Commentaries on Aristotle's Categories and De Interpretatione*, Princeton University Press, 1983.

Islamic logic has concentrated primarily on the Golden Age. While investigations into later periods look promising, most texts must still be studied.[170]

Buddhist logic in India and China and the revival of Nyaya. Buddhist logic has existed since the fifth century BCE, but it does not become of interest to our quest until the Sanskrit text *Hetucakra* ('Wheel of Reason') by the logician Dignaga in the sixth century CE.[171] Dignaga constructed a system that resembled Aristotelian syllogistics quite closely, comprising sequences of premises followed by a conclusion. The major difference is that Dignaga was primarily interested in syllogisms that can serve as instruments for debate. In his Wheel of Reason, Dignaga created a total of nine possible reasoning patterns for such reasoning schemes. Although his wheel can be defined completely in Aristotelian terms, Dignaga did not seem to have known Greek syllogistics, and his discovery was made independently of Aristotle. In the seventh and eighth centuries, Buddhist logic was also introduced into China, for instance through the translations of Xuanzang (see 3.1). Mohist logic, which had been so impressive, had long since disappeared from the scene (see 2.6).

Contrary to Dignaga's formal patterns of reasoning, the continuation of Nyaya logic remained inductive (see 2.6). This continuation—or actually revival—took place in the *Navya-Nyaya* school. This was founded in the thirteenth century by the logician Gangesa, who in his work *Tattvacintamani* ('Thought-jewel of Reality') set task the defence of Nyaya logic against attacks from other logicians as his most important task.[172] Gangesa's main importance lies in how he systemized all Nyaya concepts into four basic categories: observation, inference, comparison, and evidence (see also 2.6).

This brief overview does not do justice to the many other forms of Indian logic, which was so rich, such as *Jain* and *Catuskoti*.[173] These types of logic are parts of more encompassing philosophies that do not concern us here. However, it is fascinating to see that Jainistic and Catuskoti logic display similarities to Greek propositional logic, but it is not clear to what extent the proposed schemes can also be used as rules for new reasoning.

Parallel between Europe, Islamic civilization, and India. Medieval logic was rule-based and procedural virtually all over the world, with the syllogism as the most important reasoning pattern. If scholars believed they could discover a formal system of rules somewhere, it was apparently in the structure of human reasoning. The importance of valid reasoning on the basis of strict criteria was sensed everywhere. In logic we do not in fact find a transition from rules to examples, as we have seen in various other disciplines such as linguistics and musicology.

[170] See e.g. Khaled El-Rouayheb, 'Sunni Muslim scholars on the status of logic, 1500–1800', *Islamic Law and Society*, 11, 2004, pp. 213–32; Khaled El-Rouayheb, 'Was there a revival of logical studies in eighteenth-century Egypt?', *Die Welt des Islams*, 45(1), 2005, pp. 1–19; John Walbridge, 'Logic in the Islamic intellectual tradition: the recent centuries', *Islamic Studies*, 39(1), 2000, pp. 55–75.

[171] R. S. Y. Chi, *Buddhist Formal Logic: A Study of Dignaga's Hetucakra and K'Uei-Chi's Great Commentary on the Nyayapravesa*, Motilal Banarsidass Publications, 1990.

[172] Bhattacharya Gopikamohan, *Navya-Nyāya: Some Logical Problems in Historical Perspectives*, Bharatiya Vidya Prakashan, 1978.

[173] Piotr Balcerowicz (ed.), *Logic and Belief in Indian Philosophy*, Motilal Banarsidass Publishers, 2010.

3.7 RHETORIC AND POETICS: A MOTLEY COLLECTION OF RULES

In most regions, medieval rhetoric and poetics were so much interwoven that it makes sense to treat them together.

Christian rhetoric: St Augustine and the continuation of Cicero. We have already encountered the church father St Augustine (354–430) in his capacity as historian and art theoretician, but he had what is arguably an even greater influence as a rhetorician. St Augustine was educated in the Roman rhetorical tradition, and prior to his conversion to Christianity he spent some time as a university professor of rhetoric in Milan. In his *De doctrina christiana* (397/426) he explained how rhetoric could be employed in the dissemination of Christianity.[174] Classical rhetoric was extremely contentious among early Christians for a long time because it could be used to defend virtually any position. However, St Augustine argued that eloquence and oratory were not foreign to Christianity, and demonstrated this using the Bible itself. He encouraged his readers to imitate the persuasiveness of King David and St Paul. Following the tradition of Cicero, St Augustine spelled out the duties of a Christian clergyman: to instruct, to enthuse, and to move, but now according to the Holy Word. He illustrated his argument with passages from St Paul, St Ambrose, and other Christian writers. Through *De doctrina christiana*, rhetoric became the basis for *homiletics*, or the *art of preaching*. Despite this new application of rhetoric, we do not find any new rhetorical concepts—let alone empirically derived patterns—in the work of St Augustine.

What St Augustine did was immensely far removed from rhetoric as an empirical discipline that subjected speeches and argumentation to critical analysis and tried to infer an underlying system of rules (see 2.7). The best way to designate St Augustine's reworking of rhetoric is as the *biblical coherence principle*. Classical rhetoric was reinterpreted in terms of Christian teaching, as had already happened in historiography and would happen later in art theory. We find just as few new rhetorical ideas or principles after St Augustine, with the possible exception of Boethius who in his *In topicis differentiis* proposed a return to a more principle-based approach.[175] Boethius's proposal remained fundamentally Aristotelian, while Alcuin's manual on rhetoric was a Ciceronian pastiche.[176] Cicero was also the standard after the Carolingian Renaissance and virtually all rhetorical activities were modelled on him, from the art of writing letters (*ars dictaminis*) and the art of preaching (*ars praedicandi*) to the art of composing prose and verse (*ars prosandi/poetriae*), the last of which was part of poetics in the ancient world.[177] We do not find empirical systems of rules for these *artes*. Rhetorical treatments consisted

[174] St Augustine, *De doctrina christiana*, book IV.

[175] *Boethius's De topicis differentiis*, translated, with notes and essays on the text, by Eleonore Stump, Cornell University Press, 1978.

[176] Wilbur Samuel Howell, *The Rhetoric of Alcuin and Charlemagne*, Russell & Russell, 1965.

[177] Rita Copeland and Ineke Sluiter (eds), *Medieval Grammar and Rhetoric: Language Arts and Literary Theory, 300–1475*, Oxford University Press, 2009.

completely of normative procedures. For example, in his *Poetria nova*, Geoffrey of Vinsauf (early thirteenth century) proposed using *morae* (delays) as a method for expanding the text, as if the art of literature consisted of learning how one 'can say a lot when one has little to say'.[178]

Christian poetics: allegory and anagogy. Poetics, like rhetoric, was also recaste in Christian terms. In the same *De doctrina christiana* St Augustine argued that every text could be interpreted both literally and allegorically (see 3.5). In the case of Holy Scripture, the allegorical interpretation has to be preferred to the literal one. The Bible exegetes had a very tough job, namely to get agreement between the Hebrew Old Testament and the Christian New Testament. It emerged that allegorical interpretations were generally the only solution. Metaphorical text analyses were far from new. The oldest known example is to be found in the work of Theagenes of Rhegium, who in the sixth century BCE was able to defend Homeric mythology by providing it with a non-literal interpretation.[179] The anomalists of Pergamon also knew the ropes, but the allegorical method gained ground with the Neoplatonists and above all the influential metaphorical version of *The Dream of Scipio* by Macrobius (see 2.3). The world had to be viewed as a text, God's book, and it was full of allegorical symbols that had to be interpreted not literally but figuratively. Was it not Jesus himself who used parables to express the deeper and universal significance of a story? Slowly but surely a system was put together with which the Bible could be read at several levels. The Italian theologian Thomas Aquinas (1225–1274) defined in his *Summa Theologica* the four levels of Bible exegesis as follows (which can be traced back to St Augustine): (1) the literal, (2) the allegorical, (3) the moralistic and (4) the anagogic (the uplifting).[180] On top of the literal meaning of a passage in the Bible, the different metaphorical meanings could all be true at the same time. The only criterion that restricted possible interpretations was the 'principle of charity', according to which all explanations had to be consistent and coherent with Christian teaching. In this book we have already designated this principle as the biblical coherence principle.

Is there any system of rules, besides the permissive biblical coherence principle, to be discovered in non-literal interpretations, or was it a case of anything goes? As in the case of the anomalists of Pergamon, the interpretations by the Christian exegetes got quite out of hand. They were encouraged in this by passages in the Bible itself. In Galatians 4:21–31, for instance, St Paul's interpretation of the Old Testament story—in which Abraham's concubine Hagar was driven out at Sarah's insistence—was that 'Hagar' was the Arabic term for Mount Sinai and therefore represented the Old Testament and Moses, whereas Sarah was the symbolic mother of the Christians and therefore stood for the New Testament and Christ.[181] In other words Paul did not shrink from far-fetched etymologies and analogies

[178] C. S. Lewis, *The Discarded Image: An Introduction to Medieval and Renaissance Literature*, Cambridge University Press, 1967, p. 192.

[179] J. Tate, 'On the History of Allegorism', *Classical Quarterly*, 28, 1934, pp. 105–14.

[180] Thomas Aquinas, *Summa Theologica*, part I, question 1, article 10.

[181] Richard Harland, *Literary Theory from Plato to Barthes*, Palgrave, 1999, p. 25.

provided that the desired result was achieved—a flash-forward from the Old to the New Testament. It is illustrative in this context to recall the strict, rule-based exegesis of the Indian Vedas by the Mimamsa school which worked with precise rules rather than with free associations (see 2.7). The difference between the Mimamsa exegesis and the apparently arbitrary biblical exegesis could hardly be greater. We could almost equate the methods of Paul, and medieval biblical exegetes who came after him, to the anomaly principle, were it not the case that every interpretation had to be consistent with the church's teaching.

The allegorical interpretation method was also used for secular texts. Although St Augustine and Thomas Aquinas believed that non-literal construal was only valid for texts inspired by God, such as the Bible, others substantially reinterpreted the classical writers in accordance with the Christian pattern. Origen, for instance, believed that the fourth eclogue of the Roman poet Virgil could be understood as a messianic prophecy of the coming of Jesus Christ (see also 3.3).[182] In the sixth century CE Fulgentius even wrote an allegorical analysis of the complete *Aeneid*: every passage was given a Christian reference from the first sentence to the last.[183] In some cases this resulted in highly imaginative literature, but a critical method or underlying system is hard to find.

In his *Convivio* (Banquet), Dante Alighieri (1265–1321) also argued that Thomas Aquinas's four-level biblical exegesis was ideal for use with secular storytelling.[184] Dante showed on the basis of Ovid's works that all four forms of interpretation (literal, allegorical, moralistic, and anagogic) could be valid and did not need to exclude one another. We also find something that is new in Dante's work. In his unfinished *De vulgari eloquentia* ('On Eloquence in the Vernacular') from *c.*1302 he advocated, albeit in Latin, the use of the vernacular in poetry, which Dante put into practice himself in his beautiful sonnets and in the *Divina commedia.*[185] The *Eloquentia* is the oldest known document about using the vernacular as artistic language.

Summarizing, an analytical method based on four levels of interpretation was developed in both biblical exegesis and secular literary criticism. But except for the biblical coherence principle, it produced no underlying system that could be used to create the construal. It appeared to have been sufficient if the allegorical, moralistic, or anagogic reading led to the desired result.

Secular poetics: procedural rules in *Leys d'amors*. We find a completely different approach to poetics in the *Leys d'amors* ('Laws of Love') by Guilhem Molinier in the fourteenth century.[186] We have already referred to this discourse briefly as the

[182] Gerard O'Daly, *Days Linked by Song: Prudentius' Cathemerinon*, Oxford Univerisity Press, 2012, p. 347.

[183] Fulgentius, *The Exposition of the Content of Virgil*, translated by O. B. Hardison, in Alex Preminger, O. B. Hardison, and Kevin Kerrane (eds), *Classical and Medieval Literary Criticism: Translations and Interpretations*, F. Ungar Publishing Co., 1974, pp. 333–9.

[184] Dante Alighieri, *The Banquet*, book II, chapter 1, translated by Elisabeth Sayer, Aegypan, 2009.

[185] Dante Alighieri, *De vulgari eloquentia*, edited and translated by Stephen Botterill, Cambridge University Press, 1996.

[186] Joseph Anglade (ed.), *Las Leys d'amors*, 4 volumes, E. Privat, 1919–1920.

oldest description of Occitan (see 3.1), but in addition to being a grammar it also contained an empirical study of the poetry of Provençal troubadours and on top of that an attempt to develop a procedural system of rules. How, in the middle of an ocean of anti-empirical, allegorical interpretive practices, can this work be explained? The explanation is as fascinating as it is tragic. At the beginning of the fourteenth century the art of the Provençal troubadours was dying out as an activity. In part this was due to the terrible bloodbaths attendant on the persecution of the so-called Cathars or Albigenses in the lands of the Count of Toulouse during and after the Albigensian crusade (1209–1229). Entire towns were butchered with the encouragement: 'Kill them all. God recognizes His own.'[187] Little remained of the wonderful culture of poetry in Occitania. In order to save what could be saved, between 1332 and 1356 a gigantic work was produced under the leadership of Guilhem Molinier in which as much troubadour poetry as possible was collected and their poetry system was described as explicitly as achievable. This was the *Leys d'amors*. It was a final attempt to preserve a dying art for posterity. Besides being a grammar, which formed the basis for all later Occitan grammars, the work discusses in detail the rules of prosody, the structures of the lines, the couplets, and the poetic genres. Moreover, a few exact procedures were given for the poetry system, such as for the—relatively simple—process of finding rhyming words. Let us say that you want to find a word that rhymes with *-ori*. You start alphabetically with *a_ori* and then put each letter in the open position: *abori, acori, adori, afori* etc. From this list you next select the words that are actually in Occitan. You continue thus. It could hardly be more algorithmic, although it is extremely time-consuming and exhausting. But as we can read in the wonderful stanzas of the *Leys d'amors*: *mays dura anta que sofracha* (shame lasts longer than suffering).

Arabic rhetoric: continuation of Aristotle. Compared with Christian Europe, the continuation of Aristotelian rhetoric and poetics in Islamic civilization was almost immediate.[188] In the tenth century al-Farabi wrote an extensive commentary on Aristotle's *Rhetorica*, which was followed by commentaries by Avicenna (Ibn Sina) and Averroes (Ibn Rushd) in the eleventh and twelfth centuries.[189] In perfect Aristotelian tradition, rhetoric was defined as the syllogistic art whose goal was persuasion on the grounds of enthymemes. However, the syllogisms used varied substantially between the three great Islamic scholars, where Avicenna in particular proposed much richer, temporal modal and inductive syllogistics (see 3.6). But Aristotle's use of enthymemes, in which the premises in a syllogism do not necessarily have to be 'true' but only 'plausible' (see 2.7), was embraced by all three philosophers. This also applied to the use of Aristotle's heuristic system of rules.

As in Aristotle, but contrary to the Christian tradition, Arabic rhetoric formed a very strong unity with logic. This did not mean that there was no art of preaching

[187] '*Caedite eos. Novit enim Dominus qui sunt eius.*' This sentence is probably apocryphal and was attributed to the Cistercian church leader Arnaud Amalric by the German monk Caesarius van Heisterbach in the *Dialogus Miraculorum* (written between 1219 and 1223).

[188] Uwe Vagelpohl, *Aristotle's Rhetoric in the East: The Syriac and Arabic Translation and Commentary Tradition*, Brill, 2008.

[189] Deborah Black, *Logic and Aristotle's Rhetoric and Poetics in Medieval Arabic Philosophy*, Brill, 1990.

(homiletics) in Islamic civilization, but it was primarily a part of the theology that was *sui generis*, and historical research into which is still at a very early stage. A recently available manuscript, the *Kitab adab al-khatib* ('The Book of the Preacher's Etiquette') written by al-Dimashqi in about 1324, reveals that there was also continuity with Greek rhetoric in Islamic preaching practice. Some people even think it is possible to recognize the influence of Hermogenes's *Progymnasmata* (see 2.7).[190]

Arabic poetics: *takhyil* and literary criticism from al-Farabi to Averroes. Al-Farabi, Avicenna, and Averroes also commented on Aristotle's *Poetica.* In his *Canon of Poetry*, al-Farabi gave a thorough explanation of Aristotelian poetics and also a refined subdivision of the concept of *mimesis* (imitation).[191] Avicenna, on the other hand, had a different approach. His poetics shifted from mimesis to *oratorical evocation*, which he designated with the term *takhyil.*[192] This different approach was all about the role of poetics in Islamic civilization. Unlike the Greek world, theatre was virtually unknown as an art form in the Arab world and poetry recitation was the most important poetic activity. Poetry was considered to be one of the key sources of knowledge. After all, the Koran was written in verse.[193] Avicenna's concept of *takhyil* refers to the *poetic capacity to summon up images in the memory (the mind) of the general public.* Imagination and memory are closely linked to each other here because the evoked image does not need to contain just a stored depiction. It can also be created as a result of a complex interaction between memory, imagination, and emotion. Avicenna contrasted this capacity to summon up images, the *takhyil*, with logic. According to Avicenna a logical proof does not stir the soul, whereas *takhyil* does, and therefore the man in the street is moved more readily by *takhyil* than by logic. But after the design of these poetics, modelled on the Arabic idiom, Avicenna did not take the step of investigating whether a system of rules underlies good poetry, as Aristotle did for the concept of drama. So although Avicenna innovated and extended Aristotle's poetics in regard to a number of essential points, he did not go in search of rule-based patterns. It was the same story with the last of the great three: Averroes. He considered poetics as a way to discover universal canons that apply to all or at least most peoples. But Averroes did not take the step of actually deriving these canons. While he searched for the underlying nature of poetics, he seemed ultimately to have been more interested in defending logic and reason than in understanding the functioning of poetry.

In addition to the commentaries of al-Farabi, Avicenna, and Averroes, there was also a rich tradition in the Arab world in the field of literary criticism. The discourse *Qawa'id al-shi'r* ('The Rules of Poetry') from the ninth century by Tha'lab

[190] Philip Halldén, 'What is Arab Islamic rhetoric? Rethinking the history of Muslim oratory art and homiletics', *International Journal of Middle East Studies*, 37, 2005, pp. 19–38.

[191] Al-Farabi, *Canons of Poetry*, translated by A. J. Arberry, Revisti degli Studi Orientali, volume 17, 1938. See also Vicente Cantarino, *Arabic Poetics in the Golden Age*, Brill, 1975, pp. 109ff.

[192] Ismail Dahiyat, *Avicenna's Commentary on the Poetics of Aristotle: A Critical Study with an Annotated Translation of the Text*, Brill, 1974, pp. 8ff.

[193] Of course, the Koran itself is not poetry and even takes an explicitly anti-poetical position. It is to the Arabs' credit that they nevertheless continued to write verses.

(or al-Kufah) is particularly worthy of mention. In this work Tha'lab approached the art of poetry on a completely linguistic basis, and then primarily on the grounds of an analysis of the words rather than more poetic features such as metre or rhyme.[194] Despite the promising title, we find no system of rules at all, either declarative or procedural. A somewhat broader view of poetry was taken by Ibn Rashiq in the eleventh century as we can see from his views about diction, metre, meaning, and rhyme.[195] While Ibn Rashiq's work is much more systematic than Tha'lab's rather chaotic discourse, we find once again few if any methodical principles. The same has to be said of other, otherwise very interesting, literary critics such as the Afro-Arab scholar al-Jahiz (781–868), who made the concept of *coherence between all the parts of a poem* a cornerstone of his literary criticism.[196] While born in Basra, al-Jahiz was probably of East-African descent and also the author of one of the most remarkable works of the 'medieval' period, the *Risalat mufakharat al-sudan 'ala al-bidan* or 'Superiority of Blacks over Whites'. In it he argued that blacks had conquered and governed white countries (from Arabia to the Yemen), while whites had never overrun a single black country.[197]

We seem to have come a long way from the formalizing, musicological tradition of Islamic civilization, in which al-Farabi exposed all kinds of patterns, for example the formalization of the Arabic rhythmic cycles (see 3.4). Al-Farabi's approach would have been equally readily applicable to the metre of Arabic poetry and much else besides, but there are no indications that anything like this was ever tried in Arabic poetics.

Finally there were the encyclopaedic compilations, for instance Ibn al-Nadim's annotated catalogue of Arabic literature, the *Kitab al-fihrist*, which had entries for all the books on sale in Baghdad at the end of the ninth century.[198] The work gives an impressive overview of Arabic writings in the Golden Age and the immense wealth of medieval Baghdad. To a degree it can be compared with Photius's *Bibliotheca*, which was produced at the same time in Byzantium. But whereas Photius summarizes the *classical* Greek works in his *Bibliotheca* (which we discussed in 3.3 on philology), Ibn al-Nadim gave an annotated overview of *all* Arabic literature: grammar, historiography, poetry, law, theology, philosophy, geography, philology—anything that could be read at the time.

Indian rhetoric and poetics. In India, rhetoric and logic were virtually indistinguishable. We saw in 3.6 that both the *Navya-Nyaya* school and the Buddhist logic of Dignaga were integrated in the concept of the debate. In Indian poetics, on the

[194] Wen-chin Ouyang, *Literary Criticism in Medieval Arabic-Islamic Culture: The Making of a Tradition*, Edinburgh University Press, 1997, p. 181.

[195] Vicente Cantarino, *Arabic Poetics in the Golden Age*, Brill, 1975, pp. 141–50.

[196] G. J. H. van Gelder, *Beyond the Line: Classical Arabic Literary Critics on the Coherence and Unity of the Poem*, Brill, 1982, pp. 37–42.

[197] During the huge slave revolt in southern Iraq from 869–83 a whole province was in the hands of blacks for several years. Al-Jahiz had died just before this revolt. One can imagine how his discourse may have been received in the capital. (Thanks to Wim Raven, p.c.) For an in-depth study of al-Jahiz's genre, see Ibrahim Geries, *Un genre littéraire arabe: al-Mahasin wa-l-masawi*, G.-P. Maisonneuve et Larose, 1977.

[198] Fuat Sezgin, *Geschichte des arabischen Schrifttums*, volume 1, Brill, 1967, pp. 385–8.

other hand, we see a continuation of the Mimamsa and the Natya Shastra schools. As we were able to establish in 2.7, both schools were prescriptive. The Mimamsa school presented a system of rules for the ritual interpretation of Vedas using strict formalistic rules, and the Natya Shastra gave an extremely detailed system of procedures for theatrical performances on the basis of rasas. There are few new ideas to be found in these schools during the Middle Ages and no new principles or patterns. The most significant development was an influential commentary by Kumarila Bhatta in about 700 CE, the *Mimamsa-slokavarttika*, that enabled the Vedic tradition in India to endure for centuries.[199] And thanks to the monumental commentary of Abhinavagupta during the eleventh century, the *Abhinavabharati*, the Natya Shastra tradition was canonized into a form that was still being applied in twentieth-century poetics.[200]

A masterpiece of Chinese rhetoric and poetics: Chen Kui. In China Chen Kui (1128–1203) wrote an extraordinarily interesting work, the *Wen Ze* or *The Rules of Writing*.[201] Although this work is barely known outside China and a circle of sinologists, it is considered to be the first systematic analysis of Chinese rhetoric. The Mohist works we discussed in the previous chapter were written centuries earlier, of course (fifth century BCE), but they discussed argumentation and debating and not the art of writing itself (see 2.7). *The Literary Mind and the Carving of Dragons* by Liu Xie (fifth century CE) was also significantly earlier but it belongs primarily in the category of literary criticism and it did not attempt to deduce a system of rules for proper writing (see 2.8).

Born during the era of the Song Dynasty, Chen Kui became the registrar of the imperial library at an early age. He appears to have had an especially critical and independent mind. For example, he brought waste to the attention of the court and he asserted that there were far more civil servants than necessary. As a result Chen Kui was transferred to the provinces, but this did not prevent him from writing his masterpiece of rhetoric. We will briefly outline the historical context as an aid to correct appreciation of Chen Kui's discourse. The attitude of the ruling class with regard to government officials changed after the fall of the Tang Dynasty in 907 and the rise of the Song Dynasty in 960 (after an intervening period of five short dynasties). Officials were now chosen on the basis of a competitive examination rather than a selection system that was based chiefly on privileges, as had been normal during the Tang Dynasty. The most important part of this examination was to write an essay in which the candidate had to demonstrate his originality and skill in argumentation.[202] The examination was the first step on the road to a much sought-after government career. Chen Kui's *The Rules of Writing* was now the first

[199] Kumarila Bhatta, *Slokavarttika*, translated by Ganganatha Jha, The Asiatic Society, 1985.

[200] *Natyasastra with the Commentary of Abhinavagupta*, edited by M. Ramakrishna Kavi, 3 volumes, Oriental Institute, 1954. See especially volume 3 for the influence of the Natya Shastra tradition.

[201] The *Wen Ze* has not yet been fully translated. For an outline and discussion of the work, see Andy Kirkpatrick, 'China's first systematic account of rhetoric: an introduction to Chen Kui's *Wen Ze*', *Rhetorica*, 23(2), 2005, pp. 103–52.

[202] E. A. Kracke, *Civil Service in Early Song China*, Harvard University Press, 1953.

satisfactory handbook of Chinese rhetoric that could also serve as preparation for the examination. Chinese printing expertise did the rest.

What made *The Rules of Writing* so special (and so interesting in terms of our quest) is that Chen Kui explicitly derived the rules and principles of good writing from existing texts. These documents represented the *crème de la crème* of Chinese literature. For instance, Chen Kui distilled the rules for 'clear language' from the *Book of Rites* (one of the Confucian classics), while the rules for 'colourful language' were inferred from the *Book of Songs*. Chen Kui did not 'prescribe' rules. First they were empirically derived, then they were tested on (or compared to) new works and after that they were amended if necessary. Chen Kui also conducted a detailed study of sentence length in the classic *Tan Gong* (a section of the *Book of Rites*), in which he believed he discovered regularity, and he concluded, among other things, that 'the language of the *Tan Gong* is simple but not sparse'. It is the oldest known quantitative analysis of style in literary theory. Then Chen Kui tried to select the 'exquisitely beautiful' sentences from the *Tan Gong* and to describe them in terms of a number of properties. He compared the regularities he found to sentences in the *Spring and Autumn Annals* and the *Book of Songs*. When he discovered that his rules could not be generalized for these works, he did not immediately reject his approach (the way that Dionysius of Halicarnassus had done with his rules of natural word order—see 2.8). Instead he tried to fine-tune the rules. While this attempt did not produce an unambiguous result, it did indicate a direction that could be explored further. Chen Kui also analysed the different ways of explaining things in textual material (which we could compare to the *expositio* in Graeco-Roman rhetoric). He found three style forms: 'from detail to conclusion', 'from main point to detail', and 'from main point via detail to a reformulation of the main point'.

It is instructive to compare Chen Kui's discourse with Quintilian's influential *Institutio oratoria*, which was written at the end of the first century CE (see 2.7). The latter work lived on in fragments in the Middle Ages in Europe and—after discovery of the complete manuscript—it dominated rhetoric in the Renaissance (see chapter 4). Like Chen Kui, Quintilian made many references to the works of illustrious predecessors, from Cicero to Aristotle and Plato. But he seems to have had no pretensions to expose regularities in these texts, let alone to devise testable hypotheses about good or beautiful use of language, which is what his contemporary Dionysius of Halicarnassus—who had sunk somewhat into oblivion—had attempted (see 2.8). Chen Kui, on the other hand, combined two goals: empirical rhetoric and instruction for ambitious young men, which he consolidated into one handbook. Chen Kui showed how style analysis could go hand in hand with rules for 'good' writing, but that it was extremely tough to find general rules for 'beautiful' writing. Although this result was also reported by Dionysius (see 2.8), Chen Kui gave suggestions for further research into the rules of the 'beautiful'. While Chen Kui's work is therefore comparable with Dionysius's empirical poetics, it is miles away from the prescriptive European manuals of the *ars dictaminis* or *ars prosandi* that we discussed at the beginning of this section.

Chen Kui's *The Rules of Writing* was without doubt the most original work on rhetoric in the Middle Ages and one of the most important in the history of rhetoric. Save for a few translated fragments, this work is virtually unknown outside China—embarrassing evidence of the indifference to activities in the humanities outside Europe.[203]

[203] Cf. Andy Kirkpatrick and Zhichang Xu, *Chinese Rhetoric and Writing: An Introduction for Language Teachers*, Parlor Press, 2012, pp. 5ff.

CONCLUSION: INNOVATIONS IN THE MEDIEVAL HUMANITIES

Fragmentation. While it is relatively easy to get an overview of the humanities in Antiquity—with the Greeks towering above the rest in almost all disciplines—the balance was completely different in the Middle Ages. The most important innovations came from Islamic civilization, where historiography as a discipline rose to great heights thanks to the formal transmission theory of the *isnad*, the anthropological approach of al-Biruni and the sociological method of Ibn Khaldun. While Islamic scholars also made significant contributions to logic (the inductive, temporal syllogistics of Avicenna), linguistics (the example-based grammar of Sibawayh), and musicology (al-Farabi's rhythmic cycles), European men of learning in these fields were of at least equal stature—there were Ockham and Buridan in logic, the Modists in linguistics, and the organum formalizers in musicology. Philology (as textual criticism) was too marginal to go through life as a discrete discipline, with the possible exception of China. As regards the other humanities—art theory, rhetoric, and poetics—the centre of gravity seems to have been in China and (to a slightly lesser degree) India, for example the Chinese description of rule-breaking art, the Indian rules for foreshortening in art, and the brilliant rhetoric and poetics of Chen Kui. Activities in these disciplines in Islamic civilization were primarily a continuation of the classics, while there was a Christian reinterpretation in Africa and Europe. Although this reinterpretation was extremely influential, it has relatively little significance for the quest for empirical patterns in humanities materials.

This geographical fragmentation was matched by the fragmented nature of the principles used and patterns found. There were systems of rules to be sure, such as for source reliability in historiography, for organa in musicology, for clear use of language in rhetoric, for rhyming in poetics, and for inferences in logic, but there was no hunt for deeper, more general principles that could apply to all disciplines. We also found this fragmentation in the humanities of Antiquity. There was no coherent method for the different disciplines within the humanities. In the next chapter we will explore the early modern era, when humanists developed philological methods that were in fact used in (almost) all branches of the humanities.

Continuation of Antiquity. Despite this fragmentation, there was continuation of the patterns and metapatterns that we found in the humanities of the ancient world (see 2.9).

Pursuit of a system of rules was found on many fronts, as we have seen. The biblical coherence principle, which seldom if ever resulted in a system of rules, was an exception. The example-based tradition that started with Sibawayh was not a quest for a system of rules either, although the concept of analogical substitution could be called a rule that works through examples.

Parallel discoveries were made once again, but they were less convincing than in classical Antiquity: the discovery of example-based grammars (Sibawayh and European grammarians), foreshortening in art (India and the earlier classical Greece), congruence between the historical and theological time structures (Roman Africa,

Islamic civilization, Europe, Ethiopia, and India), and the circle of fifths (China, earlier classical Greece).

From descriptive to prescriptive: we also see this tendency in most medieval humanities, such as linguistics (for example descriptive Arabic and Latin grammars became prescriptive), art theory (the Chinese Six Principles for analysing art became normative), rhetoric, and poetics (Christian writing, poetry, and preaching). Musicology was a striking exception: the descriptive formalization of organa seemed to produce precisely the opposite of prescriptive behaviour, where the rules were broken time after time by new forms of organa.

Seldom deductive: systems of rules were seldom if ever explained on the basis of first principles, except for the continuation and extension of Aristotelian logic.

Empirical rules for the 'good', not for the 'beautiful': this pattern was also repeatedly confirmed in the Middle Ages, as we have seen in musicology, Chinese art theory, and Chinese poetics (Chen Kui).

From rules to examples. The medieval humanities might appear to be a patchwork quilt, but there are also general trends to be found. Although efforts were made in almost all disciplines to establish systems of rules, there was also an observable tendency to shift to example-based systems. We find such systems first in the work of Sibawayh, who compiled a nigh-on complete example-based grammar for classical Arabic. New sentences could be produced from earlier sentences employing only one rule—the analogical substitution of words. In other linguistic traditions this tendency resulted in more hybrid systems. Where rules could be found, they were stated. If not, a phenomenon was covered for better or worse with examples. We also see this trend from rules to examples in Islamic and above all in European musicology (where early organa could be pinned down with systems of rules, but later organa could not), and in art theory, rhetoric, and poetics. However, systems of rules were found now and again, for example by Chen Kui in poetics, but many literary studies resorted to simple lists.

Formalization and unification in the 'soft' humanities. The transition from rules to examples contrasts with the pursuit of formalization and unification in one of the 'softer' disciplines—historiography, which was 'soft' for a long time compared with the more formalistic logic, rhetoric, linguistics, and musicology. By exposing the transmission chain of a source (the *isnad* method), two earlier historical principles were formalized and unified in Arabic historiography. These were the most probable source principle and the eyewitness account principle of Herodotus and Thucydides respectively. This pursuance of both formalization and unification was exceptional in the medieval humanities, although we also see it fleetingly in European musicology in the formalization of the organum and, of course, in logic, where Aristotle's syllogistics and Abelard's *consequentiae* were unified by Buridan.

A religious revolution? If there was ever a revolution in the humanities, it was the religious and above all the Christian upheaval in the Early Middle Ages. While the Greek tradition remained intact to some degree in most Arabic humanities, in Europe the humanities were redefined according to the Christian pattern and the biblical coherence principle. Historiography became salvation history, linguistics became a quest for the universal language before Babel, musicology became a formalization of polyphonic Gregorian chants, art theory became uplifting, rhetoric became the art of preaching, and poetics became a study of biblical-allegorical interpretation. We do not see such radical changes in so short a time anywhere else in the humanities. Most of them took place during the active life of St Augustine. This Christian revolution originated in Roman Africa—the intellectual centre of early Christendom—and spread rapidly through converted Europe.

Both the subject and the method of the humanities were transformed. This method reflected an uncritical and anti-empirical attitude. The Christian humanities scholars appeared to know a priori what they would find. Whether it was literary, historical, linguistic, visual, musical, or poetic material, it emerged that it was in conformity with, or it was made to conform with, the Bible and church teachings. The wildest allegorical interpretations were admissible provided that the desired result was achieved. Although we do not find this anti-empiricism in all disciplines in equal measure—musicology and logic remained partially out of range—art theory, rhetoric, poetics, and historiography followed a virtually allegorical pattern.

It would take a second, lengthier humanistic revolt to 'undo' the Christian revolution. But contrary to what these humanists intended, or said they intended, their activities did not result in a return to the classics, as we will see.

4

Early Modern Era: The Unity of the Humanities

In the history of ideas, to which this historiography belongs, the European Middle Ages end with the fourteenth-century Italian humanist Petrarch. Humanism, which sought to revive classical Antiquity, spread out over a large area, initially in Italy and then in the rest of Europe. A line runs directly from humanism to the upheaval in the world view that is known as the Scientific Revolution. The early modern era, which roughly speaking started at the end of fourteenth century and finished at the end of the eighteenth century, is normally split into the Renaissance and the Enlightenment, but for the history of the humanities it makes sense to treat this period as a whole. It emerges that the impressive innovations in Europe had parallels in China, where the humanities likewise flourished. Activities in Islamic and Indian scholarship appeared to flag, although both reached a brief high point in the Mughal Empire. African humanities reached its highest stage of development in the Songhai Empire, from which hundreds of thousands of manuscripts have been handed down, which are still waiting to be accessed.

4.1 PHILOLOGY: QUEEN OF EARLY MODERN LEARNING

It is with some justification that we discuss philology first in this chapter about the early modern age. Rarely has a discipline brought about such major societal changes as philology did in Europe. Yet at the end of the Middle Ages this subject was in a dire state. Save for the work of a Carolingian philologist like Lupus of Ferrières (see 3.3), so much knowledge about text reconstruction had been lost that forgeries were the order of the day. However, as we will see elsewhere in the history of the humanities, the discipline that lagged furthest behind proved able to catch up the most quickly.

The beginning of early modern philology: prehumanists, Petrarch, and Boccaccio. The rise of philology was preceded by a group of prehumanists who worked in fourteenth-century Padua. The practice of collecting classical texts had never completely disappeared in Italy. For example, Lovato Lovati (1241–1309) knew Catullus, Propertius, and Tibullus long before they were 'discovered' by later humanists.[1] Giovanni de Matociis, or Giovanni Mansionario (active 1306–1320), produced the first literary criticism in Europe. On the basis of a Veronese manuscript and a text by the Roman historian Suetonius (see 2.2), he discovered that there were two Plinys rather than one, as had hitherto been believed. Giovanni published his finding in the tract *Brevis adnotatio de duobus Pliniis*.[2] Although his discovery appears not to have been based on general methodical principles, the result was historically correct.

The same was true of the first great humanist: Francesco Petrarca, or Petrarch (1304–1374), whose philological work was barely based on methodical principles, if at all.[3] However, Petrarch's influence was so far-reaching that he can justifiably be put forward as the founder of humanism.[4] Petrarch's greatest fame was as a poet, but he saw himself first and foremost as a man of learning.[5] Essentially nobody did more than he did to revive the ideals of Rome in a Christian community, which was the goal he had in mind. This ideal of revival had never disappeared in Europe, as we saw in the Carolingian Renaissance and the twelfth-century Renaissance (see 3.1), but Petrarch gave it a new, very

[1] Roberto Weiss, *The Renaissance Discovery of Classical Antiquity*, 2nd edition, Blackwell, 1988, pp. 16–29.

[2] The *Brevis adnotatio de duobus Pliniis* is published in C. Cipolla, 'Attorno a Giovanni Mansionario e a Guglielmo da Pastrengo', in *Miscellanea Ceriani*, Milano 1910, pp. 755–64. See also Elmer Truesdell Merrill, 'On the eight-book tradition of Pliny's *Letters* in Verona', *Classical Philology*, 5, 1910, pp. 175–88.

[3] For an introduction to Petrarch's philological work, see Giuseppe Billanovich, *Petrarca e il primo umanesimo*, Antenore, 1996. See also Teodolinda Barolini and H. Wayne Storey, *Petrarch and the Textual Origins of Interpretation*, Brill, 2007.

[4] Nicholas Mann, 'The Origins of Humanism', in Jill Kraye (ed.), *The Cambridge Companion to Renaissance Humanism*, Cambridge University Press, 1996, pp. 8ff. See also L. D. Reynolds and N. G. Wilson, *Scribes and Scholars: A Guide through the Transmission of Greek and Latin Literature*, 3rd edition, Clarendon Press Oxford, 1991, pp. 129ff.

[5] Cf. Victoria Kirkham and Armando Maggi, *Petrarch: A Critical Guide to the Complete Works*, University of Chicago Press, 2009.

personal form that was to become the model for later humanists. Petrarch travelled throughout Western Europe in search of hidden manuscripts in monasteries and cathedrals. He stayed for a considerable period in Avignon, which acted as a cultural exchange between north and south when it was the papal residence (1309–1377). The city became one of the leading intellectual centres in Europe, with monastery and cathedral libraries within easy reach. The papal commissions that stimulated commentaries on Roman classics, such as those of Livy and Seneca, were of immediate importance to philology. Petrarch arrived in Avignon at precisely the right time. He found a community with an interest in texts that had scarcely been read for centuries.

The large-scale reconstruction of literary, artistic, and historiographical Roman Antiquity started with Petrarch, initially still without a clear method, but gradually (during later generations) on the basis of progressively more precise principles. Petrarch's greatest philological fame is founded on his reconstruction of Livy's historical works, which was a widespread success, vulgarized in Italian and in French. Petrarch brought together different fragments from European libraries and was able to make one coherent whole of books 1–10 and 21–40 (books 41–5 were not discovered until the sixteenth century and the others are still missing without trace).[6] He corrected, annotated, and supplemented copies of Livy's work on a monumental scale. Petrarch was not the first person to try this, but he was by far the best in over a thousand years. He copied out some parts of Livy's text himself when visiting libraries. This activity instantly points to one of the most important features that identified humanistic philology: the humanists were manuscript hunters and were convinced that they made real discoveries in the world around them, which they saw as one of texts, classical and otherwise. However, their discoveries were often no more than separate or even inconsistent observations that needed considerable inventiveness before they could be fused into a coherent whole. This humanistic attitude produced a new model—the philologist's task was to bring historical Antiquity back to life by reconstructing its texts, which were waiting in medieval vaults to be unveiled.

Petrarch passed this attitude on to his admirer Giovanni Boccaccio (1313–1375).[7] Although Boccaccio remained in Petrarch's shadow as a humanist scholar, he achieved unprecedented popularity as the author of *The Decameron*. Before long Boccaccio felt ashamed of his frivolous early work and wanted above all to imitate Petrarch, but it would seem that he lacked the patience to produce sound text reconstructions. Nevertheless, we owe Boccaccio a great debt of gratitude—not least for his unbridled energy in stripping Italian monasteries of their most valuable manuscripts. After his visit to Montecassino in 1355, for instance, the Cassinese manuscripts of Apuleius and Tacitus suddenly turned up in Florence.[8] He also

[6] On Petrarch's reconstruction of Livy, see Giuseppe Billanovich, *Tradizione e fortuna di Livio tra medioevo e umanesimo*, Antenore, 1981. See also Myron Gilmore, 'The renaissance conception of the lessons of history', in William Werkmeister (ed.), *Facets of the Renaissance*, Ayer Publishing, 1971, pp. 77ff.

[7] Mann, 'The Origins of Humanism', pp. 15–16.

[8] Cornelia Coulter, 'Boccaccio and the Cassinese Manuscripts of the Laurentian Library', *Classical Philology*, 43(4), 1948, pp. 217–30.

contributed more than anyone else to popularizing humanism among the people of Florence, and his advocacy in support of studying Greek achieved more than is often assumed.

Humanism on solid ground: Coluccio Salutati and the *studia humanitatis*. Without Coluccio Salutati (1331–1406) humanism could have faded away. Salutati handed the flame lit by Petrarch and Boccaccio to fifteenth-century humanists such as Poggio Bracciolini and Leonardo Bruni. The humanist studies at the time of Salutati were often attacked as being 'pagan', for example by the influential cardinal Giovanni Dominici (1356–1419). In a series of writings Salutati defended the value of studying pagan literature to a proper understanding of the Bible and the church fathers: 'The *studia humanitatis* and the *studia divinitatis* are so interconnected that true and complete knowledge of the one cannot be had without the other.'[9] Salutati argued that the church fathers had aimed at 'Christianizing' the Graeco-Roman world, and in doing so they used pagan concepts and rhetoric (see also 3.7).

The *studia humanitatis* (or *studia humaniora*) was a term that Salutati had adopted from his great exemplar Cicero. According to Cicero, what differentiated people from animals was language, and therefore the study of languages should be at the heart of upbringing and education. Salutati's *studia humanitatis* consisted of grammar, rhetoric, poetics, history, and moral philosophy.[10] In this field of learning, the linguistic disciplines in the *artes liberales* were de facto released from the mere propaedeutic straitjacket that had served for centuries as preparatory training for theology. The *studia humanitatis* started to be taught at a number of fifteenth-century Italian universities, and students called its supporters *umanisti*, which is where the word humanist and the later nineteenth-century term humanism came from.[11]

Despite his influential programme, Coluccio Salutati still had one foot in the Middle Ages. His allegorical interpretations of the classics stand shoulder to shoulder with those of St Augustine and Thomas Aquinas. Nevertheless Coluccio was a very sound philologist—his discovery of regularities in textual corruptions attests to real insight. As a manuscript hunter, he brought Greek texts and Greek scholars to Florence, which became the centre of Greek studies in Europe—the fulfilment of Boccaccio's dream.

The century of discoveries: Poggio Bracciolini. The 'century of discoveries' began with Gianfrancesco Poggio Bracciolini (1380–1459).[12] Petrarch, Boccaccio, and Salutati had reordered the image of classical literature, but Poggio's thirst for discovery outstripped everything. He gave humanists a reputation as ruthless and

[9] Coluccio Salutati, *Epistolario* IV, edited by Francesco Novati, Fonti per la storia d'Italia pubblicate dall'Istituto Storico Italiano, 1891–1911, p. 216. See also Charles Stinger, *Humanism and the Church Fathers: Ambrogio Traversari (1386–1439) and Christian Antiquity in the Italian Renaissance*, State University of New York Press, 1977, p. 12.

[10] August Buck, 'Die "studia humanitatis" im italienischen Humanismus', in Wolfgang Reinhard (ed.), *Humanismus im Bildungswesen des 15. und 16. Jahrhunderts*, Acta Humaniora Weinheim, 1984, pp. 11–24.

[11] For the history of the word 'humanist', see P. O. Kristeller, 'Humanism and scholasticism in the Italian Renaissance', *Byzantion* 17, 1944–5, pp. 346–74.

[12] Reynolds and Wilson, *Scribes and Scholars: A Guide through the Transmission of Greek and Latin Literature*, p. 136.

even unprincipled manuscript hunters.[13] Poggio used his position as secretary to the pope to unveil the most diverse classical texts—from polemics to pornography. Poggio's expedition to Cluny Abbey in Burgundy in 1415 rewarded him with previously unseen speeches by Cicero thanks to a manuscript that had remained undisturbed there for over six centuries. Poggio's second foray was in 1416 and took him to the Abbey of St Gall. They resulted in a number of unprecedentedly influential discoveries—a complete Quintilian, Asconius's commentaries on Cicero's speeches, and the *Argonautica* of Valerius Flaccus. There was a new expedition to St Gall at the beginning of 1417, but this time with official papal sanction. After that there were even more extended trips in the summer of 1417 in France, England, and Germany. Many famous texts that until then had been completely unknown were discovered, from Silius Italicus and Manilius to a part of the renowned *Satyricon* by Petronius. Poggio's most impressive finding was the discovery of the last remaining manuscript of Lucretius's *De rerum natura* in which the Epicurean principles of atomism were explained.

By the end of Poggio's life the lion's share of the Latin literature we know today had been tracked down. Discoveries were still made after Poggio, but the best bits had already been found. However, the 'century of discoveries' was not yet at an end. The only achievement of Poggio and his contemporaries was to uncover manuscripts that had stayed untouched for centuries in the libraries of monasteries and cathedrals. The real research had yet to commence—the critical study, organization, and analysis of these texts. It is here that our story on the quest for principles and patterns in early modern philology begins.

Philology becomes an influential discipline: Valla's rebuttal of the *Donation*. The study of the classics fanned out into new areas under the fourth generation of Italian humanists (after Petrarch, Salutati, and Poggio). Thanks to the large quantity of texts it was possible to compare many versions of Latin, such as the Classical Latin of Cicero, the Silver Latin of Seneca, the Late Latin of Marcellinus and the Medieval Latin that was detested by the humanists. The introduction of printing in Europe resulted in unparalleled access to classical texts. Libraries put books into the public domain and an international forum was created for debate about text reconstruction and textual criticism.

Handbooks of Latin language and grammar were also written. The most influential was the *Elegantiae linguae Latinae* by Lorenzo Valla (1406–1457).[14] This work was both descriptive and prescriptive (see also 4.3). Valla extracted his rules from the classical texts he had studied, with which he then wanted to revive Classical Latin as the only correct version. There were no fewer than fifty-nine editions of the *Elegantiae* over the course of a few decades and it hastened the demise of Medieval Latin, which still existed as a spoken language. Classical Latin was much more

[13] William Shepherd, *Life of Poggio Bracciolini*, Longman, Rees, Orme, Brown, Green & Longman, 1837, chapter 3. See also Stephen Greenblatt, *The Swerve: How the World Became Modern*, W. W. Norton, 2011.

[14] *Elegantiae linguae Latinae*, edited by S. López Moreda, Universidad de Extremadura, 1999. See also David Marsh, 'Grammar, method, and polemic in Lorenzo Valla's *Elegantiae*', *Rinascimento*, 19, 1979, pp. 91–116.

difficult than Medieval Latin, so many authors sought refuge in local languages like Tuscan or Occitan. Together with the humanistic Latin of Valla, a new literary language that we now call Neo-Latin developed, which was used by the humanist elite.

Valla's outstanding knowledge of Classical Latin enabled him to make one of the most important discoveries in the history of humanities. In 1440 in his essay *De falso credita et ementita Constantini donatione* he showed that the document *Donatio Constantini* (the donation of Constantine) was a forgery.[15] In this document it was stated that the Roman emperor Constantine the Great (280–337) had given the Western Roman Empire to Pope Sylvester I out of gratitude for Constantine's miraculous recovery from leprosy. The document *Donatio Constantini* thus represented the most important justification for the church's worldly power. During the Middle Ages the document was widely regarded as authentic, although there had been doubts now and then. It was during the fifteenth century that humanists began to realize that the *Donatio* could not possibly be genuine. Nicholas of Cusa had already concluded that the document had to have been apocryphal in 1433,[16] but it was Valla who subjected the text to a strict critical method and identified it as a fake. Cusa's work was known to Valla and there are a few striking parallels between these two authors,[17] which suggests an early influence of northern humanism on the southern variant (according to tradition Cusa, like Erasmus, was educated at the Latin school in Deventer—see below). Nevertheless it was Valla who was the first to develop and apply strict philological principles to a text and in so doing to expose it with certainty.

Although Valla did not explicitly spell out his working methods anywhere, it is quite straightforward to distil them from his essay. It emerges that Valla employed three 'principles of consistency', namely (1) chronological consistency, (2) logical consistency, and (3) linguistic consistency.[18] I will briefly discuss each of the principles below.[19]

The principle of chronological consistency: Valla noted that the reported date of the document (as stated in the *Donatio Constantini*) was inconsistent with the content of the document itself because it referred to both the fourth consulate of

[15] *De falso credita et ementita Constantini donatione*, edited by W. Setz, Hermann Böhlaus Nachfolger, 1976. See also F. Zinkeisen, 'The Donation of Constantine as applied by the Roman Church', *English Historical Review* 9(36), 1894, pp. 625–32.

[16] Nicholas of Cusa, *De concordantia catholica*, 1433. See Paul Sigmund (ed.), *The Catholic Concordance*, Cambridge University Press, 1991, pp. 216–22.

[17] For a comparison between Cusanus's *De concordantia catholica* and Valla's *De falso credita*, see Riccardo Fubini, 'Humanism and truth: Valla writes against the Donation of Constantine', *Journal of the History of Ideas* 57, 1996, pp. 79–86.

[18] In rudimentary form, these principles can already be found in Petrarch's letter to Emperor Charles IV in 1361, in which Petrarch showed that a document claimed by Rudolf IV to be a privilege granted by Julius Caesar and Nero to Austria for being an independent state, was a forgery—see Francesco Petrarch, *Letters of Old Age/Rerum senilium libri*, translated by Aldo Bernardo, Saul Levin, and Reta Bernardo, Johns Hopkins University Press, 1992, volume 2, pp. 621–5.

[19] I have discussed these three principles in depth in 'Formalization in the humanities: from Valla to Scaliger', *First International Conference on the History of the Humanities*, University of Amsterdam, 23–5 October 2008.

Constantine (315) and the consulate of Gallicanus (317). This chronological or historical inconsistency was an indication of the corruption or forgery of the *Donatio*.

Principle of logical consistency: Valla employed an indirect, counterfactual method of reasoning to make a reasonable case that the donation did not happen. He contended that if Constantine had given the Western Roman Empire to Sylvester, this would certainly have been reported in the Acts of Sylvester. This was not the case, as Valla established. Therefore, it was extremely improbable that the donation took place.

Principle of linguistic consistency: Valla's most compelling evidence is linguistic. He observed that the document contained terms that could not have been known in Constantine's time, for example words related to the feudal system, which was not created until after the fall of the Western Roman Empire. Valla called the forger directly to account for the many linguistic inconsistencies, for example when he asserted, 'rather than *milites* you wrote *militia*, which we have adopted from the Jews, whose books neither Constantine nor his secretaries ever knew'.[20]

Using these three principles, Valla developed a method of textual criticism that complied with precise criteria. Nothing like this had ever been seen before and it gave the humanists an extremely powerful weapon. In the third century BCE Alexandrian philologists had also studied the history of classical texts in depth, but their principles were primarily aesthetic (Zenodotus) or linguistic (the analogists—see 2.3). All three criteria in Valla's method were historical through and through: the chronological, logical, and linguistic arguments were placed in a historical context. The principle of biblical coherence, which had been applied uncritically by many people less than a generation before, appeared to wither away when exposed to Valla. Of course, Valla would not have been able to develop his textual criticism without the passion of preceding humanists for collecting documents, but his new approach based on textual criticism appeared to come from such a different world that in 1440 the Middle Ages already seemed to be miles behind.

Valla's refutation was accepted virtually immediately by Pope Pius II, the humanist Enea Piccolomini (see 4.2), who recorded it in a tract (1453). Yet nothing changed in regard to the legitimation of the papal state. After Pius's death Valla's work was largely ignored. But when, during the Reformation, Martin Luther used Valla's repudiation as an argument for reforming the church, *De falso credita* was put on the list of prohibited works. Yet a few decades later the church historian and cardinal Cesare Baronio admitted in his *Annales Ecclesiastici* (1588–1607) that the *Donatio* was a forgery, and this slowly settled the matter. Valla's rebuttal was too well crafted to be contradicted.

We have thus seen that during the first century of humanism (roughly from 1350 to 1450) the attitude to text changed completely. Whereas Petrarch displayed an uncritical esteem for everything that smacked of Antiquity, by the time of Valla

[20] For an English translation of *De falso credita*, see e.g. Christopher Coleman, *Lorenzo Valla, Discourse on the Forgery of the Alleged Donation of Constantine*, Yale University Press, 1922.

this deference had changed into a critical and sceptical approach.[21] No text was sacred in his eyes. Sources could be corrupted or fabricated, and it was up to the humanist to separate the wheat from the chaff. The *Donatio Constantini* was only the beginning of Valla's cleansing. A couple of years later he demonstrated that the surviving correspondence between Seneca and St Paul was also a fake, and dated from the time of St Jerome at the beginning of the fifth century. However, when he applied his historical textual criticism to the Bible (the Vulgate translation by St Jerome—see 3.3), which he treated as a text just like all others and even dared to emend, he bungled it. It was not until 1505 that Erasmus succeeded in getting Valla's critique of the Vulgate printed, almost fifty years after Valla's death. It remains a miracle that this 'humanist with the hammer' as he has been called[22] (by way of analogy with Nietzsche, who was the 'philosopher with the hammer') was buried in the most prestigious of Rome's churches—the Papal Archbasilica of St John Lateran—where he has his own chapel.

Text reconstruction as genealogy: Poliziano's *eliminatio* or oldest source principle. Brilliant as it was, Valla's textual criticism added little to the problem of reconstructing a source from the surviving copies. Reconstructive skills were, it is true, widely employed among humanists, but theoretical underpinning was hard to find. Until about 1480 this practice was more likely to be a matter of subjective guesswork than sound emendations. And if an emendation had already been substantiated, reference was made primarily to the quantity of mutually consistent copies without investigating the *genealogical* relationship between these copies. Precise references to manuscripts were completely absent. This all changed quite radically with the work of Angelo Poliziano (1454–1494). In his *Miscellanea* in 1489 he described a method that enabled an accurate comparison and evaluation of sources.[23] Poliziano realized that a group of completely consistent sources could still be a problem. Assume that we have four sources—A, B, C, and D—which all agree on one point, and that B, C, and D are entirely dependent on A for their information.[24] Should B, C, and D nevertheless be included as extra evidence of the authenticity of A? According to Poliziano they should not: if *derived* sources were mutually consistent, they should be identified and eliminated.[25] Sources should be ranked genealogically so that their dependence in regard to an older source becomes clear. One anomalous manuscript can refute dozens of consistent manuscripts purely on the basis of its position in the genealogical ranking.

The almost self-evident preference for an older source existed long before Poliziano. Older manuscripts were more reliable than new ones because there were fewer transmission stages between the old source and the author. Poliziano's

[21] Cf. Lodi Nauta, 'Lorenzo Valla and Quattrocento scepticism', *Vivarium*, 44, 2006, pp. 375–95.

[22] Lodi Nauta, 'Valla, Lorenzo', in Anthony Grafton, Glenn Most, and Salvatore Settis (eds), *The Classical Tradition*, Harvard University Press, 2010, p. 959.

[23] Angelo Poliziano, *Miscellanea*, in *Opera omnia*, 3 volumes, edited by Ida Maïer, Bottega d'Erasmo, 1970–1.

[24] This example comes (with slight modification) from Anthony Grafton, *Defenders of the Text*, Harvard University Press, 1991, p. 56.

[25] Poliziano, *Miscellanea*, I.39.

method, however, consisted of more than establishing the oldest possible source. It also involved determining the complete genealogy of sources. Once this genealogy had been set down, a start could be made on eliminating derived sources. Poliziano's principle is therefore known as the *eliminatio codicum descriptorum*,[26] and we will refer to it as the '*eliminatio* principle' or also as the 'oldest source principle'. This principle was further developed in the nineteenth century by Karl Lachmann (see 5.2) to become the cornerstone of modern philology.

Poliziano illustrated his genealogical method using Cicero's *Epistolae ad familiares*, of which he owned a ninth-century manuscript from Vercelli and a fourteenth-century text that had originally been produced for Coluccio Salutati. Poliziano demonstrated that the fourteenth-century manuscript, in which a piece of text has been displaced by a binding error, was the source of all more recent documents because they had the same displacement but they did not have any binding error that could explain it. Furthermore he established that the fourteenth-century manuscript was a copy of the ninth-century document and that therefore all these later documents had no value for reconstructing the original text, and only the ninth-century document should be used as the starting point.[27]

In so doing Poliziano became the first to give a detailed and theoretically underpinned method for text reconstruction. According to his genealogical method, sources had to be *weighted* instead of counted. Nevertheless, Poliziano's method was not immediately greeted with open arms. Why should one old manuscript count for more than hundreds of more recent documents? A shift in philological practice did not take place until the first half of the sixteenth century, but from 1550 Poliziano's approach was applied virtually everywhere in Europe. Poliziano himself used his principle with exemplary precision and passion. His quest for the oldest surviving manuscript resulted in the most accurate and brilliant reconstructions of Terence, Virgil, Seneca, Propertius, and Flaccus.[28]

No matter how obvious Poliziano's principle may seem today, a historically-based method had not previously been advanced in European philology. While there were methodical principles in historiography—such as the most probable source, eyewitness, personal experience, and written source principles—none of them achieved the precision of the *eliminatio* procedure. Poliziano's principle perhaps most resembled the Arab transmission theory, the *isnad* method (see 3.2), in which the genealogical transmission chain was also reconstructed back to the source itself (usually a saying of the Prophet). Yet while the *isnad* set out to reconstruct the chain of verbal sources, Poliziano's principle was about reconstructing the chain of written sources. Nevertheless, the correspondence is remarkable, and it is not impossible that Poliziano was influenced by the many translated Arabic

[26] Paul Maas, *Textkritik*, 4th edition, Teubner, 1960, p. 2.

[27] Grafton, *Defenders of the Text*, pp. 59–60.

[28] Cynthia Pyle, 'Historical and philological method in Angelo Poliziano and method in science: practice and theory', in *Poliziano nel suo Tempo: Atti del VI Convegno Internazionale (Chianciano-Montepulciano 18–21 luglio 1994)*, Firenze, 1996, pp. 371–86.

works that were circulating in Christian Europe, especially in Italy, although there is no indication of this whatsoever.

Whatever the case may be, Poliziano's notion of manuscript affinity was more than just a theoretical motivation of a philological practice. His genealogical chain could be used to make predictions that could be tested out on new manuscript discoveries. A new find, an even older manuscript for instance, could support or refute earlier hypotheses about emendations, and even require a rejection or modification of the 'theory' (the methodological principle) itself. The latter was indeed the case when Erasmus discovered that a more recent but *untranslated* manuscript was more reliable than an older, but translated document—see below. This interaction between theory and empiricism, where a theory provided underpinning of empiricism and also generated verifiable predictions about that empiricism, which in turn had an impact on the theory, is one of the most fascinating aspects of early modern philology.[29] We will find this interaction in almost all other humanities, from historiography and linguistics to musicology. The humanistic tradition of the fifteenth century was to have a profound influence on the later natural sciences (see 4.4).

Collectors and printers: from Bessarion to Manuzio. The inventing of printing in Europe and the explosive growth of libraries were largely responsible for the flourishing of philology in Poliziano's time. The majority of the surviving Latin classics appeared in print between 1465 and 1475. The fall of the Byzantine Empire in 1453 brought a new stream of Greek manuscripts to Italy. Cardinal Bessarion of Trebizond (1403–1472) got his representatives to search for documents throughout the former empire.[30] He also made a significant contribution to the spread of teaching Greek in Europe, which had already been initiated by Greek humanists such as Chrysoloras and Plethon, but also by enthusiastic Italian humanists like Guarino and Filelfo.[31] The house in Rome that Bessarion lived in after his flight to Italy became an academy of humanist activities and, with its Renaissance frescos and picturesque setting, is still one of Rome's best kept secrets.

Nevertheless, the spread of printed texts in Greek lagged far behind that of documents in Latin. There were typographical difficulties, but limited knowledge of Greek was another cause. Sales of Latin translations were high enough to be profitable whereas this was not the case with Greek works. For example, in 1484 Marsilio Ficino's (1433–1499) famous Latin translation of Plato appeared in an exceptionally large edition of 1,025 copies and sold out within six years, whereas the Greek version of Plato had to wait until 1513.[32] Even so, the tone had been set. In addition to a

[29] Dirk van Miert, 'Philology and the roots of empiricism: textual criticism and the observation of the world', paper presented at *Erudition and Empiricism: The Intertwining of the Humanities and the Sciences in Early Modern Europe*, Panel for the Three Societies Conference 2008, Oxford, 4–6 July 2008.

[30] Nigel Wilson, *From Byzantium to Italy: Greek Studies in the Italian Renaissance*, Duckworth, 1992.

[31] Anthony Grafton and Lisa Jardine, *From Humanism to the Humanities*, Harvard University Press, 1987, pp. 9ff.

[32] Reynolds and Wilson, *Scribes and Scholars: A Guide through the Transmission of Greek and Latin Literature*, p. 155.

translation, a text also had to be available in the *original* language. In Venice Aldo Manuzio (1449–1515) specialized in printing Greek texts in collaboration with Greek scholars and compositors.[33] He perfected the quality of the typography and the incunables. Manuzio developed a standardized system of punctuation marks and has been credited with the semicolon and the invention of italics.

Erasmus and the original language principle. Desiderius Erasmus (1466–1536) was the first important Northern European philologist, but he was not the first major Northern humanist. The seed of Northern European humanism was planted in the Netherlands by the religious movement Brethren of the Common Life, which was created in the fourteenth century in Deventer around Geert Grote (1340–1384).[34] This movement, which was also known as the Modern Devotion, founded a fully-fledged Latin school in Deventer with illustrious pupils like Thomas à Kempis (1380–1472), Nicolas of Cusa (1401–1464), and Wessel Gansfort (1419–1489), the *lux mundi* of his time, who had acquired an excellent knowledge of Greek in Italy.[35] Among his students Gansfort could count humanists such as Rudolf Agricola and Johann Reuchlin, whose reformist ideas influenced Martin Luther. Under headmaster Alexander Hegius (*c.*1439–1498) the curriculum of the Latin school even included Greek, something that was unique in Northern Europe at the time. In the fifteenth century the school developed into the cradle of Northern European humanism and Erasmus was enrolled there in 1487. Although Erasmus rapidly became an all-round humanist, with brilliant theological, pedagogical, rhetorical, and polemic works like *De libero arbitrio, Adagia, Copia*, and *In Praise of Folly*, he also made important contributions to philology.[36]

Erasmus applied Poliziano's oldest source principle and Valla's textual criticism principles to his extremely influential edition of the New Testament. This work was based on research over many years into the oldest source of the *Greek* New Testament—which Erasmus brought back from all over Europe—after which he began to construct the best possible translation. Erasmus's translation resulted in a number of significant changes in the New Testament as compared to the existing Latin version.[37] In particular, one of these changes concerned leaving out a passage known as the *comma Johanneum*, which mentioned the Holy Trinity—one of the main doctrines of the church.[38] This led to such a major controversy that Erasmus promised he would put the words back if they could be found in another Greek

[33] Martin Lowry, *The World of Aldus Manutius: Business and Scholarship in Renaissance Venice*, Blackwell, 1979, pp. 111ff.

[34] Theodore van Zijl, *Gerard Groote: Ascetic and Reformer (1340–1384)*, Catholic University of America Press, 1963.

[35] Fokke Akkerman, Gerda Huisman, and Arjo Vanderjagt (eds), *Wessel Gansfort (1419–1489) and Northern Humanism*, Brill, 1993, p. 3.

[36] Erasmus's immensely versatile work is being collected by Manfred Hoffmann and James Tracy (eds), *Collected Works of Erasmus*, University of Toronto Press, 1974–2011 (currently 78 volumes).

[37] H. J. de Jonge, 'The character of Erasmus' translation of the New Testament', *Journal of Medieval and Renaissance Studies* 14, 1984, pp. 81–8.

[38] Bruce Metzger, *Der Text des Neuen Testaments: Einführung in die neutestamentliche Textkritik*, Kohlhammer, 1966, pp. 100–2.

manuscript of the New Testament. Such a manuscript promptly appeared, but Erasmus rightly condemned it as a forgery. That said, Erasmus put the *comma Johanneum* back in the third and later editions. Moreover, Erasmus's editorial approach was not always completely kosher. For example he sometimes amended the Greek text of the Bible—which he printed in parallel with the Latin translation—to accord with St Jerome's Vulgate, which was precisely what he claimed to be improving.

Apart from this editorial transgression, however, Erasmus adhered faithfully to Valla's standpoint that the Bible, like any other work, should be treated as a text like any other, together with Poliziano's principle that the oldest manuscript should be used. Yet it emerged that Erasmus, because he stood by the original (Greek) text of the New Testament, in this case departed from Poliziano's principle. After all, the oldest recoverable source could be a translation of the original text, in which case a source that might be not quite as old but was written in the original language would have to be preferred. This was indeed established by Erasmus with regard to a Greek manuscript of the New Testament that was less old than a Latin translation, but because it was in the original language it ultimately proved more reliable than the older Latin version. It should be pointed out here, however, that Erasmus thought his Greek manuscript was older than it actually was.[39] We shall designate Erasmus's method as the *original language principle* and contend that this method can trump the oldest source principle, but only if sources in different languages have survived. Despite considerable initial resistance, it was thanks to Erasmus that it was slowly but surely accepted that texts should be studied in their original language rather than in the form of a translation. Erasmus's approach meant that Poliziano's theory was not so much rebutted as transformed into a better one.

Although he was one of the greatest humanists, Erasmus cannot be considered as one of the leading philologists of the early modern age. In his role as a text critic, Erasmus is better known for his forewords than his actual reconstructions. In these forewords he called for the translation of the Bible into all languages and he defended the position that the Bible should be treated like any other text. Valla had advocated this last point half a century earlier, but it was not until Erasmus that textual criticism was unleashed in all its glory on the Bible. His knowledge of palaeography and genealogical text affinity might have fallen short here and there, but his argument for using philological rather than theological principles was hugely important to biblical textual criticism. Erasmus's unceasing urge to travel and his drive to found institutions were decisive in the permanent establishment of humanism in Northern Europe.[40] He injected new life into the dormant University of Cambridge and in Louvain he inspired the foundation of the famous *Collegium Trilingue*, where Latin, Greek, and Hebrew were taught on an equal footing.

[39] De Jonge, 'The character of Erasmus' translation of the New Testament'.

[40] There is a huge literature on Erasmus's life. I mention here only three quite different biographies: Johan Huizinga, *Erasmus*, Schwabe & Co., 1928. Léon-E. Halkin, *Erasmus: A Critical Biography*, Blackwell, 1993. James Tracy, *Erasmus of the Low Countries*, University of California Press, 1997.

The heyday of philology: the Scaligers. Humanistic philological attainments were whipped up to new heights in the sixteenth and seventeenth centuries by the two Scaligers. Julius Caesar Scaliger (1484–1558) was a generation younger than Erasmus and was actively involved in many fields in the humanities, for example linguistics, rhetoric, poetics, and philology.[41] His analysis of Latin grammar became one of the cornerstones of early modern linguistics (see 4.3), and his poetics laid the foundation for humanistic literary theory (see 4.7). Aside from his commentaries on the classics, J. C. Scaliger owed his fame primarily to his extraordinarily fierce attack on Erasmus, who had asserted in the *Ciceronianus* that Cicero should not be seen as the best and only model for correct Latin. However, it was J. C. Scaliger's tenth son, Joseph Justus Scaliger (1540–1609), who can be considered as one of the greatest philologists of the early modern age.[42] Thanks in part to his father, his knowledge of Latin was many times greater than that of his predecessors. This emerged early on when he was able to create something comprehensible from the surviving text *Astronomica* by Manilius (first century CE), which had become so corrupted that large parts were completely unintelligible. Scaliger turned Manilius into a readable author where others had failed (first edition 1579). Scaliger was the first to treat an author as an organic entity by considering the author's intellectual background in addition to the text itself. His fame spread rapidly and he was asked to succeed Justus Lipsius at Leiden University. Within a few years after its foundation in 1575, it had become the most prominent university in Europe, and as a Huguenot Scaliger appeared to be the ideal candidate for the vacancy. After initial hesitation and several rounds of negotiations, things got too hot for him in France and he accepted a position as a Leiden university professor without any teaching commitments. He was in charge of outstanding scholars, among them the prodigy Hugo de Groot, or Grotius (1583–1645). In a period when Europe was engulfed in horrific religious wars, during Scaliger's time Leiden was an oasis of peace and scholarship.

However, Scaliger's many reconstructions were nothing more than a limbering up for his higher objective, for which he had collected manuscripts in Syriac, Aramaic, Arabic, Hebrew, Ethiopic, and other languages. It was the reconstruction of the complete history of the ancient world on the basis of a precise scientific chronology (see 4.2 for more details), and to achieve it by using a single philological–historical principle—the oldest source principle, where Scaliger also considered the background of the author. It was the job of the philologist to reconstruct these oldest sources, in the process of which forgeries could be unmasked like the texts of Manetho and Berossus fabricated by Annius of Viterbo.[43] Once they had been restored as accurately as possible, authentic historical sources could be used to record a total history from the beginning of time to the present without falling victim to the caprice of the biblical coherence principle.

41 Vernon Hall, *Life of Julius Caesar Scaliger (1484–1558)*, American Philosophical Society, 1950.

42 For an in-depth biography of J. J. Scaliger and his works, see Anthony Grafton, *Joseph Scaliger: A Study in the History of Classical Scholarship*, 2 volumes, Oxford University Press, 1983, 1993.

43 Anthony Grafton, *Defenders of the Text*, Harvard University Press, 1991, pp. 76–103.

Scaliger's philological–historical approach was extremely successful, as we will discuss at length in the next section.

Unfortunately, Scaliger never described his complex, all-embracing methods in detail. It is clear that he employed Poliziano's oldest source principle, but his integration of many disciplines (from chronology to astronomy) and his contextual approach are not methodically explained anywhere in his work. We can perhaps describe Scaliger's philological integration as 'example based', but even this is too positive a representation of what happened. The absence of a methodical description is true of almost all early modern philology. Methodical principles usually have to be distilled from books and letters. A rare exception is Francesco Robortello, who in 1548 was one of the first to try to summarize the philological method in one manual, *De arte sive ratione corrigendi antiquorum libros disputatio* ('Lecture on the Art and Method of Correcting the Books of the Old Writers').[44] However, this work has so many unclear classifications and omissions (text reconstruction as genealogy is not discussed), that it has a very modest place in the history of philology.

The rejection of the Corpus Hermeticum: from Casaubon to Bentley. There are few new principles to be found in early modern philology after Scaliger. However, this does not mean that there were no philologists who made exceptional discoveries. After Scaliger, Isaac Casaubon (1559–1614) was seen as the most learned man of his time.[45] He was a Huguenot and a loyal friend of Scaliger, and in 1610 he fled to England after the murder of Henry IV of France. As well as many editions of the works of Greek and Roman writers, Casaubon was able to thoroughly date, and in doing so to reject, a number of texts in the *Corpus Hermeticum*. This Corpus, which was attributed to one Hermes Trismegistus, was one of the most studied works in the Renaissance and was alleged to have a biblical age.[46] It became widely known as a result of the Latin translation by Marsilio Ficino in 1471. Ficino observed agreements between the philosophy in the Corpus and Plato's dialogues, from which he believed that he could conclude that Hermes Trismegistus had lived before Plato and was even a contemporary of Moses. This generated enormous interest in the so-called Hermetic philosophy during the Renaissance. In 1614, however, Casaubon—using purely linguistic grounds—was able to date the Corpus's philosophical texts to between 200 and 300 CE.[47] It followed from this that the Corpus contained no philosophical originality and was largely eclectic. This exposé of Hermetic ideas on the basis of textual criticism so captured the imagination that philology attained an unprecedentedly high status.

[44] *De arte sive ratione corrigendi antiquorum libros disputatio*, edited and translated by G. Pompella, L. Loffredo, 1975. See also Antonio Carlini, *L'attività filologica di Francesco Robortello*, Accademia di Scienze, Lettere e Arti Udine, 1967.

[45] Mark Pattison, *Isaac Casaubon, 1559–1614*, 2nd edition, Oxford University Press, 1892.

[46] Brian Copenhaver, *Hermetica: The Greek Corpus Hermeticum and the Latin Asclepius in a New English Translation, with Notes and Introduction*, Cambrige University Press, 1992.

[47] Isaac Casaubon, *De rebus sacris et ecclesiasticis exercitationes XVI*, London, 1614, pp. 70–87. See also Anthony Grafton, 'Protestant versus prophet: Isaac Casaubon on Hermes Trismegistus', *Journal of the Warburg and Courtauld Institutes*, 46, 1983, pp. 78–93.

Many exponents of the 'New Sciences'—from Kepler to Newton—would study both nature and texts.

Scaliger had initiated a flourishing philological tradition in the Protestant Netherlands. Great numbers of editions of Latin authors were published by philologists in the Dutch Republic, for example Gronovius and father and son Heinsius (see also 4.7). Gerardus Vossius (1577–1649) was appointed as a historian at Leiden University and later as the first professor at the Athenaeum Illustre in Amsterdam, but he was also a poet and rhetorician. Vossius compiled a widely-used pedagogical Latin grammar and also overviews of literary history (see 4.7). We will meet him, and his son Isaac Vossius, in the next section on historiography. However, none of the later philologists attained the brilliant erudition of Scaliger. A possible exception is the Englishman Richard Bentley (1662–1742),[48] who was Master of Trinity College, Cambridge, and who produced exceptionally high-quality text reconstructions, based largely on the principles employed by Scaliger. For example, Bentley repeated Scaliger's reconstruction of Manilius, but did it better. Bentley's emendations were explained in such attractive Latin that his work became widely known. He found himself in the company of John Locke and Isaac Newton, and he had substantial influence on the latter's historical studies (see 4.2).

The old and the new humanism: from Gesner to Hemsterhuis. Philology in Germany was given a tremendous shot in the arm by the foundation of the *Preußische Akademie der Wissenschaften* in 1700.[49] Elector Frederick III of Brandenburg set up the academy on the advice of Gottfried Wilhelm Leibniz (1646–1716), who also became its first president. We will encounter Leibniz extensively in regard to his work on artificial languages and the impetus he gave to formal logic, but he was likewise important in establishing the foundations of the eighteenth-century philological tradition in Germany. The Prussian Academy of Sciences was the first institution to focus explicitly on both natural sciences and the humanities, each of which had its own department from 1710 onwards.

One of the most important representatives in the field of philology was Johann Matthias Gesner (1691–1761),[50] not so much because of new methodical principles or empirical patterns, but because of his different vision of the classics—sometimes referred to as the 'new humanism'—which was taught with great energy at the University of Göttingen and then elsewhere.[51] According to Gesner the old humanism had tried to create a verbal imitation of the classics and a continuation of the Latin literature of Antiquity. Around 1650 this goal was deemed to be unfeasible and was gradually abandoned. The new objective that Gesner had in mind was no longer a matter of imitating Greek and Latin style, but of mastering its substance. The classics served to form the mind and cultivate taste, and through this to create a new literature instead of reconstructing and imitating the old one.

[48] Kristine Haugen, *Richard Bentley: Poetry and Enlightenment*, Harvard University Press, 2011.

[49] Katrin Joos, *Gelehrsamkeit und Machtanspruch um 1700: die Gründung der Berliner Akademie der Wissenschaften im Spannungsfeld dynastischer, städtischer und wissenschaftlicher Interessen*, Böhlau, 2012.

[50] Friedrich Reinhold, *Johann Matthias Gesner: sein Leben und sein Werk*, Roth, 1991.

[51] John Sandys, *A History of Classical Scholarship*, volume 3, Hafner Publishing Co., 1964, p. 7.

Gesner's vision attracted a great deal of attention. It became a guiding principle for Winckelmann (see 4.5), Lessing (see 4.7), and many others. Johann Sebastian Bach even dedicated his *Canon a 2 perpetuus* (from *The Musical Offering*) to Gesner.

Late eighteenth-century German philologists, for example Ernesti and Heyne, continued on the course set by Gesner. Text reconstruction was no longer key; it was replaced by the study of Antiquity in general—from art to literature. Outside Germany, eighteenth-century philology continued to exist for a long time either as textual criticism or as the linguistic study of the classics. An example of the latter is the *analogy theory* of Tiberius Hemsterhuis (1685–1766). Hemsterhuis wanted to find underlying analogies to demonstrate that Greek word structure was regular in origin and had no anomalies.[52] Except for his immediate successors such as Valckenaer and Van Lennep, however, the *Schola Hemsterhusiana* initially remained without any appreciable influence. Hemsterhuis's ideas became, however, controversial after two crushing letters from Richard Bentley, whereupon Hemsterhuis refused to open any Greek book for months.[53] We will see, though, that Hemsterhuis's analogy theory and his ideas on underlying regularity were revived in the nineteenth century by Grimm and Bopp in their quest for the origin of Indo-European languages (see 5.3).

Evaluation of humanistic philology: rule-based or example-based? The philological methods of Valla, Poliziano, Erasmus, and Scaliger resulted in a new practice that could be used to study all humanities, as we will see in the following sections. But how did this new Queen of Learning relate to the earlier 'strict' disciplines—linguistics, logic, and musicology for instance—where we have found procedural and declarative systems of rules (see chapters 2 and 3)? Upon reflection, the strictness of textual criticism was not as rigorous as we might believe. For example, Valla's philological method had no procedural system of rules. Valla gave no unambiguous system that could reveal a forgery or derive a source reconstruction. Nor do the principles form a declarative system of rules such as we found in Aristoxenus's melodic grammar (see 2.4). Valla's principles of linguistic, historical, and logical consistency described above were not even defined by Valla or other humanists. They had to be distilled from their texts. The humanistic method was therefore more 'example-based' than 'rule-based'. By using examples of textual criticism and text reconstruction by earlier philologists, their way of working could be mastered by *imitating* this method and using it on new texts. This appears to correspond to what the majority of humanists aimed to do with the manuscripts they admired. By reproducing (*imitatio*) classical models (see also 2.8 on rhetoric), they wanted to copy the Latin of Cicero and Quintilian in terms of both grammar and style. Perhaps the only exception to this approach was Poliziano's genealogical theory of manuscript transmission. This was defined precisely enough to pass as

[52] Pieter Verburg, 'The School of Hemsterhuis', in Pieter Verburg, *Language and its Functions: A Historico-Critical Study of Views Concerning the Functions of Language from the Pre-Humanistic Philology of Orleans to the Rationalistic Philology of Bopp*, John Benjamins, 1998, pp. 445–52.

[53] Bentley's letters are contained in David Ruhnken, *Elogium Tiberii Hemsterhusii*, S. et J. Luchtmans, 1789.

'rule based' even though the theory was transmitted and probably learned on the basis of, once again, examples.

China and the Empirical School: Chen Di, Gu Yanwu, and the Jesuits. Europe was not the only region to host the creation of a critical philological tradition. Analytical study of classical texts also developed in the Chinese Ming era, and Chen Di (1541–1617) was one of the first exponents. He was a traveller and wrote the first description of the island of Taiwan and its original inhabitants.[54] His main claim to fame, though, was as a philologist. Chen Di was the first to succeed in demonstrating that Old Chinese has its own phonology (sound system) with pronunciation rules that differed fundamentally from contemporaneous Chinese. In so doing he refuted the then current practice of reading old poems in which characters were systematically changed in order to maintain the rhyme.[55] In his standard work, *Mao Shi gu yin kao* ('Investigation of the Ancient Rhymes of the Mao Shi') of 1606, Chen Di derived the pronunciation of characters in the ancient world on the basis of an in-depth analysis of the rhyme schemes in the *Shijing* ('Book of Songs') and a number of other classical works, including the *Yijing* ('Book of Changes'). In his foreword, which became famous, he wrote, 'In time, there is ancient and modern; in space, there is south and north. It is inevitable that characters undergo changes and sounds undergo shifts.' In other words, language change is of all time and space.

The late Ming and early Qing scholar Gu Yanwu[56] (1613–1682) perfected Chen Di's work and used it as the foundation for studying the Chinese classics. According to Gu Yanwu such a study had to be preceded by philological, linguistic, and historical research for which primary sources had to be used as much as possible. He is therefore considered to be the founder of the School of Textual Criticism, the *Kaoju Xue*, also known as the Empirical School, the *Kaozheng Xue*.[57] Gu Yanwu contended that a philologist and historian had to employ both internal and external proof in order to determine the authenticity of a text. A philologist ought to use an inductive method—a judgment should be given based on the highest possible probability by comparing as many sources as could be found. Knowledge had to be derived from facts and independent observations. In so doing one did not need to limit oneself to manuscripts—texts could also be compared with epigraphic remnants like stone and bronze inscriptions. Oral sources should also be taken into account. Gu Yanwu had an immense passion for finding new sources. As he stated himself, whenever he would discover an unknown or ignored source, he would be 'so overjoyed that he could not sleep'.[58]

[54] John Shepherd, *Statecraft and Political Economy on the Taiwan Frontier 1600–1800*, Stanford University Press, 1993, pp. 32ff.

[55] William Baxter, *A Handbook of Old Chinese Phonology*, Mouton de Gruyter, 1996, pp. 154–5.

[56] Although Gu Yanwu lived during the Qing dynasty of the Manchus, he never served the Manchus.

[57] Willard Peterson, 'The Life of Ku Yen-wu (1613–1682)', *Harvard Journal of Asiatic Studies*, 28, 1968, pp. 114–56.

[58] Ng and Wang, *Mirroring the Past: The Writing and Use of History in Imperial China*, p. 229.

The Empirical School of Textual Criticism had a number of important philological discoveries to its name. Yan Ruoqu (1636–1704), for instance, demonstrated on the basis of an analysis of Chinese characters and the associated pronunciation that twenty-six assumed chapters in the Book of Documents were forgeries from the fourth century BCE.[59] And Cui Dongbi, also known as Cui Shu (1740–1816), analysed the extent to which Confucius really was the compiler of the works attributed to him, such as the Spring and Autumn Annals and the Book of Songs. The result was doubt, and currently the attribution to Confucius is taken to be apocryphal. Old texts were also actively reconstructed by Jiang Yong (1681–1762) and others.

An explanation of the rather sudden creation of the philological school during the Late Chinese Empire has been sought in the prosperity in the Yangtze Delta at the end of the sixteenth century.[60] Merchants and intellectuals searched for antique works of art, early manuscripts, and rare editions. They would pay huge sums of money for a single manuscript. This paved the way for imitations and forgeries, which in turn stimulated the study of the authenticity of texts. Renewed interest also arose in reprinting the classical works, which found their way to Vietnam, Korea, and Japan in large numbers.

It could be that Chinese empirical philology received an additional impulse as a result of the arrival of Jesuits, who started to introduce Western scholarship to China in the sixteenth century. The Jesuits were obsessed by the dream of creating a Chinese-Christian civilization comparable to the Roman-Christian civilization. Although they were not particularly successful in propagating the Catholic faith in China, the Jesuits did contribute to a scholarly and cultural exchange between China and Europe. The Jesuit Matteo Ricci (1552–1610) was even convinced that Confucius's teaching contained the monotheistic concept of a Supreme Being.[61] In his view the Christian doctrine had already been embodied in the classics. Although only a very small percentage of Chinese intellectuals converted to Christianity, the influence of the Jesuits on scholarly work in China is considered by many to be substantial.

Whatever the exact contribution by the Jesuits may have been, the Chinese philological tradition corresponds surprisingly well with the European tradition—the rediscovery of classical works and their reconstruction, but also forgeries and their refutation are to be found in both regions. The principles of chronological, linguistic and logical consistency attributed to Valla were applied in both China and Europe. Although we find no formal theory of manuscript transmission in China, corresponding to Poliziano's genealogical theory, material sources such as inscriptions and epigraphs were used in both areas. As well as this striking

[59] Q. Edward Wang, 'Beyond East and West: antiquarianism, evidential learning, and global trends in historical study', *Journal of World History*, 19(4), 2008, pp. 489–519.

[60] Benjamin Elman, *From Philosophy to Philology: Social and Intellectual Aspects of Change in Late Imperial China*, Harvard University Press, 1984. See also Benjamin Elman, 'Philology and its Enemies: Changing Views of Late Imperial Chinese Classicism', in *Colloquium 'Images of Philology'*, 2006.

[61] See e.g. David Mungello, *Curious Land: Jesuit Accommodation and the Origins of Sinology*, University of Hawaii Press, 1989, p. 63.

similarity, there was also a substantial difference. In the Late Chinese Empire, textual criticism did not appear to have brought about radical social changes as it did in early modern Europe, where philological historiography led to a new secular world view (see 4.2). Yet the philological method could be just as 'destructive' for Confucius as for the Bible. However, we should remember that the Chinese Empirical School was operating in one and the same superstate, whereas European philologists were working in different countries. When the situation became too dangerous in France, Scaliger could flee to the Netherlands and Casaubon to England. Conversely, Grotius could take refuge in France when he was persecuted in the Dutch Republic for being a Remonstrant. But where could someone like Yan Ruoqu or Cui Shu go? When intellectual repression in China took on grotesque forms under the Manchu (Qing) in the eighteenth century, Chinese philologists and historians avoided all subjects that had even the slightest moral or political connotations. The Empirical School started working on annotations and commentaries, and this soon resulted in academic hairsplitting. Major social changes of the kind seen in early modern Europe were inconceivable in the Late Chinese Empire and had to wait until the twentieth century, when Confucius was desacralized.

4.2 HISTORIOGRAPHY: PHILOLOGY SPREADS AND THE SECULARIZATION OF THE WORLD VIEW

In Europe the philological refutation of historical sources contributed to one of the most drastic upheavals of the early modern age—the secularization of the world view. Philological historiography also spread in China, although the conventional court chronicles continued as well. Historiography in the Ottoman and Mughal Empires rapidly started blossoming but did not produce novel methodical principles. A new form of historiography arose in Africa in which personal experience and oral transmission were integrated.

A new historical pattern: Petrarch and (once again) Ibn Khaldun. As we did with early modern philology, we can let early modern historiography begin with Petrarch. Although only a small number of Petrarch's works can be considered historiographical (such as his *Africa* on the Second Punic War),[62] he represented a turning point in the discipline. This is because Petrarch, using his notion of the 'Dark Ages'—which he positioned between Classical Rome and his own era—pushed aside the time structure of salvation history.[63] Rather than a linear historical pattern from the Creation to the end of time, Petrarch believed he could discern a new pattern: Antiquity—Dark Ages—New Age, corresponding to the later well-known periodization Antiquity—Middle Ages—New Age. These days the Dark Ages refers primarily to the first centuries after the fall of the Western Roman Empire, but Petrarch used the term to cover all the centuries between the fall of Rome and his own time.[64] The metaphor of darkness was not new. Medieval historiographers associated darkness with the era of pagans, and light with the time of Jesus Christ. Petrarch linked darkness to the absence of classical culture and light to the revival of this culture. He was convinced that his own era was only the very beginning of this new period in world history. Thanks to humanism this new epoch was to become brilliant, and it would bring Antiquity back to life in all its beauty.

Although Petrarch put behind him the linear historical pattern that from Christ necessarily leads to the end of time, his new pattern was not a return to the cyclical structure that we have found in the work of the classical historiographers Herodotus, Thucydides, and Sima Qian. Contrary to the cyclical pattern of these classical historians, the New Age did not need to start at the beginning. Instead it continued to build on knowledge of the eras prior to the Dark Ages. In this regard Petrarch's vision looks more like Ibn Khaldun's theory of knowledge accumulation (see 3.2), although Ibn Khaldun's pattern is above all a thorough empirical study of successive civilizations that continued to build on one another, whereas Petrarch's model

[62] Thomas Bergin and Alice Wilson, *Petrarch's Africa*, Yale University Press, 1977.

[63] Petrarch, *Epistola familiaris* vi. 2, in V. Rossi (ed.), Petrarch, *Le familiari*, Sansoni, 1934, volume 2, pp. 55–60.

[64] Theodor Mommsen, 'Petrarch's Conception of the "Dark Ages"', *Speculum* 17(2), 1942, pp. 226–42. See also Bard Thompson, *Humanists and Reformers: A History of the Renaissance and Reformation*, Wm. B. Eerdmans Publishing Co., 1996, p. 13 and p. 207. Mommsen's interpretation of Petrarch's historical scheme has been criticized in Alison Brown (ed.), *Language and Images of Renaissance Italy*, Oxford University Press, 1995, pp. 66ff.

was a programmatic *revival* of a classical civilization with a 'gap' between the classical and Petrarch's epochs.

Ibn Khaldun and Petrarch were contemporaries, but it is unlikely that they knew or influenced each other. Petrarch concentrated wholly on the rebirth of Rome and his perspective was limited to Western Europe. Although he probably read Averroes in Latin translation, he did not know Arabic, unlike medieval predecessors such as Gerard of Cremona (see 3.1). Ibn Khaldun's universe, on the other hand, consisted of the huge Islamic world, which extended from West Africa to the borders of the Chinese Empire. He had little to say about Christian Europe and for him Antiquity was Greek not Roman.[65] It is moreover difficult to make a comparison between the two scholars. Ibn Khaldun was first and foremost a historian and historiographer whereas Petrarch was above all a philologist and poet. Nevertheless, their mutual ignorance is remarkable and typical of the separation of the paths. Here were two major scholars who were working at a modest geographical distance from one another—Tunisia and Italy—yet knew nothing about each other's existence. It was not until 1629 that a copy of Ibn Khaldun's work was to be found in Europe, which was brought from Istanbul to Leiden by the Arabist and mathematician Jacob Golius.[66]

Obviously we would have been equally justified in discussing Ibn Khaldun entirely in the early modern age. We only addressed his work in the preceding chapter on the Middle Ages because he was active at the end of the great Islamic civilization. The current section bears his name again in order to underline his modern ideas and the interesting parallel with Petrarch.

Application of Petrarch's pattern: Bruni as model historian. In 1410 Leonardo Bruni (*c.*1369–1444) was the first historian to use Petrarch's classification of Antiquity—Dark Ages—New Age in his *Historiae Florentini populi* ('History of the Florentine People').[67] Bruni, who had been a pupil of Coluccio Salutati, is considered by many to be the first modern historian. As chancellor of Florence he had direct access to the city's archives. His critical approach, in which—like Ibn Khaldun—he rejected myths about how the city was founded if they could not be supported by reliable sources, caused an abrupt break with previous European historiography. While Bruni's *History* was also based on earlier chronicles, he was critical of his sources and seems to have checked them by direct research in the archives. Finally, he presented his arguments in a classical model where the work was divided into 'books' alternating with dramatic rhetorical speeches. This was all written in singularly eloquent Latin, which resulted in Bruni's work being seen a humanistic model for historiography.

[65] Richard Fletcher, *The Cross and the Crescent: Christianity and Islam from Muhammad to the Reformation*, Viking, 2004, pp. 153ff.

[66] Jan Just Witkam, *Inventory of the Oriental Manuscripts of the Library of the University of Leiden*, volume 1, Ter Lugt Press, 2007, pp. 34–5.

[67] Leonardo Bruni, *History of the Florentine People*, 3 volumes, edited and translated by James Hankins, Harvard University Press, 2001–7.

Petrarch's pattern in Biondo: 'archaeology' and the material source principle. Flavio Biondo (1392–1463) used a similar tripartite periodization as Bruni did.[68] But Biondo is first of all considered to be one of the founders of 'archaeology' in the early modern age—although the term 'antiquarianism' would be more appropriate. The first 'archaeological' activities had already taken place during Antiquity. Thucydides, for instance, established that the graves on the island of Delos had to have belonged to the inhabitants of the Cyclades because they contained weapons that looked very like those of that community during his time.[69] Similarly Sima Qian wrote extensive descriptions of ancient ruins that he was able to date to the Xia Dynasty.[70] Pausanias, a Greek who lived in the second century CE, was perhaps the first to give a more or less systematic description of the monuments of classical Greece.[71] As regards Islamic civilization, Abd al-Latif al-Baghdadi (1162–1231) was called the first Egyptologist because of his treatise on Egyptian monuments.[72] Flavio Biondo fits seamlessly into this tradition of antiquarian descriptions. What Biondo added was the humanistic ideal of reviving Rome in the present.

Biondo's *De Roma instaurata* (1444) set out to establish a topography of Ancient Rome. First and foremost, though, it is a humanistic vision of the 'rebuilding' of Rome by studying its ruins. In Biondo's time these ruins were still largely undiscovered. A few years before, when Poggio climbed to the top of the Capitol, he saw little more than abandoned fields. The Forum Romanum was buried beneath meadows with cattle grazing. Together with other humanists like Leon Battista Alberti (1404–1472), Biondo started to explore and document the architecture, topography, and history of Rome. His second work, *De Roma triumphante* (1459), described pagan Rome as a model for contemporary political and military reforms. The book was particularly influential in the revival of Roman patriotism and respect for Ancient Rome. Biondo was moreover able to keep on the right side of the popes by presenting the papacy as a continuation of the Roman Empire. One of Biondo's greatest works was *Italia illustrata* (1474), in which he attempted to link Antiquity with the modern age on the basis of descriptions of each place (in a large part of Italy), the etymology of the place name and the changes over time.[73] Finally, in his work *Historiarum ab Inclinatione Romanorum Imperii*, which did not appear until 1483—years after his death—Biondo gave a historical overview of Europe using the three basic eras. This work definitively established the chronological

[68] Denys Hay, *Flavio Biondo and the Middle Ages*, Proceedings Annual Italian Lecture, Oxford University Press, 1959.

[69] Bruce Trigger, *A History of Archaeological Thought*, 2nd edition, Cambridge University Press, 1996, p. 46.

[70] Gungwu Wang, 'Loving the Ancient in China', in Isabel McBryde (ed.), *Who Owns the Past?*, Oxford University Press, 1985, pp. 175–95.

[71] Christian Habicht, *Pausanias' Guide to Ancient Greece*, University of California Press, 1985.

[72] Okasha El Daly, *Egyptology: The Missing Millennium. Ancient Egypt in Medieval Arabic Writings*, UCL Press, 2005.

[73] Flavio Biondo, *Italy Illuminated, volume 1: Books I-IV*, translated by Jeffrey White, Harvard University Press, 2005.

notion of three periods: Antiquity, Biondo's own time, and the time in between which came to be known as the Middle Ages.

As well as texts, Biondo also used material sources for his historiography, such as coins, epitaphs, and, of course, the monuments he studied. This led to new disciplines such as numismatics and epigraphy. First and foremost, however, Biondo's 'archaeology' was antiquarianism. It was part of the humanistic agenda that sought to restore the grandeur of Ancient Rome. There was as yet no question of equality between written and material sources, but Biondo was one of the first to realize the importance of non-textual sources, which leads us to attribute the *material source principle* to him.

Classical followers: Piccolomini and Sacchi. A variety of humanistic historians were followers of classical historians, for example Ennea Silvio Piccolomini (1405–1464), who later as Pope Pius II accepted Lorenzo Valla's refutation of the *Donatio Constantini* (see 4.1). In his autobiography, the *Commentarii*,[74] Piccolomini adopted the style of Caesar's autobiography *Commentarii de bello gallico*. It is still one of the very few autobiographies ever written by a pope.

The methods of Bartolomeo Platina, originally named Sacchi (1421–1481), primarily followed Plutarch (see 2.2). He wrote biographies, of the Visconti and the Sforza for instance, and a series of lives of the popes, *Vitae Pontificum* (1479), which he introduced with the life of Christ.[75] Yet Sacchi's most popular work was his *De honesta voluptate et valetudine* ('On Honourable Pleasure and Health'), a gastronomic treatise published in 1474 that described in detail the culinary art of one of the most renowned fifteenth-century chefs.[76] This chef, Maestro Martino of Como, was the source of many new recipes and discoveries ranging from *polpette* (meatballs) and *vermicelli* (dried pasta) to an improvement in cooking times. He specified the cooking time for a particular dish more accurately than before on the basis of a variable number of prayers, *Pater Noster* or *Miserere*. Moreover, Martino's theory of the right combinations of colour and taste created the foundation for scientific gastronomy. However, much it is also a product of the human mind, the study of cookery is outside the scope of our definition of the humanities.

In search of an explanation for history: Machiavelli and Guicciardini. Niccolò Machiavelli (1469–1527) and Francesco Guicciardini (1483–1540) both wrote in the vernacular. For this reason alone, they are often not considered to be 'humanists' (in Italy). Their choice of subjects and their methods have also played a role. Rather than reviving the classics, these historians gave an explanation for the major contemporaneous disasters that afflicted the Italian peninsula, such as the French incursions in 1494 and the *Sacco di Roma* (Sack of Rome) by the German army in

[74] Ennea Silvio Piccolomini, *I commentari*, translated by Giuseppe Bernetti, 2 volumes, Longanesi, 1981.

[75] Bartolomeo Platina, *Lives of the Popes*, translated by Anthony D'Elia, Harvard University Press, 2008.

[76] Bartolomeo Platina, *De honesta voluptate et valetudine*, translated by Mary Ella Milham, Tempe, 1998.

1527.[77] Yet Ancient Rome was still a role model: both wanted Italy to become a great and powerful empire again as it had been during the era of the Roman Republic.

Machiavelli's *Il Principe* ('The Prince'), written in 1513 and intended to curry favour with the Medici, was one of the most influential treatises from the Renaissance.[78] The work contains a direct application of historical knowledge to contemporary politics. Machiavelli believed he had observed a new pattern in history: it is permissible to use all means in order to survive as a ruler. The ruler's most important task is to consolidate and maintain his power. Machiavelli's analysis is still controversial, not because of his discovery of this pattern but because of the apparently uncritical acceptance of such utilitarian thinking. In Machiavelli's work the end justifies the means without an evaluation of the ethical acceptability of these means.

Even more important than *Il Principe* were the *Discorsi* (in full *Discorsi sopra la prima deca di Tito Livio* or 'Discourses on the First Decade of Titus Livy') dating from 1519 in which Machiavelli drew lessons from Rome's early history, in particular with regard to the structure of a republic.[79] He described a mixture of powers by means of a tripartite political structure and explained the superiority of a republic to a princedom. The book was full of empirical generalizations. Machiavelli appeared to be searching for the universal principles of the state and human nature—and believed that he had found them. To do this he employed the written source principle, and he made unrestrained use of the writings of earlier historians, primarily Livy but also Herodotus, Thucydides, Cicero, Sallustius, Tacitus, and even Dante.[80] Machiavelli thus utilized previous historiography directly as empirical reality, in which he found one generalization after another with great ease. For example, he felt he had found the following pattern in Roman history: 'The only things that counted in Rome were capacities, no matter under which roof they lived. . . . Incidentally age never played a role in Rome: people always chose the best candidate, whether he was young or old.' And also 'Provided the dictator was appointed in accordance with the constitutional rules and not on his own authority, his influence on the State was always salutary. No Roman dictator ever brought the Republic anything but benefit.'

It is clear that Machiavelli was on the track of something new. He wanted to find historical laws that he thought he could distil from the historiography about Rome. In so doing he did not shrink from depicting Rome as extremely bellicose. 'The

[77] Jacques Bos, 'Framing a new mode of historical experience: the Renaissance historiography of Machiavelli and Guicciardini', in Rens Bod, Jaap Maat, and Thijs Weststeijn (eds), *The Making of the Humanities, Volume I: Early Modern Europe*, Amsterdam University Press, 2010, pp. 351–65.

[78] Niccolò Machiavelli, *Il Principe*, edited by Giorgio Inglese, Einaudi, 2006. For an English translation, see e.g. Niccolò Machiavelli, *The Prince*, translated by Harvey Mansfield, University of Chicago Press, 1985.

[79] Niccolò Machiavelli, *Discorsi sopra la prima deca di Tito Livio*, edited by Francesco Bausi, 2 volumes, Salerno Editrice, 2001. For an English translation, see e.g. Niccolò Machiavelli, *Discourses*, translated by Leslie Walker, Penguin Classics, 1984.

[80] Mario Martelli, *Machiavelli e gli storici antichi, osservazioni su alcuni luoghi dei discorsi sopra la prima deca di Tito Livio*, Salerno Editrice, 1998.

Romans focused all their attention on waging war and always sought to gain advantage for themselves from both a financial and every other point of view. The wars of the Romans were almost always offensive and not defensive.' And above all, 'The Republic's whole strategy was never to pay for land, never to pay for peace.' Machiavelli initially presented his observations without value judgments, but his agenda soon became clear. He wanted to demonstrate that the best system, with the greatest chance of survival, was a republic with a mixture of powers between the consuls, senate, and tribune of the people (the last of these was instituted in Rome in 494 BCE after the people had revolted). In Machiavelli the idea of 'freedom' only exists in the absence of the subordination of one state to another state. He tried to remain as objective as possible in his generalizations, for example when he stated that 'it would appear that people either have to perpetrate aggression or submit to it'. Ultimately he arrived at 'a republic lasts longer and has more enduring prosperity than a monarchy.'

One could argue that Machiavelli's conclusions were drawn on a selective basis. Using the well-known Egyptian lists of kings, for instance, he could also have observed the opposite, namely that a monarchy has a longer existence than a republic. However, Machiavelli was interested in how one builds and stabilizes an empire, and confined himself to an analysis of Rome. His passion for finding underlying patterns was unrivalled. Rather than being a historiography, Machiavelli's *Discorsi* provides a description of human nature as found in Rome in the form of a declarative system of rules. He tried to specify the boundary conditions of everything that the Romans got up to during the republican era. We can find exceptions to the patterns he found, but Machiavelli's apotheosis only appeared in his interpretation of the generalizations he found, which can be summarized by the concepts of *necessità* (necessity) and *virtù* (virtue). By *necessità* Machiavelli meant human submission to uncontrollable forces and conditions. At the end of day the leader with *virtù* will also fail as a consequence of the necessary conditions and forces that are stronger than he is. These conditions and forces correspond to the generalizations that Machiavelli derived; they are operative everywhere and always. We do not do justice here to Machiavelli's other key texts, such as his *Istorie Fiorentine* ('History of Florence', 1525) and *Dell'arte della guerra* ('The Art of War', 1520). These works confirm Machiavelli's search for patterns, e.g. in his description of how the many oppositions in Florence were reflected in its institutional structure, in its political clans, and its social groups.[81]

Francesco Guicciardini, a friend and critic of Machiavelli, also created a historiography that could serve as a manual for a practical politician. Guicciardini was extremely sober-minded, and he exposed all pretence and imagined motives. He was much more sceptical than Machiavelli about the use of past examples for the present because of different circumstances between apparently similar historical

[81] Niccolò Machiavelli, *Le istorie fiorentine di Machiavelli: ridotte alla vera lezione su codici e stampe antiche*, edited by Pietro Fanfani, 2 volumes, Cenniniana, 1874. For an English translation, see e.g. Niccolò Machiavelli, *History of Florence and of the Affairs of Italy from the Earliest Times to the Death of Lorenzo The Magnificent*, with an introduction by Hugo Albert Rennert, Dunne, 1901.

situations.[82] In his *Ricordi* (1512–1530), Guicciardini argued that '[f]or any comparison to be valid it would be necessary to have a city with conditions like theirs, and to govern it according to their example.'[83] Guicciardini's criticism seems to anticipate later 'historicist' thinking on the impossibility on generalizing over unique historical events (see 5.1).[84] The tension between the unique and the general will fully come back in the modern era. Guicciardini's main work is *Storia d'Italia* (1537) which is considered to be a masterpiece of early modern historiography.[85] Using government sources, he analysed the Italian agony (between 1494 and 1532) with painful meticulousness.[86] Sometimes Guicciardini's details are almost perfectionistic, mainly because of his exhaustive use of documents as historical sources as well as the use of his own extensive experience in politics. In Guicciardini we find a form of realism that is absent from Machiavelli. Whereas Machiavelli was bulging with patriotism, Guicciardini could do nothing but elevate Italian corruption to an indispensable fundamental principle.

Followers elsewhere in Europe, and the historical criticism of Jean Bodin. Most European historians outside sixteenth century Italy were influenced by the Italian historians. They imitated the classics by way of an imitation of the Italian imitation of the classics. These second and third order followers were of great importance to local historiography, but they had little new to say that helps us in our quest for principles and patterns.[87]

The Frenchman Jean Bodin (1529–1596) was an exception to the humanistic followers. He was a lawyer, philosopher, and historian and, like Casaubon and Scaliger, he was a child of the French religious wars. In 1566 he wrote a sceptical historical criticism *Methodus ad facilem historiarum cognitionem* ('Method for the Easy Comprehension of History').[88] The work was a breath of fresh air after the dominant humanistic agenda. Unlike his predecessors, Bodin doubted whether it was possible to create a better today through historiography, which is what the humanists believed. His experiences during the religious wars seem to have fuelled his cynicism. He contended that a historical work could always be interpreted,

[82] Mark Phillips, *Francesco Guicciardini: The Historian's Craft*, University of Toronto Press, 1977, pp. 83ff.

[83] Francesco Guicciardini, *Maxims and Reflections of a Renaissance Statesman (Ricordi)*, translated by Mario Domandi, Harper & Row, 1965, p. 69.

[84] Woolf, *A Global History of History*, p. 189.

[85] Francesco Guicciardini, *Storia d'Italia*, edited by Silvana Seidel Menchi, 3 volumes, Einaudi, 1971. For an English translation, see Francesco Guicciardini, *The History of Italy*, translated by Sidney Alexander, Macmillan, 1969.

[86] On the effect of the Sack of Rome in Italian historiography, see Bos, 'Framing a new mode of historical experience: the Renaissance historiography of Machiavelli and Guicciardini', 2010.

[87] For example, William Camden (1551–1623) wrote the first topographical and historical overview of Britain and Ireland, the *Britannia* of 1586, but with regard to method the work does not go beyond Biondo. Camden positions himself entirely within the humanist programme when he declares 'to restore antiquity to Britaine, and Britaine to its antiquity'. The same counts for many other (e.g. Dutch and German) historiographers such as Cornelius Aurelius, Hadrianus Junius or Conrad Celtis.

[88] John Lackey Brown, *The Methodus ad facilem historiarum cognitionem of J. Bodin: A Critical Study*, Catholic University of America Press, 1939. See also Marie-Dominique Couzinet, *Histoire et méthode à la Renaissance: une lecture de la Methodus de Jean Bodin*, Coll. Philologie et Mercure, 1996.

used, and misused in different ways. By employing principles from philology, the historian's task was to write so that improper use was impossible. And the reader's task was to judge the examples they encountered when reading historical works, classifying them into 'base, honourable, useful and useless'.[89]

The *trattatisti*, *ars historica*, and the New World. Around mid-sixteenth century we find a new wave of Italian historians who are known as the *trattatisti*, such as Sperone Speroni, Francesco Robortello (see also 4.1), and Francesco Patrizi.[90] The trattatisti argued for a rhetorical historiography, an *ars historica*, analogous to the *ars rhetorica* of Aulus Gellius and Cicero (see 2.7).[91] Their examples were classical historians like Herodotus, Thucydides, Dionysius of Halicarnassus, and Tacitus (see 2.2), but also contemporary historians such as Machiavelli and Guicciardini. In addition, the trattatisti attempted to introduce critical and scholarly principles into their historical evaluations and writings. We have already seen how Jean Bodin advocated the use of critical historical principles based on philology (see above). Yet the influence of the *ars historica* went far beyond Italy and France. In the second half of the sixteenth century historians from England to Spain started to promote the new historical approach rooted in philological scholarship. And its critical shadow was felt far into the seventeenth century, for instance in the work of Gerardus Vossius (see below), John Selden (1584–1654), and Paolo Sarpi (1552–1623), who wrote one of the most admired 'impartial' histories of his time, the *Istoria del Concilio Tridentino* (1619), which was soon translated into English as the *History of the Council of Trent.*[92]

The trattatisti also proposed a new interpretation of the pattern of rise, peak, and decline. The Italian trattatisti, in particular, continued Machiavelli's work when they argued that there was a certain kind of order underlying human events. Knowledge of this order, in other words 'human motives' made it possible to explain history. To this end Francesco Patrizi (1529–1597) drew up as complete a list as possible of all human motives in his ten *Dialoghi della Historia.*[93] Ultimately he had to admit that his list was unusable for explaining specific human events, but he did find a cycle that was very reminiscent of the cycle described by Herodotus, Sima Qian, and Ibn Khaldun—a pattern of *destruction and rebirth.* Patrizi believed it was possible to find this pattern in Egyptian and Greek history and mythology. His cycle appeared to be a reversal of the existing well-known pattern of rise and fall, in that Patrizi began with decline (destruction), which was followed by rebirth and so on.

[89] Bodin, chapter 3. See also Peter Burke, *A Social History of Knowledge: From Gutenberg to Diderot*, Polity Press, p. 182.

[90] Girolamo Cotroneo, *I trattatisti dell'Ars Historica*, Giannini, 1971.

[91] Anthony Grafton, *What Was History? The Art of History in Early Modern Europe*, Cambridge University Press, 2007.

[92] David Wooton, *Paolo Sarpi: Between Renaissance and Enlightenment*, Cambridge University Press, 1983.

[93] Cesare Vasoli, 'I "Dialoghi della historia" di F. Patrizi: prime considerazioni', *Culture et societé en Italie du Moyen Age à la Renaissance. Hommage à A. Rochon*, Université de la Sorbonne nouvelle, 1985, pp. 329–52.

The discovery of the New World and unknown peoples added to the scepticism about the possibility of a linear history from the Creation to the present (see 3.2). Yet there were also historians like Bartolomé de Las Casas (1484–1566) who saw the discovery of new peoples as fulfilment of Christian salvation history. New souls could be converted, and the prediction that Christianity would become the universal religion would be borne out. In 1552 Las Casas was one of the first to write a complaint about cruel treatment of the Indians in his *Brevísima relación de la destrucción de las Indias* ('A Short Account of the Destruction of the Indies').[94] He actively condemned the mistreatment of the indigenous peoples of the Americas by trying to convince the Spanish court to adopt a more humane policy of colonization. Las Casas is often seen as the first advocate for universal Human Rights.

The end of Universal History: an about-turn in the world view, from Scaliger to Spinoza. The use of philological methods in historiography had the most dramatic impact on the Universal History and historical chronology. Thanks to humanistic textual criticism, doubtful sources, such as Annius of Viterbo's forgery of Berossus (see 4.1), could be convincingly refuted. Ultimately, chronologists thought that they could solve the problem of the exact dating of Creation, of the life of Adam and Abraham as well as of all ancient peoples. Joseph Scaliger (see 4.1) was in the best possible position to do this.[95] He had mastered more languages than anyone else, his zest for work was phenomenal and his knowledge of Syriac, Aramaic, Ethiopic, Arabic, Hebrew, and Greek sources was unparalleled. He had written the best editions of the classics, so if anyone was able to distil the correct chronology from old sources, it was Scaliger. In the philological–humanistic spirit of *precision*, *consistency*, and *documentation*, in his *De emendatione temporum* (1583) he defined a new timeframe for classical Antiquity, and using calendar comparisons he was able to place Graeco-Roman history in the context of Babylonian, Egyptian, Persian, and Jewish history.[96] To this end Scaliger developed a new unit of time—the Julian Period—so that he could unify lunar and solar calendars. This Julian Period and the Julian day derived from it are still used as a reference point in time in astronomy. As Bede had done centuries before him (see 3.2), Scaliger demonstrated how important astronomy was to historical dating and correspondingly how important historical preoccupations could be to astronomy.

Perhaps Scaliger's most important innovation was that he was able to reconcile the divergent chronological systems of different peoples. To do this he used one historical principle, which was based on the oldest source principle advanced by Poliziano. *The source that is as close as possible in time to the date of the event described is the most reliable. If only derived sources exist, the original must be reconstructed as well as possible.*[97] This principle was not without problems, but it gave Scaliger a

[94] Bartolomé de Las Casas, *An Account, Much Abbreviated, of the Destruction of the Indies With Related Texts*, edited by Franklin Knight and translated by Andrew Hurley, Hackett Pub Co., 2003.

[95] Grafton, *Joseph Scaliger: A Study in the History of Classical Scholarship*, 1983, 1993.

[96] Philipp Nothaft, *Dating the Passion: The Life of Jesus and the Emergence of Scientific Chronology (200–1600)*, Brill, 2011, pp. 2–9.

[97] Anthony Grafton, 'Joseph Scaliger and historical chronology: the rise and fall of a discipline', *History and Theory*, 14(2), 1975, pp. 156–85.

tool for separating the wheat from the chaff. He applied the principle in an exemplary fashion during the remaining twenty-four years of his life, primarily in his *Thesaurus temporum* of 1606.[98] In this work he collected, restored, and ordered virtually every surviving historical fragment. Scaliger reconstructed a few extremely important historical texts, among them Manetho's history of the earliest Egyptian dynasties (see 2.2). Using the information from these sources, particularly about the duration of the different dynasties, Scaliger was able to date the beginning of the first Egyptian dynasty to 5285 BCE. To his dismay this date was nearly 1,300 years before the generally accepted day of Creation, which according to biblical chronology had to be around 4000 BCE. However, Scaliger did not draw the ultimate conclusion from his discovery, which would have meant that either the Bible or his own method was incorrect. In order to 'save the phenomena', Scaliger introduced a new concept of time—the *tempus prolepticon*—a time before time.[99] He placed every event that occurred before the Creation, such as the early Egyptian kings, in this proleptic time. Scaliger's solution may come across as artificial, but for a Protestant in around 1600 it was inconceivable to cast doubt on the Bible. Yet at the same time Scaliger was too consistent to give up on his philological method just like that. He preferred to introduce an imaginary era rather than abandon the oldest source principle.

Scaliger's chronological dating of the earliest Egyptian dynasties, which is currently thought to be largely correct, was barely accepted in his own time. Even his immediate followers Ubbo Emmius and Nicolaus Mulerius did not go along with Scaliger in his dating, simply because it flatly contradicted the Bible. The meticulous Gerardus Vossius (1577–1649) thought he could solve the problem by assuming that the Egyptian dynasties were not successive but simultaneous (and occurred in different places).[100] However, apart from an analogy with Babylonian history, he had no evidence whatsoever to support his position. Vossius's proposal almost appeared to be a return to the biblical coherence principle (see 3.2), according to which every historical fact had to be brought into line with Christian biblical teaching. Others, the theologian Jacob Revius for instance, argued that everyone was wrong, referring to the usual biblical fragments, whereas in 1654 in his *Annalium pars posterior* the Irish archbishop James Ussher again determined that the creation of everything had taken place on Sunday, 23 October 4004 BCE.[101]

Within a year, though, all hell broke loose. In 1655 the French theologian Isaac La Peyrère (1596–1676) asserted that people had lived before the creation of Adam and Eve—the so-called *pre-Adamites*.[102] For the time being his claims appeared to

[98] Joseph Justus Scaliger, *Thesaurus temporum*, Joannem Janssonium, 1658 [1606]. See also Nothaft, *Dating the Passion: The Life of Jesus and the Emergence of Scientific Chronology (200–1600)*, p. 274.

[99] Scaliger, *Thesaurus temporum*, p. 278.

[100] Grafton, 'Joseph Scaliger and historical chronology: the rise and fall of a discipline', 1975.

[101] Charles Richard Elrington, *The Whole Works of the Most Rev. James Ussher, D.D.*, volume 11, Hodges and Smith, 1847, p. 489.

[102] The pre-Adamite hypothesis had a long history before it was made famous by La Peyrère—see Richard Popkin, *Isaac La Peyrère (1596–1676): His Life, Work and Influence*, Brill, 1987, pp. 26ff.

have been created out of thin air. For example, La Peyrère contended that the Egyptian kings had ruled for millions of years. However, Isaac Vossius (1618–1689), the son of Gerardus, provided philological and historical underpinning. Rather than contending that people had lived before the Creation, he showed in *De vera aetate mundi* (1659) that the earth had to be at least 1,440 years older than had been hitherto assumed.[103] Isaac substantiated his argument with additional evidence from geographical studies and Chinese and Ethiopian texts. His work became widely known in scholarly European circles and it had a profound effect on radical biblical criticism in the second half of the seventeenth century.

Baruch Spinoza (1632–1677) elevated biblical criticism to a secular political philosophy. In his *Tractatus theologico-politicus*, which was published anonymously in 1670, he argued with a passion not previously displayed that books of the Bible were texts written by people that had grown historically and were transmitted in a specific time.[104] The biblical criticism that Spinoza employed for his purposes was based on the historically underpinned textual criticism of his illustrious philological predecessors.[105] In Spinoza's hands the destructive power of philology led to an eruption—no text was absolute. He took the results of philologists and historians and extrapolated them to the ultimate implication, and then demanded the right to the free use of reason, without interference from theologians, with democracy emerging as the preferred form of government. Spinoza was able to use the historical–philological paradigm for a new, secular world view, which represented the de facto beginning of the Enlightenment.[106]

In this context, Scaliger's discovery that world history conflicted with biblical chronology had far-reaching implications. What he had found stood at the beginning of a chain of sweeping changes that resulted in a world view in which the Bible was no longer taken to be a serious historical source and freedom of thought was necessary for the welfare of citizens and the state.[107] These were the ideas that the eighteenth-century 'rationalist' Enlightenment thinkers would use to create a craze. However, right at the beginning of this long chain were the humanists of the fifteenth and sixteenth centuries, of whom Valla was the first relevant scholar and Scaliger was the greatest—a *sceptical view* of everything, including the Bible, and the *precision*, the *consistency*, and the *empirical approach* together with sound *theoretical underpinning*. This method influenced all scholarly activities, not just biblical criticism.[108] Although we must not forget that many humanists had the

[103] Isaac Vossius, *Dissertatio de vera ætate mundi: qua ostenditur natale mundi tempus annis minimum 1440 vulgarem æram anticipare*, Adriaen Vlacq, 1659. See also Eric Jorink and Dirk van Miert (eds), *Isaac Vossius (1618–1689): Between Science and Scholarship*, Brill, 2012.

[104] Benedict de Spinoza, *Theologico-Political Treatise*, edited and translated by Jonathan Israel and Michael Silverthorne, Cambridge University Press, 2007.

[105] Piet Steenbakkers, 'Spinoza in the history of biblical scholarship', in Bod, Maat, and Weststeijn, *The Making of the Humanities, Volume I: Early Modern Europe*, 2010, pp. 313–26.

[106] Jonathan Israel, *Radical Enlightenment*, Oxford University Press, 2002.

[107] According to Eric Jorink, *Het Boeck der Natuere*, Primavera Pers, 2007, p. 429, there is an unbroken line running from Scaliger via Saumaise and Isaac Vossius to Spinoza.

[108] See also Cynthia Pyle, 'Text as body/body as text: humanists' approach to the world around them and the rise of science', *Intellectual News*, 8, 2000, pp. 7–14.

sole goal of letting Antiquity live again, it also led to a critical selection of surviving sources, when the most critical exponents, for example Valla, Poliziano, Erasmus, and Scaliger, cast doubt on every text.

Nevertheless, Scaliger's dating of the Egyptian dynasties was disputed until well into the eighteenth century. Even the icon of the Scientific Revolution, Isaac Newton, subordinated historical facts to biblical stories when he asserted that the pharaohs lived simultaneously. 'Ye kingdoms of Egypt were at first like those of Greece, many in number.'[109] However, these opinions slowly but surely died out,[110] after which the Enlightenment reached full maturity. It was also the era in which philology and the new natural sciences grew apart. The disciplines became so specialized that it was almost impossible for them to be combined by one person. We see this in Newton's philological–historical work and also in Scaliger's earlier 'humanistic mathematics'. Although Scaliger was outstanding in philological historiography and calendar comparisons, he did not really have much of a clue when it came to mathematics. For example, he maintained through thick and thin that he had solved the quadrature of the circle—but with philological methods rather than mathematical ones.[111]

As a chronologist, through, Scaliger was unequalled. He was the first person who did not subordinate his chronology to theology. A century and a half earlier, Lorenzo Valla had valued pagan manuscripts over ecclesiastical documents. Now Scaliger valued the historical record of the world de facto over the record offered by the Bible, with all that this implied.

The search for ever more precise methods: Mabillon and Le Clerc. The quest for a philological historiography based on precise methods became ever more urgent when exponents of the 'New Sciences' dismissed historiography for its bias and inaccuracy. In his *Discourse de la méthode* ('Discourse on Method', 1637), René Descartes argued that even the most reliable histories would always omit the lowest or least striking circumstances, so that what is left does not represent the truth.[112] As a response, some, like Pierre-Daniel Huet (see also 4.7), tried to apply the geometrical method to historiography but without much success. Others emphasized with more success the importance of using original sources, such as the French Benedictine scholar Jean Mabillon (1632–1707). In his *De re diplomatica* ('On the Science of Diplomatics') Mabillon dealt with the various methods of dating charters on the basis of handwriting, formulae, style, seals,

[109] Cited in Anthony Grafton, 'Joseph Scaliger and historical chronology: the rise and fall of a discipline', 1975, p. 180. See also Frank Edward Manual, *Isaac Newton, Historian*, Harvard University Press, 1963.

[110] Yet several historians still ventured an attempted salvation historiography that unified world history with the Bible—from Georgius Hornius, who incorporated the New World in his Universal History, to Walter Raleigh, who assimilated the insights of Copernicus and Galileo. But in the course of the eighteenth century, historians simply came to treat Christian history as a part of general history.

[111] Rienk Vermij, *The Calvinist Copernicans*, Edita KNAW, 2003, p. 20.

[112] René Descartes, *Discourse de la méthode*, Part I. See also Peter Burke, 'History, myth, and fiction: doubts and debates', in José Rabasa, Masayuki Sato, Edoardo Tortarolo, and Daniel Woolf (eds), *The Oxford History of Historical Writing, Volume 3: 1400–1800*, Oxford University Press, 2012, p. 265.

signatures, and other intrinsic and extrinsic factors.[113] Mabillon's work became the standard for source-critical method in historiography by which forgeries could be detected and authenticity established. While there was a clear continuity from Valla, Erasmus, and Scaliger to Mabillon (see 4.1), Mabillon used the technique in such a systematic way that it could be turned into a handbook of rules for historical source criticism, which was indeed carried out by the theologian Jean Le Clerc in his *Ars Critica* (1697).[114]

The last humanist or first historicist? Vico and the principle of *verum factum*. The Neapolitan historian-philosopher Giambattista Vico (1668–1744) is sometimes referred to as the last humanist historian, but he also had one foot in the modern era.[115] The concept of culture as a 'systematic whole' has been attributed to Vico.[116] He proposed a strongly anti-Cartesian vision of man, arguing in favour of a science of the human rather than a science of nature. In 1725 in *Scienza Nuova* he introduced a new scholarly discipline that was meant to shed light on the developments relating to all human existence (not to be confused with the *Nuove Scienze* of Galileo, which refer to the new natural sciences).[117] Vico's new science investigated the human past in the form of life cycles. These cycles all followed one and the same pattern, the *storia eterna ideale*, consisting of three periods—the age of the *gods*, the age of the *heroes*, and the age of *men*. The age of the gods comprised myths and oral narratives, followed by an age of bards and poets, who narrate the deeds of heroes, and finally the age of stories about ordinary men. Unlike previous adherents to the cyclical pattern, we find no preference in Vico's work for one of the three ages. Each period is equally valuable. In the third age there is an unavoidable decline from the moment that human achievement reaches an apex. A civilization then returns to barbarism or is conquered and absorbed by another culture.

According to Vico, the Cartesian assumption that nature would be more accessible than human affairs was fundamentally wrong. Vico argued that because God created nature, only He could really know it, whereas men could know about what *they* have created, to wit their own civilization. The *factum* (that which man creates) is the *verum* (the truth). In other words, people have a better understanding of what they themselves have made (*factum*) than what confronts them (nature created by God). Human history was inherently understandable because all people experience hope, fears, desire, etc., while they would always remain outsiders when

[113] Jean Mabillon, *On Diplomatics*, in Peter Gay, Gerald Cavanaugh, and Victor Wexler (eds), *Historians at Work, Volume 2: Valla to Gibbon*, Harper & Row, 1972, pp. 161ff.

[114] Maria Cristina Pitassi, *Entre croire et savoir: le problème de la méthode critique chez Jean Le Clerc*, Brill, 1987.

[115] Peter Burke, *Vico*, Oxford University Press, 1985. See also Joep Leerssen, 'The rise of philology: the comparative method, the historicist turn and the surreptitious influence of Giambattista Vico', in Rens Bod, Jaap Maat, and Thijs Weststeijn (eds), *The Making of the Humanities, Volume II: From Early Modern to Modern Disciplines*, Amsterdam University Press, 2012, pp. 23–36.

[116] Breisach, *Historiography: Ancient, Medieval and Modern*, pp. 211ff.

[117] Giambattista Vico, *New Science*, translated by David Marsh with an introduction by Anthony Grafton, 3rd edition, Penguin, 2000.

it came to nature.[118] In his anti-Cartesianism, Vico contended that the proper study of man was and had to be man. Here Vico laid the foundations of the humanities as an integrated area of learning that would be built on by Wilhelm Dilthey and others (see 5.1).

It would take almost a century before the implications of Vico's history would fully register, initially in the work of Johann Gottfried Herder (see below) and then among nineteenth-century historians. It was Jules Michelet (see 5.1) who fully recognized Vico's understanding of the creative power of ordinary people (and which he used for his own history of France). Vico also seems to have had a major influence on Comte and Marx: they saw a new interpretation method in Vico's ideas according to which history inevitably develops in line with set laws. However, this new historical approach had not yet been fully applied by Vico. His own work mainly concerned the cyclical pattern, the *storia eterna ideale*, from which in Vico's view there was no escape.

The spiral pattern of progress: Voltaire, Turgot, Condorcet. Whereas Vico saw history as a perpetual cyclical record, the French Enlightenment thinkers interpreted the course of history as a pattern of *progress*. Few historical interpretations have had as much influence as this pattern. According to the French statesman Turgot (1727–1781) the whole of humanity advances, in alternating periods of peace and turmoil, towards ever greater perfection.[119]

Progressive thinking became the maxim of the Enlightenment. But what was the evidence for this pattern? There was a growing conviction among eighteenth-century 'philosophes' that man was able to take control of his own fate. The world proved to have been ordered according to scientific principles that had been discovered by a rational mind free of superstition. There was of course the problem of other peoples and civilizations which displayed a less convincing pattern of progress, but a way could be found to get round that too. In his *Essai sur les mœurs et l'esprit des nations* ('Essay on the Manners and Spirit of Nations', 1754),[120] for example, Voltaire (1694–1778) comprehensively discussed the Chinese, Indian, Persian, and Islamic civilizations, but he stood by the pattern of progress by emphasizing the superiority of Europe in his own time. Voltaire sidestepped the problem of diversity by introducing the concept of 'unequal development'. European peoples did not differ intrinsically from other peoples—they were just ahead in the area of 'rationality'. The fact that other peoples were lagging behind was caused only by irrational forces like religion, laws, and customs. Sooner or later, however, all would share in the greatest merit of rationality: happiness.

[118] Gino Bedani, *Vico Revisited: Orthodoxy, Naturalism and Science in the Scienza Nuova*, Berg Publishers, 1989, pp. 196ff.

[119] Ronald Meek (ed.), *Turgot on Progress, Sociology and Economics: Three Major Texts Translated with an Introduction by the Editor*, Cambridge University Press, 1973, p. 41.

[120] François-Marie Arouet de Voltaire, *An Essay on Universal History: The Manners, and Spirit of Nations, from the Reign of Charlemaign to the Age of Lewis XIV*, translated into English, by Mr Nugent, 1759. See also Ben Ray Redman (ed.), *The Portable Voltaire*, Penguin Books, 1977, pp. 547ff.

This new order of progress was accompanied by total silence on the multiplicity of peoples and historical periods. The Christian Middle Ages were therefore seen as no more than a temporary interruption in the liberation of reason, which had been set in motion during Antiquity. Whereas Montesquieu (1689–1755) was still tolerant towards Christian learning,[121] Nicolas de Condorcet (1743–1794) thought the Middle Ages were a spectre of superstition and fanaticism.[122] On the other hand Condorcet saw many patterns of progress in his own time—improvements in communication, greater freedom, more equality, and enlightened governments. Condorcet admitted that the historical pace sometimes went hand in hand with delays, such as temporary medieval decline, but at the end of the day there was only a path forwards.

It was therefore not the case that Enlightenment thinkers denied the cyclical pattern of peak and decline, but they integrated it with, and made it subordinate to, the pattern of progress. Turgot even thought he had detected a new historical pattern that did justice to both the eternal cycle and the idea of linear progress. He described the course of history as a *spiral* path to the desired goal.[123] As far as Turgot was concerned progress in knowledge and rationality was not held back by the coming and going of civilizations. Arts and sciences were indeed delayed and then their development was accelerated, but the net effect of this was advancement. The spiral pattern joins the cyclical and the strict linear pattern (as in salvation history) as the third formal model that we have met for the description of the 'shape of the past'.

Yet this pattern was found in only one civilization—the European—and even here it had to be nuanced. The Enlightenment thinkers observed no pattern of progress whatsoever in other civilizations, yet it was deemed to apply. The concept of progress among the 'philosophes' was therefore more of an a priori imposed regularity than an a posteriori observed one. It has therefore been associated more with philosophy than historiography (as has been asserted from time to time, philosophy is 'wisdom beforehand' and history is 'wisdom afterwards'). Centuries before, however, Herodotus had been able to verify his cyclical pattern on the basis of *several* civilizations (see 2.2). This does not mean that Herodotus's generalization would not have been affected by his Greek world view, but it does mean that it can show more historical evidence.

The pattern of progress was also criticized by the Enlightenment thinkers themselves. Jean-Jacques Rousseau (1712–1778) expressed mainly moral objections whereas Montesquieu contended that the idea of universal progress was not supported by historical evidence.[124] Ultimately only the cyclical pattern enjoyed

[121] Charles de Secondat Baron de Montesquieu, *L'Esprit des lois*, Chatelain, 1748, Book 25. For an English translation, see e.g. Montesquieu, *The Spirit of the Laws*, edited by Anne Cohler, Basia Miller, and Harold Stone, Cambridge University Press, 1989.

[122] Nicolas de Condorcet, *Esquisse d'un tableau historique des progrès de l'esprit humain*, Agasse, 1795. For an English translation, see e.g. Nicolas de Condorcet, *Outlines of an Historical View of the Progress of the Human Mind*, reprint of first (anonymous) English translation, G. Langer, 2009.

[123] Meek, *Turgot on Progress, Sociology and Economics: Three Major Texts Translated with an Introduction by the Editor*, p. 41.

[124] For a brief overview, see Breisach, *Historiography: Ancient, Medieval and Modern*, pp. 209–10.

general recognition while the progress pattern had dubious status at best, even though many Enlightenment thinkers supported it.

Reactions to the French Enlightenment historians: Hume, Gibbon, Herder. In contrast to the rather theoretical Italian and French historiography, in eighteenth-century England and Scotland there was a preference for a more empirical way of writing history. In *The History of England* (1763), David Hume (1711–1776) reacted against speculative interpretations during the course of history.[125] In complete agreement with his radical-empirical philosophy, Hume saw an endless variation in the many separate events.

We see an emphasis on empiricism and less concern with theory in the work of Edward Gibbon (1737–1794) too. In his much-praised six-volume *Decline and Fall of the Roman Empire* Gibbon wondered what ultimately caused Rome to fall, and above all how it was possible for it to last so long.[126] He did not give his answer on the grounds of the simple formula of rise, peak, and decline but compiled extremely detailed step-by-step pieces of evidence. The result of Gibbon's labour had immediate success. He became popular in the world of men and women of learning, and his work also refuted the simplistic progress pattern. And yet Gibbon was also too much a child of his time not to believe in a form of progression. However, it was never described as a general pattern but only as a unique phenomenon.

As a grandiose finale to the European eighteenth century—and in effect the beginning of nineteenth (see chapter 5)—the work of Johann Gottfried Herder (1744–1803) was of great importance, above all his notion of *Volk* (people).[127] To Herder a *Volk* was the central unit of history, an organic entity that individuals could unite through common language, art, and literature.[128] According to Herder all peoples on earth had a unique, unrepeatable identity. This diversity was also designated *Volksgeist* (people's spirit). As an organic entity, a people could grow, develop, and die. Similar to Vico, Herder stressed that every period in the cycle of a people was equally valuable.[129] He rejected the French concept of progress (and the associated spiral pattern) because it denied the life process. This process consisted of growth as well as decline and death. Herder himself was a great devotee of popular literature and he collected enormous quantities of folk songs. He has often been seen as one of the spiritual fathers of nineteenth- and twentieth-century nationalism, but also as a source of inspiration for multiculturalism.[130]

[125] David Hume, *The History of England*, 6 volumes, Fili-Quarian Classics, 1985.

[126] Edward Gibbon, *Decline and Fall of the Roman Empire*, edited by Betty Radice, The Folio Society, 1998.

[127] Tilman Borsche (ed.), *Herder im Spiegel der Zeiten: Verwerfungen der Rezeptionsgeschichte und Chancen einer Relektüre*, Fink, 2006.

[128] Johan Gottfried Herder, *Ideen zur Philosophie der Geschichte der Menschheit*, Hartknoch, 1784–91, translated by T. Churchill, *Outlines of a Philosophy of the History of Man*, J. Johnston, 1800, IX, ii.

[129] We encounter this view (of equally valuable periods) also among earlier *Aufklärung* historians like Johann Ernesti (1707–1781) and Johann Semler (1725–1791).

[130] Joep Leerssen, *National Thought in Europe: A Cultural History*, Amsterdam University Press, 2006.

China: from the relativism of Li Zhi to the historical criticism of Zhang Xuecheng. As in previous periods, the dynastic chronicle represented the official historiographical activity during the Ming Dynasty (1368–1644). The way this court historiography was produced by the 'historiographical bureau' (*Shi guan*) was fully regulated and formalized (see 3.2). The Ming period also displayed an unprecedented number of private histories. Here one historian clearly came to the fore—Li Zhi (1527–1602). As well as many biographies grafted onto the dynastic histories, Li Zhi penned in his *Cangshu* ('A Book to be Concealed') pointed criticism of earlier historical works. According to him, these works were subject to the standards and values of the then compilers.[131] Consequently, judgments from the past had to be amended. He demonstrated how people who had been described as depraved in previous works could be quite simply and completely convincingly depicted in his own biographies as heroes. Based on his method we can characterize Li Zhi as a historical relativist. Li Zhi even dared to cast doubt on the position of Confucius, which led to his imprisonment and, in 1602, to his suicide. Li Zhi was not rehabilitated until the time of the People's Republic.

During the Manchu or Qing Dynasty (1644–1912) 5,478 historical works appeared. The transition from the Ming to the Qing era (roughly 1640–1700) was a period of great social unrest and many wars. At the same time it was one of the most fruitful phases in Chinese historiography. Many historians, including private ones, tried to understand who was on the right or wrong side, why the Ming fell and why the peasant rebels failed to establish a new dynasty after the Ming was beaten in 1644. Eleven hundred works were written in under sixty years.[132] Generally speaking, the methods used in them were imbued with the traditional *annalistic pattern* (year-month-day) of the official court chronicles. The biggest innovation, however, was to be found in the creation of the Chinese history of ideas.[133] In his *Mingru xue'an* ('Records of Ming Period Confucian Scholars'), Huang Zongxi (1610–1695) assembled the first overview of Chinese philosophy.[134] Huang discussed more than three hundred thinkers of the Ming era, whom he classified into nineteen different schools. Huang's overview contained biographies and quotations from characteristic passages together with his own comments. In another work, the *Mingyi daifang lu* ('Waiting for the Dawn: A Plan for the Prince'), Huang rejected the rise and fall of dynasties as a causal factor in history. He maintained that it was the condition of the masses that counted, which had been continuously repressed since the Qin dynasty (221–206 BCE). Huang thus

[131] Carrington Goodrich and Chaoying Fang (eds), *Dictionary of Ming Biography*, Columbia University Press, 1976, p. 811.

[132] On-cho Ng and Q. Edward Wang, *Mirroring the Past: The Writing and Use of History in Imperial China*, University of Hawai'i Press, 2005, pp. 223ff.

[133] Around the same time we also find in Europe the first histories of ideas, such as the *Historia critica philosophiae* by Johann Jakob Brucker (1742). See Wouter Hanegraaff, 'Philosophy's shadow: Jacob Brucker and the history of thought', in Bod, Maat, and Weststeijn (eds), *The Making of the Humanities, Volume II: From Early Modern to Modern Disciplines*, 2010, pp. 367–84.

[134] Thomas Wilson, *Genealogy of the Way: The Construction and Uses of the Confucian Tradition in Late Imperial China*, Stanford University Press, 1995.

found a new historical pattern that took Chinese history as a process of progressive deterioration.[135]

The philological practices of the Empirical School of Textual Criticism (*Kaozheng Xue*, see 4.1) also had an impact on historiography in China. Gu Yanwu (see 4.1) and to a lesser extent Wang Fuzhi[136] (1619–1692), as well as many others, applied the textual criticism method to the verification or refutation of historical sources. Yet these philological–historical activities did not have major social consequences in China as they did in Europe. On the one hand this was because the Taoist/Confucian body of thought was less vulnerable to refutation. Dates for the creation of the world cannot be deduced from it (as was attempted in Europe with regard to the Bible—see above). On the other hand the repressive literary inquisition under the Manchus made it quite simply impossible to use philology for any type of critical historiography. Only the distant past could be criticized, and then only in small measure. Still, the Empirical School continued to flourish under the Manchus, who were great defenders of the Confucian orthodoxy. The immense Imperial Encyclopaedic compilation projects, such as the *Siku quanshu* ('Complete Library of Four Treasuries') which consists of 2.3 million pages constructed by 3,826 copyists,[137] made use of the improved philological methods invented by the Empirical School. Yet, the major changes that philological methods brought about in early modern Europe were out of the question in Late Imperial China and had to wait until the intellectual revolution after 1900 when the classics in China were desecrated.

The verification of historical sources by the Empirical School by means of textual criticism was itself criticized by Zhang Xuecheng (1738–1801) in his *Wenshi tongyi* ('General Principles of Literary and Historical Criticism').[138] According to Zhang, the philological and linguistic analysis of historical sources was too static. On the contrary, the most important characteristic of history was change. Admittedly this change had its own features in every epoch, but in Zhang's view it was possible to discover the general principles of change. They would be universally applicable and could serve as a guideline for the present. Zhang also produced a historiographical-methodical scheme that was reminiscent of Sima Qian's historical genres (see 2.2): (1) chronological overviews, (2) biographies and treatises about institutions, and (3) maps, tables, graphs, and statistics to supplement the first two parts. The content of these three parts had to be strictly objective and based on the author's own investigation of the factual material found. There was barely any interest in Zhang's work during his own time. The philological source criticism of

[135] Ng and Wang, *Mirroring the Past: The Writing and Use of History in Imperial China*, p. 226.

[136] Wang Fuzhi mainly commented on existing historical works, from which he believed to have found a single universal principle: the first responsibility of a ruler was to guarantee the welfare of the people. See Ian McMorran, 'Wang Fu-chih and Neo-Confucian tradition', in W. Theodore de Bary (ed.), *The Unfolding of Neo-Confucianism*, Columbia University Press, 1975.

[137] It should be added, though, that almost a third of the manuscripts, i.e. those that were considered to be anti-Manchu, were destroyed—see R. Kent Guy, *The Emperor's Four Treasuries: Scholars and the State in the Late Ch'ien-lung Era*, Harvard University Press, 1987, p. 198.

[138] David Nivison, *The Life and Thought of Chang Hsue-ch'eng*, Stanford University Press, 1966.

the Empirical School drew by far the most attention. It was not until the twentieth century that Zhang's ideas about the importance of historical 'change' were taken up again when they were rediscovered by Chinese and Japanese scholars.

Africa: Kati and the integration of personal experience and oral tradition. Ibn Khaldun, who was a contemporary of Petrarch, was probably the first African historian in the period from roughly 1400 to 1800. However, his *Muqaddimah* (see 3.2) was not the only historical work from Africa in Arabic. Over the last fifty years a huge wealth of manuscripts from the Niger Valley has surfaced. They were written in Arabic but also in Tuareg (Tamasheq), Songhai, and Fulani.[139] The two greatest chronicles are the *Tarikh al-fattash* from Djenné, containing a history of the Songhai Empire,[140] and its continuation, the *Tarikh al-Sudan* from Timbuktu.[141] Djenné and Timbuktu were among the major intellectual centres in Africa. Djenné was well known because of its architectural opulence (including its world famous mud brick mosque), while Timbuktu had the biggest mosque schools and libraries south of the Sahara. In 1550 in the *Descrittione dell'Africa* the Moroccan-Andalusian traveller and merchant Leo Africanus described the fabulous wealth of Timbuktu.[142] The city maintained contacts with book markets in Morocco and Spain, and the work of Ibn Khaldun as well as many other writings were in stock.

The chronicle *Tarikh al-fattash* was written by three generations of the Kati family in Djenné, whose family library was recently rediscovered.[143] Mahmud Kati started the chronicle in around 1519 and it was completed by his grandson in 1591. It gave a summary history of the Songhai Empire up to the Moroccan conquest in 1591. Like the work of Polybius (see 2.2), the contemporaneous part of the chronicle is based on the personal experience principle, whereas the descriptions of earlier historical periods are based on centuries-old oral transmission which, as was normal in this region, was kept alive by the family itself. The *Tarikh al-Sudan* by Abderrahman al-Sadi in Timbuktu, also covered the later history of the Songhai Empire up to 1655 in a similar fashion.

This form of writing chronicles, based on a combination of personal experience and oral transmission, spread from Djenné and Timbuktu to the south and west. And there, during the course of the eighteenth century, the long-existing orally transmitted lists of kings, biographies, tribal genealogies, and local chronicles were written down in Arabic or one of the local languages. For instance the *Kitab al-Ghunja*, a chronicle of the Kingdom of Gonja in the northern Gold Coast

[139] Kevin Shillington (ed.), *Encyclopedia of African History*, Routledge, 2004, volume I, p. 640.

[140] *Ta'rikh al-fattash: The Timbuktu Chronicles 1493–1599*, English translation of the original work by Kati, edited by Christopher Wise, translated by Christopher Wise and Haba Abu Taleb, Africa World Press, 2011.

[141] John Hunwick, *Timbuktu and the Songhay Empire: Al-Sadi's Tarikh al-Sudan down to 1613 and other Contemporary Documents*, Brill, 1999.

[142] Leo Africanus (al-Hasan ibn Mohammed al-Fasi), *The History and Description of Africa: And of the Notable Things Therein Contained*, edited by Robert Brown, translated by John Pory, Cambridge University Press, 2010.

[143] *Ta'rikh al-fattash.*

(present-day Ghana), was one of the most compelling examples of this tradition.[144] The manuscript can currently be seen at the IFAN library in Dakar.

There was a similar tradition in the field of chronicle writing in East Africa. The language used was Arabic, or Swahili using Arabic script. Most chronicles were about the history of individual coastal towns and cities, such as the *Pate Chronicle*[145] and the *Kilwa Chronicle*.[146] Pate is in present-day Kenya and it was a major trading post in the eastern part of Africa. Until the end of the eighteenth century it competed with the Portuguese. Kilwa, or Quiloa as the Portuguese called it, is off the coast of Tanzania and it was one of the biggest ports on the Indian Ocean from the ninth to the nineteenth centuries. The city is currently renowned for its extended medieval ruins such as the Husuni Kubwa palace with a public market, housing complexes, mosques, city walls, and cemeteries. The Pate Chronicle has been orally transmitted (the original was lost during the British conquest, but has recently been reconstructed), while the Kilwa Chronicle has been saved as a manuscript. The latter was written between 1520 and 1530, and it was translated into Portuguese as early as 1552, possibly because of the city's strategic importance. John Milton's famous seventeenth-century poem *Paradise Lost* refers to Kilwa (i.e. Quiloa).[147] In terms of methodical principles, the chronicles of both Pate and Kilwa strongly resemble those from western regions of Africa. They combine oral transmission of centuries-old lists of kings and genealogies with the personal experiences of the authors. They demonstrate how rich and prestigious the oral culture—alongside the written one—was in Africa and still is (see also 5.1).

There was also a very long tradition of chronicle writing in Ethiopia, as we have seen in 3.2. From the accession of the Solomonic Dynasty to the throne in 1270, the royal chronicles formed an unbroken line, initially written in Ge'ez and later in Amharic. According to tradition the genealogy of this Solomonic Dynasty went back to Menelik I, the son of King Solomon and the Queen of Sheba, as described in the *Kebra Nagast* (see 3.2). We have not found any new principles or patterns here.

The chronicles discussed above are only the tip of the iceberg. Many manuscripts from Senegal, Ghana, Nigeria, Cameroon, and other regions have not yet been inventoried. The number of manuscripts, generally kept by families, in the area around Timbuktu alone is estimated at 700,000. A few thousand documents have been catalogued by the Ahmed Baba Institute[148] but they are currently under

[144] Joseph Ki-Zerbo (ed.), *General History of Africa I: Methodology and African Prehistory*, Heinemann, 1981, p. 130.

[145] Marina Tolmacheva, *The Pate Chronicle: Edited and Translated from MSS 177, 321, 344, and 358 of the Library of the University of Dar es Salaam*, Michigan State University Press, 1993.

[146] Arthur Strong, 'The History of Kilwa, edited from an Arabic MS', *Journal of the Royal Asiatic Society*, January, 1895, pp. 385–431.

[147] John Milton, *Paradise Lost*, Book XI, line 399.

[148] Sidi Amar Ould Ely and Julian Johansen (eds), *Handlist of Manuscripts in the Centre de Documentation et de Recherches Historiques Ahmed Baba, Timbuktu* (Handlist of Islamic Manuscripts Series V: African Collections Mali, volume 1, publication no. 14), Al-Furqan Islamic Heritage Foundation, 1995.

threat.[149] The vast majority of manuscripts are still waiting to be accessed, and the most urgent matter is to rescue and conserve the ones already known to exist.[150] Slowly but surely it is becoming clear that African written culture has been underestimated for centuries. One reason for this, and not the least important, can be ascribed to European colonial prejudices.[151] Although only a fraction of the historical works are currently accessible, it can be established from the African chronicles that have been studied that they virtually never employed the formal *isnad* method of source transmission (see 3.2), no matter how much the regions concerned were under Islamic influence. Instead we find in these chronicles a combination of the personal experience principle (for contemporaneous historiography) enriched by even more intensive use of the oral transmission principle (for earlier historiography). It is not yet entirely clear to what extent the oral sources were tested for reliability.

Historiography in the Ottoman Empire and the Arab world: from Peçevi to Cantemir. For four centuries the Arabic-speaking world was part of the Ottoman Empire, with Istanbul as its cultural centre. Although the Ottoman period is often seen as a time of scholarly decline, the humanities during this epoch are still largely unstudied. The documents that are accessible show the reverse of decline. There was a rapidly developing historiographical culture.

While the first 'Ottoman' historians were Byzantine historiographers (such as Laonicus Chalcondyles and Michael Critobulus—see 3.2), in the sixteenth century the Turks had European historical works translated in order to acquire greater insight into Europe. Ottoman historiography therefore had to be built up from virtually nothing. Take the case of Ibrahim Peçevi (1572–1650) who, when he wanted to write his history of the Ottoman Empire, had to resort to Hungarian 'enemy' historians for some periods, because not all battles had been documented in Turkish.[152] Others, like Mustafa Ali (1541–1600) continued the tradition of Islamic universal history that echoed that of Rashid al-Din (see 3.2).[153] Similar to the work of his mediaeval precursors, there is a strong tension in Mustafa Ali's writing between Islamic universalism and the Ottoman regionalism.

The historian Hüseyin Hezarfen (who died in 1691) wrote an almost complete world history based on European sources.[154] Other historians, in contrast, combined Arabic and European examples. One such was Ibn Lütfulla, also known as

[149] On 28 January 2013, Islamists set fire to the Ahmed Baba Institute. Although many of the manuscripts seem to have been saved from destruction, their exact status is currently unknown.

[150] Mary Minicka, 'Towards a conceptualization of the study of Africa's indigenous manuscript heritage and tradition', *Tydskrif vir Letterkunde*, 2008, 45(1), pp. 143–63.

[151] For a discussion on colonial prejudices, see Ki-Zerbo (ed.), *General History of Africa I: Methodology and African Prehistory*, pp. 1–6.

[152] Bernard Lewis, *Islam in History*, Alcove Press, 1973, pp. 107–8. See also Franz Babinger, *Die Geschichtsschreiber der Osmanen und ihre Werke*, Otto Harrassowitz, 1927, pp. 192ff.

[153] Cornell Fleischer, *Bureaucrat and Intellectual in the Ottoman Empire: The Historian Mustafa Âli (1541–1600)*, Princeton University Press, 1986.

[154] M. Fatih Çalisir, 'Decline of a "myth": perspective on the Ottoman "decline"', in *The History School*, IX, 2011, pp. 37–60. See also Babinger, *Die Geschichtsschreiber der Osmanen und ihre Werke*, pp. 228ff.

Müneccimbaşi (1631–1702), whose universal history is based on that of Rashid al-Din, enriched with contemporary European sources.[155] The first 'official' historian of the Ottoman empire was Mustafa Naima (1655–1716) who wrote one of the most cited chronicles for that period. He also devised a number of rules for writing history which are strikingly similar to those proposed by the *trattatisti*.[156]

The historian in the Ottoman Empire most interesting to our quest was Dimitrie Cantemir 1673–1723), Prince of Moldavia, who was also active in other fields, such as musicology (see 4.4). During his long exile in Istanbul from 1687 to 1710, Cantemir wrote an incredibly detailed Ottoman history using many local sources, the *Historia incrementorum atque decrementorum aulae othomaniae*.[157] Cantemir connects the Ottoman decline to that of the Indians, Assyrians, Persians, Egyptians, Greeks, and Romans. All these empires displayed the pattern of rise and fall. Already in 1734, Cantemir's *Historia* was translated into English, and translations in French and German followed soon after. Cantemir is probably the first Ottoman historian who had an impact on Western European historians. Gibbon expressed a general admiration for Cantemir in his *Decline and Fall of the Roman Empire* (see above), and Voltaire, Montesquieu, William Jones, and others took Cantemir's *Historia* as the standard reference of Turkish history.[158] Being shaped both by Ottoman traditions and Western European learning, Cantemir's work is particularly fascinating, the more because his work shows the impact of Ottoman scholarship before Western Europe became dominant.

India, Mughal Empire: the present as a variation on the past. Chronicles were also written in the Islamic Mughal Empire in the form of both Universal Histories and court chronicles. An official historical tradition was created under Akbar (1556–1605), the third Mughal emperor, which resembled the genealogical practice of the Safavids in Persia.[159] Abul Fazl (1551–1602), a minister to Akbar, brought together a large number of sources in his *Akbarnama* ('Book of Akbar'). He enriched his work with thoughts about the nature of history, and traced Akbar's pedigree back to Adam in fifty-two biographies.[160] The book was illustrated with Mughal miniatures by at least forty-nine different artists from Akbar's studio. The hagiographic character of the Mughal court chronicles is especially present in the *Padshahnama* ('Chronicle of the King of the World') on the fifth Mughal emperor Shah Jahan (died 1666), best known as the builder of the Taj Mahal. The chronicle

[155] Lewis, *Islam in History*, p. 109.

[156] Lewis Thomas, *A Study of Naima*, New York University Press, 1972.

[157] For an English translation, see Dimitrie Cantemir, *The History of the Growth and Decay of the Othman Empire*, translated into English by N. Tindal, 2 volumes, Gale ECCO Print Editions, 2010.

[158] For the impact of Cantemir on these and other historians, philosophers, and linguists, see Michiel Leezenberg, 'The Oriental origins of Orientalism: the case of Dimitrie Cantemir', in Bod, Maat, and Weststeijn (eds), *The Making of the Humanities, Volume II: From Early Modern to Modern Disciplines*, 2012, pp. 243–63.

[159] Harbans Mukhia, *Historians and Historiography during the Reign of Akbar*, Vikas Publishing House, 1976. See also John Dowson (ed.), *The History of India, as Told by Its Own Historians: The Muhammadan Period*, The Posthumous Papers of the Late Sir H. M. Elliot, Calcutta, 1959.

[160] *The Akbarnama of Abul Fazl*, translated by Henry Beveridge, 3 volumes, Asiatic Society 1902–39, reprinted 2010.

consists of almost 3,000 pages and gives accounts of almost every public activity of the emperor, all sumptuously illustrated by the finest imperial artists of the time.[161]

Clearly, Mughal historiography was primarily normative, but earlier texts were typically 'included' in descriptions of the present—usually in the form of summaries. In doing so, the present was somehow presented *as a variation on the past*. The past served as a primordial model that could be used to interpret the present. Unfortunately this historical method, and the cyclical variation arising from it, got little or no chance to crystallize out into a more descriptive method. After the death of Emperor Aurangzeb in 1707 the Mughal Empire collapsed, and Indian court chronicle writing collapsed along with it.

The patterns in early modern historiography. What was the status of historiography at the end of the early modern age? In Europe, Petrarch's classification of Antiquity—Middle Ages—New Age was generally embraced. The past as a source for empirical generalizations was also broadly accepted, from Machiavelli to Condorcet. At a global level, the linear pattern from Creation to the end of days seems to have given way in most places to the 'ancient' cyclical pattern. This had already been rediscovered by Ibn Khaldun, who was followed by some Mughal and Ottoman historians, and it achieved full glory once again under the *trattatisti*, Vico, Cantemir, and Herder. The pattern was also adopted by the 'philosophes' in their spiral pattern of progress, which showed some similarity to Ibn Khaldun's cyclic pattern with cultural augmentation. However, progressive thinking was criticized, first in France itself (Montesquieu) and then in England/Scotland (Hume) and Germany (Herder). There were also many 'unique' developments like Li Zhi's historical relativism and the combination of personal experience and oral transmission by African historians (Kati).

There is nevertheless a second remarkable parallel in early modern historiography. *The philological–historical approach led to a turning point in historical practice in both China and Europe.* A document could be completely negated even though it had been thought to be genuine for centuries. Now one single historian could make or break a source provided that the strict philological method was used. In Europe this method resulted in establishing that world history was older than biblical history, and this broke through the metapattern created previously, namely that the time structure of history corresponded with that of the Holy Book (see 3.2).

[161] Milo Beach, Ebba Koch, and Wheeler Thackston, *The King of the World: An Imperial Manuscript for the Royal Library Windsor Castle*, Thames & Hudson. 1997.

4.3 LINGUISTICS AND LOGIC: UNDER THE YOKE OF HUMANISM

Whereas humanism was extremely fertile for philology and historiography, it was less beneficial to the disciplines that had risen to great heights during the Christian Middle Ages, such as linguistics and logic. No humanist wanted to be associated with Modist linguistics (see 3.1) or scholastic logic (see 3.6). Instead, the ancients became the standard. The result was an initial impoverishment of the linguistic and logical body of thought. Rather than move forward with the creative insights of the medieval scholars, many early modern linguists and logicians had to rediscover their discipline. Linguistics was dominated by philology in China, too. Chinese logic, which had also flourished at one point, appeared to be virtually dead in the water.

Humanistic grammars for the vernacular: Alberti. In the fifteenth century, European linguistics largely coincided with philology. Lorenzo Valla's handbook of Latin grammar, the *Elegantiae* (see 4.1), was in fact a philological exercise, and the Latin grammars of Thomas Linacre[162] (1460–1524) and Julius Caesar Scaliger[163] (see 4.1) were on the humanistic agenda in order to bring Antiquity back to life. At the same time, of course humanism, with its *studia humanitatis*, wanted to emancipate the linguistic disciplines (the *trivium*) in the general university curriculum (the *artes liberales*). Yet the frequency with which medieval achievements were suppressed remains striking. Linacre and Scaliger, for example, peppered their grammars with Modist terms, although they said their work was based on Varro and Priscian. There were no references to Speculative Grammarians like Thomas of Erfurt or Roger Bacon (see 3.1).

Vernacular grammars also underwent drastic change under the influence of humanism. Usually the most important goal was to show the degree to which these grammars corresponded to Latin, the rules of which were deemed to be 'universal'. Leon Battista Alberti's *Grammatica della lingua toscana*, dating from 1437 to 1441, was one of the first examples of a humanistic study of everyday language.[164] There was a typical humanistic background to Alberti's grammar: a controversy had arisen between Poggio, Valla, Guarino, and Filelfo about the question of whether classical Latin was or was not spoken by everyone in the ancient world.[165] In other words, was classical Latin the language of the male elite in Antiquity, or was it also used by women, children, and slaves? In view of the fact

[162] Thomas Linacre, *De emendata structura latini sermonis*, 1524. See also Francis Maddison, Margaret Pelling, and Charles Webster (eds), *Essays on the Life and Work of Thomas Linacre c.1460–1524*, Oxford University Press, 1977.

[163] Julius Caesar Scaliger, *De causis linguae latinae*, 1540. On Scaliger's social view on language, see Robert Hall Jr, 'Linguistic theory in the Italian Renaissance', *Language*, 12(2), 1938, pp. 96–107.

[164] Giuseppe Patota (ed.), *L. B. Alberti, Grammatichetta e altri scritti sul volgare*, Salerno Editrice, 1996.

[165] Mirko Tavoni, 'The 15th-century controversy on the language spoken by the Ancient Romans: An inquiry into Italian Humanist concepts of "Latin", "grammar", and "vernacular"', *Historiographia Linguistica*, 9(3), 1982, pp. 237–64.

that the humanists themselves had not found it easy to master the rules of Latin, it seemed to them unlikely that this language could be spoken by everyone. The vast majority would have spoken a simpler vernacular with 'rules' of an arbitrary nature, whereas the higher language of classical Latin had a complex system of rules for which learning was required.

No text was able to resolve this issue so Alberti came at the problem from a different direction. He showed that the conjugation rules of verbs in *spoken* Tuscan were just as consistent as those of Latin, including essentially the same categories for temporal designations such as *praesens*, *imperfectum*, *perfectum*, *plusquamperfectum*, *futurum*. So although the basis of ordinary language appeared to be arbitrary, upon closer inspection it was found to be as precise as Classical Latin. There was no reason whatsoever to assume that knowledge of Latin would have been limited to a small group of erudite speakers. In the best tradition of the early Roman grammarians, Alberti used the procedural system of rules principle as his proof, which—again like his predecessors—he restricted to the inflection of words.

The word structure of Hebrew: Reuchlin. While it was still possible to base local everyday languages on Latin, this turned out to be impossible with Hebrew. Humanists aspired to the learning of Hebrew because it was one of the *tres linguae sacrae*.[166] However, many got bogged down in the unfathomable word structure of this language until, that is, the German Hebraist Johannes Reuchlin (1455–1522) was able to establish that in Hebrew the *stem* of a verb, from which all verb forms are derived, did not correspond to the *first person* indicative mood (as is the case in most European languages) but the *third person* singular in the past tense.[167] This was a surprising and unexpected insight. Reuchlin also ascertained that, unlike Latin, a word in Hebrew could inflect at the beginning (the prefix), at the end (the suffix), and in the middle of the word (the infix). The study of Hebrew resulted in a new formal articulation of the concept of a 'word'. A word was now seen as being built up from a sequence of basic units—the stem, prefix, infix, and suffix. These terms proved to be immensely useful for describing both the vernacular and the many non-European languages that Europeans came into contact with.

A new syntactic theory based on four operators: Sanctius. With hindsight, Francisco Sánchez de las Brozas (1523–1600), known as Franciscus Sanctius in Latin, who worked in Salamanca, emerges as one of the most influential linguists of the early modern period. In his work we see, for the first time in centuries, a profound interest in syntax, primarily from a logical perspective. In his *Minerva seu de causis linguae latinae* ('Minerva, or the Underlying Principles of the Latin Language') in 1587 he contended that 'the purpose of grammar is the sentence or syntax'.[168] In so doing Sanctius forged together the grammars of Linacre and Scaliger to form a new entity, while the influence of Sibawayh's descriptive grammar (see 3.1) also seemed to be present—the Arabic heritage was still tangible in sixteenth-century Spain. But Sanctius's grammar was more than the sum of the

[166] Vivien Law, *The History of Linguistics in Europe*, Cambridge University Press, 2003, p. 247.
[167] Johannes Reuchlin, *De rudimentis hebraicis*, Anshelm, 1506, pp. 582ff.
[168] Franciscus Sanctius, *Minerva seu de causis linguae latinae*, J. et A. Renaut, 1587, I ii.

parts of earlier grammars. In his *Minerva* Sanctius gave a new syntactic theory on the basis of four operators (including Sibawayh's *substitution*) with which all sentences can be constructed.[169] Sanctius's operators were:

(1) *Substitution.* Sanctius illustrated this operator using Psalm 8 from the Bible. In the sentence *lunam et stellas quae Tu fundasti* ('the moon and the stars, which you have set in place') the neuter plural *quae* ('which') needs a neuter plural head, such as *negotia* ('objects'), by which we can say that *quae* can substitute for *quae negotia* ('which objects').

(2) *Deletion.* This operator makes it possible to omit words or groups of words (ellipsis), as in the shortened sentence *Fire when ready*, which can be derived from *Fire when you are ready.* The deletion operator can also be used to deduce the sentence discussed previously in 2.1, *John ate an apple and Peter a pear*, from the sentence *John ate an apple and Peter ate a pear.*

(3) *Addition.* This can be used for declensions (of nouns and adjectives) and conjugations (of verbs), in so far as they concern prefixes and suffixes. Instead, the substitution operation is necessary where an infix is concerned (see Reuchlin above).

(4) *Permutation.* Using this operator, words or groups of words can be put in a different sequence, which occurs, for example, in compound words such as the Latin *mecum* that means *cum me* (with me), and where the composition results in a reversal of the order followed by a deletion of a space between words.

These four operators were subject to rules. Not everything could be replaced, omitted, added, or permutated. The rules varied from language to language, but the four operators that the rules employed were the same for all languages. Sanctius gave rules for a number of syntactic phenomena in Latin. Although he did not strive for completeness, the question nevertheless arises as to what extent Sanctius's system was adequate, in particular whether his system of four operators was rich enough or perhaps too rich. Could Sibawayh's system with only one operator (*substitution*) not bring about the same as Sanctius's more complex system of four operators?

This question was not only empirical but 'aesthetic' too and was about how many basic operations and rules one wanted to allow anyway and how many irregularities and constructions one wanted to put in what was normally called the 'lexicon'. Using Sanctius's four operators, one could get away with a smaller lexicon than with Sibawayh's single operator. For example, in Sanctius's system elliptical constructions could be derived from non-elliptical ones thanks to the *deletion* operator, as in Panini (see 2.1). In a Sibawayhian grammar, on the other hand, both elliptical and non-elliptical constructions had to be classified as 'examples'. In these examples words could be substituted for other words, but

[169] Manuel Breva-Claramonte, *Sanctius' Theory of Language: A Contribution to the History of Renaissance Linguistics*, John Benjamins, 1983.

there was no deletion operator that could eliminate a verb in its entirety—something that occurs with ellipsis. The question, therefore, was what the right balance was between rules and examples. Panini was virtually completely rule based, whereas Sibawayh was essentially all example based. Sanctius came between these two extremes. Based on considerations of simplicity, one could contend that all language constructions that cannot be derived from rules should be incorporated in an external lexicon, while all regular constructions should be justified using basic operators. However, this causes us problems in the case of so-called exception rules. Panini put these in his grammar, while exceptions could also be accommodated in a lexicon. But if several exceptions 'resemble' one another it might be better to justify them by means of a new rule. In principle, every linguistic phenomenon could be justified with a 'rule', including the most notorious exception. The balance between rules and examples was to become a recurring theme in linguistics (see 5.3).

Sanctius's theory catches on: Port Royal grammarians. Sanctius's *Minerva* remained unnoticed for a long time, until it was rediscovered in France in 1650 by Claude Lancelot (*c.*1615–1695), a *Port Royal* linguist. Port Royal (Port-Royal-des-Champs) was a thirteenth-century monastery near Paris, which became a centre of Jansenist philosophers, theologians, and linguists in the mid-seventeenth century.[170] The most important among them were Claude Lancelot and Antoine Arnauld (1612–1694). Sanctius's influence on the Port Royal group can be recognized primarily in Lancelot's pedagogical grammar, the *Nouvelle méthode pour facilement et en peu de temps comprendre la langue latine* (1644), which even Louis XIV used to learn Latin when he was six. The grammatical rules were written in rhyming form and everything was targeted at making learning as easy and pleasant as possible. When, after the second edition, Lancelot came across Sanctius's *Minerva*, he decided to completely rewrite his grammar, which as a result became four times as long. To his great surprise he had largely overlooked Latin syntax—which could also be said of most Roman but not medieval (Modist) grammarians. In the foreword of the longer, third edition of his *Nouvelle méthode* Lancelot wrote, 'Sanctius has dwelt particularly on the structure and connection of speech, by the Greeks called "syntax", which he explained in the clearest manner imaginable, reducing it to its first principles.'[171]

It was thanks to Lancelot that the Port Royal linguists became aware of the idea that 'first principles' were the basis of both the word forms and the sentence forms of language. It is therefore remarkable that Sanctius's influence decreased again dramatically in the later Port Royal work. The subsequently famous *Grammaire générale et raisonnée* by Lancelot and Arnauld written in 1660, for instance, was a conventional work (so too was *La logique* by Arnauld and Nicole in 1662—see

[170] Augustin Gazier, *Histoire générale du mouvement janséniste depuis ses origines jusqu'à nos jours*, volume 1, Librairie Ancienne Honoré Champion, 1924. See Seuren, *Western Linguistics: An Historical Introduction*, pp. 46ff.

[171] See Robin Lakoff, 'Review of Herbert Brekle (ed.), *Grammaire générale et rassonnée ou La grammaire du Port Royal*, Fromann, 1966', *Language*, 45(2), 1969, pp. 343–64, p. 356.

below).[172] This grammar concentrated mainly on spoken and written letters, word classes, and their inflection, and auxiliary verbs in everyday language. It was not until the last chapter that syntax was briefly addressed, where the 'rules of agreement', such as between subject and verb, were the primary topic. The reason for omitting Sanctius's syntactic method is not clear. It has been suggested that the absolutist policy of Louis XIV was to blame.[173] What happened was that in 1653, thanks to the Jesuits, the Jansenists were excommunicated for being a 'sect', and nobody wanted to attract attention with unconventional ideas about grammar, logic, or any other discipline. References to Ramism (see below) were already completely taboo. However, it was of no avail. In 1709 Louis XIV decreed the closure of the monastery at Port Royal, and in 1712 it was destroyed. Meanwhile, however, the Port Royal grammarians became well known inside and outside France. Thanks to Lancelot, Sanctius's approach—no matter how diluted—spread throughout Europe, and as a result linguists became au fait with the idea that both word construction and sentence construction could be covered by a system of rules. The name of Sanctius was soon forgotten.

The humanistic yoke (under which everything was subordinated to reviving the classics) appeared to have been largely cast off at Port Royal. Nevertheless, the medieval linguists were also swept under the carpet by the Port Royal grammarians. For example, Arnauld and Lancelot argued that they had discovered the central function of the linking verb as well as the analysis of a verb as an auxiliary verb plus an adjective, whereas both were Modist achievements (see 3.1). It is extremely unlikely that the Port Royal grammarians did not know the work of the Modists. Thomas of Erfurt's *Grammatica speculativa* was still being reprinted in Paris in 1605.[174] The humanistic prejudice that only Antiquity possessed useful knowledge was therefore still being propagated. References to the Middle Ages were beyond the pale and had to await the later eighteenth century.

Comparing languages: from Sassetti and De Laet to Jones. During the period after the Port Royal school there was a huge increase in descriptive grammars that continued building on the work of Lancelot and Arnauld. If there was one starting point that emerged from these works, it was the endeavour to stop modelling vernacular languages on Latin. As early as 1647 Claude de Vaugelas argued that his grammar was not based on Latin or logic,[175] and in 1709 the Jesuit Claude Buffier asserted that every language was autonomous and that an analysis of one language could not be applied to another.[176] In so doing he reinvigorated the discussion about universal grammar, on which he cast serious doubt because all languages were different from one another at a profound level.

[172] Jacques Rieux and Bernard Rollin, *General and Rational Grammar: The Port-Royal Grammar*, Mouton, 1975.

[173] Seuren, *Western Linguistics: An Historical Introduction*, p. 48.

[174] Esa Itkonen, *Universal History of Linguistics*, Benjamins, 1991, p. 264.

[175] Claude Favre de Vaugelas, *Remarques sur la langue française: utiles à ceux qui veulent bien parler et bien écrire*, Éditions Champ Libre, 1981.

[176] Claude Buffier, *Grammaire française sur un plan nouveau, avec un Traité de la prononciation des e et un Abrégé des règles de la poésie française*, Le Clerc, 1709.

The first comparative studies of languages had already taken place by then. They started with the Florentine merchant Filippo Sassetti, who while in India in 1585 had observed a number of similarities between Sanskrit and Italian, for example *deva/dio* for 'God', *sapta/sette* for 'seven' and *nava/nove* for 'nine'.[177] It was another couple of generations, though, before methodical principles for comparing languages were introduced by the Dutch polyglot Johannes de Laet (1581–1649). De Laet was one of the founders of the Dutch West India Company, but he spent his time on his ethnological and linguistic interests rather than running the company.[178] Earlier Hugo de Groot (Grotius) had argued that the American Indian languages had to display Hebrew influences because all people descended from Adam and Eve. In his *Notae ad dissertationem Hugonis Grotii* (1643) De Laet refuted Grotius's view and stated that there was no relationship whatsoever between American Indian languages and Hebrew, Greek, Latin, or any modern European language.[179] It was not unusual in De Laet's time to compare languages, but it was done only associatively and on the basis of a small stock of words. Like Grotius, De Laet had been trained under Joseph Scaliger (see 4.1), who in his posthumously published *Opuscula varia* (1610) had divided European languages into eleven language families—four major and seven minor.[180] In fact, his four major language families corresponded to the present-day Romance, Greek, Germanic, and Slavic groups. However, Scaliger's comparisons of languages were based on only a few words in each one, and then primarily the word for 'God'. He therefore talked about *Deus, Theos, Godt,* and *Boge* languages respectively. Now De Laet advocated stricter methodical principles in word comparisons that also had to be extended to other linguistic levels. To this end he developed a precise interpretation of the existing notion of *permutatio litterarum,* which permitted equating words from different languages with each other. De Laet formulated two principles:

(1) A quantitative principle that linguistic affinity could only be established if a sufficiently large number of words were involved in the comparison.

(2) A qualitative principle that any claim to linguistic affinity had to be supported not just at a phonological and lexical level but at a syntactic level too.

Using these two principles he was able to demolish Grotius's position that all languages were descended from Hebrew. The origin of American peoples now became a subject of discussion throughout Europe, and De Laet's 'proof' was eagerly quoted. For example, La Peyrère referred to De Laet's work when he defended his renowned contention that people must have lived before the creation of Adam and Eve (see 4.2). Isaac Vossius was likewise happy to use De Laet's insights to underpin his argument that the earth had to be older than could be

[177] Jean-Claude Muller, 'Early stages of language comparison from Sassetti to Sir William Jones (1786)', *Kratylos,* 31, 1986, pp. 1–31.

[178] Peter Burke, *A Social History of Knowledge: From Gutenberg to Diderot,* Polity Press, pp. 164–5.

[179] Johannes de Laet, *Notae ad dissertationem Hugonis Grotii De origine gentium americanarum, et observationes aliquot ad meliorem indaginem difficillimae illius quaestionis,* 1643. See also Eric Jorink, *Het Boeck der Natuere,* Primavera pers, 2007, pp. 307ff.

[180] Joseph Scaliger, *Opuscula varia antehac non edita,* Hieronymus Drovart, 1610.

deduced from the Hebrew Bible. So, besides Scaliger's philological–historical work, La Peyrère's speculative work and Vossius's geographical work, De Laet's linguistic work forms part of the complex chain of seventeenth-century transformations that resulted in a new secular world view where theologians no longer had the last word—see 4.2.

De Laet's method was followed up successfully in the eighteenth century. While Leibniz was still mainly propagating lexical comparisons between languages, Nicolas Beauzée (1717–1789) used both lexical and syntactic comparisons when he investigated linguistic affinity. He studied Semitic, Romance, Germanic, and Celtic languages in addition to Latin, Greek, Basque, Japanese, Chinese, and Peruvian (Quechua).[181] Beauzée was a philosopher-grammarian, and he was searching for the principles of universal grammar, which in his case, however, essentially boiled down to the principles of thinking. Alongside this descriptive work there was also a non-empirical discussion about the origin of languages. Giambattista Vico, for instance, argued on the basis of his cyclic world view that language was created from poetry (the era of myths and oral narratives—see 4.2). Further speculation was not long in coming, and in his prize-winning essay in 1772 Johann Gottfried Herder asserted that language was created from the primitive songs of the first people.[182] Jean-Jacques Rousseau believed that Oriental languages were the oldest known and that they belonged to the world of poetry, whence all languages stemmed.[183]

This early Romantic speculation about the origin of language, together with empirical comparative linguistic study, sowed the seed for *comparative philology*. It originated at the end of the eighteenth century with the famous lecture from the amateur linguist William Jones (1746–1794).[184] Jones assumed that Sanskrit, Latin, Greek, Gothic, Celtic, and Persian were rooted in a common source language, which is currently known as (proto-)Indo-European, but it was not convincingly explained until after the nineteenth-century discovery of the sound shift laws (see 5.3).

Sign language and the creation of phonetics: from Bonet to Holder. In the seventeenth century the empirical tradition also led to a study of sign language. In 1620 the Spanish priest Juan de Pablo Bonet published an early work on speech therapy through which deaf people could be taught in sign language.[185] Although sign language had existed as long as there had been deafness, Bonet's work was intended to improve communication with the deaf. In *The Deaf and Dumb Man's Tutor* (1680), the Scot George Dalgarno devised a finger alphabet in which letters

[181] Nicolas Beauzée, *Grammaire générale, ou Exposition raisonnée des éléments nécessaires pour servir à l'étude de toutes les langues*, J. Barbou, 1767.

[182] Johann Gottfried Herder, *Abhandlung über den Ursprung der Sprache*, 1772. For an English translation, see John Moran and Alexander Gode, *On the Origin of Language: Two Essays*, University of Chicago Press, 1966.

[183] Jean-Jacques Rousseau, *Essai sur l'origine des langues*, 1781. For an English translation, see Moran and Gode, *On the Origin of Language: Two Essays*.

[184] William Jones, *Discourses delivered before the Asiatic Society: and miscellaneous papers, on the religion, poetry, literature, etc., of the nations of India*, C. S. Arnold, 1824.

[185] Juan de Pablo Bonet, *Reduction de las letras y Arte para enseñar á ablar los Mudos*, Abarca de Angulo, 1620.

could be conveyed using fingers and palm. This enabled the expression of words or names for which there were not yet any signs.[186] Dalgarno also developed artificial languages, which we will discuss below. The interest in studies of deafness went so far that the Englishman William Holder (1616–1698) succeeded in getting a deaf-mute to speak 'plainly and distinctly, and with a good and graceful tone'.[187] However, this achievement was also claimed by the mathematician John Wallis, which resulted in a bitter dispute within the Royal Society. Despite this controversy, the result of the new phonetic activities was a flourishing *School of Phonetics*, where phonetics was studied from a linguistic point of view but also from physiological and articulatory perspectives (which had already been initiated in the eleventh century by Avicenna). Holder's *Elements of Speech* published in 1669 was the most important work, in which we find a detailed articulatory diagnosis about the distinction between voiced and unvoiced consonants.[188] However, Holder's articular phonetics remained unnoticed for over a century, and after that it was overshadowed by the discovery of Indian phonetics. Holder's activities struck a more sympathetic chord in music theory (see 4.4).

Formal versus natural logic: Valla, Ramus, and Port Royal. As they had done with linguistics, humanists ignored medieval logic—not because it was incorrect but because of the so-called barbaric Latin in which it was recorded. Who still wanted to read Buridan if he could read the recently discovered *De rerum natura* by Lucretius? The classical authority was no longer Aristotle but Cicero and Quintilian, who still used logic as an auxiliary discipline in their rhetoric manuals. Lorenzo Valla and Rudolf Agricola maintained that logic had no independent status and should stay within the bounds of rhetoric.

In his *De falso credita* Valla demonstrated substantial dialectical competence (see 4.1) but it was completely subordinate to rhetorical argumentation. In his *Repastinatio dialectice et philosophie* (*c.*1431) Valla maintained that all that mattered was whether an argument 'worked', in other words whether it convinced the listeners.[189] He criticized Aristotle and the scholastic logicians for the artificiality of their syllogisms. The syllogistic pattern of reasoning did not correspond with the *natural* method of argumentation. What was the use of a syllogism that deduced that all Greeks are mortal if one had already postulated that all men were mortal and that all Greeks were men (see 2.6)? According to Valla such reasoning was a pedantic exercise and was not worthy of the qualification *ars*. This criticism showed how strongly Valla argued on the basis of a rhetorical framework. If a logical rule was not useful in a line of reasoning, it was superfluous. Valla was also the first to analyse a number of paradoxes and dilemmas linguistically, and he stated that common sense solutions existed for them. Valla's criticism revealed

[186] David Cram and Jaap Maat (eds), *George Dalgarno on Universal Language: The Art of Signs (1661), The Deaf and Dumb Man's Tutor (1680), and the Unpublished Papers*, Oxford University Press, 2001.

[187] Jonathan Rée, *I See a Voice*, Flamingo, 1999, pp. 107–8.

[188] William Holder, *Elements of Speech*, reprinted by AMS Press, 1975.

[189] Lorenzo Valla, *Repastinatio dialectice et philosophie*, edited by Gianni Zippel, 2 volumes, Antenore, 1982.

that he was *anti-formalist*—he cast doubt on the utility of *formal* argumentation and advocated *natural* lines of reasoning.[190] Valla's own work—such as the refutation of the *Donatio*—is punctuated by principles relating to logical consistency but they remained implicit and subservient to rhetoric. Although Valla's contribution to logic was modest, his invigorating criticism made him one of the most interesting sceptics—and in any event one of the greatest philologists—in the history of the humanities.

There was also a fiercely anti-Aristotle attitude in the work of Peter Ramus (1515–1572). In 1543 in his *Aristotelicae animadversiones* Ramus appeared to have been motivated by pedagogical motives—he wanted to bring order and simplicity to the teaching of philosophy and to that end he employed dialectics as a methodical basis for the disciplines. In *Dialecticae partitiones*, also published in 1543, Ramus foresaw a new structure for all human knowledge, for which he proposed summaries, headings, citations, and examples.[191] Ramus's technique for presenting and categorizing was indeed innovative, but in terms of content his work was not much more than an erudite pedagogical manual and was not a treatise on logic.[192] Ramus was not able to substantiate his claim to have replaced Aristotle with a new independent system, and all he did was make a marginal modification in syllogism that in fact corresponded to Aristotle's original scheme. He also went along with Valla in his criticism of syllogism. These days Ramus has been largely forgotten, but in his own time he was extremely influential and founded Ramism, a movement of almost the same order as Cartesianism. It was strictly forbidden in Catholic Europe but was broadly accepted by Protestants. It could be that his tragic death during the massacre of St Bartholomew (23/24 August 1572) had something to do with this. As a murdered Huguenot, Ramus was seen by Protestants as a martyr.

Like the humanists, the exponents of the 'New Science' also lashed out at logic. While humanists primarily contested scholastic logic, in 1637 René Descartes made mincemeat out of all logic in his *Discours de la méthode*. He advocated the new mathematics and logic, and in particular his own analytical geometry, which was not artificial and deductive like Euclid's *Elements*, but heuristic and problem solving. Nevertheless, 1662 saw the appearance of another complete overview of 'traditional logic'—the Port Royal logic—*La logique ou l'art de penser* by Arnauld and Nicole (see above).[193] There are barely any new logical ideas to be found in it. Until far into the nineteenth century the work was used primarily as an introduction to logic. Yet, if we put together the Port Royal logic and the Port Royal grammar discussed earlier—the *Grammaire générale*—one can discern an agenda in the form of an overarching logical theory of language and thinking. In it, sentences

[190] Lodi Nauta, *In Defense of Common Sense: Lorenzo Valla's Critique of Scholastic Philosophy*, Harvard University Press, 2009.

[191] Peter Ramus, *Dialecticae Partitiones: ad celeberrimam & illustrissimam Lutetiae Parisorum Academiam*, Iacobus Bogardus, 1543.

[192] Walter Ong, *Ramus, Method, and the Decay of Dialogue: From the Art of Discourse to the Art of Reason*, Harvard University Press, 1958.

[193] Antoine Arnauld and Pierre Nicole, *La logique ou l'art de penser*, edited by Charles Jourdain, Gallimard, 1992.

expressed thoughts using words, where the words were nothing other than symbols for thoughts. Writing subsequently consisted of symbols for words, and the syntactic categories of these symbols were equated with logical units like subject and predicate. Using these logical primitives, sentences were produced by means of syntactic rules, as in Panini's grammar. In present-day linguistics this process is called *language production*. A language production theory takes a given logical form (the 'meaning') and generates the corresponding syntactic form (the 'sentence'). The reverse process is called *language comprehension*. Here an observed syntactic form is converted into its meaning in terms of logical units. This interpretation of Port Royal as a 'generative grammar'—although the term did not exist at that time—was made popular primarily in the twentieth century by Noam Chomsky (see 5.3).[194] However, we have already encountered a logical language theory in the work of Sanctius, and of course much earlier in the work of Greek and Indian linguists.

The declining prestige of Latin and the ascent of artificial languages: Bacon, Dalgarno, and Wilkins. The most important contribution to early modern logic lay in the development of universal languages. The aspiration for a universal language goes as far back as the Modists, but it became a serious subject in the seventeenth century when the invulnerability of Latin came under discussion. The grammars of everyday languages that had been compiled showed that these languages were a match for Latin in terms of precision (as Alberti had demonstrated for Tuscan). The influence of Latin, and with it humanism, steadily declined and works were increasingly written in the vernacular. Many exotic languages from the New World were also discovered, and thus there was a growing realization that the *ideal language* was not Latin but a new, artificial language that would have to be constructed from the best parts and properties of existing languages—like a sort of neo-Platonic 'ideal selection', which we will also come across in art theory (see 4.5).

Francis Bacon (1561–1626), the great champion of the new vision of nature through experimental investigation, advocated the development of a universal language for knowledge sharing that expressed exactly what someone meant, so that misunderstandings would be eliminated for once and for all.[195] His most radical proposal was the construction of a completely new language for the whole civilized world. This language would communicate knowledge, thoughts, and ideas in symbols, in a direct and universal way. Some believed that the Chinese pictogram system could serve as an example here. This was because ever since knowledge of Chinese reached Europe in the late sixteenth century, it had been assumed that Chinese characters were direct representations of ideas. Although this assumption was incorrect—like all other languages Chinese is based on a complex grammatical system—the construction of pictograms seemed to be semantically transparent. It was therefore no coincidence that the first artificial languages were inspired

[194] Noam Chomsky, *Cartesian Linguistics: A Chapter in the History of Rationalist Thought*, Harper & Row, 1966.

[195] John Robertson, *The Philosophical Works of Francis Bacon*, Routledge, 1905, pp. 119ff.

by Chinese script, for example the *Ars signorum* by George Dalgarno (1661)[196] and the *Essay towards a Real Character and a Philosophical Language* (1668) by John Wilkins,[197] who was a co-founder of the Royal Society in 1660.

The works of Dalgarno and Wilkins resemble each other from several points of view.[198] The two scholars even worked together for a while, but their views were too divergent so that each soon went his own way. Dalgarno was a 'minimalist' and he believed that a universal language should be built up using a minimum number of words for the basic concepts. All other words would have to be derived from these basic concepts by means of precise rules or operations. Wilkins on the other hand was an 'encyclopaedist' who wanted to use a classification scheme to express all human knowledge in symbols, including abstract relationships, actions, processes, logic concepts, natural species, or varieties of both living and dead things, as well as the relations between people in families and society. Apart from these differences, though, both were aiming at a scientific universal language that could produce an infinite number of statements on the basis of a finite lexicon.

The characters of Wilkins's lexicon, for instance, are wholly semantically transparent and every symbol can be translated into a word in a natural language. The symbol for 'father' is made up of the basic symbol for 'interpersonal relationship', from which specific instantiations can be derived by adding sub-symbols, resulting in representations for father, mother, sister, brother, etc.[199] Wilkins moreover proposed a universal phonetics that was built from the known languages in the world. The syntactic rules of his universal language were reduced to a minimum, and the word class and grammatical relationships between words were also depicted graphically using sub-symbols. Wilkins compared his 'philosophical language' with Latin, which he called the second most universal language, and Chinese, which he praised for its character system. Whereas Latin was full of lexical redundancy and irregularities, according to Wilkins his artificial language was an oasis of order and simplicity. And while semantic analysis of Chinese characters was often impossible, the character system in his language was completely transparent.

Nothing of practical value came out of all this wonderful work. One could almost describe the seventeenth-century construction of universal languages as a useless exercise. Yet later on, eighteenth-century scientists did make use of Dalgarno's and Wilkins's classification schemes, culminating in 1735 with Linnaeus's classification of the plant world in his *Systema naturae*. The idea of expressing human knowledge in a system of symbols would also be successfully implemented later in formal linguistics and symbolic logic.

Beginning of symbolic logic: Leibniz and later. It is no exaggeration to state that the early modern age produced only one really great logician—Gottfried Wilhelm Leibniz (1646–1716). Leibniz was at the leading edge in many disciplines of

[196] Cram and Maat, *George Dalgarno on Universal Language*.

[197] John Wilkins, *Essay towards a Real Character and a Philosophical Language*, 1668, reprinted (facsimile) by Thoemmes Press, Works in the History of Language, 2002.

[198] Jaap Maat, *Philosophical Languages in the Seventeenth Century: Dalgarno, Wilkins, Leibniz*, Kluwer, 2004.

[199] R. H. Robins, *A Short History of Linguistics*, Longman, 1989, pp. 128ff.

learning and science, and he made important discoveries in a vast number of areas. Like Dalgarno and Wilkins, he was fascinated by symbolic artificial languages and he designed plans himself for such a language, which he called *lingua characteristica universalis*. His most important contributions, though, were in the field of logic. In the sixteen-eighties he designed a symbolic binary logic that was virtually identical to the famous Boolean logic dating from 1847 (see 5.3). He discovered that computing processes could be done more easily with a binary coding.[200] However, Leibniz kept almost all of his discoveries on the drawing board and none of this was published until 1903, when symbolic logic was already in full bloom (see 5.3), with the result that Leibniz's insights in this field had barely any influence.

In 1666 Leibniz wrote the work *De arte combinatoria*, which was considered to be a continuation of Llull's project to discover truths through the exhaustive combination of concepts (see 3.6).[201] Leibniz's plans for a *lingua characteristica universalis* also built on the compositional idea that complex concepts are constructed from simple concepts that everyone could understand in a 'natural' way. Leibniz's most important activity in logic was his design for the *Calculus ratiocinator*.[202] He wanted to mechanize the reasoning process by manipulating symbols so that new truths could be discovered. To achieve this, the reasoning process had to be executed by means of algorithms and machines so that it was no longer susceptible to human error and shortcomings. To this end Leibniz developed a *logical calculus* with which syllogistic premises (see 2.6) could be written as formulas. For example 'All As are Bs' was written as 'A = AB'. This equation stated that the concepts included in the concepts of both A and B are the same as those in A. A syllogism, 'All As are Bs; all Bs are Cs; therefore all As are Cs' becomes the sequence of equations 'A = AB; B = BC; therefore A = AC'. This conclusion ('A = AC') can be deduced from the two premises (A = AB and B = BC) by simple *substitution*—by substituting B in the first premise A = AB by BC (from the second premise) we get A = ABC, which according to the definition is equivalent to A = AC. Leibniz's great insight was to realize that underlying the 256 different types of Aristotelian syllogisms there was a deeper pattern, which he believed he could formalize as a logical calculus with precise operations. These operations resemble both seventeenth-century mathematics and the sixteenth-century linguistics of Sanctius.

In his logical calculus Leibniz employed the procedural system of rules principle, as Aristotle had done before him (and like Panini and Sanctius in linguistics). But where Aristotle had to assume that the different types of syllogisms were primitives, Leibniz tried to reduce these syllogisms to a more fundamental system. Yet Leibniz was only partially successful. He was the first to realize that his calculus was not able to tackle any syllogisms with *negative* sentences. The problem of accounting for such syllogisms was not solved until centuries later (see 5.3). The fact remains that

[200] Louis Couturat, *La Logique de Leibniz*, Felix Alcan, 1901, p. 115.
[201] Gottfried Wilhelm Leibniz, *Dissertatio de arte combinatoria*, 1666, *Sämtliche Schriften und Briefe*, Akademie Verlag, 1923.
[202] G. H. R. Parkinson, *Leibniz: Logical Papers*, Oxford University Press, 1966.

Leibniz's attempt was unmatched in the early modern age. More than two centuries would pass before European logic would take a step of the same order.

As far as we know, Leibniz had no direct influence on other logicians. This was the tragedy of a man who left his scribblings on his desk for too long.[203] The most important contributions after Leibniz were made by the Jesuit Giovanni Saccheri (1667–1733), who resolved a number of ambiguities in Euclid, and Leonhard Euler (1707–1783), who could express syllogisms in the form of diagrams using his famous Euler circles.

Linguistics and logic in China. In Late Imperial China linguistics was dominated by philology,[204] as it was in Europe. Chen Di's study of sounds was inspired by his philological interest in Chinese poetry in the Book of Songs (see 4.1). There was also research into language variation. Pan Lei (1646–1708) travelled throughout China in order to record the sound properties of the different dialects. Yet in no case did these investigations lead to an attempt to produce an overarching Chinese grammar. It was recognized that the phonological element of a Chinese character, its pitch and intonation, largely dictated the meaning of a character. Given this, one would expect there to have been interest in a quest for underlying rules between phonological form and meaning, but we have found nothing of this kind. The study of Chinese by the Jesuit Matteo Ricci resulted in not much more than a dictionary with transcriptions. The absence of internal word structure in Chinese is put forward now and again as a possible obstacle to start searching for Chinese lexical and syntactic categories.

Chinese logic likewise appears to have undergone virtually no new developments during the early modern age. For a long time it was based on the Buddhist art of argumentation, which had been brought to China in the seventh century CE (see 3.6). Mohism and its inductive logic had disappeared and been forgotten centuries before, after the repression by the legalistic Qin regime (see 2.6). China was introduced to deductive logic for the first time when Ricci and Xu Guangqi translated Euclid's *Elements* into Chinese in 1607. However, this work did not trigger a new logical tradition in China—let alone a deductive one. It was seventy years before Fang Zhongtong wrote the first Chinese mathematical commentary on Euclid.[205] The only thing about logic in this commentary was a trivial digression. Later commentators had similarly little interest in Euclid's method of reasoning and proving.

The introduction of logic as an independent discipline by the Jesuit Giulio Alenio (the 'Confucius of the West') in 1623 made even less impression in China, if that were possible. The first logic manual, by Francisco Furtado in 1631 and based

[203] Maria Rosa Antognazza, *Leibniz: An Intellectual Biography*, Cambridge University Press, 2008.

[204] Benjamin Elman, *From Philosophy to Philology: Intellectual and Social Aspects of Change in Late Imperial China*, Harvard University Press, 1984.

[205] Peter Engelfriet, *Euclid in China: A Survey of the Historical Background of the First Chinese Translation of Euclid's Elements (Jihe yuanben; Beijing, 1607), an Analysis of the Translation, and a Study of its Influence up to 1723*, PhD thesis, University of Leiden, 1996, p. 296 (also published as monograph with Brill, 1998).

on Aristotle and Porphyry, suffered the same fate.[206] For over two centuries European logic was looked on in China as an eccentric intellectual activity. In the twentieth century the decanonization of the Confucian classics sparked off a public debate about the discipline of 'Chinese logic', which had been unknown until then.[207] Mohism was rediscovered and the first comparative studies were set up in which the three major logic traditions—European, Indian, and Chinese—were put under the spotlight.

Evaluation of early modern linguistics and logic. The procedural system of rules principle made a resplendent return in early modern linguistics and logic—from Sanctius to Leibniz. Methodical principles were also defined in comparative linguistics (by De Laet), but they were not procedural and only specified the boundary conditions with which a language comparison would have to comply (declarative system of rules principle). Alongside such formal linguistics and logic there was no lack of scepticism—humanists like Valla and Ramus even disputed the independent status of logic. However, there was an observable general tendency towards order and simplification. Humanists and seventeenth-century scientists alike wanted to order the often incomprehensible scholastic logic and restore it to clear arguments. The combination of empirical and theoretical approaches, which was already being used by humanist philologists and flourished in the seventeenth century, ultimately appeared to be the most successful. This resulted in new research areas that were inconceivable in the fifteenth century—sign language, a finger alphabet, the study of linguistic affinity, and the construction of artificial languages.

It is difficult to find new patterns in linguistics and logic outside Europe, with the exception of the Chinese phonology of Chen Di and Pan Lei. No new turns seem to have been taken in linguistics and logic in either India or Islamic civilization. However, it should be remembered that many sources are not yet accessible.[208]

[206] Needham and Harbsmeier, *Science and Civilization in China: Volume 7*, pp. 164ff.

[207] Joachim Kurtz, 'Coming to terms with logic: the naturalization of an Occidental notion in China', in Michael Lackner, Iwo Amelung, and Joachim Kurtz (eds), *New Terms for New Ideas: Western Knowledge and Lexical Change in Late Imperial China*, Brill, 2001, pp. 147–76.

[208] See for example Khaled El-Rouayheb, 'Sunni Muslim scholars on the status of logic, 1500–1800', *Islamic Law and Society*, 11, 2004, pp. 213–32.

4.4 MUSICOLOGY: THE MISSING LINK BETWEEN HUMANISM AND NATURAL SCIENCE

While many 'new' works from Antiquity were discovered by philology and historiography, there were no comparable discoveries in the field of music. Nothing of the music of Antiquity had survived and the classical texts from music theory had already been known since the Early Middle Ages thanks to Boethius. Yet Boethius's *De institutione musica* appears only rarely to have been read accurately.[209] His treatise was therefore one of the first musicological works to be studied by humanists. Other Greek musical discourses (from Aristoxenus to Ptolemy) soon followed. Two traditions then developed in Europe—a quest for the laws of consonances and a search for systems of rules for musical analysis and composition (*musica poetica*). Ottoman musicology continued to build on the Arab rules of melodic and rhythmic cycles, whereas Indian music theory reached a state of near perfection in the systems of rules for *raga*s. African musicology is still largely contained in sources that are waiting to be accessed, and China concentrated on historical overviews during the Ming and Qing Dynasties.

The battle of the consonances: from Ramis to Vincenzo Galilei. Thanks to the study of Greek music theory, renewed interest arose in 'harmonious' or consonant intervals. According to Pythagorean musicologists, consonant intervals corresponded to simple ratios of the first four whole numbers, such as 1/2 for the octave, 2/3 for the fifth, and 3/4 for the fourth—but that is about as far as it went (see 2.4). However, this law resulted in controversies about intervals that sounded 'more or less' consonant, such as the third and the sixth, but were classified as dissonant by the Pythagoreans.[210] Aristoxenus declared that it was not theory but empiricism (human hearing) that should have the last word. The controversy was expressed in Boethius's standard work, but the thread was not taken up in the Middle Ages. We do, however, see extensive musical systems of rules, such as for the organum (see 3.4), but they were based entirely on the Pythagorean consonances. There was no deeper investigation into the assumed laws of harmonies.

However, when the musicological classics were put under the spotlight again in the fifteenth century, new interest arose in the laws of harmony. Meanwhile, the third and the sixth had been introduced in music composition, particularly in the work of the composer John Dunstaple (*c.*1390–1453), who created elegant harmonies with thirds and sixths. An unthinking embrace of Pythagorean music theory had become dubious, to put it mildly, and from the second half of the fifteenth century the humanists reinvigorated their search for theoretical and empirical foundations of harmony. Almost fifteen hundred years after Ptolemy (see 2.4), new life was breathed into one of the oldest questions in learning and

[209] This can be derived from the contents of thirteenth- and fourteenth-century musical treatises, according to Claude Palisca, *Humanism in Italian Renaissance Musical Thought*, Yale University Press, 1985.

[210] See <http://en.wikipedia.org/wiki/List_of_musical_intervals> for links to sound fragments of intervals.

science—*are consonances based on an underlying system*? Ptolemy's vision, which can be traced back to Aristoxenus, became the focus of attention. According to Ptolemy, music was a human experience and therefore opinions about consonances had to be developed on the basis of hearing, *only* with help of human reason. In the Pythagorean view, on the other hand, the ultimate opinion should be given based on reason *alone* because senses can too easily be led up the garden path.

Bartolomeo Ramis de Pareia (*c*.1440–1491), one of the first humanistic music scholars, experimented with the monochord, a single string instrument. Initially Ramis worked in Salamanca (like the linguist Sanctius—see 4.3) and later in Bologna and Rome, where he encountered Italian humanists. In his *Musica practica* (1482) he reported the problems he found in Boethius—who had apparently not read Ptolemy—and then addressed the 'imperfect consonances' that corresponded to intervals other than the Pythagorean ones.[211] Pythagoras based the consonances on simple ratios between only the first four whole numbers, whereas Ramis went as far as the number eight. This laid him open to fierce criticism from the dyed-in-the-wool Pythagorean Franchino Gaffurio (1451–1522), who considered these intervals to be irrational. At the same time, however, in the light of existing musical practice, Gaffurio could not deny that there was a practical problem contained in the Pythagorean system, because—related to the consonances issue—there was the problem of *tuning* instruments. Tuning could likewise be based either on hearing or on mathematics. However, actual practice was so recalcitrant that in his *Practica musicae* (1496) Gaffurio acknowledged that instruments could only be tuned acceptably if one let the fifth deviate slightly from the pure mathematical ratio.[212] Gioseffo Zarlino (1517–1590), on the other hand, the chapel master at St Mark's in Venice, appeared to help the Ptolemaic vision gain ascendancy until in 1585 Giovanni Battista Benedetti demonstrated empirically that it was impossible to use it for polyphonic singing. This meant that this way of tuning was of no practical value for contemporary music. The controversy was made an even bigger issue when Vincenzo Galilei (1520–1591) argued that equal temperament tuning, where an octave was divided into twelve equal parts, was the only solution for instrumental music.[213] Vincenzo based his claim on a large number of experiments with strings of different lengths, materials, and tensions. In doing so, Vincenzo refuted Zarlino by means of experiment rather than by theory. The Pythagorean number magic was done away with by empiricism and had, according to Vincenzo, to be substituted by equal temperament.

Now, however, there was a colossal theoretical problem—no whole numbers were found with which an octave could be divided into twelve equal parts (there are indeed twelve tones, but the ratios between them correspond to the twelfth root of two, which cannot be expressed as the ratio of two whole numbers). Anyone who

[211] Bartolomeo Ramis de Pareia, *Musica practica*, commentary and translation by Clement Miller, Hänssler-Verlag, 1993.

[212] Irwin Young, *The Practica musicae of Franchinus Gafurius*, University of Wisconsin Press, 1969, pp. xxiiff.

[213] Stillman Drake, 'Renaissance music and experimental science', *Journal of the History of Ideas*, 31, 1970, p. 496.

did not want to work on a purely practical basis but also wanted to base empiricism on a theory using ratios of whole numbers was aware of this problem. People could not accept that there was no deeper theoretical principle underlying a practical solution. Although music theory was in this resepect still under the numerological yoke, the search for a *theoretical underpinning of empiricism* proved to be extraordinarily fruitful. This desire for unity between theory and empiricism is perhaps the most singular feature of the early modern study of music and seemed to have had a huge effect on later scholars—not least on Vincenzo Galilei's own son, Galileo Galilei.[214] We have already encountered this tendency towards a theoretical foundation in other fields, such as fifteenth-century philology, where a basis was developed for manuscript selection and emendation by Poliziano (see 4.1). The singular feature of Poliziano's genealogical theory is that it not only justified text reconstruction practice—it transcended it. New unexpected problems could be tackled using his *eliminatio* principle. This marvellous synergy between theory and empiricism, where the theory proved able not only to underpin facts that already existed but also to account for or predict unsuspected, new facts (which in turn could have an influence on the theory) was a characteristic of both early humanistic disciplines and the later natural sciences ('New Sciences').

Thus people also sought a theoretical foundation for the ratios between consonances within the empirical-Aristoxenian approach to music. In *Le istitutioni harmoniche* (1558) Zarlino proposed putting reason above hearing.[215] He introduced a limit for consonance, called the *scenario*, corresponding to the first six whole numbers—one to six—which could cover all 'natural' intervals. Now the thirds and sixths could also be reproduced on the basis of the acceptable numerical ratios 4/5 and 3/5 respectively. He justified the number six on the grounds of extensive theoretical and practical arguments, but the most important was that six was the first 'perfect' number because the number was equal to the sum of its divisors (1 + 2 + 3). Once again, however, it was obstinate reality that dealt summarily with Zarlino's proposal. It turned out that his *scenario* was unworkable in practice and was soon swept aside. In his *Discorso intorno all'opere di Messer Gioseffo Zarlino* (1589) Vincenzo Galilei tried to theoretically underpin his empirical approach by contending that *all* intervals were 'natural', and he asserted that an infinite number of consonant intervals existed.[216] A sound theory that could distinguish consonances from dissonances seemed further away than ever.

Seventeenth-century practice comes to the aid of theory: from Mersenne to Huygens. In the seventeenth century, performing musical practice rushed to the aid of music theory. It emerged that intervals that had previously been perceived as completely dissonant were becoming increasingly widely accepted in music

[214] For an in-depth study of the relation between the music theory and the scientific revolution, see H. Floris Cohen, *Quantifying Music: The Science of Music at the First Stage of the Scientific Revolution, 1580–1650*, Reidel Publishing Company, 1984.

[215] Gioseffo Zarlino, *Istituzioni armoniche*, translated by Oliver Strunk, in Source Readings in Music History, W. W. Norton, 1950.

[216] Vincenzo Galilei, *Discorso intorno all'opere di messer Gioseffo Zarlino da Chioggia, et altri importanti particolari attenenti alla musica*, G. Marescotti, 1589, reprinted by University of Rochester Press, 1954.

composition. Even the (diminished) seventh was no longer shunned by a composer like Claudio Monteverdi. This brought up the question of whether harmony theory was universal, or whether it depended on time and place, like the theory of organum construction (see 3.4).

Although there was no absolute distinction between consonant and dissonant intervals, it could be that there was a gradual progression from consonant to less consonant up to dissonant intervals. But complications remained here too. According to Marin Mersenne (1588–1648), one of the staunch defenders of the New Sciences, the biggest problem in musicology was the question as to whether or not the fourth was more consonant than the third because reason and sensory perception appeared to contradict each other.[217] Mersenne was the first to revive the study of overtones. Although Aristotle had made a vague observation that a tone always contained its higher octave (which had been postulated as a real phenomenon by Oresme—see 3.4), it was Mersenne in his *Harmonie universelle* who was able to distinguish the first four overtones, which also corresponded with the first four numbers. Mersenne had no idea where these overtones came from, but he assumed that a source of sound was perceived as more harmonious and pleasant as the number of overtones it contained increased. Mersenne compiled tables that gave the degree of 'sweetness' and 'agreeableness' of the different intervals. However, even then the theoretical underpinning could still not account for sensory observation.

Virtually all seventeenth-century New Scientists grappled with the problem of the theoretical or empirical explanation of the degree of 'pleasantness' of consonance—from Galileo, Beeckman, Descartes, Wallis, Holder, and Huygens to Euler. We describe a few of their ideas here.[218] The most important new and shared insight was that *consonance was no longer linked to abstract ratios of numbers but to the physical category of the vibration frequency* (where, to be sure, the ratios remained the same).[219] For example, according to Galileo's 'coincidence theory', consonances occurred when vibrations often coincide, which pleases our hearing in an agreeable way. But this theory delivered incorrect predictions more often than correct ones. Isaac Beeckman argued that dissonance was caused by pulses in the sound vibrations. His perception was to be worked out in the nineteenth century with more success by Hermann von Helmholtz (see 5.4). In 1677 John Wallis gave a physical analysis of overtones but he did not get much further than Mersenne in accounting for the continuity between consonances. William Holder (who was also a phonetician and linguist—see 4.3) and Christiaan Huygens designed micro-intervals, and even completely new scales, but neither of these led to a sound theory of consonances. It emerged ultimately that it was musical practice, and the

[217] Marin Mersenne, *Harmonie universelle*, 1636–7, translated into English by Roger Chapman, M. Nijhoff 1957.

[218] For an in-depth overview, see Cohen, *Quantifying Music: The Science of Music at the First Stage of the Scientific Revolution, 1580–1650.*

[219] The ratios stayed the same because a chord that is *n* times longer has a vibration frequency that is *n* times smaller—other things being equal.

associated gradual acceptance of 'new' consonances, that largely nullified all the theoretical work.

Although it failed to find an absolute law, the quest during the fifteenth, sixteenth, and seventeenth centuries for a theory of consonances was not fruitless. Firstly, it became clear that there was no hard and fast distinction between consonances and dissonances—and this de facto rebutted the centuries-old Pythagorean cosmic harmony (see 2.4).[220] Secondly, this quest heralded interaction between theory and empiricism, in which empiricism was to have the last word, no matter how fine the underlying theory was. We dare to contend that it was the fifteenth- and sixteenth-century humanistic musicologists who unearthed the synergetic interaction between theory and empiricism and passed it on to the New Scientists of the seventeenth century, who elaborated it further.[221] Galileo, Kepler, Beeckman, and Huygens all adopted this extremely fruitful synergy and applied it to music, acoustics, and other aspects of studying nature. Musicology—even more than philology—seemed to be the missing link between the humanities and the New Sciences.

Rameau's eighteenth-century harmonic grammar. Despite the wholesale failure of the search for the universal law of music, musicology was far from dead. On the contrary, harmony theory became more and more focused on musical practice, and the great French composer and theoretician Jean-Philippe Rameau (1683–1764) was its most important exponent. Now it was the turn of the New Sciences to exert influence on musicology, rather than the other way round. Rameau took his inspiration from Descartes, Kepler, and Newton. In *Traité de l'harmonie reduite à ses principes naturels* (1722), he ascribed his method to the Cartesian concept of an 'evident and clear principle'.[222] All notes and chords had to be derived from one single source tone. Like Zarlino before him, Rameau succeeded in producing both the consonant and dissonant intervals using the first six divisions of a string. However, he ran into problems generating the basic chords, the so-called triads. A breakthrough appeared to come when he heard about the existence of overtones. In the same way as Newton had demonstrated that white light consisted of a spectrum of separate colours, so, in *Nouvelle système de musique théorique* (1726), Rameau wanted to show that a single tone comprised a spectrum of harmonic overtones.[223] It gave him an even more fundamental 'first principle' than dividing up a string. Yet he ran into problems again with a number of chords (the minor triad and the diminished seventh chords). Initially Rameau tried to

[220] H. Floris Cohen, 'Music as science and as art: the 16th/17th-century destruction of cosmic harmony', in Bod, Maat, and Weststeijn (eds), *The Making of the Humanities, Volume I: Early Modern Europe*, 2010, pp. 59–71.

[221] Penelope Gouk, 'The role of harmonics in the Scientific Revolution', in Thomas Christensen (ed.), *The Cambridge History of Western Music Theory*, Cambridge University Press, 2002, pp. 223–45. See also Claude Palisca, *Studies in the History of Italian Music and Music Theory*, Oxford University Press, 1994, pp. 200–37.

[222] Jean-Philippe Rameau, *Traité de l'harmonie reduite à ses principes naturels*, 1722, reprinted by Nabu Press, 2011. See also Thomas Christensen, *Rameau and Musical Thought in the Enlightenment*, Cambridge University Press, 1993, pp. 26ff.

[223] Jean-Philippe Rameau, *Nouvelle système de musique théorique*, Ballard, 1726.

manage the phenomena by assuming there were non-existent 'undertones', but in his *Nouvelles réflexions* (1760) he finally admitted that only the major triad (e.g. the C-E-G chord) could be generated directly and that *all* other chords could only be derived from the overtones on the basis of analogy.[224]

Although Rameau's attempt to develop a Cartesian explanation of the consonances and chords was in vain, he did succeed in analysing chords using harmonic progressions (sequences of chords). This part of his work appeared to be aimed primarily at musical practice. By the end of the eighteenth century, Rameau's harmony theory formed the dominant pedagogical paradigm in Europe. While he did not succeed in creating an overarching music theory on the basis of first principles, his attempt was sufficiently convincing to attract the attention of the best scholars of his time—from Euler to Diderot.[225] His many schemes for the sequence of chords could be read as a system of rules, a harmonic grammar that reproduced the underlying chord systems of pieces of music. However, his grammar was not of the same order as Panini's language grammar, which established all possible linguistic utterances using a recursive procedural system of rules. Rameau's system seemed to be more like a declarative system of rules in the style of Aristoxenus's melodic grammar. It spelled out the limits of the class of possibilities within which the composer still had immense freedom to write a piece of music as he wished. Rameau's work became the standard for the following two centuries and his approach is still the basis for studying harmony today.

Musical grammars: from Dressler to Koch. It is remarkable that although Aristoxenus was one of the major sources of inspiration for the early modern study of harmony, his melodic laws (which we discussed in 2.4) received no attention. I know of no early modern attempt to construct a melodic grammar in Aristoxenus's *deductive* style. However, the much greater interest in harmony theory is understandable from a humanistic perspective. Aristoxenus did not realize that melody theory was dependent on time and place, whereas the humanists probably did. The universality of music was at its best in harmony, in the law of consonances. Although we now know that the perception of consonant intervals is also largely dependent on time and place, there is indeed something universal in harmony theory. As far as we know, the octave and possibly the fifth are considered to be consonant in all tonal cultures, in addition, of course, to the trivial unison (see 2.4 and 3.4).[226] What comes after the fifth is completely open. Harmony therefore appears to be partly objective (the mathematical ratios for the few universal consonances) and partly subjective (the time- and place-related perception of dissonances—but see 5.4).

All this is not to say that no systems of rules were devised for the melodic composition of music in the early modern age (from the Renaissance to baroque

[224] Christensen, *Rameau and Musical Thought in the Enlightenment*, pp. 242ff.

[225] Michaela Maria Keane, *The Theoretical Writings of Jean-Philippe Rameau*, Catholic University of America Press, 1961, p. 170, p. 180. See also Philip Gossett, *Preface to Jean-Philippe Rameau: Treatise on Harmony*, Dover, 1971.

[226] Philip Ball, 'Facing the Music', *Nature* 453, 2008, pp. 160–2.

and classicism). There had been a continuous tradition of composition manuals since the Middle Ages—from the tenth-century *Musica enchiriadis* and the twelfth-century *Ad organum faciendum*, which we discussed in the preceding chapter (3.4), to the fourteenth-century *Ars contrapunctus secundum Phillippum de Vitriaco*.[227] The search for rules of composition was given a shot in the arm by the humanistic programme *musica poetica*, which sought unification with rhetorical composition (a similar unification with rhetoric was also proposed by Alberti for art analysis—see 4.5). The *musica poetica* was worked out in great detail by Gallus Dressler (1533–1580/9) in his *Praecepta musicae poeticae* (1563).[228] This treatise contained the oldest known system of rules for Renaissance motets in the style of the composers from the Low Countries such as Josquin des Prez and Orlandus de Lassus. Although Dressler was born in Thüringen, he appears to have undergone his musical training in the Netherlands.[229] His manual applied two principles:

(1) the rhetorical principle of *exordium* (introduction), *medio* (middle part), and *finis* (end) for the structure of a motet;
(2) the grammatical principle of *clausula* (sentence) to progressively smaller musical parts, where the combination of sentences, phrases, and tones was specified both horizontally (melodic line) and vertically (polyphony).

Because of the relationship to the linguistic concept of 'sentence', Dressler's system of rules is also referred to as *musical syntax* or *musical grammar*. As the title of his treatise makes clear, he was mainly interested in the rules ('praecepta') of *musica poetica*. Although his work can be seen as a prescriptive composition manual, Dressler appeared to want to give his grammar a primarily descriptive approach for existing motet practice in the Netherlands. And if the work was in fact intended to be prescriptive, Dressler's rules did not last long. Polyphonic music changed so rapidly that new manuals for new musical styles appeared in quick succession.

In 1606, for example, Joachim Burmeister published his *Musica poetica*, which was also very much based on notions taken from rhetoric and linguistics.[230] Now, however, Burmeister was describing the musical idiom of his own time—which was still primarily vocal and polyphonic. He recorded, for example, the different rules of vertical combination of tones, the sequences of vertical harmonies, and the overall plan of beginning, development, and cadence.[231]

One of the most influential musical grammars from the early modern era was *Gradus ad Parnassum* by Johann Joseph Fux (1660–1741), in which the polyphony of Renaissance music—in particular of Palestrina (1525–1594)—was presented as

227 Additionally, 'rules' were developed for the performance of music, the *musica ficta*, that described the notes that lay outside the system of the *musica recta* as defined by the hexachords of Guido of Arezzo (see 3.4).

228 Robert Forgács, *Gallus Dressler's Praecepta Musicae Poeticae: The Precepts of Poetic Music*, University of Illinois Press, 2007.

229 Walter Blankenburg, 'Dressler, Gallus', *Grove Music Online*, Oxford University Press, 2006.

230 Joachim Burmeister, *Musical Poetics*, translated by Benito Rivera, edited by Claude Palisca, Yale University Press, 1993.

231 Harold Powers, 'Language models and musical analysis', *Ethnomusicology*, 24(1), 1980, pp. 1–60.

one large system of rules.[232] Essentially all the later work on Renaissance music is indebted to Fux's *Gradus*. Yet his system of rules was not procedural but declarative. The rules do not represent a procedure but give the boundary conditions or constraints within which polyphonic composition could play out. Fux's rules look, for instance, like this:

- A piece must begin and end on a perfect consonant interval.
- Opposite movements (of the voices) must dominate.
- Avoid movements in parallel fourths, fifths, or octaves.
- Avoid lengthy movements in parallel thirds or sixths.

These rules were of a different order from those in the *Musica enchiriadis* for polyphonic organa (see 3.5), where the procedure is specified step by step. The rules in the *Gradus ad Parnassum* consist of restrictions or preferences but not of algorithmic steps. A high degree of freedom remained for the composer. Nevertheless, Fux showed that Palestrina's famous contrapuntal style was subject to a declarative system of rules.

The fugue had to follow suit. In *Der wohlklingende Fingersprache* Johann Mattheson (1681–1764) developed fugues of different forms of complexity, for which he spelled out the underlying system of rules.[233] Johann Sebastian Bach took up the challenge and developed new fugue forms thereby superseding Mattheson's work. Heinrich Christoph Koch went furthest with the linguistic analogy in his three-volume *Versuch einer Anleitung zur Composition*, the first part of which appeared in 1782.[234] Koch introduced the concept of melodic phrase structure, which was to play an important role in nineteenth-century musicology (see 5.4), and which established direct links between linguistic and musical syntax, where linguistic terms were applied to the melodic idiom. Koch's notions of punctuation, *Vordersatz* (antecedent phrase) and *Nachsatz* (consequent phrase) are still being used in music analysis.[235]

Just as in harmony theory, we can therefore talk about interaction between theory and empiricism in the composition manuals too. However, unlike harmony theory, which was deemed to be at least partly universal (the octave and the fifth), it was patently obvious that musical practice, with its successive styles, was time- and place-dependent. Here the interplay between theory and empiricism was less far-reaching than in the study of consonances. The systems of rules in the composition manuals were not so much *refuted* as *superseded* by new styles. The rules for consonances on the other hand were rebutted. Hypotheses like the coincidence theory and the overtone theory (see above) were invalidated in practice, and were

[232] Johann Joseph Fux, *The Study of Counterpoint from Johann Joseph Fux's Gradus ad parnassum*, translated and edited by Alfred Mann, with the collaboration of John Edmunds, Dent, 1965.

[233] Birger Petersen-Mikkelsen, *Die Melodielehre des 'Vollkommenen Capellmeisters' von Johann Mattheson*, Eutin, 2002.

[234] Heinrich Koch, *Versuch einer Anleitung zur Composition*, 3 volumes, Böhme, 1782–93, reprinted by Olms, 1969.

[235] Nancy Kovaleff Baker, *From 'Teil' to 'Tonstück': The Significance of the 'Versuch einer Anleitung zur Composition' by Heinrich Christoph Koch*, PhD thesis, Yale University, 1975.

subsequently improved or rejected. While the quest for regularities in consonances did influence the New Sciences and even partially shaped them, this seems not to have been the case in the search for patterns in composition practice. However, this does not detract from the great value of these systems of rules to an understanding of the structure of these musical forms—from motet to fugue—or in other cultures from *gamaka* to *raga* (see below).

Initial impetus for the history of music. The early modern age also saw the appearance of the first European musical histories. The oldest—the three-volume *Storia della musica* by the Italian priest Giovanni Battista Martini (1706–1784)—dates from the middle of the eighteenth century. This work was organized in a way that was reminiscent of the medieval salvation histories (see 3.2). Martini began his overview with the hypothetical music from Adam to the Flood, then from the Flood to the birth of Moses, after that from the birth of Moses to his death and so on. It is only in the third volume that we read about the music of Classical Greece, about which little was known, whereupon we reach the end of this colossal work. It was a curiosity in the Age of Enlightenment.[236]

It was not until the end of the eighteenth century that the history of music was approached more descriptively, among others by the German theologian Martin Gerbert. In his *De cantu de musica sacra* published in 1774 he gave an overview of church music from about the third century to the invention of the printing press. The first history of music written in England was by Sir John Hawkins in 1776, but his work was relegated to oblivion by Charles Burney's impressive *General History of Music* (1776–89).[237] Burney approached his task extremely systematically—he travelled through France, Italy, Germany, and the Netherlands in order to acquire an accurate picture of contemporaneous music. The first two volumes of his musical history were based on Martini's *Storia della musica,* but the last volume consisted of his own researches. Although his treatment of Bach and Handel was not very good, his history of music got an excellent reception and within a few years was translated into German and Dutch. Burney's work was probably the first musical history based on fieldwork, but he did not use the material he had gathered for an empirical quest for regularities or patterns in the history of music. We do not find such an approach until the nineteenth century (see 5.4).

The history and the technical aspects of musical instruments—*organology*—were also studied alongside musical history. These activities already existed in India (Bharata Muni, see 2.4) and Islamic civilization (al-Farabi, see 3.4), but the first European work did not appear until early in the sixteenth century—the *Musica getutscht und ausgezogen* (1511) by Sebastian Virdung.[238] However, Michael Praetorius ranks alongside Marin Mersenne as one of the most important

236 Elisabetta Pasquini, *Giambattista Martini*, L'Epos, 2007, pp. 31ff.

237 Maria Semi, 'An unnoticed birth of "musicology" in eighteenth-century England', in Bod, Maat, and Weststeijn (eds), *The Making of the Humanities, Volume II: From Early Modern to Modern Disciplines*, 2012, pp. 93–102.

238 Sebastian Virdung, *Musica getutscht*, edited by Klaus Wolfgang Niemöller, Bärenreiter Verlag, 1970.

seventeenth-century organologists. He gave an overview of all known Renaissance musical instruments in his *Syntagma musicum* (1618).[239] The *Musurgia universalis* (1650) by the Jesuit Athanasius Kircher also belongs in this category.[240] Although some of Kircher's results turned out afterwards to have been contrived, such as his infamous 'deciphering' of hieroglyphs (see 5.2), his encyclopaedic collector's mania was beneficial. He was one of the first to collect and record birdsong in musical notation.

China: history of music remains a government matter. Whereas a fierce debate raged in Europe about the nature of consonances and little was done about musical history, the opposite happened in China. The history of music and of music theory was extensively documented while empirical musicology came to an almost complete standstill. The most impressive work from the Ming era was the 5,000-page *Yuelü quanshu* ('Collected Works of Music Theory', 1584) by Zhu Zaiyu, a prince of the Ming dynasty.[241] The work gave a historical overview of the different music theory attainments from China, including the famous and probably oldest theory of equal temperament tuning by Cai Yuanding during the Song Dynasty (see 3.4). Zhu Zaiyu did not shrink from expressing extensive criticism of his fellow music theoreticians. He also innovatively described equal temperament via a more accurate calculation of the twelfth root of 2, which was brought to Europe via Italian merchants in the seventeenth century. But this was a mathematical rather than a musicological result.

The *Shenqi mipu* ('Mysterious and Marvellous Tablature') by Zhu Quan, the seventeenth son of the founder of the Ming dynasty, contained an anthology of sixty-four annotated pieces of Song music.[242] This work, which was of great importance to musical historiography, set down detailed instructions for performing the Song pieces as well as discussions about their historical and theoretical import. The transition to the Qing Dynasty in 1644 changed little in Chinese musicology. While the Jesuits produced several musical pieces in Chinese, their influence made itself felt only in the course of the nineteenth century. There were many descriptions of popular operas and, in the mid-eighteenth-century gigantic anthologies of pre-Qing melodies appeared containing dozens of volumes, but there is no harmonic or stylistic analysis of these illustrious pieces of music to be found.

India: declarative systems of rules for *gamaka* and *raga*. The Indian musicological tradition remained rule-based and declarative (see 3.4). The boundary conditions of possible pieces of music were defined, but there was no procedure for new compositions. This can be seen from the many sixteenth- and seventeenth-century treatises, for example the *Sangita-Parijata*, which gave rules for

[239] Michael Praetorius, *Syntagma musicum*, 3 volumes, Bärenreiter Verlag, 1958–9.

[240] Athanasius Kircher, *Musurgia Universalis*, facsimile, Olms 1970.

[241] Fritz Kuttner, 'Prince Chu Tsai-Yü's life and work: a re-evaluation of his contribution to equal temperament theory', *Ethnomusicology*, 19(2), 1975, pp. 163–206.

[242] L. Carrington Goodrich (ed.), *Dictionary of Ming Biography*, volume 1, Columbia University Press, 1976, p. 305.

ornaments—the *gamaka*.[243] However, this work pales by comparison with the 1609 discourse *Raga-Vibodha* by Somanatha, which specified the endless variations of the *gamaka* using a new, almost pictorial, notation system. We also find manuals with comprehensive procedures for musical composition in Carnatic music (see 3.4). For example the seventeenth-century *Sangita-Sudha* explained how motifs were developed into melodic units, how they could be extended or shortened and could then be combined with rising or falling patterns.[244] It described the complex structure of the *raga*, including the first and second expositions of motifs, and the ways in which these motifs can return and expand into longer and broader chains (the *brikka* or *phirukka*), and ended by addressing the way in which a virtuoso piece could then be brought back down again step by step, so that with a few phrases it was possible to end back on the fundamental tone. These discourses are unbelievably complex and fascinating. They appeared to define virtually the whole span of a musical idiom.[245] Although the treatises had a clear musical pedagogical purport, it is not clear whether the systems of rules were intended to be descriptive or prescriptive.

Africa: little accessible musicology. A huge wealth of music has been produced in Africa, but little musicology has survived. However, it has since become known that the extremely delicate Songhai manuscripts from Timbuktu also discuss the music of their time (see 4.2). Unfortunately, most Songhai manuscripts are not accessible and the ones that are consist primarily of prayer books and chronicles. With the exception of a few brief references to music in Ibn Khaldun, the only contemporaneous African musical histories that are currently available come from external sources. For example, when he visited Mozambique in 1596 on the way to India, Jan Huygen van Linschoten gave one of the earliest descriptions of an African musical instrument complete with a picture of a mouth bow.[246] In the seventeenth century Italian missionaries who were active in the African Kingdoms of Congo and Matamba described local musical practice. Girolamo Merolla's *Breve e succinta relatione del viaggio nel Regno di Congo nell'Africa Meridionale* published in 1692 is one of the most important sources for African musical history. It became clear from this and other works, including Peter Kolb's description of South African music in 1719, that polyphonic singing culture was not unique to Europe.[247] There is centuries-old, very rich polyphonic music in Africa—from the Pygmies to the Khoikhoi.[248]

Ottoman Empire: heir to Arabic musicology. Ottoman musicology has often been dismissed as pedagogical and non-theoretical, but this is incorrect. While we

[243] Anupam Mahajan, *Ragas in Indian Classical Music*, volume 1, Gyan, 2001, pp. 200ff.

[244] *The Sangita Sudha of King Raghunatha of Tanjore*, Music Academy, Madras, 1940.

[245] Harold Powers, 'The structure of musical meaning: a view from Banaras (a metamodal model for Milton)', *Perspectives of New Music*, 14(2), 1976, pp. 308–34.

[246] Jan Huyghen van Linschoten, *Voyage to Goa and Back, 1583–1592, with His Account of the East Indies*, reprint, AES, 2004.

[247] Peter Kolb, *Caput bonae spei hodiernum. Das ist: Vollständige Beschreibung des Vorgebürges der Guten Hofnung*, 1719, reprint Marktredwitz, Volkshochschule, 1975.

[248] John Gray, *African Music: A Bibliographical Guide to the Traditional, Popular, Art and Liturgical Musics of Sub-Saharan Africa*, Greenwood Press, 1991.

do not find any musicological investigations of harmony, there were many studies of the underlying schemes of the Turkish melodic idiom. These schemes defined the class of possible melodies and could—as in Indian music—be looked on as a declarative system of rules for Turkish music. The fourteenth-century Ibn Kurr and the fifteenth-century al-Ladhiqi, for instance, defined the different tonal steps that are needed to create a melodic contour.[249] A later seventeenth-century anonymous work, the *Shajara*, gave both melodic and rhythmic cycles for Turkish music. This work resembled what Safi al-Din and al-Farabi had done centuries earlier for Arabic musical cycles (see 3.4).

The most important musicological work in the Ottoman tradition is a treatise from around 1700 by Dimitrie Cantemir, Prince of Moldavia, who we have already encountered as an important historian in 4.2. During his exile in Istanbul, Cantemir studied Turkish music and wrote the work *Kitab-ı ilm al-musiki ala vech al-hurufat* ('The Book of the Science of Music through Letters').[250] In it he addressed the different classes of melodies using 'simplified' tone schemes without any interval designations. One could interpret these shortened tone schemes as a notational limitation since musical notation was not generally known among the Turks. In the case of Cantemir, however, it is more plausible that he intended to reproduce the schematic system underlying the Turkish music idiom. In fact, he had an excellent command of musical notation and even developed a specific script for representing Turkish instrumental music, the *ebced* notation. He used this notation to preserve hundreds of seventeenth-century pieces of Ottoman music for posterity.

Relationship between musicology and the New Sciences. If we try to take stock of European, Chinese, Indian, African, and Ottoman musicology, we see that systems of rules were developed for musical composition in most regions, but it was only in Europe that the interaction between theory and empiricism matured in the field of consonance studies. This interaction was characterized by a cycle of theoretical underpinning and empirical verification, which seemed to strengthen each other through rejection or improvement. One idea was barely refuted before an alternative had been developed. How different it was from the Aristotelian approach, in which all phenomena could be traced back deductively to first principles, which were considered to be self-evident and inviolable (as was also the case with Aristoxenian melody theory—see 2.4). The early modern European tradition put these so-called self-evident first principles under the spotlight and tried to advance new principles, first of all in the humanities and later in natural sciences.

This brings up the question of what new methodical achievements remained for the new natural sciences, which also emerged in early modern Europe. It has often been said that the mathematical description of reality was one of the greatest achievements of the New Scientists. Yet this already existed in humanistic musicology,

[249] Kurt Reinhard and Ursula Reinhard, *Musik der Türkei, Band 1: Die Kunstmusik*, Wissenschaftliches Buchgesellschaft, 1984.

[250] Owen Wright, *Demetrius Cantemir: The Collection of Notations, Volume 2: Commentary*, Ashgate, 2001. See also Eugenia Popescu-Judetz, *Prince Dimitrie Cantemir: Theorist and Composer of Turkish Music*, Pan Books, 1999.

from Gaffurio to Zarlino. Even the new concept of experimentation, where a phenomenon was put on the rack under controlled conditions, which has often been attributed to Galileo, had already been implemented by Ramis de Pareia and Galileo's father Vincenzo. However, the use of both mathematics and experimentation remained at a fairly basic level. We do not find in musicology the form that experimentation took during the seventeenth century with Robert Boyle, for instance, and the mathematical virtuosity that we see with Isaac Newton, for example. It is therefore going too far to contend that the Scientific Revolution, with its new understanding of the laws of the universe, arose out of the study of music, as has been put forward by some historians.[251] It is clear, though, that all scholars (humanists and 'natural scientists' alike) tackled the theory of consonances, resulting in the new field of acoustics. And it remains a surprising and most significant feat that the fruitful interaction between theory and empiricism ended up in the New Sciences thanks to the fifteenth- and sixteenth-century humanists. It was the natural scientists, though, who can take credit for its elaboration and application to the study of nature in all its finesse.

[251] Drake, 'Renaissance music and experimental science'.

4.5 ART THEORY: A TURNING POINT IN THE REPRESENTATION OF THE VISIBLE WORLD

It is almost impossible to imagine a greater contrast than that between the medieval and the early modern study of art. Apart from a few technical manuals, for over a thousand years nothing was produced in the field of European art theory. Yet, as we have seen in philology, the discipline with the biggest 'backlog' is able to catch up the quickest. This has everything to do with the ubiquitousness of classical art in Italy. All that was needed was a spark to ignite interest, for the material was there for the taking.

A magnificent start: Alberti and linear perspective. Art theory was consequently able to get off to a flying start in Italy. And what a start it was! The first work was at the same time one of the best discourses in the history of the humanities—*De pictura* ('On Painting') written in 1435 by Leon Battista Alberti (1404–1472).[252] Virtually every idea in this book was picked up and elaborated on in subsequent centuries. Born the illegitimate son of a wealthy Florentine banker, Alberti had a classical education at the outstanding school of Gasparino Barzizza.[253] After reading law in Bologna, Alberti proved to excel in everything—from athletics to horsemanship, from poetry to music, from architecture to painting, from pedagogy to philology, from cryptography to art theory. And we already came across Alberti in linguistics, with his brilliant analysis of Tuscan (see 4.3). In the history of the humanities he is seen as an outstanding example of *homo universalis*.[254] Alberti's *De pictura* was the first theoretical work about the visual arts in Europe. As he himself emphasizes in Book II, his work was not a history of art like Pliny's (see 2.5), but a 'theoretical' work. What did Alberti mean by this? In the best humanistic tradition he wanted to develop a model for the ancient rules of art, which he believed he could find in *illusionism* (2.5). However, while elaborating these rules he came up with a model that went far beyond classical art. He developed a completely specified method for the *illusionistic reproduction of three-dimensional objects on a two-dimensional surface.*

In so doing Alberti provided a theoretical underpinning to a practice that had existed for at least ten years in Florentine art—*linear* (or *mathematical*) *perspective*, which he attributed to the sculptor and architect Filippo Brunelleschi, to whom he also dedicated the Italian version of his work. To an extent Alberti was *prescriptive* in his treatment: for example, when he argued that painting had to comply with the laws of linear perspective and not the rules of thumb developed in the studios of the Late Middle Ages (section 19 of *De pictura*). However, he was also *descriptive* when he explained and substantiated an existing tradition—linear perspective had already been used in the relief *St George and the Dragon* by Donatello in 1417 (Bargello,

[252] There are several translations of Alberti's work, e.g. Leon Battista Alberti, *On Painting*, translated by Cecil Grayson, with an introduction and notes by Martin Kemp, Penguin Classics, 1991.

[253] Anthony Grafton, *Leon Battista Alberti: Master Builder of the Renaissance*, Harvard University Press, 2002.

[254] Joan Kelly-Gadol, *Leon Battista Alberti: Universal Man of the Early Renaissance*, University of Chicago Press, 1969.

Figure 4. Masaccio, *The Holy Trinity*, *c.*1425, Santa Maria Novella, Florence.

Florence) and in *The Holy Trinity* fresco by Masaccio in *c.*1425 (Figure 4).[255] As well as a description of perspective, Alberti also supplied a geometrical analysis and a theoretical foundation for the technique. This foundation could—as was the case with the description of musical consonances as numerical ratios—be seen as the underlying, hidden rule that an artist like Masaccio had already used and that could now be defined with precision.

Alberti's analysis of perspective is one of the clearest explanations in the literature. According to Alberti a depiction of reality should always be made such that it

[255] Martin Kemp, *The Science of Art: Optical Themes in Western Art from Brunelleschi to Seurat*, Yale University Press, 1990, pp. 17–20.

looks like a 'view out of a window'. This view corresponds to the pictorial image and the window to the picture plane (the painting). Alberti elucidates his method using imaginary lines which link the artist's eye with the subjects in the depiction and which, when they are intersected by the picture plane (the window, the painting), result in the portrayed composition. Here Alberti applied optical ideas from the eleventh-century Arabic polyglot Alhazen (Ibn al-Haytham), who was known in Europe through the work of thirteenth-century Franciscans like Roger Bacon.[256] In Alberti's thinking, the illusion of a 'view out of the window' was achieved by getting all lines that cut the picture plane to meet at one single point on the horizon (the *vanishing point*). This method enables artists to establish the relative sizes of the objects in the picture. Alberti's famous example is the problem of the illusion of a tiled floor with objects such that the gradual diminution of the tiles and the objects is credible (Figure 5 gives a visual representation following Alberti's description).

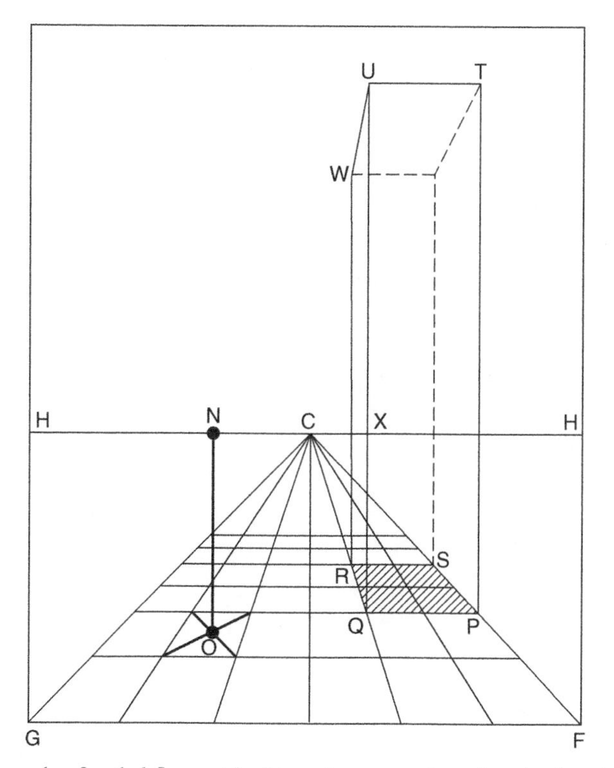

Figure 5. Example of a tiled floor with objects in perspective, after the description in *De Pictura*, Leon Battista Alberti, 1435. (Alberti's original work had no illustrations.)

[256] David Lindberg, *Roger Bacon and the Origins of Perspectiva in the Middle Ages*, Clarendon Press, 1996.

Alberti's impact, and Leonardo's empirical perspective. Alberti's theory of perspective was a resounding success and was immediately adopted by the artists of his time.[257] In around 1470 the painter Piero della Francesca wrote an important follow-up treatise, *De prospectiva pingendi* ('On the Perspective for Painting'), in which he made Alberti's method easier to use in practice by selecting a number of key points in the picture that were projected one for one on the surface.[258] He appeared to have applied his own point-by-point method in 1450 in *The Flagellation of Christ* (see Figure 6).

Figure 6. Piero della Francesca, detail of *The Flagellation of Christ*, *c.*1450, Urbino.

[257] Though the first impulse to using linear perspective was given by Brunelleschi—see Kemp, *The Science of Art: Optical Themes in Western Art from Brunelleschi to Seurat*, p. 9.

[258] Piero della Francesca, *De prospectiva pingendi*, reprinted by Aboca Edizioni, 2008.

All the Italian artists who were abreast of the times very soon switched to linear perspective. As had been the case 2,000 years before with the Pythagorean laws of music, initially Alberti's perspective laws seemed to be a success story at which no one demurred. However, anyone who looks closely at paintings based on purely linear perspective cannot avoid the perception that there is something artificial about this type of portrayal. The precise application of the laws of perspective appears to produce rather sterile and laboured pictures that make one think of the technical illustrations in a present-day popular science magazine. It is somewhat like the situation with basic intervals in consonance theory (see 4.4). These are defined mathematically as ratios of whole numbers, but they sound better—or should we say 'more exciting'—if the pure ratios are adjusted slightly upwards or downwards (depending on the instrument and the interval).

The same proved to be true of mathematical perspective. What Alberti could not have surmised was discovered by Leonardo da Vinci (1452–1519): an image is not necessarily 'harmonious' if it complies scrupulously with mathematical laws. Leonardo still believed that there were laws underlying the discipline of visual depiction, it was just that they were 'other' laws. To a degree this was true. Alberti's approach was demonstrably inadequate for wide angles, as had been shown by Piero della Francesca.[259] This problem could still be solved mathematically, though. However, what Leonardo undertook was *a quest for the underlying laws of perspective in which it was not mathematics but empiricism that had the last word.* Leonardo set to work extremely systematically, as we can observe from his *Trattato della pittura* (published posthumously).[260] He investigated all possible perceptual changes that manifest themselves when the relative positions of the objects, the picture plane, and the observer are varied. During this process he worked with differences in colour and shape, he developed machines that could draw in perspective and he tried to take the working of the eye into account. The result of all this was that Leonardo no longer embraced Alberti's linear perspective because he had become aware of the deceptions and complications of the visual process. Leonardo did not, though, completely turn his back on the laws of mathematical perspective. Indeed, he tried to resolve the shortcomings by incorporating small, experimentally derived changes so that the result looked even more credible. He also added a number of advances in *non*-linear perspective, such as the gradual shift of light and shade as the objects were further removed. He did not, however, succeed in developing an alternative consistent system for perspective as a whole. A theoretical basis for visual depiction seemed further away than ever.

A parallel with Pythagorean and Aristoxenian music theory comes to mind. While the musicologists wanted to base their work on pure Pythagorean ratios, perceptual practice was so recalcitrant that these ratios had to be adapted, and for tuning people had to go in search of other systems that also did justice to musical

[259] Kirsti Andersen, *The Geometry of an Art: The History of the Mathematical Theory of Perspective from Alberti to Monge*, Springer, 2007, p. 40.

[260] Leonardo da Vinci, *A Treatise on Painting*, translated by John Francis Rigaud, Kessinger Publishing, 2004.

perception (see 4.4). This primarily resulted in hybrid systems, no matter how hard people tried to find a sound theoretical basis. As with consonance in music, perspective in painting was partially objective (based on mathematical principles) but also partially subjective (founded on perception, which is time- and place-dependent). It is true that Alberti's mathematical perspective gave a guideline, but it was also necessary to use the empirical and aesthetic 'corrections' described by Leonardo. Some humanists must have been aware of this remarkable but annoying parallel between musicology and the study of visual art. For example in the sixteenth century Giovanni Battista Benedetti sought a definitive mathematical theory for both consonances (see 4.4) and perspective,[261] but did not succeed in either.

In Northern Europe the new insights into linear perspective only became known during the course of the sixteenth century. In the fifteenth century painters in the Southern Netherlands (Flanders) like Jan van Eyck and Rogier van der Weyden were already applying an approximation of linear perspective, but without a mathematical foundation. Thanks to the German artist Albrecht Dürer, the theoretical justification of the technique of perspective was brought from Italy to the north in 1506. However, in his book *Underweysung der Messung* ('Instructions for Measuring', 1525) Dürer did not adhere to Alberti's explanation in all regards,[262] and one could even say there was a design error that worked out well and by chance resulted in a more empirical perspective. Above all it was Dürer's invention of the perspectograph which had the greatest impact on art output and art theory, and which was included in all later manuals on perspective (Figure 7). Studies of perspective reached great heights in the seventeenth century in Dutch painting in the works of Pieter Saenredam and Samuel van Hoogstraten. However, the perspectival results of these studies are the paintings themselves.

Alberti's *disegno* and Ficino's Neoplatonism: canon of *ideal selection*. Over and above the introduction of perspective, Alberti was the leading figure in the theory of *disegno* ('design'), which is the subject of the second book of *De pictura*. Here, too, Alberti was both descriptive and prescriptive. He analysed the existing methods of Italian artists in designing an image—the general outline, the composition and the objects. However, he also added a number of guidelines with which he created a new, prescriptive canon.

Alberti divided the art of painting into three components:

(1) *Circumscriptio* (Sketch): the drawing of outlines of all objects and parts of objects on the basis of detailed descriptions.

(2) *Compositio* (Composition): the hierarchical arrangement of the constituent parts of the image; the way that the *historia* (the depicted story) is divided into forms, which are subdivided into parts, which in turn are broken down into the smallest constituent parts (e.g. hand, fingers, and nails).

[261] J. V. Field, 'Giovanni Battista Benedetti on the Mathematics of Linear Perspective', *Journal of the Warburg and Courtauld Institutes*, 48, 1985, pp. 71–99.

[262] For an English translation, see Albrecht Dürer, *De Symmetria Partium In Rectis Formis Humanorum Corporum/Underweysung Der Messung*, translated by Silvio Levy and Walter Strauss, with an introduction by David Price, Octavo, 2003.

Figure 7. Albrecht Dürer, artist with mechanical grid, in *Underweysung der Messung*, 1525.

(3) *Receptio luminum* (Reception of light): the contribution of light and shades.

As far as we know, Alberti's art analysis was the first of its type in Europe. We have, however, come across theories about the structure and organization of a painting in the Indian and Chinese traditions (such as the 'limbs theory' and Xie He's Six Principles—see 2.5), but these were unknown to Alberti. Alberti's hierarchical notion of *compositio* therefore almost appears to have been plucked out of thin air. Yet we have already seen a comparable concept in classical rhetoric in the structure of a text or speech (see 2.7) and also in Vitruvius's architectural theory with regard to the organization of a building (see 2.4).[263] In the fifteenth century the concept of *compositio* was known to everyone who had been taught humanistic rhetoric—words make up a phrase, phrases a clause, and clauses a sentence. Alberti extended this concept to painting. As in rhetoric, he characterized the composition of a painting as a hierarchically layered structure in which the tiniest elements of an image are combined in the right way to create ever greater parts, which between

[263] Michael Baxandall, *Giotto and the Orators: Humanist Observers of Painting in Italy and the Discovery of Pictorial Composition 1350–1450*, Oxford University Press, 1971, p. 130.

them make up a coherent organization of the whole work. Using his new concepts, Alberti was able to analyse a fresco by Giotto as if it were a sentence by Cicero.[264] As we have already seen, this analytical method also had repercussions in other disciplines, including the analysis of music (see 4.4). The hierarchically segmented approach was to become a constant in virtually all modern humanities (see chapter 5).

In addition to a new art analytical method, Alberti also made a number of normative recommendations that have become indicative for generations after him. He rejected the medieval use of gold in paintings. The illusion of gold should on the contrary be achieved by light and colour techniques. Alberti moreover argued that the representation of a *historia*—a narrative painting—was at the top of the hierarchy of subjects that a painter can choose. Other subjects were also integrated into the *historia*, such as portraits, landscape (the background to the story), the correct depiction of emotions and the reactions to them. And the *historia* concerned more than just the narrative aspects of an image—the allegorical qualities were part of it too.

In the third and last book of *De pictura* Alberti addressed the artist's training. It had to correspond to that of an all-round humanist trained in the liberal arts. In making this assertion, Alberti definitively departed from the medieval idea of an artist as an artisan. An earlier work, the *Libro dell'arte* ('Book of Art'), written by Cennino Cennini in about 1399, had already contained a plea for elevating painting to the status of poetry, but the elaboration became bogged down in technical instructions.[265] Alberti gave art practice a complete intellectual repertoire based on empirical and theoretical insights. For example, according to Alberti the artist should not just imitate nature. He must also *improve* it in order to achieve true beauty, although at the same time he warned against wanting to better nature without having studied it sufficiently beforehand. Alberti recommended a method (described in Pliny) that was developed by Zeuxis, who had created an idealized portrayal of Helen of Troy by selecting the most perfect character traits of different models and combining them to create a new whole. This approach of *ideal selection* became the new canon for painting for the coming centuries. Like Polykleitos's classical guideline, Alberti's canon is prescriptive (see 2.5), but at the same time both canons arose out of an existing practice—in Alberti's case the assumed method of Zeuxis.

Alberti's canon of ideal selection received a philosophical foundation from Marsilio Ficino (1433–1499). Ficino, a translator of Plato and the Corpus Hermeticum (see 4.1), was in pursuit of a universal philosophical system that combined all fundamental truths from all other systems—from classical to modern and from Christian to non-Christian.[266] This form of *synthetic Neoplatonism* struck a chord

[264] This analogy comes from Baxandall, *Giotto and the Orators*, p. 131.

[265] D. V. Thompson Jr, *Cennino d'Andrea Cennini da Colle di Val d'Elsa: Il Libro dell'Arte*, 2 volumes, Yale University Press, 1933.

[266] Michael Allen and Valery Rees (eds) with Martin Davies, *Marsilio Ficino: His Theology, His Philosophy, His Legacy*, Brill, 2002.

with humanist artists.[267] It gave them justification for harking back to the classical past as an example. After all, absolute truth or beauty could only be obtained by combining the best of what existed. In this regard ideal selection is more example-based than rule-based (see the discussion of examples vs. rules in Sanctius in 4.3).

Finally, Alberti dealt with the rules of *decorum*. Old men must employ restrained gestures in contrast to young men. Philosophers, soldiers, and saints had to be dressed in accordance with specific guidelines and so on. These rules complied with the Aristotelian recommendation to gather as much knowledge as possible about different types of people in order to achieve the desired rhetorical and poetic effect (see 2.7 and 2.8). All these facts contributed to knowledge of the visible world.

Art historical patterns: from Vasari and Zuccari to Bellori. Art theory was reanimated, and so too was art historiography. We see this in *Le Vite de' più eccellenti pittori, scultori, ed architettori* ('The Lives of the Most Excellent Painters, Sculptors and Architects') by Giorgio Vasari (1511–1574).[268] The first edition of this work appeared in 1550 and it was enlarged in 1568. Vasari's book consists of a collection of biographies—over 150 artists arranged in chronological order from the thirteenth century to Vasari's own time.[269] This sequence of lives resulted in a magnificent story of art which, after the decline in medieval painting, improved again bit by bit from humble beginnings with the *primi lumi* such as Cimabue and Giotto and which, via Masaccio and Donatello, ended up with the great Leonardo and achieved the ultimate perfection in the art of Michelangelo—said Vasari. Vasari 'discovered' a *pattern of progress* as Pliny had observed in classical art, namely an increase in the *illusionistic* portrayal of reality. In addition to this growth in illusionism, Vasari also saw a pattern of increase in *beauty*, *persuasiveness*, and *expression of abstract ideas*.[270] Vasari assumed that his assessments of quality were completely objective and universal. For example, Michelangelo was the greatest artist of all time because he had perfected painting, sculpture, and architecture individually and also because he had demonstrated the unity between these three on the basis of the *disegno* principle—the hierarchical cohesion between the parts of the whole.

After Vasari, the number of art theory works started to increase apace. Gianpaolo Lomazzo gave as complete an overview as possible of the subtleties of painting in practice,[271] and Federico Zuccari attempted to base *all* arts—from the liberal to the

[267] Marieke van den Doel, 'Ficino, Diacceto and Michelangelo's Presentation Drawings', in Bod, Maat, and Weststeijn, *The Making of the Humanities, Volume I: Early Modern Europe*, 2010, pp. 107–32.

[268] For an English translation, see Giorgio Vasari, *The Lives of the Most Eminent Painters, Sculptors and Architects*, translated by G. de Vere, 3 volumes, Abrams, 1979 [1912–15]. On the significance of Vasari for European art history, see Einar Rud, *Giorgio Vasari: Vater der europäischen Kunstgeschichte*, W. Kohlhammer Verlag, 1964.

[269] For a critical view on Vasari's authorship, see Patricia Lee Rubin, *Giorgio Vasari: Art and History*, Yale University Press, 1995. See also Charles Hope, 'Can you trust Vasari?', *New York Review of Books*, 5 October 1995, pp. 10–13.

[270] Williams, *Art Theory: An Historical Introduction*, pp. 71–2.

[271] Gianpaolo Lomazzo, *Scritti sull'arte*, edited by R. P. Ciardi, 2 volumes, Marchi & Bertolli, 1973–4.

visual—on the one principle of *disegno*.[272] According to Zuccari *disegno* was nothing less than the fundamental principle of all thinking. He created a hierarchy of all knowledge, in which the visual arts came in second highest place—just below theology.

In the seventeenth century, art theory treatises were used as manuals at art academies.[273] Charles LeBrun and Giovanni Bellori diversified the theme of ideal selection, and Bellori also spotted a new pattern in art history. In his *Vite* of 1672 Bellori described the sharp decline in art after Raphael (from 1520 onwards).[274] According to Bellori we do not find a revival of the fine arts until the late sixteenth-century painter Annibale Carracci. Carracci was able to once again achieve perfect beauty by adapting and correcting natural forms. Bellori was disparaging about Caravaggio, whose purely realistic painting fell grossly short, no matter how much his naturalistic style was imitated. The *recurring cycle of rise, peak, and decline* had now also appeared in art historiography.

An alternative canon? The 'picturesque' among Dutch art theorists. The canon of ideal selection was more theoretical than empirical. With the exception of the academies, only a few artists followed the Neoplatonic lead. Earlier we mentioned Caravaggio, whose non-idealizing style was widely copied throughout Europe, but it was primarily in the Netherlands, with its divergent 'realistic' painting culture from Frans Hals to Rembrandt, where ideal selection was in many cases completely absent (see Figures 8 and 9).

Although a Dutch classicism with many renowned exponents co-existed alongside Dutch realism,[275] some northern art theoreticians sought to justify the realistic bent of Dutch art. The first, Karel van Mander, modelled his 1604 *Schilder-boeck* ('Book of Painting') on Vasari's *Vite*.[276] After an outline of the Italian art theory he mentioned a new concept, for which he used the Dutch word *schilderachtig* which is commonly translated as 'picturesque' (albeit the latter word came into use only in the late eighteenth century). Although Van Mander did not use this concept consistently, in an art theoretical sense it appeared to mean 'from life', 'true to life', and 'characteristic'.[277] In any event it did not signify 'from the classics' or 'ideal selection'. Later theorists like Joachim von Sandrart and Samuel van Hoogstraten, as well as writers like Bredero, adopted the concept of picturesque and gave it their own interpretation. Von Sandrart used the term to make it clear that Rembrandt's

272 Federico Zuccaro, *Scritti d'arte di Federico Zuccaro*, edited by D. Heikamp, Olschki, 1961.

273 Carl Goldstein, *Teaching Art: Academies and Schools from Vasari to Albers*, Cambridge University Press, 1996.

274 Giovanni Bellori, *Le Vite de' pittori, scultori e architetti moderni*, edited by E. Borea, Einaudi, 1976. See also Francis Haskell, *Patrons and Painters: A Study in the Relations between Italian Art and Society in the Age of the Baroque*, Yale University Press, 1980.

275 Albert Blankert, 'Classicism in Dutch history painting', in A. Blankert, J. Giltaij, and F. Lammertse (eds), *Dutch Classicism*, NAi Publishers, 2000.

276 Karel van Mander's *Schilder-boeck* has been translated into English as *The Lives of the Illustrious Netherlandish and German Painters*, translated and edited by Hessel Miedema, 6 volumes, Davaco, 1994–9.

277 Caroline van Eck, Jeroen van den Eynde, and Wilfred van Leeuwen (eds), *Het schilderachtige*, Architectura & Natura Pers, 1994, pp. 12–15.

Figure 8. Frans Hals, *Malle Babbe*, *c*.1630.

Figure 9. Adriaen Brouwer, *The Bitter Tonic*, *c*.1635.

art was not based on learning but on the agreeable aspects of nature. The picturesque seemed to offer an alternative to Alberti's canon.

This alternative 'canon' did not acquire much prestige though. The art theoretician Jan de Bisschop asserted that painters could lose themselves in the picturesque to such an extent that they gave preference to ugly subjects.[278] Yet De Bisschop also contended that all elements of the visible world contain a certain beauty. And if nature is already beautiful, why would one still strive for idealization? After all, art was about the whole of the visible world, as Van Hoogstraten argued. There was a hierarchy in the many objects, but the value of *imitatio* conferred a right to exist on even the ugliest things. The art theorist Willem Goeree stated that beauty was only a relative concept. It depended on the context and the opinion of the beholder. Some even reasoned that ugliness has its charms. Van Hoogstraten for example wrote about the 'likeable ugliness' in the work of Adriaen Brouwer (see Figure 9 for instance), and Goeree determined that 'in art some ugliness is also beautiful'. As an explanation for this Dutch concept, the picturesqueness of the ugly has on occasion been associated with the Calvinist preoccupation with the temporary, the transient. However, what interests us above all here is that these art theoreticians tried to give some foundation to Dutch painting practice. And that is why they found some common pattern in Dutch art—the desire to create lifelike, and therefore sometimes ugly, images—in contrast to the academic pursuit of classical and universal 'beautiful' portrayals.

Yet this descriptive approach in art theory did not appear to set the tone. The more normative and prescriptive theorists held all the cards. In his *De pictura veterum* ('On the Painting of the Ancients', 1637) the philologist Franciscus Junius, who was active in Holland and England, contended that the perfect art of painting that had existed in Antiquity was so brilliant and pure that it could never be matched.[279] On the grounds of Junius's discourse many went in search of the laws of perfect art, and at the end of the century people believed they had found them. Perfect painting had the greatest simplicity and clarity that one could achieve with a guideline based on rules. Gerard de Lairesse's *Groot Schilderboek* ('Great Book of Painting') is one of the best known examples of such a guideline, which in our terminology corresponds to a declarative system of rules.[280] A painter, De Lairesse went blind at the age of fifty and started to teach classes in his home in Amsterdam. His children made notes of his lessons, which were published and translated into

[278] For the full citations by De Bisschop, Goeree, and Van Hoogstraaten, see Thijs Weststeijn, *De zichtbare wereld: Samuel van Hoogstratens kunsttheorie en de legitimering van de schilderkunst in de zeventiende eeuw*, PhD thesis, Universiteit van Amsterdam, 2005, pp. 214ff. For the English translation, see Thijs Weststeijn, *The Visible World: Samuel van Hoogstraten's Art Theory and the Legitimation of Painting in the Dutch Golden Age*, Amsterdam University Press, 2008.

[279] Junius's *De pictura veterum* was highly influential: it was translated into English as *On the Painting of the Ancients* in 1638, and into Dutch as *De Schilder-konst der Oude begrepen in drie boecken* in 1641.

[280] De Lairesse's *Groot Schilderboek* was translated into English as *A Treatise on the Art of Painting in all its Branches, Accompanied by Seventy Engraved Plates, and Exemplified by Remarks on the Paintings of the Best Masters, Illustrating the Subject by Reference to their Beauties and Imperfections*, translated, revised, corrected, and accompanied with an essay by W. M. Craig, London, Edward Orme, 1817.

French, German, and English from 1707 onwards. At an art theoretical level De Lairesse's work most resembles Alberti's ideal selection. In the more practical realm, however, De Lairesse provided a list of rules that could be used to produce consummate art. He specified the proportions of all parts of the male and female bodies, and gave an exhaustive overview of drawn postures, positions, and compositions that should be used in all conceivable situations and actions. The artist could demonstrate his erudition in the knowledge and execution of these rules, but the constraints within which he was free to move became more and more restrictive. And, needless to say, Rembrandt was condemned as a representative of the wrong way to do things.

Winckelmann's new art history and the sublime in Burke. Vasari's art historiography remained dominant for centuries. This situation did not change until Johann Joachim Winckelmann wrote his *Geschichte der Kunst des Altertums* ('History of Ancient Art', 1764).[281] In response to the rather anecdotal biographies before him, he tried to describe art history as a continuing narrative line through which he wanted to arrive at the essence of art. Winckelmann was the source of the division of Greek art into three development stages—from *archaic* via *classical* to *hellenistic*—which is still being used by art historians. Winckelmann considered the classical period in Greek art to be the most perfect of the three stages. His reverence for Antiquity was conventional, but his explanation for it was not. Winckelmann attributed the superiority of Greek classical art to the character of the Greeks in the fifth century BCE and their major attainments in democracy and philosophy. In other words Winckelmann saw the social and intellectual context rather than the artist as the key.

The eighteenth century also saw the creation of a new discipline—philosophical aesthetics—of which Alexander Baumgarten, who wrote *Aesthetica* (1735/1762), was one of the founders. Although philosophical aesthetics is largely outside the scope of our quest, Edmund Burke's *A Philosophical Enquiry into the Origin of our Ideas of the Sublime and Beautiful* (1757) is worthy of mention because of his assertion that both the beautiful and the ugly can lead to sublime experiences. Burke was clearly inspired by the classical treatise about the sublime in the literature of Longinus (see 2.8), but extended his ideas to painting and to human emotions in general. His philosophy has been very valuable in regard to the insight that the non-beautiful and things that break the rules can evoke aesthetic emotions.[282] We already came across this perception in the work of Dutch art theoreticians, for example Van Hoogstraten's 'likeable ugliness' and Goeree's contention that some ugliness in art may also be appealing. Outside the Netherlands their ideas had no impact whatsoever on classicistic art theories. However, the fact that Burke's philosophical art theory struck the right chord in Europe half a century later—from Denis Diderot to Immanuel Kant—showed how unbearable the classicistic straitjacket had become in the course of the eighteenth century.

[281] Johann Joachim Winckelmann, *History of Ancient Art*, translated by G. Henry Lodge, 2 volumes, F. Ungar Publishing Company, 1969. See also Alex Potts, *Flesh and the Ideal Winckelmann and the Origins of Art History*, Yale University Press, 1994.

[282] Koen Vermeir and Michael Funk Deckard (eds), *The Science of Sensibility: Reading Burke's Philosophical Enquiry*, Springer, 2012.

The laws of architecture: Alberti again and his continued impact. The dominance of Alberti in the art theory of the early modern age can barely be overestimated. His formulation of linear perspective represented a turning point in the portrayal of the visible world and prompted a torrent of new discourses. His concept of *historia* determined all narrative painting in Europe. And his canon of ideal selection moulded all art theorists except those few who wished to justify 'wrong' local painting practice. Yet Alberti's most voluminous work was not about painting but architecture. His *De re aedificatoria*, written in *c.*1452, was the first architectural treatise in Europe since Vitruvius.[283] In this work too Alberti harked back to the classics, but whereas many elements in Alberti's *De pictura* were new, his architectural theory is primarily a more systematic explanation of Vitruvius's rather obscure *De architectura* (see 2.5).

Alberti wanted to base architecture on the 'laws of nature', by which he meant the universal principles of correct proportions. In keeping with the classics, Alberti argued that the harmony of the cosmos was expressed in mathematical terms that could be imitated in architecture. An architectural design therefore had to be based on three principles: (1) number, (2) proportion, and (3) distribution. The correct application of these three principles resulted in *concinnitas*: the right harmony between the parts of the whole that would otherwise remain separate from one another. Alberti's harmony was based on Pythagorean music theory, where the consonances are expressed in perfect mathematical ratios such as 1/2, 1/3, 2/3, and 3/4.[284] He used these ratios in his own designs, for example Palazzo Rucellai in Florence, where he used a carefully proportioned façade to make a harmonious whole from a row of houses (Figure 10).

Alberti brought forward an extremely influential concept, which was that if the 'laws of nature' were complied with, the harmonic relationships between the parts resulted in a situation where nothing could be removed or added without making the outcome worse. The most important difference from Vitruvius is Alberti's use of musical consonances as the basis for architectural ratios, whereas Vitruvius saw the human body as the primary source for the correct proportions (see 2.5). There was also a subtle rhetorical distinction between the two. While Vitruvius described how buildings *were* designed, Alberti wrote how they *should be* designed. As has happened so often in the history of the humanities, a transition from description to prescription can be observed here.

We also find an emphasis on the correct architectural proportions in most of the discourses after Alberti, although the ratios were not always based on musical consonances. For example in 1490 Francesco di Giorgio stressed, as had Vitruvius, that the human body was the prime wellspring for the correct proportions.[285] Michelangelo also used the human body as the foundation for architecture because,

[283] Leon Battista Alberti, *On the Art of Building in Ten Books*, translated by Joseph Rykwert, Neil Leach, and Robert Tavernor, with an introduction by Joseph Rykwert, The MIT Press, 1991.

[284] Rudolph Wittkower, *Architectural Principles in the Age of Humanism*, A. Tiranti, 1952, p. 102.

[285] L. Reti, 'Francesco di Giorgio Martini's treatise on engineering and its plagiarists', *Technology and Culture*, 4, 1963, pp. 287–98.

Figure 10. Leon Battista Alberti, *Palazzo Rucellai*, 1446–51, Florence.

as he explained in a surviving letter, buildings had to imitate the human body in their symmetry and openings.[286] He was even opposed to the use of a priori proportions, at which point he departed from Vitruvius's system of thought. Leonardo on the other hand went along with Vitruvius when he was able to depict in unparalleled fashion the Vitruvian descriptions of the human poses in the two basic cosmic forms, *homo ad circulum* and *homo ad quadratum* (see 2.5), in one figure: *Vitruvian Man* (Figure 11).

The use of musical consonances in architectural theory resurfaced in the second half of the sixteenth century with the famous treatise by Andrea Palladio *I quattro libri dell'architettura* (1570).[287] Palladio criticized Michelangelo and once again based architectural ratios on Pythagorean harmony theory. Architecture was restrained and reduced to a declarative system of rules of strict boundary conditions. Sebastiano Serlio's *Tutte le opere dell'architettura* (published between 1537 and 1575) is an example of such a system of rules that defines architecture as a form of *grammatical syntax* that went beyond the painting syntax of De Lairesse and the

[286] James Ackerman, 'The Tuscan/Rustic order: a study in the metaphorical language of architecture', *Journal of the Society of Architectural Historians*, 42(1), 1983, pp. 15–34.

[287] Andrea Palladio, *The Four Books on Architecture*, translated by Robert Tavernor and Richard Schofield, The MIT Press, 1997.

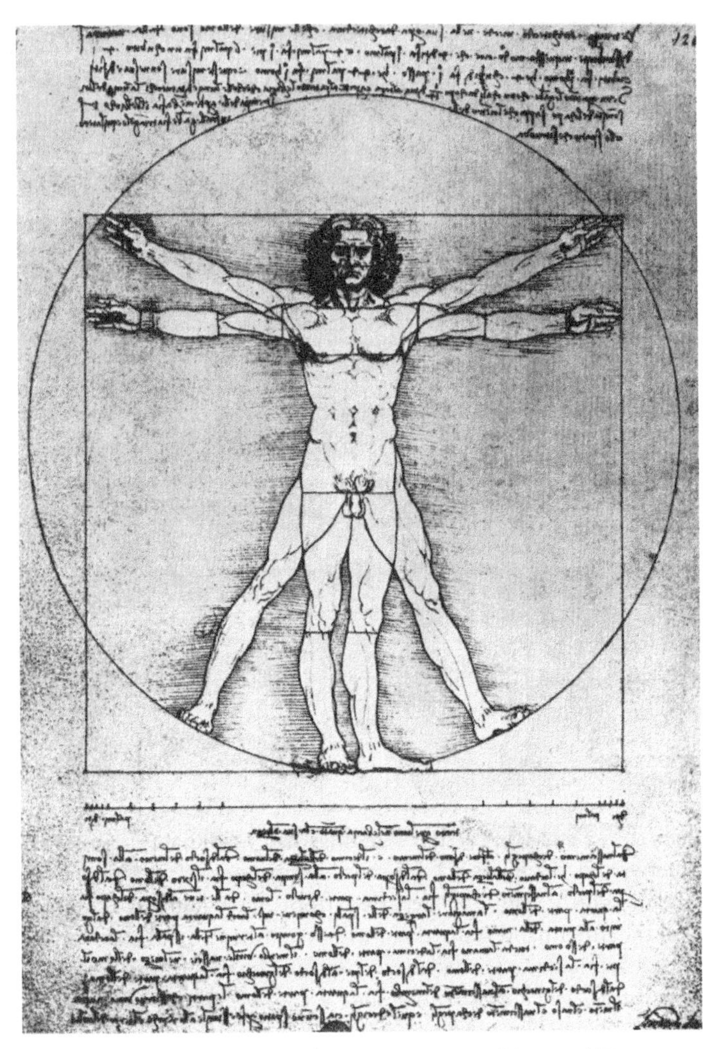

Figure 11. Leonardo da Vinci, *Vitruvian Man, c.*1487.

musical syntax of Dressler in terms of precision.[288] When Palladio's pupil Vincenzo Scamozzi moreover gave rhetorical underpinning to his master's architectural theory, we can say that there was a new *architectural canon.*[289] This canon was to form the basis of the strict seventeenth- and eighteenth-century classicistic theory of architecture. It spread extremely widely in Protestant Europe, while Italian architecture set off once again in a different direction with the much freer

[288] Vaughan Hart and Peter Hicks (eds), *Sebastiano Serlio on Architecture*, 2 volumes, Yale University Press, 1996–2001.

[289] Rudolph Wittkower, 'Review of Franco Barbieri, Vincenzo Scamozzi', *The Burlington Magazine*, 95, 1953, p. 171.

baroque of Francesco Borromini (a very similar thing happened with the classicist theory of poetics—see 4.7). However, baroque architecture did not get any theoretical underpinning for the time being.

Classicistic architectural theory also remained dominant in the eighteenth century, although as a result of the direct study of Greek architecture it was given a new interpretation that became known as *neoclassicism*. Greek architecture had not been directly accessible for a long time, but when Robert Wood published his engravings of Palmyra (Syria) and Baalbek (Lebanon) in 1751 he caused a real sensation in Europe.[290] Inspired by Winckelmann's study of Greek art, Wood's work resulted in a new definition of classical orders and their correct proportions. Whereas Alberti made the proportions depend on musical consonances, they were now determined *empirically* on the basis of the historical study of Greek art and architecture.

To put it in a nutshell, from the fifteenth to the eighteenth centuries early modern architecture theoreticians sought and found underlying patterns in architecture that were descriptive in some cases, mostly prescriptive and finally historical. These patterns consisted predominantly of mathematical ratios that were underpinned by musical, physical, or empirical knowledge.

China and India: towards a historicization of style. As in the Tang and Song Dynasties, during the relatively short Yuan Dynasty (1279–1368) Chinese art critics used the principles of Xie He in their assessment of painting (see 2.5 and 3.5). During the course of the fourteenth century a large number of technical manuals were produced that associated different forms of depiction (for example flowers and branches) with specific emotions. The two manuals by Wang Yi (*c.*1333–*c.*1368) addressed the portrayal of colour and facial expression.[291] These works were the forerunners of even more specific manuals in the seventeenth century. During the Ming period there were growing numbers of art connoisseurs or *literati* who commented on Chinese art history. Initially the opinions of these cognoscenti circulated as collected notes and in some cases were put in the form of manuscripts.[292]

This great interest in Chinese art among literati was comparable to the interest in classical Chinese manuscripts (see 4.1). The emancipation of these connoisseurs even led to new art theories, such as we find in the study notes of Wang Shizhen (1526–1590).[293] Wang described the history of art as a succession of turning points that were marked by pre-Tang and Tang *figure* painters and draughtsmen, followed by Yuan *landscape* painters. In the Chinese court, on the other hand, there was a preference for the Song style of art, which was also historically documented.

At the beginning of the Qing era no Chinese artist could still paint without taking account of the huge rich artistic past from which he could choose. The

[290] Hanno-Walter Kruft, *Geschichte der Architekturtheorie*, Verlag C. H. Beck, p. 235.

[291] David Sensabough, 'Images of the man of culture in fourteenth century China', in Perry Link (ed.), *The Scholar's Mind*, Chinese University of Hong Kong, 2009, pp. 1–16. See also Susan Bush, *The Chinese Literati on Painting: Su Shih (1037–1101) to Tung Ch'i-ch'ang (1555–1636)*, Harvard University Press, 1971.

[292] Lin Yutang, *The Chinese Theory of Art: Translations from the Masters of Chinese Art*, Putnam Sons, 1967. See also Craig Clunas, *Empire of Great Brightness: Visual and Material Cultures of Ming China, 1368–1644*, Reaktion Books, 2007.

[293] Man Leung Tang, *Wang Shizhen (1526–1590): A Study of Patronage in Art*, PhD thesis, Chinese University of Hong Kong, 2006.

eclectic practice of art, with its historicizing 'neo-styles', began in China as early as the seventeenth century, while a similar development did not occur in Europe until the nineteenth century. There was nothing that seventeenth-century Jesuits could teach the Chinese about style awareness and historicization. On the contrary, the few Jesuit artists working in China, such as Giuseppe Castiglione (1688–1768), started to paint in the styles described by Chinese literati, although they continued to display traces of European influence.[294]

As in previous periods, perspectival renderings were widely used during the Yuan and Ming periods (see 2.5). But save a corrupted fragment discussed in 2.5, a search for the underlying principles is not found for this type of 'ruled-line' perspective.[295] Clearly artists had knowledge about it, but this knowledge was not theorized as in Alberti's work. Rather, as with Northern-European art (see above), the works on perspective were the actual paintings themselves. A foundation for the practice of Chinese perspective is not found before the modern era.

In India it appeared that the rich eleventh- and twelfth-century tradition of art discourses with the earliest description of foreshortening (see 3.5) did not develop further—in any event not with regard to our quest. Similarly, it is extremely difficult to find anything about art history or art theory in the Mughal Empire, with its dazzling miniatures (see 4.2).[296] It is of course possible that appearances are deceptive, and there are still many manuscripts waiting to be unveiled or accessed.

Is art theory empirical? On the one hand, from Alberti onwards European art theory provided a foundation for the existing practice of linear perspective, which Leonardo made empirical. On the other hand, the same art theory gave highly normative rules that could not be derived empirically, such as the *historia* as the best achievable painting design and the a priori proportions in architecture (as well as 'ideal selection' in painting). After Alberti virtually all art theory became prescriptive and anti-empirical. That was different from the world we find in music theory, where one hypothesis had no sooner been devised than it was replaced by another—although we have seen prescriptive elements in composition theory too. With the exception of a few Dutch theorists, the 'classical' starting point in early modern art theory remained largely intact until we get to Burke. Art theory seemed to be entangled in its predilection for exact rules and its urge for classical imitation. Although little art had survived from Antiquity—and that was mainly in the form of Roman copies—art theory laboured more under the yoke of humanism than music theory did. We do not see a return to a more descriptive approach until the eighteenth-century studies of Greek art and architecture. The only other region with which we could have compared European art theory is China, where a historicizing approach made its debut in the seventeenth century—far before anything similar in Europe occurred.

[294] Fred Kleiner, *Garner's Art Through the Ages: A Global History*, Thompson Wadsworth, 2009, p. 729.

[295] Michael Sullivan, *The Arts of China*, 4th edition, University of California Press, 1999, pp. 166ff.

[296] Milo Cleveland Beach, *Early Mughal Painting*, Harvard University Press, 1987. See also Daniel Ehnbom, 'A leaf of the *Qissa-i Amir Hamza* in the University of Virginia Art Museum and some thoughts on early Mughal painting', in Rosemary Crill, Susan Stronge, and Andrew Topsfield (eds), *Arts of Mughal India: Studies in Honour of Robert Skelton*, Mapin, 2004, pp. 28–35.

4.6 RHETORIC: THE SCIENCE OF EVERYTHING (OR NOTHING)?

During the Middle Ages rhetoric barely rose above the status of preaching and writing letters, but during humanism it towered above all the other *artes*. According to Lorenzo Valla rhetoric was the *ars* that integrated logic and linguistics and even made them superfluous (see 4.3). With its notion of hierarchical stratification, rhetoric acted as theoretical underpinning for the practice of musical, visual, and literary composition. It endeavoured to bring together all knowledge and was to become one of the most significant elements of the Renaissance.[297] As Cicero said, a competent rhetorician should know about everything, provided that this knowledge can be put to practical use.

Assimilation of classical oratory. Despite rhetoric's elevated status, we return home with virtually empty hands if we go in search of new insights, principles, or patterns in early modern rhetoric. As the historian of rhetoric Brian Vickers remarks, the history of humanistic rhetoric is the story of the assimilation and synthesis of classical treatises that were discovered one by one by manuscript hunters.[298] This story began with the discovery of previously unknown Ciceronian speeches and letters by Petrarch and Salutati.[299] In Cicero's case, humanists could use his personal writings to get to know the great Roman orator, who also emerged as a philosopher and statesman. One discovery was swiftly followed by another. Poggio Bracciolini found a complete version of Quintilian's *Institutio oratoria* in 1416 while travelling to the Council of Constance (see 4.1). He copied out the manuscript in a very short space of time and a few years later forty copies were circulating. Dozens of editions had been printed by the end of the fifteenth century. This gives us an idea of the demand for discourses on rhetoric. The discovery of the complete text of Cicero's *De oratore* created a sensation that was at least as great.[300] Finally people could read Cicero in the form of a superbly composed dialogue. The text became a model of *suavitas* (charm) and an example for imitation. What the humanists really wanted to do was to copy not solely the grammar of the classics but also and especially the eloquence.

After the fall of Constantinople in 1453 many Greek scholars found their way to Italy, where they introduced the humanists to Hermogenes's *Progymnasmata* (see 2.7), which was printed in large numbers by Aldo Manuzio.[301] Aristotle's *Rhetorica* also finally became available. 'New' rhetorical works were created, such as the *Rhetoricorum libri quinque* by George of Trebizond, but this was swept aside the

[297] Heinrich Plett, *Rhetoric and Renaissance Culture*, Walter de Gruyter, 2004.

[298] Brian Vickers, *In Defence of Rhetoric*, Oxford University Press, 1988, p. 255.

[299] Jerrold Seigel, *Rhetoric and Philosophy in Renaissance Humanism: The Union of Eloquence and Wisdom, Petrarch to Valla*, Princeton University Press, 1968.

[300] Steven Lynn, *Rhetoric and Composition: An Introduction*, Cambridge University Press, 2010, p. 217.

[301] Jonathan Harris, *Greek Émigrés in the West, 1400–1520*, Porphyrogenitus, 1995. See also Annabel Patterson, *Hermogenes and the Renaissance: Seven Ideas of Style*, Princeton University Press, 1970.

moment that Hermogenes became known in Italy.[302] With the exception of a few parts of Aristotle's work, nearly all rhetorical writings were normative manuals. New studies did appear in specific branches of rhetoric: in no fewer than 206 chapters, Erasmus's extremely popular *De copia* concentrated on the rhetorical art of embellishment and variation.[303] By way of illustration, in chapter 33 he gave 195 variations on the sentence 'your letter pleased me greatly'. Other rhetorical subfields that flourished were the *ars mnemonica* (the art of memory) and *ars inveniendi* (the art of invention) in the studies of Giordano Bruno[304] (1548–1600) and Gottfried Leibniz (see 4.3), but the authors themselves did not consider these 'arts' to be part of rhetoric.

Manuals of logic were also written that were so close to rhetoric they were used as handbooks for the art of persuasion, for example *De inventione dialectica* by Rudolf Agricola and the *Dialectique* by Peter Ramus (see 4.3). Gerardus Vossius, who based his work primarily on Aristotle and Hermogenes, branded Ramus as anti-Aristotelian. However, even in Vossius's hands humanistic rhetoric was first and foremost a synthesis of existing texts.[305] There was at most some change in emphasis.

In the work of Francis Bacon (1561–1626), whom we already came across as a champion of a universal language for sharing knowledge (see 4.3), we find for the first time a new vision of rhetoric that can be summed up primarily as anti-rhetorical. Yet Bacon himself was a master of oratory. When he spoke in the English parliament there was not a cough to be heard and nobody's gaze wandered.[306] And Thomas Hobbes feared the rhetoric of demagogues, but his own rhetorical passion was no less intense for all that.[307] In his early *History of the Royal Society* (1667), Thomas Sprat wrote that all oratory should be banned. One should convince others not with words but with facts. After Sprat, early modern rhetoric never really recovered.

Bernard Lamy's rhetorical grammar: sympathetic kinaesthesia. Nevertheless, there were a few serious attempts to expose the underlying systems of rules for persuasive speaking. Here the Cartesian mechanistic ideas of the New Sciences left a considerable mark on rhetorical research. The best known work in this area is the *Art de parler* ('Art of Speaking', 1675) by Bernard Lamy (1640–1715).[308] Lamy was an eminent rhetorician as well as a gifted and influential mathematician and physicist, who represented a dying race with his *Traité de mécanique* in 1679.

[302] John Monfasani, *George of Trebizond: A Biography and a Study of his Rhetoric and Logic*, Brill, 1976.

[303] Desiderius Erasmus, *Copia: Foundations of the Abundant Style (De duplici copia verborum ac rerum commentarii duo)*, Collected Works of Erasmus, volume 24, edited by Craig Thompson, University of Toronto Press, 1978.

[304] Ingrid Rowland, *Giordano Bruno: Philosopher/Heretic*, Farrar, Straus and Giroux, 2008.

[305] Thomas Conley, *Rhetoric in the European Tradition*, University of Chicago Press, 1990, p. 160.

[306] Ben Jonson, *Timber: Or Discoveries Made Upon Men And Matter*, edited by F. Schelling, 1892, reprinted by Kessinger Publishing, 2007, p. 30.

[307] Conley, *Rhetoric in the European Tradition*, p. 163.

[308] John Harwood, *The Rhetorics of Thomas Hobbes and Bernard Lamy*, Southern Illinois University Press, 1986.

Lamy's *Art de parler* began where Port Royal's *De l'art de penser* left off (see 4.3). The work started with an explanation of the principles of speech and meaning, but went on to focus mainly on the rhetorical effects of pronunciation and the use of words. Unexpected pronunciation and word order can pique listeners' interest and make them more attentive. Lamy's treatment of the composition of sounds and emotion is a splendid example of *rhetorical grammar*. He gave an overview of the combinations of individual sounds and the ranking of words, which can be interpreted as a linguistic exercise. Yet when Lamy defined rules for the 'correct' sentence length, pauses, and ends of phrases and sentences, he was back in the world of rhetoric. His analysis of the rhetorical communication process appeared to be based on the New Sciences. Lamy described this process as a physical phenomenon that he called *sympathetic kinaesthesia* where vibrations in the speaker's soul are transferred by sounds that produce vibrations in the listener's soul. While this hypothesis is the most interesting part of Lamy's rhetorical analysis, it is alas unverifiable. However, as far as we know Lamy's work is the only seventeenth-century rhetorical theory that was able to go beyond the classical standard texts in several regards.

Vico and Gottsched: historical analysis and example-based rhetoric. Descartes, Bacon, and Hobbes were hostile to rhetoric. So too were many scientists and scholars after them. Nevertheless, in the eighteenth century we find a classicistic rhetorical study once again in Giambattista Vico's *Institutiones oratoriae* (1711). For this reason one could call Vico old-fashioned, but his self-confident historicizing approach, which permeates his entire work (see 4.2), was ahead of its time and he stood at the beginning of the nineteenth-century humanities, in which historical analysis was central (see 5.1).[309] It is true that rhetoric ended up on the dunghill of history, but Vico was the first to recognize its historical value.

We find another attempt to pin down both German rhetoric and poetics in the work of Johann Gottsched (1700–1766). Where rhetoric was concerned, Gottsched did this in the form of a *combined rule-based and example-based system* (his poetics will be discussed in 4.7).[310] Gottsched's *Ausführliche Redekunst* ('Complete Rhetoric', 1736) was published in two volumes and ran to more than 700 pages altogether. The first volume contained an attempt to set down the general principles of good speaking. In the second volume he filled hundreds of pages with examples to illustrate the general principles by imitations in German of classical speeches and instances from a variety of genres. However, there was no overarching example-based theory such as we saw earlier in linguistics.

Explanation of humanistic rhetoric. How can we understand the unprecedentedly rapid rise and the equally swift decline of humanistic rhetoric? The initial popularity of rhetoric has on occasion been explained by the cultured alternative

[309] Joep Leerssen, 'The rise of philology: the comparative method, the historicist turn and the surreptitious influence of Giambattista Vico', in Bod, Maat, and Weststeijn, *The Making of the Humanities, Volume II: From Early Modern to Modern Disciplines*, 2012, pp. 23–36.

[310] Uwe Möller, *Rhetorische Überlieferung und Dichtungstheorie im frühen 18. Jahrhundert: Studien zu Gottsched, Breitinger und G.Fr. Meier*, Wilhelm Fink, 1983.

that it offered to the uncivilized use of violence.[311] That is to say that rhetoric was the art of convincing without compulsion. The humanists saw this art as the most humane form of persuasion, which stood in strident contrast to the bloody oppression, persecution, and religious wars in early modern Europe. But this explanation, as laudable a goal as it is, is not completely satisfactory. The discovery and assimilation of classical culture had also become a goal in itself. Many wanted the grandeur of ancient Rome to be revived in all its facets, and language was one of the most important elements of this revival. However, it was not enough simply to write grammatically correct Latin. The humanists also wanted to match the eloquent style with all its power of persuasion, and who could serve as better models than the great orators who were philosophers and statesmen? They represented the ideal exemplars for the ambitions of cities like Florence. It is in this 'art of imitation'—and one step further, the 'art of emulation'—where the greatest contribution of humanistic rhetoric lies (see 4.1). It contributed little or nothing in terms of new principles or patterns, but in *imitatio* it was unmatched.[312]

Imitation played a less significant role in the other humanities. In text reconstruction, philologists did not imitate the Alexandrines (see 2.3) but went in search of the best principles and methods themselves (Valla, Poliziano). The same applied to musicology (with its new theories of consonance) and art theory (with the development of linear and empirical perspective). Although one found oneself under the classical yoke here too—particularly in art theory with Alberti's ideal selection—there was less imitation of the classics simply because ancient paintings and music had not survived. When the prestige of Latin decreased in the course of the sixteenth century and above all in the seventeenth century (see 4.3), a large part of rhetoric fell short as art of imitation and its decline set in. Only its concepts continued to live on in the study of music, literature, and art.

[311] Conley, *Rhetoric in the European Tradition*, p. 110.

[312] On the rhetorical use of *imitatio*, see Brian Vickers, 'Rhetoric and poetics', in Charles Schmitt and Quentin Skinner (eds), *The Cambridge History of Renaissance Philosophy*, Cambridge University Press, 1988, pp. 719ff.

4.7 POETICS: CLASSICISM *IN EXTREMIS*

As was the case with rhetoric, humanistic poetics was subjected to classical hegemony, and it barely rose above the prescriptive character it had taken on since Roman Antiquity. Initially people fell back on Donatus's commentary on Terence and Horace's *Ars poetica* (see 2.8), but these texts had been known since the Middle Ages. It was not until Aristotle's *Poetica* was translated in 1549 that new studies and commentaries followed hard on one another's heels.

The dramatic unities by Minturno, Scaliger, and Castelvetro. The prime leading poetical studies from the sixteenth century were Antonio Minturno's *De poeta* (1559), J. C. Scaliger's *Poetices libri septem* (1561), and Lodovico Castelvetro's *Poetica d'Aristotele vulgarizzata e sposta* (1570).[313] One would expect that these scholars brought Aristotelian poetics back to life in its empirical glory. But that could not be further from the truth. Whereas Aristotle had tried to derive general principles from Greek tragedies and epic, the Italian humanists transformed these principles into a priori goals. The new literary theoreticians concentrated above all on specifying the 'desirable' form of literature and not on describing the existing literature of their own time. All aspects of Aristotle's *Poetica* were taken to be normative. Even the notion of *catharsis* was reinterpreted as a form of moral improvement. In this regard little seemed to have changed since medieval European poetics—humanists simply replaced the Bible with the classics, in this case Aristotle.

Nevertheless, humanists also came up with something new. The unity of action that Aristotle had found in Greek tragedies was extended to the 'unity of time, place, and action'. This expansion took place in two steps—the addition to include unity of time was proposed by Minturno and Scaliger, and the inclusion of unity of place was put forward by Castelvetro.[314] This 'triad' was brought about on purely theoretical grounds. Minturno and Scaliger dismissed an interruption in the timeline, such as the passage of days between different scenes, by invoking *mimesis* (the imitation of reality, see 2.8), which was taken to the ultimate consequence. The timeline of the theatrical performance and that of the action it purported to show had to coincide exactly. In any case the events in a play must never span more than twelve hours. The third unity, that of place, was justified by Castelvetro thus. Since a person can never observe more than one place at a time, the action must remain constant. That is to say that it needs to be limited not just to only one town or one home but even to one place that corresponds to a space that can be seen by a single person. Everything is devoted to the goal of creating maximum clarity.

In so doing the Italian trio had forced drama into an almost impossible straitjacket. Like their fellow theorists in art and architecture, they believed that they had

[313] Baxter Hathaway, *The Age of Criticism: The Late Renaissance in Italy*, Cornell University Press, 1962, p. 173, pp. 182–5.

[314] Lodovico Castelvetro, *Poetica d'Aristotele vulgarizzata e sposta*, in Bernard Weinberg, *A History of Literary Criticism in the Italian Renaissance*, volume 1, University of Chicago Press, 1961. See also Roberto Gigliucci (ed.), *Lodovico Castelvetro: filologia e ascesi*, Bulzoni, 2007.

grasped the universal rules of the theatre thanks to the thoroughly thought-through implications of the mimesis principle. Their 'triad', however, was miles away from Italian theatrical practice. Even so, the ideas of these sixteenth-century humanists had an unexpectedly large following, but primarily outside Italy. Castelvetro's rules developed into the standardized rules of Dutch, French, English, and German classicism. And so it was that classical poetics became the canon for the theatre in Northern Europe (similar to what happened with classical architectural theory—see 4.5), whereas the blossoming baroque theatre in Italy (Giambattista Marini for instance) and Spain (Luis de Góngora and others) was largely devoid of a theoretical foundation.

Baroque poetics? Tesauro and Gracián. One of the very few treatises written about baroque theatre and literature was Emanuele Tesauro's *Il cannocchiale aristotelico* (1654).[315] At its heart is the concept of the *metaphor*—which could link divergent phenomena through analogy. Whereas classicists wanted to leave no uncertainty about anything, Tesauro advocated the polysemy of the metaphorical figure of speech that generated enjoyment and surprise. The discovery of a metaphor breached conventional rules and led according to Tesauro to a concept of poetics that was beyond Aristotle's horizon.[316] The title of Tesauro's work ('The Aristotelian Telescope') expresses an analogy with the New Sciences. Aristotle's shortcoming lay in the lack of a telescope, which meant that many phenomena remained invisible to him.

The Spanish Jesuit Balthasar Gracián (1601–1658) also extolled the pleasure of clever metaphors, astute plays on words, and complex word orders in his *Agudeza y arte de ingenio* ('Wit and the Art of Inventiveness').[317] The core of Gracián's work was the notion of *concepto*, which he defines ambiguously as 'an insight that reveals the relationships between concepts'. Gracián used *concepto* to indicate both nimble-mindedness (or wittiness) and a comparative metaphor. He contended that 'understanding without wit or *concepto* is like sun without light, without beams.' Contrary to what the classicists argued, sharp-wittedness and metaphor were integral parts of the comprehension of a literary or theatrical work by the readers or audience. Gracián put his ideas into practice in his masterpiece *El Criticón* and his popular collection of aphorisms *Oráculo manual y arte de prudencia* ('The Oracle, a Manual of the Art of Discretion'), which preached the art of circumspection and is still being reprinted.[318]

We do not find in the work of Tesauro or Gracián a deeper theoretical underpinning in the way we do with Castelvetro, who was able to fit poetics into a system of precise rules with the universal unity of time, place, and action. The baroque theorists remained marginal in early modern poetics. However, they were much closer to poetical practice, and their work was more descriptive than prescriptive.

[315] Emanuele Tesauro, *Scritti*, Edizioni dell'Orso, 2004.

[316] Eugenio Donato, 'Tesauro's poetics: through the looking glass', *Modern Language Notes*, 78(1), pp. 15–30.

[317] Evaristo Correa Calderón (ed.), *Agudeza y arte de ingenio*, 2 volumes, Castalia, 1969. For a French translation, see *Art et figures de l'esprit*, translated by Benito Pelegrín, Seuil, 1983.

[318] Virginia Foster, *Balthasar Gracián*, Irvington Publishers, 1975.

The scope of Classicism: Heinsius, Vossius, Boileau, and Gottsched. In the Netherlands the ideas of classical poetics were disseminated by Daniël Heinsius (1580–1655), the favourite pupil of the great philologist Joseph Scaliger (son of J. C. Scaliger). Over and above numerous philological commentaries and editions, in 1610 Heinsius edited Aristotle's *Poetica*, followed in 1611 by his influential discourse *De tragica constitutione* ('How to Make a Tragedy'), in which he explained Aristotle's ideas and poetic classicism in an easily understandable way.[319] Heinsius added barely any new ideas, but his procedures were adopted by tragedians in the Dutch Republic (e.g. Vondel), England (e.g. Ben Jonson), France, and Germany.

One of the most systematic overviews of humanistic literary theory was produced by Gerardus Vossius in 1647 in his impressive *Poeticae institutiones*.[320] Using works in his own library, Vossius described all classical literary genres and their characteristics. Vossius asked the States-General for financial support to help pay for the costs of printing this extremely voluminous work, and a grant of 800 guilders—over six months' salary—was forthcoming. This demonstrated the status that classicistic poetics enjoyed. Like rhetoric, it was part of the classical education of any ambitious young man, even if he ultimately went into trade.

The introduction of the Italian Aristotelians was also highly influential in seventeenth-century France. In his famous *Essais*, Michel de Montaigne (1533–1592) advocated clarity and simplicity and rejected each and every embellishment.[321] However, it is hard to find new poetical ideas in French theoretical works, although there was a proliferation of procedures for *decorum*, with many specific rules for social standings. Even the most important critic, Nicolas Boileau (1636–1711), in his *L'art poétique* of 1674, did not get much further than a renewed interpretation of the mimesis principle and the unity of time, place, and action.[322] Yet Boileau's ideas also met quite some resistance, culminating in what is known as the *Quarrel of the Ancients and the Moderns*. The Ancients (*Anciens*), led by Boileau, contended that a poet could do no better than imitate the ancient authors, whereas the Moderns (*Modernes*), led by Charles Perrault's *Le siècle de Louis le Grand* ('The Century of Louis the Great', 1687) supported the great achievements of the authors during the reign of Louis XIV. The 'Quarrel' was foremost a dispute between authority and innovation, and it triggered analogous debates in several other European countries.[323] Perhaps the only real surprise in Boileau's work is the sentence 'je ne sais quoi'. Even with the strictest application of the classicistic rules of mimesis, decorum, and the unities it still takes an unpredictable ingredient ('I don't know what') to elevate a literary work to art.[324] Alexander

[319] J. H. Meter, *The Literary Theories of Daniel Heinsius: A Study of the Development and Background of his Views on Literary Theory and Criticism during the Period from 1602 to 1612*, Van Gorcum, 1984.
[320] Gerardus Vossius, *Poeticarum institutionum libri tres/Three Books on Poetics*, edited and translated by Jan Bloemendal in collaboration with Edwin Rabbie, 2 volumes, Brill, 2010.
[321] Michel de Montaigne, *The Essays of Michel Eyquem de Montaigne*, translated by Charles Cotton, edited by W. Carew Hazlit, Encyclopaedia Britannica, 1952, p. 114.
[322] Marcel Hervier, *L'Art Poétique de Boileau, étude et analyse*, Mellottée, 1948.
[323] Joan DeJean, *Ancients against Moderns: Culture Wars and the Making of a Fin de Siècle*, University of Chicago Press, 1997.
[324] Richard Harland, *Literary Theory from Plato to Barthes*, Palgrave Macmillan, 1999, p. 41.

Pope interpreted this 'anomalist' principle later in his *Essay on Criticism* in 1711 as 'A Grace beyond the Reach of Art'.[325] It was recognition of what came outside the theory. As we have seen in the classical poetics of Dionysius and the art theory of Pliny, apparently there are rules for good art but not for beautiful art (see 2.8).

Johann Gottsched went the furthest in creating one overarching system of rules in 1730 in his *Versuch einer critischen Dichtkunst vor die Deutschen* (see also Gottsched's work on rhetoric, 4.6).[326] As a reaction to the disordered baroque popular theatre, where social standings were mixed up and verse forms were disharmonious, Gottsched taught the actors to speak Alexandrine lines and to stimulate catharsis. Everything that was not rationally understandable was prohibited. Rather than an application of the mimesis principle, Gottsched seemed to be endeavouring to completely dehumanize the theatre. It is telling that Gottsched's wife, 'die geschickte Freundin' (the skilful friend) Luise Adelgunde Kulmus, broke all her husband's rules in her own popular plays.

The end of Classicism: Doctor Johnson. It was not until the second half of the eighteenth century that the most important critic of his time, Samuel Johnson (1709–1784), also known as Doctor Johnson, launched a successful attack on classicistic poetics. The sickly Johnson, son of a bookseller, had an encyclopaedic knowledge of almost all the literature of his time and had learned large parts of English literature by heart. It took him only a few years to produce one of the best dictionaries ever written, but he nevertheless complained constantly about his laziness. He had a strong social conscience and he opposed slavery. He once proposed a toast to 'the next rebellion of the negroes in the West Indies'. His life was chronicled in the minutest detail by James Boswell in the unforgettable *Life of Samuel Johnson*.

In 1765, in his preface to the collected works of Shakespeare, Johnson opened the attack on one of the maxims of classicistic poetics—the unity of time and place.[327] Although many doubted this unity, it was Johnson who was able to demolish the fundamental premise theoretically. According to Castelvetro the unity of time and place was necessary to make the drama credible (in view of the associated corresponding structure of reality). Johnson simply refuted the existence of such credibility. He argued that 'the spectators are always in their senses, and know, from the first act to last, that the stage is only a stage, and that the players are only players.'[328] Plays were not judged on the grounds of how well they corresponded to real life but on their dramatic content. Castelvetro had erroneously assumed that someone in the audience would be completely absorbed in a play. According to Johnson this led him to the mistaken conclusion that the audience sees a reality that it must be possible to believe is reality and not a drama. In

[325] Samuel Holt Monk, 'A Grace Beyond the Reach of Art', *Journal of the History of Ideas*, 5(2), 1944, pp. 131–50.

[326] Phillip Marshall Mitchell, *Johann Christoph Gottsched, 1700–1766: Harbinger of German Classicism*, Camden House, 1995, chapter 5.

[327] Samuel Johnson, 'Preface to Shakespeare', in *Yale Edition of the Works of Samuel Johnson*, volume 7, Yale University Press, 1968.

[328] Johnson, 'Preface to Shakespeare', p. 77.

Johnson's view a play is like a narrative in which different times and places can be given shape. The things that are observed are never ultimately realities, but only symbols of something else. This was an apposite argument from someone who described himself as a harmless drudge.

Despite his devastating criticism of the unity of time and place, Johnson still wrote within a classicistic framework. However, classicism was completely abandoned by theorists after him. In England this happened with Burke's concept of the sublime that can also be evoked by the ugly and the uncontrolled (see 4.5). Classicism in Italy was abandoned by Vico's historicizing approach (though Vico's work was only very slowly accepted), while in Germany in 1766 Gotthold Lessing put the mechanical rules of classicism behind him in *Laocoon*.[329] In France, Diderot went in 1773 so far as to argue that a poetic imitation of reality (mimesis) simply cannot exist—after all, a verbal utterance is an expression of thought and observation, not of reality.[330]

First steps to literary history: Bembo and Huet. More than 1,500 years after Longinus's masterpiece *On the sublime* (see 2.8), literary history again became a subject of study in Europe. J. C. Scaliger's work *Poetices libri septem* contains a historical classification of Greek poetry in which he, like Aristotle, assigned a key position to Homer. Vossius's *Poeticae institutiones* also gave an overview of classical poetry, but there was no mention of contemporary literature. On the other hand in 1525 the Venetian humanist Pietro Bembo (1470–1547) produced an outline of Italian literature from its origin to Bembo's own time in his *Prose della volgar lingua*.[331] He identified Dante, Petrarch, and Boccaccio as the new founders of the Italian language and its literature. In Bembo's view Petrarch was the model for the purest poetry and Boccaccio for the best prose. This was the first time that non-classical writers were honoured thus. Bembo's editions of the Italian classics, published by Aldo Manuzio, were now being read in large parts of Europe. Bembo's work extended to many other fields of the humanities—his most important achievement besides poetics was his analysis of the madrigal, the secular musical genre that expanded enormously with the work of the composer Monteverdi.

Literary historiography was also taken up in Northern Europe. One of the most remarkable works, which appeared in 1670, was the *Traité de l'origine des romans* by the Frenchman Pierre-Daniel Huet.[332] Most humanists brushed the novel aside as being anti-classicistic, but Huet devoted an entire treatise to it. He defined a novel as a fictional story written completely in prose and intended for the enjoyment of the reader. Huet surveyed the novel from Antiquity to his own time. He also analysed the fictional aspects of religious works like the Bible and the Koran,

[329] Gotthold Lessing, *Laocoon, Nathan the Wise, Minna von Barnhelm*, translated by William Steel, Dent, 1930.

[330] Denis Diderot, *The Paradox of Acting*, translated by Walter Pollock, Kessinger Publishing, 2007.

[331] Pietro Bembo, *Prose della volgar lingua*, edited by Pasquale Stoppelli, Zanichelli, 2010.

[332] Pierre Daniel Huet, *Lettre-traité de Pierre Daniel Huet sur l'origine des romans*, edited by Fabienne Gégou, Nizet, 2005.

including the parables of Jesus in the New Testament. While at around the same time Spinoza proposed a 'correct method for interpreting Scripture' in his *Tractatus theologico-politicus* with far-reaching consequences (see 4.2), Huet limited himself to the evidently fictional aspects in the Bible. However, there is no direct line from Huet to modern literary historiography. Although there were a few local literary historiographies, like the *Lives of the Most Eminent English Poets* (1779–81) by Doctor Johnson, the field did not become a discipline until the nineteenth century (see 5.6).

Poetics and the other humanities. There was an observable common trend in Europe in the development of poetics, art theory, architectural theory, and, to a lesser extent, musicology. All began prescriptively, but at the end of the early modern age they turned away from a priori rules and proportions. The instigation of historicizing approaches in the eighteenth century may have accelerated the abandonment of such prescriptive rules, but it was the criticism of people like Doctor Johnson, Burke, and even the 'mistaken' Dutch art theoreticians (see 4.5) that set this reversal in motion. It was therefore not the nineteenth-century new humanities scholars but the seventeenth- and eighteenth-century theorists and historians who were able to free the humanities from the classicistic yoke.[333]

Aside from this common trend in the European humanities there were also manifest differences. Classicistic poetics based itself on imitating reality (mimesis), whereas art theory was founded on improving reality through ideal selection, and musicology and architectural theory were based on universal proportions. In other words, *classicistic poetics was primarily Aristotelian*, *classicistic art theory was Neoplatonic*—except for the Aristotelian concept of decorum—and *musicology and architectural theory were primarily Pythagorean*. The classics had degenerated into a catch-all in which every humanist who wanted to could find something.

Running parallel to this trend there was also an empirical tradition in the study of the arts, but for a long time this was limited to explorations of perspective by Leonardo and a few 'mistaken' theorists or historians who were bold enough to study the literature, art, or architecture of their own time. With the exception of musicology, the empirical quest was subordinate to the classicistic view. It is possible that the study of art, literature, and architecture—and to a lesser extent music—was so close to their output (most theoreticians were practitioners), that for many the pursuit of reviving these classical arts coincided with their theory.

Chinese and other poetics: from Hu Yinglin to Zhao Zhixin. We have bypassed Islamic civilization, India, and Africa in our discussion of early modern poetics. There is hardly any trace of Arabic poetics in the early modern age. However, just as in India, appearances may be deceptive, as we discussed with regard to Arabic logic and linguistics (see 4.3). There is likewise little known about an African poetic

[333] See also Rens Bod, 'Introduction: the dawn of the modern humanities', in Bod, Maat, and Weststeijn, *The Making of the Humanities, Volume II: From Early Modern to Modern Disciplines*, 2012, pp. 9–19.

theory, although the creation and transmission of the great epics, such as the Sunjata epic, were in full swing at this time.[334] As far as we know the only poetic tradition that went in search of theoretical underpinning was in China.[335]

Since Antiquity, principles had been developed (and patterns had been found) in China for creating poetic language and good literature,[336] as we have seen in the composition methods of Liu Xie's *The Literary Mind and the Carving of Dragons* (see 2.8) and in the systems of rules of Chen Kui's *The Rules of Writing* (see 3.7). The most important movement in Chinese literary criticism during the Ming period (1368–1644) was represented by Hu Yinglin (1551–1602), a protégé of the scholar and art critic Wang Shizhen (see 4.5).[337] Hu asserted that the essence of poetry consisted of not more than two principles:

1. Formal style and musical tone
2. Imagination and personal spirit

According to Hu this dichotomy was crucial because while there were rules that people could follow for the first principle, there were none for the second. We came across a similar contrast in Chinese art theory too, namely Xie He's Six Principles (see 2.5 and 3.5). Here Xie He's first principle of 'Spiritual Resonance' corresponds roughly to Hu Yinglin's second poetic principle of 'Imagination'. No rules can be devised for either of them, whereas this is possible for the others. Yet without spiritual resonance or imagination there is no point in spending any more time looking at a work of art (see 2.5) or listening to a poem. We therefore once again find the pattern of good versus beautiful art that we also came across in European poetics, musicology, and art theory—good art can be formalized, beautiful art cannot.

During the Qing Dynasty we find both moralistic and empirical visions of literature. The *Handbook of Tonal Patterns* written by Zhao Zhixin during the first half of the eighteenth century is the most important example of the empirical vision. This work explained the prosodic (tone) rules for classical Chinese poetry.[338] It provided one of the most thorough studies of a poetry based on a tone language. This brought poetics very close to the philological studies of classical Chinese poetry (see 4.1).

[334] Ralph Austen, *In Search of Sunjata: The Mande Oral Epic as History, Literature, and Performance*, Indiana University Press, 1999.

[335] James Liu, *Chinese Theories of Literature*, University of Chicago Press, 1975.

[336] Leonard Chan Kwok-kou, *A Critical Study of Hu Ying-lin's Poetic Theories*, PhD thesis, University of Hong Kong, 1982.

[337] Peter Bol, 'Looking to Wang Shizhen: Hu Yinglin (1551–1602) and late Ming alternatives to Neo-Confucian learning', *Ming Studies*, 53, 2006, pp. 99–137.

[338] John Wang (ed.), *Chinese Literary Criticism from the Ch'ing Period (1644–1911)*, Hong Kong University Press, 1993, pp. 47–53.

CONCLUSION: WAS THERE PROGRESS IN THE EARLY MODERN HUMANITIES?

Having arrived at this point, I will take a step back and try to survey the history of the humanities from Antiquity up to the early modern age as a whole. Looking at this huge time span, is there a continuity in the humanities with regard to the questions asked or problems solved? Additionally, can we talk about 'progress' or 'growth' in the humanities? And if so, what does this progress consist of? Before we can address these questions, we first need to describe the common patterns in the early modern humanities.

I. General patterns in the early modern humanities

Unity in the humanities and a new world view. Through the revival of the classics it sought, humanism gave the European humanities a unity unequalled before or since. All humanistic activities, from historiography to musicology, were cast in a classical mould. In the process, the philological method of *precision*, *consistency*, and *documentation* soon became the standard for all other disciplines. While this classical revival was beneficial to historiography, musicology, and philology, it was less fruitful for the disciplines that had already blossomed in the Middle Ages, such as linguistics and logic. In many cases one could talk of a humanist yoke: the medieval achievements were suppressed, and everything was governed by the emancipation of the *studia humanitatis*. In this respect the medieval Christian humanities were not as different from the early modern humanistic humanities as is often assumed. The Bible was the authority for the former and the classics for the latter. Nevertheless, humanistic activities led to a much more critical and sceptical attitude than we found in the Christian Middle Ages. The explanation for this outlook appears to be the endeavour to distinguish between genuine classical sources and the many forgeries and corruptions. At the same time there was a dawning awareness that the Bible could also be approached philologically. The critical study of the Bible as a text rather than a sacrosanct book resulted in the discovery that biblical history could not be made to agree with profane history, which in turn led to the new secular vision of man and state.

Relationship between man and cosmos. According to traditional scientific historiography, it was foremostly the discoveries of New Scientists such as Galileo, Kepler, Bacon, and Descartes that toppled the Aristotelian cosmos. But this view has proved to be incomplete: it was the sum total of all early modern scholarly activities—from philology to art theory and from anatomy to natural philosophy—that caused a revolution in the world view. Within the humanities it was first of all philology which initiated the much-discussed interaction between theory and empiricism and influenced all scholarly activities, from the study of music to the study of nature. The realization that both the humanities and natural sciences contributed to the new world view, and that the humanities not only preceded the natural sciences but also shaped them to a large extent, is currently widely shared.[339]

So while the original goal of humanism was the revival of Antiquity, this happened in such a way that, possibly unintentionally, the Christian biblical world view was refuted in a historical sense. This also called into question the Aristotelian world view, which had formed a package deal together with Christendom since the Middle Ages (the integration of Aristotle's theory of celestial spheres and Christian theology—see 3.5).

Increasing formalization: logical, mathematical, and procedural. One of the characteristics of the humanities of Antiquity was the pursuit of systems of rules (see 2.9). During the early modern age a revival of these systems of rules was accompanied by growing formalization, which we find in the form of underpinning for empirical practice, for example in philology, linguistics, musicology, and part of art theory (perspective), and also as normative and prescriptive rules in architectural theory, rhetoric, poetics, art theory, and musicology. In the natural sciences, formalization is usually equated with a mathematical description, as carried out by Kepler, Huygens, and Newton, but in the humanities three types of formalization can be identified.

For example Poliziano's theory was not founded on mathematical principles but on logical consistency principles for eliminating manuscripts (see 4.1). We will designate his philological method, which was employed throughout Europe in the sixteenth century, as *logical formalization* of philological practice. The linguistics of Sanctius and Port Royal, as well as Leibniz's logic, come into this category (while they adhered to the procedural system of rules principle, their formalization is based on logical operations). On the other hand, the theory of harmony and the theory of perspective embraced a form of *mathematical formalization.* Attempts were made to get to grips with the perception of harmonies and the illusion of depth by using numerical and geometric laws. Finally, we can designate the normative, almost procedural, instructions in literature, art, and musical composition—where artistic freedom was reduced to the 'initial conditions' of a play, work of art, or piece of music—as *normative* or *procedural formalization.*

These three types of formalization are not mutually exclusive. For instance, Alberti originally formalized perspective theory mathematically, but it was then made more accessible by Piero della Francesca, who extended it in the form of a step-by-step plan for achieving linear perspective that corresponded to procedural formalization (see 4.4). Architectural theory was also formalized partially mathematically (the arithmetic proportions) and partially procedurally (the way in which the parts are brought together to form a whole). And there were of course also intellectual activities where formalization was barely, if at all, possible, such as in historiography. However, where systems of rules could be found, there was

[339] See e.g. Anthony Grafton, *Defenders of the Text,* Harvard University Press, 1991, pp. 1–22; Steven Shapin, *The Scientific Revolution,* University of Chicago Press, 1998, p. 229; Eric Jorink, *Het Boeck der Natuere,* Primavera pers, 2007 (translated into English as *Reading the Book of Nature in the Dutch Golden Age, 1575–1715,* Brill, 2010).

formalization. What is striking here is that the humanities made use primarily of procedural and logical and only rarely mathematical formalization.

The advantage of formalization is obvious: a theory can be recorded and evaluated with precision, whether it is descriptive or prescriptive. We have seen that in so doing, theories in philology, linguistics, logic, and parts of musicology (consonance) and art theory (perspective) were refuted or adapted, whereas formalization of poetics and music and art composition resulted mainly in a poetical, compositional theory being followed by a new stylistic form. The function of formalization was therefore not the same in the different humanities.

From prescriptive back to descriptive. Distaste of prescriptive systems of rules defined a priori arose at the end of the early modern era. This dislike was followed by a return to a more descriptive quest for rules and proportions in humanistic material (particularly in art, architecture, theatre, literature, and music). We will describe this pattern as a *tendency away from prescriptive and towards descriptive.* This trend was opposite to the one in Antiquity, where we found a pattern from descriptive to prescriptive. Early modern humanities therefore started as being primarily prescriptive, but in the course of the seventeenth century and particularly the eighteenth the direction changed and prescription was replaced by a descriptive approach to material based on observation.

It appeared that in the course of the early modern age people realized that the existing 'descriptions' of art, architecture, music, language, and literature did not correspond to the empirical material, after which they focused on a more historicizing approach (Vico, Winckelmann, Wood) or a renewed descriptive one (the Dutch art theorists, the baroque critics, and also linguists like De Laet). Although this renewed approach was never completely descriptive—it remained normative to a degree—we can by and large distil a long-term pattern in the history of the humanities, that is a *cycle from descriptive to prescriptive and then back again to descriptive.*

Parallel versus unique developments. As we did in Antiquity and the Middle Ages, we find a number of fascinating parallels between regions in the early modern era. Poliziano's genealogical theory, for example, is comparable in a number of regards with the Arabic *isnad* method, and Petrarch's historical pattern fits into Ibn Khaldun's culture enhancement cycle. There were also parallel developments in China and Europe, particularly in philology. Manuscripts were rediscovered and reconstructed, but also forged and unmasked in both regions. Linguistic, poetical, and historical methods were employed for the dating and verification of texts—but these philological activities only brought about a revolution in the world view in Europe. There were also parallels in historiography in virtually all regions. During the course of the early modern age the cyclical pattern was widely embraced, from Europe and China to the Ottoman and the Mughal empires.

Over and above these parallel developments, there were also unique developments. For example, the fruitful interaction between theoretical speculation and empirical validation only developed in Europe (see 4.4). Equally unique were logical formalization in philology and mathematical formalization in the study of consonances and perspective theory. In contrast, we find procedural formalization in other regions too, particularly in musical composition systems. The historical relativism of Li Zhi and the historicizing 'neo-styles' were unique to China for the time being, whereas integration of the personal experience principle and the oral transmission principle appeared to be unique to Africa.

II. Was there progress in the humanities?

Continuity and the notion of progress. The early modern developments discussed above give rise to the question of whether we can speak of 'progress' in the humanities. No historical concept has been criticized as fiercely as that of cumulative progression (see for example 4.2 about the French Enlightenment thinkers). The first thing we should ask ourselves is what we mean by 'progress'. Do we mean progress in formalization, in achieving scientific and scholarly unity, in the trend towards empirical description, in the secularization of the world view, or something else again? There is extensive literature about scientific progress in the philosophy of science, but it does not deal with the humanities.[340]

On the face of it, it seems that any choice of a definition of progression is arbitrary. However, as we are concentrating in this book on a quest for patterns, it makes sense to define 'progress' on the basis of the patterns that were found. As we remarked above, in many cases these patterns correspond to systems of rules. However, we have seen that systems of rules never stand alone. They always serve to solve a concrete problem, such as establishing the reliability of a source, reconstructing a text, determining a language's possible word forms and sentence structures (often as a grammar), depicting reality, or verifying the validity of a line of reasoning. We can therefore compare different systems of rules in terms of the degree to which they solve a particular problem. This means we can use the increase or decrease in the problem-solving capacity of the systems of rules as a yardstick for 'progress', provided that the problem concerned is comparable in the different periods (and preferably also regions).

Problems in different periods are seldom exactly the same, but there proves to be a great deal of coherence in humanistic problems. This strikes one immediately in comparison to the study of nature. For example, where Aristotle and his followers defined the problem of movement in terms of the 'teleological' goal of movement (where things in the sublunary world tend to the natural place in the centre of the earth), Galileo and his followers defined the problem of movement by using a mathematical description without any notion of a goal.[341] In other words during Antiquity and the Middle Ages kinetics was primarily part of natural philosophy, whereas in the early modern era this field of study developed into a mathematical–empirical discipline. Such a sharp split was much less evident in the humanities.

[340] For an overview of theories of progress, see John Losee, *Theories of Scientific Progress: An Introduction*, Routledge 2004. See also Philip Kitcher, *The Advancement of Science: Science without Legend, Objectivity without Illusions*, Oxford University Press, 1993.

[341] Helen Land, *The Order of Nature in Aristotle's Physics: Place and the Elements*, Cambridge University Press, 1998. H. Floris Cohen, *How Modern Science Came into the World: Four Civilizations, One 17th-Century Breakthrough*, Amsterdam University Press, 2010.

Instead there was substantial continuity in the *artes liberales*—even after the rise of humanism—and in many cases the problems barely vary between different periods (though the solutions do). For instance, in Antiquity and the early modern age—and often in the Middle Ages too—scholars wanted to compile grammars in the form of systems of rules, reconstruct sources using a system of rules, and likewise verify reasoning by means of, once again, systems of rules. While the solutions were different, the problems were often the same.

The concept of problem-solving capacity as a yardstick for the success and progress of a theory stems from Thomas Kuhn's *The Structure of Scientific Revolutions*.[342] However, whereas Kuhn deemed his notion of progress to apply only in a period of relative rest ('normal science'), when problems are solved on the basis of a particular paradigm, and not in periods that are separated by scientific revolutions, I contend that the idea of problem-solving capacity can also be employed between periods.[343] I believe I can do this because of the substantial coherence between the problems in grammar, logic, rhetoric, poetics, historiography, and musicology.

Below I will try to identify in each of the humanities a problem which was more or less central and check whether there was *progress in the degree to which the particular problem could be solved*. Obviously, my concept of progress is circumscribed, but it is precise enough to determine whether there is any progression at all. My analysis does not rule out the possibility that another selection of problems would create a different picture (and it goes without saying that my analysis says nothing about the status of the problems concerned in present-day humanities—see the conclusion to chapter 5). If there appear to be several central problems in a discipline, I will therefore try to include them as far as possible.

Linguistics

Problem: Determining a language's word forms and sentence structures.

Solution: Panini was the first linguist to produce an essentially complete system of rules, or grammar, for a language's word forms and sentence structures and their meanings. Moreover, for a very long time he was the only linguist—and according to some people he still is—who succeeded in doing this (but see 5.3). His grammar appears to have been unsurpassed up to the present day. With the exception of Sanskrit, no complete grammar for a natural language has yet been constructed. There therefore does not seem to have been cumulative progression. Yet, as we remarked in chapter 1, Panini's grammar *cannot* be tested outside the finite corpus

[342] Thomas Kuhn, *The Structure of Scientific Revolutions*, University of Chicago Press, 1962.

[343] For an account of progress in terms of theory comparison with respect to problem-solving capacity in science, see Larry Laudan, *Progress and Its Problems: Toward a Theory of Scientific Growth*, Routledge, 1977. A similar plea is found in Nicholas Jardine, *The Scenes of Inquiry: On the Reality of Questions in the Sciences*, Oxford University Press, 1991. Yet, as noted by Bart Karstens (in his review of the Dutch version of this book, *De vergeten wetenschappen*, Prometheus, 2010), these ideas about progress did not find much resonance among historians. According to Karstens my account is one of the first attempts in the historiography of science to put these ideas to work on a grand scale (see Bart Karstens, 'Recursion, rhythm and rhizome: searching for patterns in the history of the humanities', in *Beiträge zur Geschichte der Sprachwissenschaft*, 21, 2011, pp.153–62).

of classical Sanskrit. It consequently makes sense to evaluate the achievements of the Greek, Roman, and above all more recent linguists, particularly in regard to languages that have not died out and for which there is an arbitrary amount of verification information. If we do this, some cumulative progress can be demonstrated in resolving the determination of the word forms and sentence structures. For example, Apollonius Dyscolus's grammar continued in a cumulative fashion from Dionysius Thrax's, and can cover a greater proportion of the possible linguistic utterances. Whereas Dionysius focused primarily on the word forms, Apollonius increasingly addressed the structure of sentences or syntax (see 2.1). Medieval linguists continued to concentrate on determining word forms, but also extended their investigation to sentence structures, as seen among the Modists (see 3.1). Sanctius and other early modern linguists broadened out to more complex phenomena, such as ellipsis (see 4.3). There seemed to be a slowdown in the quest for a solution for determining word forms and sentence structures after Sanctius, and at Port Royal we could even say that, after initial growth, there was decline. Yet after the Port Royal grammarians, there was another way forward, with increasingly specific grammars—for French, for example—that generally speaking could cover progressively more sentence structures. These grammars were never complete, although they did strive to define the official written language. This line of progress is less easy to detect in Arabic linguistics, where in the eighth-century Sibawayh had already reached an undisputed apogee that has never been matched. And similarly with Chinese linguistics, where these sorts of systems of rules were not produced until the modern age.

Historiography

Problem: Dating historical events.

Solution: Alongside establishing the reliability of a (written or oral) historical source and the quest for moral examples from the past, dating a historical event was one of the central problems in historiography. However, dating was not always considered relevant. Herodotus barely dated his histories and Thucydides remained largely vague, but starting with Berossus and Manetho, and in particular with Timaeus (see 2.2), chronological rankings and dating of foundations of cities and lists of kings were deemed to be very significant. But was there any improvement in the accuracy of this dating? Berossus gave a fairly arbitrary chronological arrangement, whereas Timaeus tried to date the most critical historical events as precisely as possible on the grounds of Olympiads as units of time in conjunction with a list of the Pythian Games. There was a further increase in the accuracy of dating thanks to the Roman annals, but this remained limited to contemporaneous events. Moreover, this type of dating was only usable in the Roman calendar system. The Roman annals became inadequate as soon as the problem of dating older Jewish, Babylonian, Persian, or Greek events arose. Eusebius, al-Tabari, and Bede were also often wide of the mark with the dating of events that were outside their own calendar systems (see 3.2). It was not until the major calendar harmonizations by Joseph Scaliger and others (see 4.2) that the histories of different peoples could be brought together into one chronological scheme. In doing this Scaliger continued to build on much local dating by classical historians. In other words, a solution for local dating was created in the course of Antiquity and Islamic civilization on the basis of internal calendar calculations, such as the Greek Olympiads, Chinese dynasties, Roman *ab urbe condita*, or Islamic *hijra*, but the problem of global dating was not solved until after the harmonization of these different calendar systems during the sixteenth century, on which others could continue to build.

Philology

Problem: Determining the reliability of a textual source.

Solution: The problem of the reliability or authenticity of a text has commonly been associated with establishing the original source. The early Alexandrian philologists had drawn up analogical comparisons between words that could be used to discover and emend corruptions in inconsistent sources (see 2.3). There was a break in the solving of this problem during the Christian Middle Ages—the only source that was regarded as reliable was the Bible, and more particularly the Vulgate (see 3.3). Essentially, all learning had to be made to agree with biblical revelation (see 3.2). Techniques for unmasking forgeries or tracking down corruptions were virtually lost. From the time of the first humanists, but in particular Lorenzo Valla, onwards, there was a resurgence of the philological criteria of precision, consistency, and documentation, which were successfully employed in revealing fakes (see 4.1). The historical–philological method became a powerful weapon, although it remained subjective for a long time. This changed with Poliziano's approach of deriving the genealogical relationships between texts, which brought greater precision, and hence progress, in determining the reliability of sources (this resembled the Islamic *isnad* method, but it is not clear whether Poliziano knew this). Poliziano's method was extended by Erasmus and perfected by Scaliger, Bentley, and others. When Casaubon was able to refute (and date) the Corpus Hermeticum on purely philological grounds, the field acquired a well-nigh unassailable status. Philology also flourished in China during the Ming Dynasty, with Gu Yanwu and his Empirical School as its leaders (see 4.1). Here we can speak about progress too if we make a comparison with the problem-solving capacity of earlier practices (see 2.3, 3.3).

Musicology

Problem: Determining the consonance of intervals.

Solution: In Greek Antiquity Pythagoras's solution for consonance held sway as the received view for a long time (2.4), and we find a similar solution in Chinese and Indian music theory. However, Aristoxenus placed empirical observation above mathematics, which signalled the start of a lengthy controversy. This controversy went into hibernation during the Middle Ages, but it was reawakened during humanism. Many theoretical and empirical treatises on consonance appeared, but no progress was made in solving the problem—unless it was that all consonance models were refuted (see 4.4). This refutation and the subsequent improvement of models nevertheless brought about one of the most important innovations in early scientific methodology—the stimulating interaction between theoretic speculation (including mathematical formalization) and empirical testing. It was this methodology that resulted in major progress in problem-solving capacity in the New Sciences—but it meant less in musicology. However, determining consonances was not the only problem in the field. Musicologists in Antiquity (Aristoxenus), the Middle Ages (the *Musica enchiriadis*), and the early modern age (Dressler, among others) worked on developing systems of rules for melodies (4.4). But it is difficult to speak of progress in terms of problem-solving capacity because these systems of rules always applied to specific musical styles. All the same, there was progress in other areas, such as the problem of notation, which went from very inaccurate neumes to precise musical notation (see 3.4 and 4.4), the problem of twelve-tone tuning (see 2.4, 3.4, and 4.4), and the problem of calculating the Pythagorean comma (2.4 and 3.4). However, around 1800 the consonance problem was still unresolved.

Art theory

Problem: Portraying three-dimensional objects on a two-dimensional surface.

Solution: This problem is not found in all periods or regions. Pliny developed a guideline in Roman Antiquity for the 'illusionistic depiction' of reality, but a specific solution was not given (although in practice many fresco painters could portray such three-dimensional objects). Principles for 'correct observation, size, and structure' and 'resemblance to the subject' were given in the Indian Sadanga theory and Xie He's Chinese art theory respectively (see 2.5), but without a system of rules. We find the first surviving treatment of three-dimensional projections in Indian twelfth-century painting discourses (see 3.5), in which rules for 'foreshortening' were explained. The European Middle Ages did not produce solutions for the depiction of three-dimensional objects, but Alberti's linear perspective provides us with a solution that covers both foreshortening and progressive diminution from the viewpoint. This appeared to have solved the problem of portraying three-dimensional objects on a two-dimensional surface. Nevertheless, Leonardo subsequently demonstrated experimentally that the functioning of the visual process was much more recalcitrant than Alberti had described it (see 4.5). His treatment of *empirical* perspective built cumulatively on linear perspective but led to a hybrid theory that was partially mathematical (Alberti's mathematical formalization) and partially experience based (Leonardo's procedural formalization).

Logic

Problem: Determining the validity of a line of reasoning.

Solution: From Antiquity onwards there were three solutions for testing the logical validity of a line of reasoning. These were the Greek, Chinese, and Indian solutions. The first was deductive, the second analogical, and the third inductive (see 2.6). Explicit logical systems of rules were only developed in Greece, and it was only in Europe and the Islamic civilization that there was an increase in the problem-solving capacity of these systems of rules, above all in Avicenna's Arabic logic and Buridan's scholastic logic, which continued to build on Aristotle. In both cases a few shortcomings were removed from Aristotelian syllogistics, or it was formalized further (see 3.6). In the early modern age Leibniz tried to provide Aristotelian syllogistics with a deeper 'logical language', although negations (among other things) could not be formalized in Leibniz's system (see 4.3). The tradition of logic in China disappeared from the scene for many centuries after the legalistic Qin regime, while in India there was a proliferation of logical systems but no observable progress in the problem of determining the validity of lines of reasoning (see 3.6). Summarizing, there was only cumulative progression in the syllogistic tradition in Islamic civilization and Europe. However, as we explained in 2.6, this logic had little practical value. No Euclidian assertion could be tested using it, and Valla rejected it because of its artificiality (see 4.3).

Rhetoric

Problem: Constructing convincing argumentation.

Solution: Rhetoric as the study of convincing argumentation began with Aristotle, in particular with his concept of an enthymeme (see 2.7). His solution was barely if ever surpassed after him. Even more than in the Graeco-Roman world, in China and India rhetoric formed a unity with logic and, as in Europe, it experienced a high point in Antiquity. During the Christian Middle Ages rhetoric was subordinate to preaching (3.7), but in the early modern era it reached a new zenith in the works of Valla, Agricola, and others (4.7). In Valla's rhetoric the concept of natural argumentation is compared and contrasted with Aristotelian argumentation (see also 4.3), but Valla did not come up with a concrete solution, let alone a system of rules for developing a rhetorical proof. At the end of the seventeenth century Lamy tried to write a grammar of rhetoric and an underpinning of what he called 'sympathetic kinaesthesia' (4.7), but argumentation cannot be constructed using his system (only rhetorically 'good sentences'). The best argumentation systems in the early modern age were based on Aristotle's rhetoric, in some cases extended to incorporate a few clarifications and shifts in emphasis. Progress in terms of problem-solving capacity in rhetoric is in fact not noticeable in the Antiquity–early modern era period.

Poetics

Problem: Composing a poetic work.

Solution: Since Antiquity, systems of rules have been developed for creating a good play, poem or novel, or simply for creating good or beautiful sentences (see 2.8). However, a general solution to this problem was not found, basically because 'good' art is time- and place-dependent. Many tried to transcend this time- and place-dependence by generalizing the empirical regularities they found into absolute, prescriptive rules. For example, the unity of action in the Greek theatre that Aristotle established empirically was advanced as an absolute rule by Horace (2.8). In the Middle Ages Aristotelian poetics languished in Europe whereas it was the subject of extensive commentaries in Islamic civilization (3.7). During the era of humanism the unity of action described by Aristotle was extended to prescriptive unity of time and place, with which poeticists like Castelvetro and Boileau believed they had solved the problem of composing a poetic work (4.7). Yet as we have also seen in the many melodic grammars for music, these poeticists only solved the problem for one specific style, and there was no cumulative progression. The same was true of the systems of rules for Indian and Chinese poetics (see 2.8 and 3.7). However other problems were solved, for example the interpretation problem of the Vedas with a Paninian model (3.7), the problem of the Occitan art of rhyming in Europe (3.7), and above all the impressive quantitative style analysis of Chen Kui (3.7).

Progress in most humanities. The humanistic palette is too complex to be summed up in one simple conclusion, but by and large we can contend that with the exception of rhetoric, poetics, and (surprisingly enough) musicology, there was clear progress to be observed in the problem-solving capacity of the humanities over the long period of time from Antiquity up to and including the early modern era in terms of our postulated central problem. Yet this progression did not advance at the same rate in all periods, nor was it present in all regions. However, the problem-solving capacity of the systems of rules of most disciplines in the eighteenth century was greater than it had been in Antiquity. And even in the disciplines where no progress with our postulated central problem was apparent, there was at least one other problem where there *was* progress to be identified, such as in poetics and musicology—with rhetoric as the only exception. The question that we asked at the beginning of this conclusion can therefore largely be answered positively: *except for rhetoric there was progress in the humanities in the sense of 'problem-solving capacity'.*

We should continue to bear in mind that we have only investigated one concept of progress in the humanities, namely the degree to which a particular problem could be solved. Nevertheless, it emerges that in spite of the received view, the idea of progress and the associated notion of scientific growth are applicable to the humanities.

5

Modern Era: The Humanities Renewed

During the nineteenth century a large part of the humanities became 'new'. While the subject of the humanities remained the same (texts, music, art, language, literature, theatre, the past), the way of investigation changed. In historiography for instance there was a real 'philologization', which had started in the early modern era but now became the paradigm at all Western universities. Philology itself also underwent a transformation: from being a purely classical philology the discipline was converted into a national one. These changes did not appear out of a clear blue sky. During the course of the eighteenth century the response to the glorification of the classics became progressively more critical. Moreover, the aspiration to create nation states resulted in a growing interest in national history. Starting with the French Revolution, the past was made more accessible. Monastery archives were nationalized and museum collections became public. A nation's interest in its own past was matched by a growing appetite for popular literature and folklore. Johann Gottfried Herder (1744–1803) was a pioneer in this field. In the previous chapter we described Herder as the successor of Vico, but he could equally well be considered as coming from the modern age. For example, Herder was the source of the notion of a people that can grow and die, as well as the concept of national spirit. Moreover, Herder was a spiritual father of the nationalism that was to play an overpowering role in the nineteenth century.[1] It was not until far into the twentieth century that the humanities were able to free themselves from the nationalist yoke. Once again we see a continuing line of principles and patterns in the modern humanities and also—surprisingly enough—in the postmodern humanities. The sciences and humanities in the other regions became more and more subject to European domination, which led to widespread postcolonial criticism in the second half of the twentieth century.

[1] Joep Leerssen, *National Thought in Europe: A Cultural History*, Amsterdam University Press, 2006.

5.1 HISTORIOGRAPHY: THE HISTORICIZATION OF THE WORLD

All of the past became relevant in the nineteenth century. The classical yoke was cast off and all historical periods were proclaimed to be of equal value. History, which was now also used for nationalist purposes, was institutionalized. Positivist historiography as developed by Comte and Marx had a profound effect on early twentieth-century historiography. Post-war historiography featured an embrace of patterns on the one hand and their rejection on the other. European views largely determined historiography in the other regions, with the striking exception of African historiography, with its principle of oral transmission, which in fact appeared to have influenced Europe in turn.

The claim to objectivity: Leopold von Ranke and historism. The humanistic ideal of a changeless classical Antiquity had already come under heavy fire in the eighteenth century. The first major historian who followed the line of Vico and Herder (see 4.2) and who, in flagrant opposition to earlier humanist scholars, *wanted to treat all historical periods as having equal status* was Leopold von Ranke (1795–1886). After a career as a grammar school teacher, he joined the University of Berlin, following the success of his first great work, *Geschichte der romanischen und germanischen Völker von 1494 bis 1514* ('History of the Latin and Teutonic Nations from 1494 until 1514'), published in 1824.[2] In this work, Ranke used an arsenal of written texts, including memoires, diaries, national archives, and diplomatic sources. He subjected them all to the strict methodical principles of philology. Ranke's work led to the creation of a new type of history, which became known as *historism*. This movement did not seek to make pronouncements about the past but merely to show 'wie es eigentlich gewesen' or 'how it really was'.[3] Ranke combined humanistic philology with a narrative historiography in order to achieve this. His students were dispatched to the many recently opened state and church archives, where they had to apply in-depth philological source criticism.[4] Both the content of the source and the external facets, such as the form and the carrier, were subjected to a critical analysis. The use of this philological method was intended to guarantee the objectivity of the historian, so that Ranke's goal—establishing facts—was achieved.[5]

The way Ranke put his historiography into practice brought much praise and imitation in Germany. For example, he refused to let his abhorrence of the French

[2] Leopold von Ranke, *History of the Latin and Teutonic Nations from 1494 to 1514*, translated by Phillip Ashworth, George Bell and Sons, 1887, reprinted by Kessinger Publishing, 2004.

[3] Leopold von Ranke, *Fürsten und Völker*, edited by Willy Andreas, E. Vollmer, 1957, p. 4. See also Rudolf Vierhaus, 'Rankes Begriff der historischen Objektivität', in Reinhardt Koselleck, Wolfgang Mommsen, and Jörn Rüsen (eds), *Objektivität und Parteilichkeit in der Geschichtswissenschaft*, Deutscher Taschenbuch Verlag, 1977, pp. 63–76.

[4] Kasper Eskildsen, 'Leopold Ranke's archival turn: location and evidence in modern historiography', *Modern Intellectual History* 5, 2008, pp. 425–53.

[5] Leonard Krieger, *Ranke: The Meaning of History*, University of Chicago Press, 1977. See also Siegfried Baur, *Versuch über die Historik des jungen Ranke*, Duncker & Humblot, 1998.

Revolution and the papacy have any influence on his historical findings. These days, however, Ranke's objectivistic endeavour is considered to be unachievable. Firstly, sources are seldom 'objective' (they could have been written by clergymen with a coloured world view), and secondly, every period has its implicit assumptions that nip any pursuit of objectivity in the bud. It looks as though Ranke was aware of this because he embraced Herder's notion of *Zeitgeist* (the spirit of the times)—every age is controlled by a few guiding ideas that set the tone for the period in question. Perhaps Ranke believed that the historian could expose these guiding principles, including those of his own time. However, it emerged that Ranke was not able to do that either. It was obvious to him, for instance, that there was a divine plan in the history of humanity, but he never made this assumption explicit, let alone bring it up for discussion.

Ranke's great merit was his comprehensive source criticism. Never before had such a high standard been set with regard to the use of historical documents. While the 'philologically underpinned source principle' had been employed centuries before by humanists like Joseph Scaliger (see 4.2), Ranke was the first who was able to give philological–historical practice as a method a place in the university curriculum. This made history more systematic and orderly as an independent discipline, complete with auxiliary subjects and a standardized methodology.[6] The new Humboldtian universities, with their freedom of research and teaching, guaranteed a stable environment and academic independence. The school that Ranke left behind was therefore impressive. From about 1880 to 1940 the Rankean method represented the most important historiographical movement in Europe and the United States. The upshot was that oral sources, which had played such an important part in the work of Thucydides and Sima Qian (see 2.2) and in the *isnad* tradition (see 3.2), were kicked into the long grass for nearly a century.

Despite their official scholarly independence, many nineteenth-century Rankeans started to dance to a nationalist tune. After his illustrious standard work on Rome,[7] for example, Theodor Mommsen (1817–1903) went into politics and became a fervent supporter of Bismarck's pursuit of national unification. Others used historiography to provide foundations for a specific national identity. Robert Fruin (1823–1899), the first holder of the chair of national history at Leiden University, was a prominent example. Although he was a self-declared Rankean—his inaugural lecture was entitled 'The impartiality of the historian'—he gave a biased and over-simplified picture of the seventeenth-century diplomat Lieuwe van Aitzema, whom he declared to be a secret Catholic intriguer.[8] Yet it is to the credit of Ranke's disciples, in particular Georg Waitz, Heinrich von

[6] For the notion of academic discipline, see e.g. Hubert Laitko, 'Disziplingeschichte und Disziplinverständnis', in V. Peckhaus and Ch. Tiel (eds), *Disziplinen im Kontext: Perspektiven der Disziplingeschichtsschreibung*, Fink, 1999, pp. 21–60.

[7] Theodor Mommsen, *Römische Geschichte*, 1854–1885, reprinted by Deutscher Taschenbuch Verlag, München 2001.

[8] Jonathan Israel, *The Dutch Republic: Its Rise, Greatness, and Fall, 1477–1806*, Oxford University Press, 1995, pp. 731–2.

Sybel, and especially Johann Gustav Droysen, that Ranke's ideas became institutionalized in nineteenth-century Germany.

The Whig interpretation of history: from Macaulay to Bancroft. While German historians were striving for impartial historiography, initially the British took the opposite course. They searched for support for the assumed English history of gradualness without revolutions. In his *Reflections on the Revolution in France* Edmund Burke (4.5) had already argued that a state could not structure itself on the basis of abstract ideas and was better advised to follow tradition.[9] The Whigs (liberals) saw the fact that 1848—the Year of Revolution—had passed them by virtually unnoticed as confirmation of the gradual English development.

Thomas Macaulay (1800–1859) elevated the gradual development route to a historical pattern, albeit imposed. In his *History of England from the Accession of James II* Macaulay believed he could demonstrate that nineteenth-century Britain was the result of centuries-long step-by-step development to ever higher forms of civilization with more and more freedom.[10] In Britain change meant continuity, and this continuity showed nothing but progress. First of all the British had freed themselves of their superstition, and then of their autocratic system, after which they also succeeded in constructing a well-balanced constitution with freedom of religion and freedom of speech. Even the only revolution that Britain had ever experienced, the Glorious Revolution in 1688 (when the Dutch stadholder Willem III became King William III of England after a coup d'état),[11] was interpreted as a part of the steady development of constitutional institutions. Rarely has a history book been more popular than Macaulay's *History of England.* Hundreds of thousands of copies in no fewer than eleven languages were printed and, as he himself wrote, replaced 'the last fashionable novel on the tables of young ladies'.[12]

Macaulay's story of nothing but continuous progress is currently known as the Whig interpretation of history.[13] In such an interpretation, historical complexity is reduced to a success story in which every event is given its necessary place. No matter how simplistic such a story was, it struck a chord with the general public and resulted in a view of history that was thereafter almost impossible to eradicate. In fact, the pattern of continuous progress had already existed among eighteenth-century French Enlightenment thinkers like Turgot and Condorcet (see 4.2), but whereas they worked in a more or less universalist way and asserted that all civilizations exhibited progress—sometimes with temporary periods of relapse—Macaulay limited his Whig history to the lot of Britain. Macaulay even dismissed

[9] Edmund Burke, *Reflections on the Revolution in France*, J. Dodsley, 1790. See also Crawford Brough Macpherson, *Burke*, Hill and Wang, 1980.

[10] Thomas Macaulay, *History of England from the Accession of James II*, 5 volumes, 1848, reprinted by Kessinger Publishing, 2003.

[11] Lisa Jardine, *Going Dutch: How England Plundered Holland's Glory,* HarperCollins Publishers, 2008.

[12] Thomas Macaulay, 'History', in *The Complete Writings*, 10 volumes, Boston & New York, 1901, volume 1, p. 276.

[13] This qualification comes from Herbert Butterfield, *The Whig Interpretation of History*, W. W. Norton, 1931.

the history of other countries, particularly the French nation with its bloody revolutions.

Nevertheless, the Whig interpretation also enjoyed popularity in nineteenth-century France for a while. For example, in his *Histoire de France* (1846) Jules Michelet explained that the French Revolution arose out of the unique French national spirit dedicated to freedom and fraternity.[14] And in the New World, George Bancroft was not afraid of arguing in 1882 in *The History of the Formation of the Constitution of the United States* that the United States was created as a necessary outcome of the progress in universal freedom.[15] As Anglo-Saxon descendants of Teutonic warriors, the Americans were possessed of a natural freedom-loving national spirit.

However, when the Rankean method found its way to European and American universities, the Whig historians lost favour and short shrift was given to uncritical use of sources and biased interpretations. Nevertheless, the idea of continuous progress to ever higher levels of civilization lived on in the popular vision of the past.

The upcoming social sciences and positivist historiography: Comte and Buckle. In the nineteenth century history acquired very considerable prestige, but this was eclipsed by the tremendous success of natural science. The up and coming social sciences therefore looked primarily to natural science for their methods. The sociology of Auguste Comte (1798–1859) is an example of a study of society and its history based on a *positivist* method.[16] According to Comte this method was based solely on observable facts that were subject to laws that were verifiable using deductive rules. Religious or metaphysical explanations were rejected. However, positivism was linked to the belief in progress, as could be seen in Comte's classification of the history of man into three stages: the *theological, metaphysical,* and *positive* stages, where according to Comte the third stage never ends.[17] While Comte's ideas set the tone for early social theorists and anthropologists such as Harriet Martineau, Herbert Spencer, and Émile Durkheim, few historians were impressed by Comte's periodization. Yet his positivist approach did have a few followers. The Englishman Henry Thomas Buckle (1821–1862) took it the furthest in his *History of Civilization in England.*[18] He tried to postulate laws for the course of human progress. Using primary causes like climate, soil, food, flora, and fauna, he believed he could deduce, for instance, that 'in Europe man is stronger than nature and elsewhere nature is stronger than man'. Although none of his 'laws' have withstood the test of time, Buckle was one of the first historians to try to explain why the Scientific Revolution took place in Europe and not elsewhere.

[14] Jules Michelet, *History of France*, translated by G. H. Smith, 2 volumes, D. Appleton, 1892.

[15] George Bancroft, *The History of the Formation of the Constitution of the United States*, D. Appleton, 1882.

[16] Auguste Comte, *Cours de philosophie positive*, 6 volumes, Rouen/Bachelier, 1830–42, reprinted by Hermann, 2 volumes, 1975. For a (condensed) English translation, see Harriet Martineau, *The Positive Philosophy of Auguste Comte*, J. Chapman, 1853.

[17] Pierre Macherey, *Comte: la philosophie et les sciences*, Presses Universitaires de France, 1989.

[18] Alfred Henry Huth, *The Life and Writings of Henry Thomas Buckle*, volume 1, D. Appleton, 1880.

A positivist 'salvation history': Marx. The nineteenth-century Industrial Revolution also produced a new field—*economic history*, for which Comte's positivism was used as a model. The most important exponent of this historiography was Karl Marx (1818–1883), who is known first and foremost as a political philosopher and economist, but who also earned an eminent place as a historian. Marx contended that in order to understand history one had to study how people learn to survive, which goods they produce to that end, and how they do so.[19] No religious or other intangible matters should be used as explanations. Marx called the way people produce goods a *means of production*. Such means controlled every historical change and represented an explanation in itself. By focusing on an analysis of the means of production throughout history, Marx believed he could break human history down into four stages, which would ultimately lead to a final fifth stage:[20]

(1) *Primitive communal society*: no private ownership of the means of production, for example tribal cultures.
(2) *Slavery*: means of production in the hands of the slave-owning class, for example the Roman Empire.
(3) *Feudalism*: means of production in the hands of the aristocracy, for example the European Middle Ages.
(4) *Capitalism*: means of production in the hands of the bourgeoisie; this stage goes together with a growing class struggle, for example industrialized Europe.
(5) *Socialism and communism*, no private ownership of the means of production; this stage is only reached after a revolution against capitalism.

Using European history, Marx gave an overview of the means of production that were employed in each stage, from primitive tools to the hand-mill and from the windmill to the steam engine. He showed that there were tensions between social classes at every stage, which resulted in a class struggle and revolution, after which a new stage commenced. The English Glorious Revolution, for instance, ushered in emancipation of the bourgeoisie, and the French Revolution and the Paris Commune heralded emancipation of the working class. We will designate Marx's method as a separate principle—the 'analysis of means of production principle'.

The historical pattern that Marx believed he could deduce with his methodical principle was that of a recurring class struggle and subsequent revolution—until the last stage, which has no end, was reached. In this regard Marx's historiography resembles that of Comte, whose last *positive* stage was also endless. Marx was moreover influenced by the philosophy of history and dialectic of Georg Wilhelm

[19] Karl Marx, *Das Elend der Philosophie*, with preface and notes by Friedrich Engels, J. H. W. Dietz, 1885. For an English translation, see *The Poverty of Philosophy*, Twentieth Century Press, 1900. See also Karl Marx, 'Der 18te Brumaire des Louis Napoleon', *Die Revolution. Eine Zeitschrift in zwanglosen Heften. Erste Hefte*, 1852.

[20] Karl Marx and Friedrich Engels, *Collected Works, 1845–47, Volume 5: Theses on Feuerbach, The German Ideology and Related Manuscripts*, International Publishers, 1976.

Hegel (1770–1831).[21] Hegel had a cosmic concept of history in mind, where the development of being coincided with the development of the reasonableness of 'spirit' (*Geist*). This grew from a subjective spirit (the individual person) to an objective spirit (the 'world spirit') to the absolute spirit. But Marx was influenced above all by Hegel's *historical dialectic*, according to which all historical changes emerge from a synthesis of two extremes. Marx extended Hegel's dialectic to include the *materialism* of the philosopher Ludwig Feuerbach (1804–1872), who contended (like Comte) that historical changes should only be explained on the basis of material and not metaphysical causes.[22]

Although Marx thought he could determine a new course of history, his pattern was more ideological than empirical—particularly the last stage. Nevertheless, Marx claimed universality for his pattern. All societies went through the same sequence from a common beginning to an ultimate goal. This vision, which thanks in part to Friedrich Engels (1820–1895) had a huge effect on the socialist movement, bears a resemblance to Christian salvation history, in which a universal line is also thought to extend from a common beginning to a final goal (see 3.2 and 4.2). Marx's assertion rapidly became controversial and, moreover, easy to refute, which was indeed successfully done.[23]

Marx's salvation history was therefore ahistorical, but his dialectic interpretation of history served another purpose—as support to Marx's political vision, which condemned the exploitation of the working class and aspired to a society without private ownership of the means of production. In his *Thesen über Feuerbach* ('Theses on Feuerbach'), written in 1845 and posthumously published by Engels in 1888, he argued that '[t]he philosophers have only interpreted the world, in various ways; the point is to change it'.[24] Although virtually no historians take Marx's historical pattern seriously any more, his perception that the forces of production represent the real motive power in history, with a double superstructure of the means of production first with ideology on top of it, has been extremely influential. As we shall see below, this methodical principle affected social–economic historiography and also the *Annales* school.

Explaining versus understanding: Dilthey and Windelband. A response to the ever growing influence of the positivist method was inevitable. The first systematic philosophical reflection on history and the humanities in general came from Wilhelm Dilthey (1833–1911), who in 1883 started developing a new theory of history.[25] He argued that whereas natural scientists were concerned with explaining

[21] Sidney Hook, *From Hegel to Marx: Studies in the Intellectual Development of Karl Marx*, Humanities Press, 1950.

[22] Warren Breckman, *Marx, the Young Hegelians and the Origins of Social Theory: Dethroning the Self*, Cambridge University Press, 1999.

[23] Karl Popper, *The Poverty of Historicism*, Routledge, 1957.

[24] Friedrich Engels, *Ludwig Feuerbach und der Ausgang der klassischen deutschen Philosophie*, J. H. W. Dietz, 1888, pp. 69–72. For an English translation, see Marx and Engels, *Collected Works, 1845–47, Volume 5: Theses on Feuerbach, The German Ideology and Related Manuscripts*.

[25] Wilhelm Dilthey, *Einleitung in die Geisteswissenschaften: Versuch einer Grundlegung für das Studium der Gesellschaft und der Geschichte*, 1883, reprinted by Teubner, 1959. For an English translation, see Wilhelm Dilthey, *Selected Works*, volume 1, translated and edited by Rudolf

(*erklären*), humanist scholars concentrated on understanding (*verstehen*). Counting and making measurements was of no use to historians. They had to search for the inner motives and intentions of historical figures. The term *Geisteswissenschaften* (literally: 'sciences of the spirit', nowadays equated with 'humanities') was given substance for the first time by Dilthey in 1883. Yet as I explained in the Introduction to this book, I generalize this concept to all humanistic disciplines, from Antiquity to the present day.[26]

The philosopher Wilhelm Windelband (1848–1915) brought the distinction between *Naturwissenschaften* (natural sciences) and *Geisteswissenschaften* closer to a head by contending that natural science used a *nomothetic* approach that focused on what was *general* and *systematic* (*nomos* is Greek for law), while the humanities employed an *idiographic* method that aimed at understanding the *unique*, *individual* event.[27] Such a pointed antithesis between these two fields of scholarship had never been postulated before. However, despite its conceptual clarity, Windelband's distinction is not historically correct. As is emerging from our quest, for centuries humanistic activities have displayed a nomothetic tradition, such as the search for laws in philological text reconstruction, the search for grammatical rules in linguistics, perspective in art theory, and consonance theory in musicology. Yet when it comes to historiography, Windelband may well have had a point. The search for general historical laws had produced little, except for the rather trivial pattern of rise, peak, and decline. Windelband therefore stated that historians should not try to formulate major explanatory systems, let alone laws, about the past. They should be satisfied with studying parts of the past, in which every component was created in a unique social and cultural context. And what applied to history applied by extension to all disciplines in the humanities, from philology to musicology. Although Windelband was of the mark here, the difference between *nomothetic* and *idiographic* and Dilthey's distinction between *erklären* and *verstehen* were extraordinarily influential, and even today they still dominate thinking about the disparity between natural science and the humanities. Dilthey's and Windelband's ideas gave the humanities a clear identity, but it did not always correspond to the *practice* of the humanities.

Twentieth century: neo-Rankean historicism. Almost without noticing we have entered the twentieth century. One of the movements within which the Diltheyan *verstehen* and the Windelbandian *idiographic* approach were implemented was

Makkreel and Frithjof Rodi, Princeton University Press, 1991. See also Rudolf Makkreel, *Dilthey: Philosopher of the Human Studies*, Princeton University Press, 1993.

[26] The neo-Kantian philosopher Heinrich Rickert called the *Geisteswissenschaften* also *Kulturwissenschaften* ('cultural sciences') which included sociology and anthropology—see *Die Grenzen der naturwissenschaftlichen Begriffsbildung*, Mohr, 1896. But the social sciences had already taken their own path and came to be seen as a specific set of disciplines. As explained in the Introduction, I will not go into the history of the social sciences, as this has been excellently dealt with elsewhere (see references on p. 4 n. 11). Yet I will discuss some of the influence of the social sciences on the modern humanities.

[27] Wilhelm Windelband, *Geschichte und Naturwissenschaft*, 3rd edition, Heitz, 1904. There seems to be no English translation available of Windelband's text.

neo-Rankean historiography, also referred to as *historicism*.[28] Although neo-Rankeans did not search for patterns or generalizations, they did assume that *there was a truth about the past that could be known*. Access to the past could be obtained through the critical study of sources, where the maximum distance from *the present* should be practised. Neo-Rankean historicism is an elaboration of Ranke's idea of objectivity, i.e. that there was a 'true' past that could be reconstructed, but now without Ranke's concepts of zeitgeist and people.[29] Historicism's most important method was the greatest possible immersion in a particular historical period in order to understand (*verstehen*) the human world, in which every phenomenon is unique (*idiographic*) and not a variation of something general. According to this method, which rejected patterns, every history was dependent on its individual context, and no historical knowledge could be derived by generalizing over other contexts. Patterns did not do justice to the unique, smaller contexts. The historian was therefore left with nothing more than a *multitude of unrelated contexts*. The idea of a cause and effect relationship, let alone a universal explanatory force, was rejected. Among the neo-Rankeans we see to a degree renewed use of the time-honoured anomaly principle, under which every phenomenon is considered separately and generalizations do not exist (see 2.3). *Relativism* was bound to receive considerable attention.

In its most extreme form, neo-Rankean historicism appeared to be more of an offshoot of the philosophy of history than a historiographical practice. Complete exclusion of external contexts and influences from present-day concepts is unfeasible. For some people the 'presentism' discussed earlier (chapter 1) was therefore an accomplished fact. As the Italian philosopher and historian Benedetto Croce (1866–1952), and 'follower' of Vico, aptly stated: *all true history is contemporary history*.[30]

The *Annales* school and the hierarchical pattern of time spans: Bloch, Braudel. Alongside the historiography of the neo-Rankeans, which rejected patterns, the approach of seeking patterns remained unprecedentedly strong. Instead of the ahistorical methods of Comte and Marx, though, at the beginning of the twentieth century a new social–economic school of historians deeply rooted in the tradition of critical source analysis developed. Initially this *Annales* school was active in Strasbourg, then in Paris, and after 1945 it came to dominate French historiography.[31] In contrast to the neo-Rankeans, these historians went in search of large-scale structures and forces, together with cause and effect relationships and historical generalizations. The *Annales* historians plunged into the economic and social

[28] There are several definitions of *historicism*: according to *Collins English Dictionary* (2003) historicism is the view that history is subject to natural laws. We follow the definition in Breisach, *On the Future of History: The Postmodernist Challenge and its Aftermath*, p. 325 and others who link it to the view that there is a truth about the past that can be known.

[29] Wolfgang Mommsen, 'Ranke and the Neo-Rankean School in imperial Germany', in G. G. Iggers and J. M. Powell (eds), *Leopold von Ranke and the Shaping of the Historical Discipline*, Syracuse University Press, 1990, pp. 124–40.

[30] 'Ogni vera storia è storia contemporanea', in Benedetto Croce, *Teoria e storia della storiografia*, Laterza, 1917, p. 4.

[31] Peter Burke, *The French Historical Revolution: The Annales School 1929–89*, Polity Press, 1990.

history of all aspects of human life. From the very beginning *Annales* historians such as Henri Berr, Lucien Febvre, and Marc Bloch believed it was possible to create a *total history*, as Febvre did in 1911 with his *Philippe II et la Franche-Comté*.[32] This type of historiography included geographical, psychological, social–cultural, and economic factors in explanations of historical phenomena.

The *Annales* historians were first and foremost historians, and the works they wrote were less theoretical. 'Whenever a historian reflects on the theory of history, his work stands still,' believed Lucien Febvre.[33] In his only manuscript on historical theory, written in prison in 1944 just before he was executed by the Gestapo, Marc Bloch argued that the task of the historian was to *go beyond the written and oral sources*.[34] Historians should also consider other sources, in fact all possible ones, in order to understand historical phenomena. Ernest Labrousse, for instance, elucidated the French Revolution from a social–economic perspective by considering eighteenth-century price trends and economic crises in his analysis. After his renowned total history of the Mediterranean, the 'prince' of the *Annales* school, Fernand Braudel (1902–1985), wrote a history of capitalism from the fifteenth to the eighteenth century, *Civilisation matérielle, économie et capitalisme*, in which he included not just all possible sources, but moreover posited risky generalizations, which he then held up to the light.[35] For example, Braudel found such laws as 'A dominant capitalist city always lies at the centre' and 'There is always a hierarchy of zones within a world economy'.[36] Braudel reached his generalizations by investigating a long time span—a *longue durée*—and abstracting from local contexts. He identified three hierarchically ordered *time spans*:

(1) *Structures*: occurrences that happen over a long time span (the aforementioned *longue durée*) over which generalizations can be found.

(2) *Conjunctures*: occurrences that happen over a medium time span that recur with a certain degree of regularity, but are less important to the changes in society.

(3) *Events*: short-term occurrences; according to Braudel these barely influence the course of history, and they are subordinate to and are affected by the higher time spans.

This sort of stratification makes one think of the hierarchical analysis of structural linguistics (see 5.3). We also came across hierarchical strata in the early modern humanities, such as *compositio* in Alberti's *disegno* theory (4.5), and even earlier in the *constituent structure* of Modist linguistics, in which *lower levels are subordinate to*

[32] Lucien Febvre, *Philippe II et la Franche-Comté: étude d'histoire politique, religieuse et sociale*, Honoré Champion, 1911, reprinted by Perrin, 2009.

[33] Lucien Febvre, *Pour une histoire à part entière*, S.E.V.P.E.N., 1962, p. 852.

[34] Marc Bloch, *Apologie pour l'histoire ou métier d'historien*, Cahier des Annales 3, Librairie Armand Colin, 1949. Translated into English as *The Historian's Craft*, by Peter Putnam, Vintage Book, 1953.

[35] Fernand Braudel, *Civilisation matérielle, économie et capitalisme, XVe-XVIIIe siècle*, 3 volumes, Armand Colin, 1986. Translated into English as *Civilization and Capitalism, 15th–18th century*, 3 volumes, by Siân Reynold, University of California Press, 1992.

[36] Braudel, *Civilization and Capitalism, 15th–18th century*, volume 3, pp. 26–36.

higher levels (3.1). But in historiography the hierarchical pattern of structures, conjunctures, and events was first worked out by the *Annales* historians. In broad terms this stratification corresponds to the distinction that is often made between macrohistory, mesohistory, and microhistory.[37]

Braudel showed that there could be historical generalizations at the highest level of the *longe durée*. His vision contrasted with the starting points of the neo-Rankeans, who maintained that generalizations were out of the question because contexts always remained specific. Nevertheless, it is not impossible to integrate the two traditions of the neo-Rankeans and *Annales* school, provided that we assume a gradual transition between the macro and micro levels. It is then possible to argue as follows. The more we zoom in, the more specific the history, and the more we zoom out, the more general the history. The extremes of this continuum are given an asymptotic shape by historicism and the *Annales* school, but one cannot draw a boundary line anywhere. Both specificity and generality play a role at every point. Although justice can be done in this way to both the idiographic and the nomothetic characteristics of historiography, the question remains as to whether exponents of these two extremes could go along with this.

We came across the *Annales* historians' practice of considering all possible sources (social, economic, cultural, literary, and geographical) many centuries before in the historical critique of Liu Zhiji (661–721)—see 3.2. However, it is improbable that the *Annales* historians knew the work of Liu Zhiji. This was not because of the long interval between the two (after all the Graeco-Roman historiographers were even further in the past), but because the *Annales* school's field of interest was primarily in the area of Western and Mediterranean history. However, we know that the sociological history of Ibn Khaldun (see 3.2) had an effect on the *Annales* historians thanks to the discussion of his method by Marc Bloch.[38] It is also recognized that Marx's historical–materialistic historiography had an influence. The social–economic factors employed by Marx were generalized by the *Annales* historians to *all* factors and *all* possible sources. We will consequently designate the method of the *Annales* historians as the 'all possible sources principle', and we give Liu Zhiji the honour of having introduced it.

We therefore find two main approaches in early twentieth-century historiography, both of which claimed to be 'objective'. The first was idiographic and sought the unique and individual (the neo-Rankeans), and the other was primarily nomothetic and searched for the general and systematic (the *Annales* school). While it is true that the influence of Dilthey and Windelband was substantial, it was certainly not dominant. We find both methods that seek patterns and reject patterns. The fact that their work nonetheless appeared to represent the accepted view of the humanities was largely because, together with the work of Croce, at the beginning of the twentieth century it was virtually the only philosophical reflection concerning the humanities.

Pattern-seeking historiography in the later twentieth century. The second half of the twentieth century displayed a multiplicity of historical approaches that we

[37] Peter Burke (ed.), *New Perspectives on Historical Writing*, Polity Press, 1991, pp. 93ff.

[38] See e.g. Marc Bloch, *Feudal Society*, volume 1, Routledge, 1975, p. 54.

can still characterize as pattern-seeking or pattern-rejecting.[39] We will begin by discussing the nomothetic, pattern-seeking group.

Social–economic history. This movement was typified by the desire to find generalizations in social and economic history. It actually began in the nineteenth century, but was not given sound underpinning until the *Annales* historians. The American New Left Historians and German *Historische Sozialwissenschaft* continued to build on it.[40] Social–economic historiography was dominated by a quantitative approach to deriving historical trends and patterns.[41] *Historical demography*, which tried to trace long-term developments on the basis of population figures and migration balances from the past, also arose out of this tradition.

Cliometrics. Originating from economic historiography, cliometricians used mathematical–economic models of historical data to interpret the past. As had been attempted earlier in musicology and art theory (see chapter 4), it now seemed to be the turn of history to be approached mathematically. This was done primarily by using a *counterfactual* approach (which had also been applied in a very preliminary form by Lorenzo Valla—see 4.1). Robert Fogel and Albert Fishlow, for instance, speculated about what would have happened to the nineteenth-century American economy if the railways had not been built. Using econometric models, they tried to measure the difference between the actual costs of railway construction and the hypothetical costs of an economy without railways.[42] However, the cliometric method became controversial with Robert Fogel and Stanley Engerman's book *Time on the Cross: The Economics of American Negro Slavery* (1974). Once again mathematical–economic models were applied to a historical question, in this case whether nineteenth-century slavery in the United States had been a lucrative means of production. Flying in the face of accepted views, Fogel and Engerman believed they could demonstrate that slavery was profitable. However, the way they applied their mathematical analyses resulted in a great deal of criticism. In his *Slavery and the Numbers Game: A Critique of 'Time on the Cross'* (1975), for example, Herbert Gutman objected that (1) Fogel and Engerman's analysis was based on only one plantation that was not representative, (2) mistaken criteria were used to measure the cruelty of slavery, and (3) systematically more reliable data were disregarded. Despite this criticism, the conclusions in *Time on the Cross* remained largely intact, but not until a further thirty years of study had been conducted.

[39] Additionally, there was also a totalitarian historiography that was communist, fascist, or National-Socialist. These ideological writings followed state ideology: the history is (re)interpreted in terms of Marxist stages, Roman heroes, or German racism. Dissident historians were persecuted (like Antonio Gramsci in Italy) or fled (like Erwin Panofsky, who fled from Germany). Benedetto Croce was just about being tolerated in fascist Italy, but entirely isolated. While the situation is somewhat more complex (under Lenin there was more tolerance than under Stalin), there is little to say about principles and patterns, except that ideological historiography is not based on critical analysis but on supporting established doctrines.

[40] Reinhard Rürup (ed.), *Historische Sozialwissenschaft: Beitraäge zur Einführung in die Forschungspraxis*, Vandenhoeck und Ruprecht, 1977.

[41] See e.g. Rondo Cameron and Larry Neal, *A Concise Economic History of the World: From Paleolithic Times to the Present*, 4th edition, Oxford University Press, 2003.

[42] Robert Fogel, *Railroads and American Economic Growth: Essays in Econometric History*, Johns Hopkins Press, 1964.

Neo-positivist history (and its failure). Neo-positivism—an even more extreme vision of history, which was related to cliometrics and the positivist approach—caught on in the 1950s. According to this view, all historical processes and events can be derived deductively from general laws. This notion should not be confused with the cliometric approach, but it fits well into the historiographical movement that employed logical and mathematical derivations. Carl Hempel's deductive-nomological (D-N) explanation model is the best known.[43] As in natural science, according to Hempel historians, had to first of all describe the *initial conditions*, after which the historical events could be derived on the basis of logical deduction and universal laws. A given set of initial conditions would always produce the same effects. However, Hempel soon realized that not all historical explanations were deductive, and he extended his model to include induction and statistics. Nevertheless, the neo-positivist movement was short-lived and died an inglorious death.

Cultural history. A nomothetic approach developed from an unexpected angle on the basis of the models of early cultural historians such as Oswald Spengler (1880–1936) and Arnold Toynbee (1889–1975). While these models are not mathematical or quantitative, they do follow the tradition of Vico's and Herder's systematic pattern of rise and fall. In his *Der Untergang des Abendlandes* ('The Decline of the West') Spengler argued in 1918 that the nineteenth-century positivist pattern of progress was not supported empirically.[44] In its place he proposed a history of civilization in which cultures develop, mature, and die. He underpinned his views on the grounds of eight major world civilizations—Indian, Babylonian, Chinese, Egyptian, Arabic, Mexican, Graeco-Roman, and Western. Spengler believed it was possible to deduce that Western culture was on the threshold of its destruction.

In his twelve-volume work *A Study of History* (written between 1934 and 1961), Toynbee also used a model of the rise and fall of cultures. He based his analysis on an even bigger synthesis of world history. According to Toynbee civilizations began to flourish when they tried to face up to a series of serious challenges. If a culture was no longer able to do this, it began to decline. Civilizations died as a result of suicide, not murder. Toynbee expressed his great admiration for Ibn Khaldun's *Muqaddimah*, in which the systematic trends of history were exposed (see 3.2). Yet Toynbee's work was fiercely criticized because of his frequent use of myths and metaphors (in this regard he ignored Ibn Khaldun's criticism of myths).[45]

Not all cultural historians were obsessed with the cyclical model. Johan Huizinga (1872–1945), for example, at first instance maintained an aesthetic approach to

[43] Carl Hempel, 'The function of general laws in history', *Journal of Philosophy*, 39(2), 1942, pp. 35–48.

[44] For an English translation, see Oswald Spengler, *The Decline of the West*, Oxford University Press, 1991.

[45] Pieter Geyl (1887–1966) claimed that Toynbee left out evidence that went against his assumed cycle. He described Toynbee's approach as 'metaphysical speculations dressed up as history'—see Pieter Geyl, Arnold Toynbee, and Pitirim Sorokin, *The Pattern of the Past: Can we Determine It?*, Greenwood, 1949.

history, in which art and the reconstruction of celebrations and performances played an important part. Yet Huizinga also applied the pattern-seeking paradigm: in his famous *Herfsttij der Middeleeuwen* ('The Autumn of the Middle Ages', 1919) he revealed on the one hand the 'cultural rules' of the Late Middle Ages, and on the other he established a link to the systematic pattern of rise and decline when he believed he could demonstrate that the fourteenth and fifteenth centuries had to be considered as a period of decadence instead of the beginning of a revival. We can already find the preoccupation with the concept of revival in the work of Jacob Burckhardt (1818–1897) who, because of his importance for art history, will be dealt with in 5.5.

There were also studies of the few times in the course of history that a cyclical model appeared to have been breached. One of the works in which we find examples of these is Max Weber's historical–sociological series *Die Wirtschaftsethik der Weltreligionen* ('The Economic Ethic of World Religions'). In *Die protestantische Ethik und der Geist des Kapitalismus* ('The Protestant Ethic and the Spirit of Capitalism', 1905) in particular, he contended that the West took a unique turn with the introduction of modern capitalism. According to Weber there was a causal connection between the ascetic life lived by Calvinists 'in the world', where virtually all profit was reinvested, and the creation of capitalism. The studies of the Industrial Revolution also assumed that there was a departure from a cyclical pattern. We also find this in the history of natural science, where the Scientific Revolution broke out of the pattern of rise, peak, and decline.[46] In their comparison of several cultures, these studies were *searching for patterns*, but at the same time were *culture-specific* in their focus on the unique in one culture (see also below, 'Between pattern-seeking and pattern-rejecting historiography').

Summary of pattern-seeking history. The picture of historiography that searches for patterns is more complex than Dilthey and Windelband could have suspected. The *Annales* school and the *Sozialgeschichte* sought *social–economic patterns* in historical processes, the cliometricians used *mathematical models* for analysing possible historical processes, the neo-positivists looked for *logical derivations* of historical events, while the cultural historians hunted for *general and specific cultural generalizations* in the historical material. Only cliometrics was based on mathematical formalization. The neo-positivist method employed logical formalization, whereas the cultural and *Annales* historians did not actually use any formalization (although the latter group also regularly utilized statistical analyses). If there was any quest for a 'procedural system of rules' in pattern-seeking historiography, it was in neo-positivist history, which failed pitifully in this regard.

Rejecting (universal) patterns in later twentieth-century historiography. While the pattern-rejecting tradition did criticize and dismiss patterns, this refutation made way for a quest for *different* patterns that were culture-specific, ideological, colonial, or post-colonial. It therefore makes sense to define 'pattern-rejecting' exclusively as the refutation of *universally* assumed, world-historical patterns.

[46] H. Floris Cohen, *The Scientific Revolution: A Historiographical Inquiry*, University of Chicago Press, 1994; and also H. Floris Cohen, *How Modern Science Came into the World: Four Civilizations, One 17th-century Breakthrough*, Amsterdam University Press, 2010.

Narrativism. In the second half of the 1950s a group of historians threw doubt upon the ideals and practices of nomothetic history and contended that it was not causal explanation but the story that formed the essence of historiography. History's autonomy had to be restored again by means of the narrative. More than a return to Dilthey's *verstehen*, *narrativism* represented a plea for a narrating historiography that was largely independent of how sources were obtained. Narrativists could still use economic data, but causal connections did not need to be made between the facts. What a historical narrative *did* have to comprise became the subject of in-depth discussion that is still continuing. According to the early narrativists, the structure or order of reality also served as the structure of the narrative. In his *Histoire et vérité* ('History and Truth', 1955) Paul Ricoeur talked about the insurmountable temporality of existence, which gave the narrative a predetermined time dimension.[47] David Carr's starting point was a human existential experience, without which the historical narrative could not even exist or be understood.[48] In his *Narrative Logic: A Semantic Analysis of the Historian's Language* (1983) Frank Ankersmit went one step further and asserted that coherence and order in the past were the result of the historian's storytelling. By means of the narrative itself, the historian created meaning as a replacement for a past reality that was by definition absent. If patterns were found at all in history, they were not necessarily intersubjective and consequently not verifiable. So on the one hand there were narrativists who contended that *the structure of reality determined the structure of the account*, while on the other there were narrativists who believed the opposite, i.e. that *the structure of the story gave structure to the (past) reality*. The work of Hayden White occupies an interesting position here. In his *Metahistory* (1973) White discusses analyses of the major nineteenth-century historiographies on the basis of the notion of figures of speech. Every historiographical period turned out to have its own rhetorical device. For example, synecdoche is considered a rhetorical device for Rankean historiography, while irony is characteristic for Burckhardtian historiography. In White's work the relationship between historical and literary narrative became so close that they essentially coincided.

Critical school. Like narrativism, the Frankfurt philosophical school rejected a history based on positivist starting points. This neo-Marxist school became famous for its so-called Critical theory.[49] In their *Dialektik der Aufklärung* ('Dialectic of Enlightenment', 1947) the two most important exponents, Theodor Adorno and Max Horkheimer, expressed their opposition to all interpretative methods that were grafted on to rationality—whether they were scientific, progress-based, or otherwise. This was because such approaches did not do justice to the Marxist concept of 'alienation' in Western civilization—the feeling of no longer being part

[47] Paul Ricoeur, *History and Truth*, translated by Charles Kelbley, Northwestern University Press, 1965.

[48] David Carr, *Time, Narrative and History*, Indiana University Press, 1986. See also Edward Casey, 'David Carr on history, time, and place', *Human Studies*, 29(4), 2006, pp. 445–62.

[49] Zoltán Tar, *The Frankfurt School: The Critical Theories of Max Horkheimer and Theodor W. Adorno*, Wiley, 1977.

of society. The Critical school called into question the apolitical attitude of the pattern-hunting approach, which implicitly puts all knowledge at the service of the existing power and production relationships. Myths could only be defused through a totally critical approach. Above all, the Frankfurt school produced historical criticism rather than history. Yet the concept of criticism, which was elevated by this school to a basic principle, resulted in an awareness of many hidden assumptions in historiography.

Postmodernism. The horrors of the world wars caused many historians to definitively reject the idea of progress. Some even wanted to rid themselves of the designation 'modernity', which they associated with the catastrophes of the twentieth century. And so the term 'postmodernism' was adopted. It was first used in historiography by Toynbee, who stated that the postmodern era started at the end of the First World War. The criticism of modernity only really got under way in the work of French philosophers such as Michel Foucault, Jacques Derrida, and Jean-François Lyotard, who refuted all claims to universality.[50] The centuries-old historical problem of change versus continuity was solved by attributing any reality exclusively to change. In this vision no stable or absolute truth can exist. Interpretations of texts, people, and events had to be 'decentred' or stripped of stability so that plurality triumphed over artificial unity. According to this revolutionary vision, historical works have no special status and they are texts just like all others (see also 5.6). Metahistories and claims to any truth were analysed, deconstructed, and rejected. Nevertheless, the postmodernists actually had their own universal historical pattern when they talked about 'continuous change'. This produced an inconsistency in their historical critique, which on the contrary wanted to refute all patterns, absolute or otherwise. The ambition of the postmodernists went a step further than that of the Critical school—even Marxist interpretations had to face up to it. Responses were not long in coming, and critics brought up the fact that not a single historical work could be in agreement with the postmodernist vision. The upshot was that postmodernist thinking in the narrow sense was brushed aside by most historians.[51]

Summary of pattern-rejecting history. In simplified terms we can describe the non-nomothetic movements as follows. According to (one of the visions within) narrativism, only *the 'narrative' gives meaning to an absent past,* according to the Critical school *only general criticism can demythologize the past,* and according to postmodernism *any claim to historical truth is subject to deconstruction.* The most extreme form of history that rejected patterns produced little historiography, as did the most extreme form of pattern-seeking history.

Between pattern-seeking and pattern-rejecting historiography. There has been a strong revival of cultural history since the 1980s. In it, all fields of human life are considered to be expressions of culture. This so-called *new cultural*

[50] Perry Anderson, *The Origins of Postmodernity*, Verso, 1998. See also Ernst Breisach, *On the Future of History: The Postmodernist Challenge and its Aftermath*, University of Chicago Press, 2003.

[51] Sigurdur Magnusson, 'The singularization of history: social history and microhistory within the postmodern state of knowledge', *Journal of Social History* 36(3), 2003, pp. 701–35.

history, which is also associated with the history of mentality and *nouvelle histoire*, appears to stem partially from the *Annales* school and partially from previous cultural history (see above).[52] Yet contrary to the *Annales* school, the new cultural history focuses more on specific periods or regions, or even a single village such as the famous *Montaillou, village occitan de 1294 à 1324* ('Montaillou, an Occitan Village from 1294 to 1324') published in 1982 by the French historian Emmanuel Le Roy Ladurie. Clifford Geertz's 'thick description', which had become fashionable in cultural anthropology, was cited approvingly by new cultural historians.[53] Unlike previous cultural history, however, the new one presents itself as not explicitly searching for or rejecting patterns. In fact, it transcends the movements discussed above and is better characterized on the basis of its subjects.

Cultural history concentrates primarily on phenomena that are shared by non-elitist social groups as well as on social movements like nationalism, feminism, and post-colonialism. New cultural history can perhaps best be characterized by the concepts that it uses, for example power, ideology, culture, identity, race, and perception, as well as its 'new' historical methods, including oral narratives and biographies. Although cultural historiography does not associate with its nomothetic predecessors such as Spengler and Toynbee—only Huizinga and Burckhardt (see 5.5) are quoted with approval—it has appeared more than once to be in search of cultural rules and schemes.[54] This has involved trying to find patterns in a historical epoch by employing categories and principles *from that period.* If a historian knows the rules of fifteenth-century art theory or rhetoric, for example, he can use them to analyse works of art, texts, and other, even less obvious objects, dating from that time.[55] The patterns and analyses obtained in this way are normally culture specific, whereas the quest for patterns across different eras generally has a more universal pretension.

In the course of the later twentieth and the early twenty-first century, increasing specialization has led to further fragmentation of historiography. Current directions include such diverse fields as *women's history, new oral history, intellectual history, psychohistory, indigenous history*, and *subaltern history* (see below under 'India'), to name a few.[56] At the same time there is a renewed tendency towards a more universal view as exemplified by *global history* and *world history.*[57] The question arises whether there remain any common principles in this plethora of historical subdisciplines—an issue to which we will come back at the end of this section.

China: the advance of Western and Marxist views of history. Starting in the twentieth century, European historical methods made significant headway in

[52] Peter Burke, *What is Cultural History?*, 2nd edition, Polity Press, 2008.

[53] For an introduction, see Clifford Geertz, *The Interpretation of Cultures: Selected Essays*, Basic Books, 1973, pp. 3–30.

[54] Burke, *What is Cultural History?*, p. 11 and p. 41.

[55] Michael Baxandall, *Giotto and the Orators: Humanist Observers of Painting in Italy and the Discovery of Pictorial Composition 1350–1450*, Oxford University Press, 1971. See also Caroline van Eck, *Classical Rhetoric and the Visual Arts in Early Modern Europe*, Cambridge University Press, 2007.

[56] See Woolf, *A Global History of History*, pp. 457ff.

[57] See e.g. Patrick Manning, *Navigating World History: Historians Create a Global Past*, Macmillan, 2003.

China. In the nineteenth century, however, Chinese historiography was still tied to the tradition of dynastic chronicle writing. During this last phase of Imperial China (the Qing Dynasty lasted until 1912) there was still an enormous amount of historiographical output. Over and above the normal annalistic court chronicles, we find descriptions of border areas like Mongolia, Xinjiang, and Tibet which once again adopted the standard historiographical layout. He Qiutao (1824–1862), the author of the first study of Sino-Russian relations, could not be accused of new historical ideas when he wrote his *Shuofang beisheng* ('Historical Sources on the Northern Regions').[58] Sometimes we do find new patterns, such as the one introduced by Liu Fenglu (1776–1829) who divided the past into three large epochs: first the three ancient dynasties of Xia, Shang, and Zhou; next the period of Confucius and finally that of succeeding generations.[59] The most important late Qing historian was Wei Yuan (1794–1856).[60] An immensely influential intellectual figure, Wei wrote *Shengwu ji* ('A Military History of the Qing Dynasty') and a narrative work on the Opium War, which is one of the first accounts on Sino-Western relations. But the work for which he is best known today is *Haiguo tuzhi* ('Illustrated Treatise on the Maritime Kingdoms') where he investigates the history and geography of Europe. Two hundred and fifty years after the Jesuits studied the history of China (see 4.2), Wei published the first Chinese study of the history of the European countries.[61]

The dramatic revolutions in twentieth-century China brought substantial changes in Chinese historiography. The centuries-old state exams for civil servants were scrapped in 1905, as a result of which the strict annalistic style of Sima Qian (see 2.2) also disappeared. It was replaced by European views of history, particularly during the Republican Era (1912–1949). Liang Qichao (1873–1929) was one the most important reformists who called for a historiographical revolution (*shijie geming*) inspired by European, especially German historiography.[62] Such a history should no longer be focused on rulers and officials but should examine all aspects of the past. Liang also investigated a new periodization scheme for Chinese history, modelled on the Western pattern of ancient, medieval, and modern.[63] This was the final blow to Chinese dynastic historiography and the ideal of a harmonious society with a mandate from heaven.

Marxist salvation history became dominant when the communist People's Republic was founded in 1949, though it was already initiated by earlier works of Guo Muruo (1892–1978) and Fan Wenlan (1893–1969). Such a Marxist history usually covers the time span from proto-communism via a period of slavery,

[58] Endymion Wilkinson, *Chinese History: A Manual, Revised and Enlarged*, Hardvard University Asia Center, 2000, p. 946.

[59] Benjamin Elman, *Classicism, Politics, and Kinship: The Ch'ang-chou School of New Text Confucianism in Late Imperial China*, University of California Press, 1990, pp. 227ff.

[60] Ng and Wang, *Mirroring the Past: The Writing and Use of History in Imperial China*, p. 255.

[61] Jane Kate Leonard, *Wei Yuan and China's Rediscovery of the Maritime World*, Harvard University Press, 1984.

[62] Hao Chang, *Liang Ch'i-Ch'ao and Intellectual Transition in China*, Oxford University Press, 1971.

[63] Ng and Wang, *Mirroring the Past: The Writing and Use of History in Imperial China*, pp. 262–3.

a feudal age, and a capitalist stage to a socialist era (see discussion of Marx above). However, it proved to be no easy job to make this periodization correspond to the many Chinese dynasties. Under Mao Zedong the complex dynastic history was forced into the Marxist straitjacket to which an extra period was added in order to make it plausible for Chinese history. A period of semi-colonialism between 1840 and 1919 was assumed, which was seen as a transition phase from feudalism to capitalism.[64] Every proposal to amend Mao's outline was life-threatening in the People's Republic, with the result that critical history was completely absent. However, here and there we find fascinating historiographies of peasant life, admittedly interpreted within the official framework of the class struggle.[65] Even after the Cultural Revolution (1966–1976) the Marxist model continued to be present in Chinese historiography, although it was pushed further into the background.

In present-day Chinese historiography we once again find Western views of history, but certainly others too. The use of scientific methods for dating Chinese dynasties is booming. There is moreover a more traditionalist school in Chinese historiography, particularly in the work of Qian Mu (1895–1990), who because of his writings had to flee to Hong Kong and Taiwan. Contrary to Marxist historiography, Qian Mu emphasized the importance of Confucian philosophies to the understanding of Chinese history.[66] After the economic reforms by Deng Xiaoping, Marxist historiography in the People's Republic also became more inspired by Confucius. 'Pattern-rejecting principles', such as in Western postmodernism, are hard to find in China.

India: the reconstruction of the past. Western views of history were dominant in other regions too, initially as a result of European colonialism and later through the international scholarly culture that was controlled by the West. As we have seen, historiography in India did not start until quite late with the eleventh-century chronicle of Kashmir (see 3.2) and Mughal historiography from the sixteenth century to the eighteenth century (see 4.2).

From the nineteenth century onwards, Indian historiography was grafted onto the European movements we discussed above.[67] Itihasacharya Rajwade (1863–1926), for instance, applied Rankean source criticism when he wrote his history of the Marathas in twenty-two volumes and corrected many errors in earlier historiographies.[68] One of the most prominent Indian historians was Jadunath Sarkar (1870–1958), whose work covered essentially all periods of Indian history.[69] His reconstruction of India's past was the result of a search lasting over fifty years for first-hand material in India and Europe—an almost superhuman achievement if we consider that no Indian historiography prior to the chronicle of Kashmir existed or has survived (see 2.2).

[64] Albert Feuerwerker, *History in Communist China*, The MIT Press, 1968.

[65] James Harrison, *The Communists and Chinese Peasant Rebellions; A Study in the Rewriting of Chinese History*, Atheneum, 1969.

[66] Jerry Dennerline, *Qian Mu and the World of Seven Mansions*, Yale University Press, 1988.

[67] R. C. Majumdar, *Historiography in Modern India*, Asia Publishing House, 1970.

[68] R. S. Sharma, *Rethinking India's Past*, Oxford University Press, 2009.

[69] Kiram Pawar, *Sir Jadunath Sarkar: A Profile in Historiography*, Books & Books, 1985.

Nevertheless, we find no new ideas in the area of methodology. The same can be said of the Indian independence movement and socio-economic historiography. Damodar Kosambi (1907–1966) and Ram Sharma (born 1919) described peasant life, in which fascinating parallels were drawn between European and Indian feudalism. Their methods were based on Marxist and neo-Marxist historiography.

The neo-Marxist approach is fully embraced by Ranajit Guha (born 1923) who is the founder of Subaltern Studies.[70] The term 'subaltern' is taken from Antonio Gramsci (1891–1937), who introduced the influential distinction between hegemonic (dominating) and subaltern (subordinate) cultures.[71] Inspired by Edward Said's *Orientalism* (1978), Guha and his fellow historians are especially interested in postcolonial societies, like India, and their approach is one of *history from below*, focusing on what happens among the masses rather than among the elite. Guha's classic is *Elementary Aspects of Peasant Insurgency in Colonial India* (1999).

Africa: oral transmission and the effect on European oral history. African historiography appears to have been an exception to the European intellectual dominance that we find in China and India. Contrary to Rankean textual source criticism, African historiography has been largely based on oral transmission (see also 4.2).[72] One example is *The History of the Gold Coast and Ashanti*, which was originally written in Ga by the Ghanaian Carl Christian Reindorf (1834–1917).[73] Reindorf's historiography utilized oral sources (he interviewed more than 200 men and women) and is very important to our knowledge of the Ashanti Empire and its fall in the nineteenth century. The work that put African historiography permanently on the map was *The History of the Yorubas* by the Nigerian Samuel Johnson (1846–1901)—not to be confused with the eighteenth-century Samuel Johnson (see 4.7).[74] This historiography, which was not published until 1921, was based mainly on the Oyo oral tradition. Oyo was an empire of the Yoruba people that existed from about 1400 to 1836 and covered an area from Nigeria to present-day Ghana. Johnson wrote a history of the Oyo Empire up to and including the British Protectorate on the basis of the phenomenal orally transmitted history of the Yoruba (including complete genealogies and chronicles). The work covered kings, heroes, the rise, peak, and oppression of the Yoruba, the arrival of the British, wars, and finally the dramatic disintegration. It was only in the last part—about the nineteenth century—that Johnson based his historiography on his own personal experiences.

Contrary to current opinion, the African historiographical principles of oral transmission combined with personal experience thus continued to be employed

[70] Ranajit Guha, *Subaltern Studies Reader, 1986–1995*, University of Minnesota Press, 1997.

[71] See e.g. Tom Bottomore, *The Dictionary of Marxist Thought*, Blackwell, 1992.

[72] In addition, there are also charges against slavery written by freed slaves, such as the autobiographies by Ottobah Cugoano and Olaudah Equiano at the end of the eighteenth century. These are based on the personal experience principle.

[73] Raymond Jenkins, *Gold Coast Historians and Their Pursuit of the Gold Coast Pasts, 1882–1918*, PhD thesis, University of Birmingham, 1985.

[74] Samuel Johnson, *The History of the Yorubas: From the Earliest Times to the Beginning of the British Protectorate*, African Books Collective, 1997. See also Toyin Falola (ed.), *Pioneer, Patriot and Patriarch: Samuel Johnson and the Yoruba People*, University of Wisconsin Press, 1993.

(as we saw in the sixteenth century with the historiography of the Kati family—see 4.2). There appears to have been no trace of dominance by European historical methods. It has been suggested that this steadfastness of the oral transmission principles can be explained by the lack of written African sources. But this view is based on an incorrect assumption: there are hundreds of thousands of known manuscripts in Timbuktu alone—see 4.2. Moreover, this view underestimates the huge prestige that oral transmission had and still has in all parts of Africa. Centuries-old knowledge about the past is retained by 'specialists of the word' (the *griots*) as well as by broad sections of the population.[75] Western anthropologists who visited the African continent were immediately struck by the value that had been and was being associated with oral history in Africa.[76] It resulted in a scholarly discussion in Europe and the United States about the value of oral history and life histories in an era in which the Rankean paradigm of textual sources was dominant. It is not implausible that the discovery of oral history outside Europe, in particular in Africa, stimulated interest in oral history in the West, with the Chicago school of sociologists in the 1930s as a good example.[77] In any event, the study of life histories and oral history was fully adopted by the new cultural history that we discussed above. The oral transmission principle may thus be an intriguing example of an African influence on European humanities.

The personal experience and oral transmission principles continued to survive in Africa in the twentieth century too. At the same time Western principles got a foot in the door, but an African approach remained present. The work of the Senegalese historian Cheikh Anta Diop (1923–1986) is an interesting example. Diop grew up in Senegal and studied for his research degree in Paris, where he argued on the basis of archaeological and anthropological research that the Pharaohs were of Negroid origin. Back in Senegal, Diop developed a scientific method for determining the ethnic descent of mummies. Although his method was very controversial, Diop's criticism of Western historiography on Africa has been broadly accepted. In his *Antériorité des civilisations nègres: mythe ou vérité historique?* (1967) he showed that both before and after decolonization European historians grossly underestimated the possibility of black civilizations.[78] Diop's historiographical work has influenced post-colonial studies all over the world.[79]

The diversity of the past reality. Summarizing, we can state that modern historiography shows a proliferation of methodical principles that contradict and exclude one another more than once. How can we explain such enormous fragmentation? Perhaps the answer lies in *the diversity of 'past reality' too immense to take in.* There

[75] Jack Goody, *The Interface between the Written and the Oral*, Cambridge University Press, 1987, chapter 3.

[76] Joseph Ki-Zerbo (ed.), *General History of Africa I: Methodology and African Prehistory*, Heinemann, 1981.

[77] Many thanks to Daniela Merolla for this suggestion.

[78] For an (abridged) English translation, see Cheikh Anta Diop, *The African Origin of Civilization: Myth or Reality*, translated by Mercer Cook, L. Hill, 1974.

[79] Chris Gray, *Conceptions of History in the Works of Cheikh Anta Diop and Theophile Obenga*, Karnak House, 1989.

are no 'absolute' principles for all forms of historiography because history gives no boundaries to its subject. For example, the history of both ice skating and the Enlightenment belong to historiography, but the historical methods used for them are not easily interchangeable. And the historian who studies an oral culture will not be able to use the philological methods that have been developed for a written culture. History therefore differs in an essential way from other humanistic disciplines, such as musicology or linguistics. One can study everything from the past, but by definition musicology studies only music.

Despite the abundance of contrasting historical principles, there is also a common approach—source criticism. This is a constant factor in historiography, from Herodotus's naive source selection and the formal *isnad* method for oral transmission to the Rankean philological approach and the general critical method of the Frankfurt school—and more. Whatever history is, without source criticism there can be no history writing.

5.2 PHILOLOGY: A COMPLETED DISCIPLINE?

Like historiography, from the nineteenth century onwards philology also focused on all periods. The tried and tested humanistic method of text reconstruction was perfected and applied to people's 'own' national literature. The deciphering of script systems also received renewed attention, with the decoding of hieroglyphs as its greatest achievement. The term 'philology' was associated more and more with the study of linguistics and literature in the broadest sense of the words. The term fell into disuse during the course of the twentieth century.

Philology as decipherment: from hieroglyphs and Linear B to Maya. The deciphering of manuscripts, in particular Homeric, had been part of philology since the Alexandrians. However, no decoding appealed more to the imagination than that of Egyptian hieroglyphs. Many people were defeated in their attempts, and deception and fraud were commonplace. The first serious effort came from Islamic civilization. In the ninth century Dhul-Nun al-Misri and Ibn Wahshiyya were able to understand a part of the hieroglyphs by relating them to contemporary Coptic.[80] In the seventeenth century Athanasius Kircher spared neither expense nor time to sell his decoding of hieroglyphs.[81] He added hieroglyphs that he had designed to the obelisks that had been brought from Egypt to Rome, but nobody understood them. Although Kircher's deciphering was extremely controversial, he argued correctly that Coptic was not a separate language, but the last stage in the development of Ancient Egyptian.

It is not until we get to Thomas Young (1773–1829) and Jean-François Champollion (1790–1832) that we find real decoding of hieroglyphics. This was made possible by an opportune discovery in 1799: the *Rosetta Stone*, which was unearthed by French military engineers near Rosetta (now El Rashid) in Egypt.[82] The stone bears inscriptions in three identical texts in three languages and script systems: Ancient Egyptian hieroglyphs, late Egyptian demotic script, and Greek script (which was used by the Ptolemaic rulers of Egypt). The text is an expression of gratitude from the priests of Memphis to King Ptolemy V and is dated 27 March 196 BCE.

One of the biggest subjects for debate with regard to hieroglyphs was whether they represented sounds or concepts.[83] In 1821 Champollion assumed the latter, probably because this had already been assumed since the seventeenth century and also appeared to be the case for Chinese (and was moreover applied in the artificial languages of Dalgarno and Wilkins, see 4.3). Yet less than a year later, Champollion published his hieroglyphic alphabet in which hieroglyphs could also represent phonetic symbols. He was almost certainly prompted in this regard by the earlier

[80] Okasha El Daly, *Egyptology: The Missing Millennium: Ancient Egypt in Medieval Arabic Writings*, UCL Press, 2005.

[81] Steven Frimmer, *The Stone that Spoke: And Other Clues to the Decipherment of Lost Languages*, Putnam, 1969, pp. 38ff.

[82] Richard Parkinson, *The Rosetta Stone*, British Museum Press, 2005.

[83] Lesley Adkins and Roy Adkins, *The Keys of Egypt: The Obsession to Decipher Egyptian Hieroglyphs*, HarperCollins Publishers, 2000.

discovery by Thomas Young (published in 1818) that certain hieroglyphs were phonetic. However, Champollion always maintained that he had made the discovery himself. In any event both of them used the names of kings on the Rosetta Stone to disentangle alphabetical symbols. The names of these Ptolemaic kings were spelled out in alphabetical symbols because they were of Greek origin. And because the names of the kings in the Egyptian texts were in cartouches (oblong figures), the phonetic part of the hieroglyphic script could be interpreted fairly simply.

Champollion realized what Young had failed to understand. He suspected that certain hieroglyphs could also represent syllables and that other hieroglyphs could even refer to whole words. His suspicion proved to be correct. Hieroglyphic script recorded virtually the entire history of the transition from concepts to sounds. Every symbol was saved, probably because the script was holy to the Egyptians. The perception that the hieroglyphic script was a hybrid—and gradual—gave Champollion the key with which to embark on the actual deciphering. He set to work like someone possessed by a demon, and in 1824 he came up with his famous *Précis du système hiëroglyphique*.[84] Champollion's decipherment caused a real sensation. Finally, the many Egyptian inscriptions and papyruses, which covered more than 3,000 years of history, could be read. The decoding was successfully verified by the Italian Ippolito Rosellini on the basis of an expedition to Egyptian monuments along the Nile.

Yet this sensational decoding, on which Champollion literally worked himself to death, did not result in a general deciphering method. Every script system appeared to have its own problems. While it is true that philologists could learn from the insights of predecessors (such as Champollion's discovery of the *hybridity of symbols*), the actual decoding itself often depended on fortuitous archaeological finds.

For instance, the unravelling of the oldest script system in the world, the cuneiform script from Mesopotamia (*c.*3400 BCE), was made possible by the discovery of the *Behistun inscriptions* in Persia by Henry Rawlinson in 1835.[85] This inscription consists of identical texts in the then three official languages—Old Persian, Babylonian, and Elamite. After the discovery of these inscriptions it was only a matter of time before cuneiform script was definitively deciphered (1857).

An archaeological find was likewise the breakthrough in the decoding of Linear B (the Mycenaean that was discovered in 1900 near Knossos by Arthur Evans).[86] For many years there had been nothing more than a hypothetical sound pattern for Mycenaean. There was no further progress until 1953, when Michael Ventris found a tablet from Ugarit (Syria) that bore the names of local towns and cities. In view of the fact that the names of towns and cities usually remain unchanged for a

[84] Jean-François Champollion, *Précis du système hiëroglyphique*, Adamant Media Corporation, 2001 (reprint of 1828).

[85] Lesley Adkins, *Empires of the Plain: Henry Rawlinson and the Lost Languages of Babylon*, St Martin's Press, 2003.

[86] John Chadwick, *The Decipherment of Linear B*, Cambridge University Press, 1990.

long time, Ventris went in search of similar tablets in Linear B. By then filling in the assumed sound patterns, Ventris obtained a few words that, while they had hiatuses, could be compared with the names of towns and cities near Knossos. One symbol after another could consequently be decoded, after which it was even possible to unravel the case system of Linear B. Many had assumed that Mycenaean was related to Etruscan, but it proved to be an early form of Greek!

Spectacular deciphering is taking place now too. Currently people are working on decoding the Maya script, and over 90 per cent of Maya texts can now be read with great accuracy, something that has produced recent discoveries in Maya historiography and musicology (see chapter 6).[87] However, there are also languages for which complete decoding is still not in sight, for example Etruscan, Linear A and Rongorongo, which was once spoken on Easter Island. A fortunate archaeological find can still work wonders.

No matter how compelling the history of deciphering may be, it has emerged that there are no general methodical principles for unravelling script systems. However, one can find a few patterns in the history of decoding between 1780 and 2000 that could serve as guidelines. It appears that the *number of symbols in a script system* is a crucial factor. If this number is lower than thirty, the symbols are probably sounds. If there are over thirty, they are syllables, and if there are many thousands, they are concepts. Symbols can be hybrid though, and there are different types of symbols, for example punctuation marks and characters. Proper names (kings, towns, and cities, for instance) are extremely important clues because to a degree they are language-independent. Yet over and above such patterns there are virtually no methodical principles for practical philological deciphering. The only method to which there is occasional reference relates to Etruscan and is known as the 'combinatorial method'.[88] It propagates the use of a combination of all possible knowledge sources—from archaeological–antiquarian, epigraphic, and paleographic to linguistic fonts of information. If we recall the historical principle of all possible sources (3.2, 5.1), we see that there is nothing new under the sun here.

Philology as textual criticism: Karl Lachmann's formalization. Unlike the field of deciphering, it is possible to find general principles for text reconstruction. The role of Karl Lachmann (1793–1851) was crucial. He contributed more than any other to a text reconstruction theory that is currently known as the *stemmatic theory* or *stemmatology*.[89] In this method a family tree (a stemma) of surviving texts is built that can be used to reconstruct the original text. Some elements of the stemmatic theory had already been in use for centuries, such as the concept of an archetype of a

[87] Dennis Tedlock, *2000 Years of Mayan Literature*, University of California Press, 2010. See also Michael Coe, *Breaking the Maya Code*, Thames & Hudson, 1992.

[88] Ambros Josef Pfiffig, *Einführung in die Etruskologie: Probleme, Methoden, Ergebnisse*, Darmstadt Wissenschaftliche Buchgesellschaft, 1972.

[89] For the fundamentals of Lachmann's theory, see Karl Lachmann, *Kleinere Schriften zur deutschen Filologie*, G. Reimer, 1876 (reprinted by Oxford University Press, 2007). See also Winfried Ziegler, *Die 'wahre strenghistorische Kritik': Leben und Werk Carl Lachmanns und sein Beitrag zur neutestamentlichen Wissenschaft*, Kovač, 2000.

text (see 2.3) and the genealogical method (employed by Poliziano as well as in the earlier Arabic transmission theory, the *isnad*, see 4.1 and 3.2). Lachmann put these separate elements into one systematic whole.[90] First of all he divided the philological method into three separate phases:

(1) *Recensio*. In this stage the philologist collects all surviving versions of a text, inventories the differences (*variants*) and determines the genealogical relationship between the surviving texts—a *stemma codicum*, a sort of family tree (see below). This phase is executed as mechanically as possible in order to keep it separate from the interpretation of the text.

(2) *Examinatio*. After the 'primitive' text has been established by the *stemma*, the philologist has to decide whether or not it is authentic.

(3) *Emendatio*. If the primitive text is judged not to be authentic, the philologist has to emend it in order to reconstruct the lost archetype from the oldest surviving accurate version.

Lachmann did not completely formalize any of these phases. The well-informed guess of the philologist remained an inherent part of text reconstruction. Once the family tree of the *stemma* of text variants had been put together, though, Lachmann asserted that a number of very precise rules could be applied to it. The concept of the *stemma* is therefore one of the showpieces of philology. The first published genealogical tree for a classical text is attributed to Carl Zumpt in his edition of Cicero's *Verrine Orations* in 1831, but it was Lachmann who spelled out which rules applied to a stemma and how they could be used in his editions of Lucretius (1850) and the New Testament (1842–1850).

In the following I will explain the three phases of *recensio, examinatio*, and *emendatio* with a concrete example that is taken from Ben Salemans.[91] In the first phase, the *recensio*, a stemma is built up on the basis of the differences between the variants. The underlying principle is as follows: *if an error is created in a version of a text—for example a missing paragraph—it is probable that descendants of that text contain the same common error*. Lachmann assumed that all texts with a particular missing paragraph (or missing word, line, etc.) lead back to the same common ancestor where this error is found for the first time. If there are several common errors, they can be used to draw the genealogical family tree. We can compare deducing such a family tree with establishing a unique hereditary disease or a DNA sequence that children inherit from their parents and then pass on to their own children. The incidence of the disease or DNA sequence can be used to find the genealogical familial relationships. A tree showing familial relationships was first

[90] Sebastiano Timpanaro, *La genesi del metodo del Lachmann*, Le Monnier, 1963, pp. 5–13. For an English translation, see Sebastiano Timpanaro, *The Genesis of Lachmann's Method*, edited and translated by Glenn Most, University of Chicago Press, 2005.

[91] Ben Salemans, *Building Stemma's with the Computer in a Cladistic, Neo-Lachmannian, Way: The Case of Fourteen Text Versions of Lanseloet van Denemerken*, PhD thesis, Radboud University Nijmegen, 2000, pp. 3–4. I am most grateful to Ben Salemans for allowing me to reproduce his example with explanation here.

developed in textual philology, and it was not employed successfully in genetics until many years later.[92]

We will explain the construction of a family tree using the following imaginary example.[93] Assume that we have six versions of a text, which we designate as A, B, C, D, E, and F and which are shown in Figure 12 (the Greek letters after these designations will be explained below). One or more common errors are identified in certain passages in A and B, whereas the other versions—C, D, E, and F—have other words in common but they are *different* from the common errors. We can therefore conclude that A and B have a common ancestor in which the errors occurred for the first time. We will designate this lost common ancestor as 'a'. However, text 'a' cannot have been the ancestor of text C, for instance, because one or more common errors in A and B do not occur in C. Let us also assume that we find unique common errors in C and D (which lead to a common ancestor 'b'), and that other mistakes occur in E and F (leading to 'c'). We then assume for the purposes of our example that C, D, E, and F have common errors (from which one can deduce that they all have one common ancestor, which we will designate 'd'). We moreover assume that these six texts can be traced back to one common original text O. All these common errors in this imaginary example result in the stemma in Figure 12, which is the result of the *recensio*.

A stemma can be considered as a *historical depiction of the relationships between texts*. Using this stemma we can now start the reconstruction of O, the *emendatio*, but not until we have established that none of the surviving texts can be the source text (the *examinatio*), which indeed emerges from the stemma above. We will clarify the text reconstruction once more with an example.[94] Let us assume that the first line of text A begins with α, whereas in the same place text B has a β, C and D have a γ, E begins with δ and F with α. We therefore have four different variants—α occurs in two texts, β in one text, γ in two texts and δ in one text (Figure 12). If we consider variant γ we see that it occurs in texts C and D, and therefore it is likely that the common ancestor of C and D (text 'b') also had variant γ. Let us now consider variant α, which is found in A and F. The first common ancestor of both the texts is O, the lost source text. According to Lachmann's method we can now assume that the source text also had the reading α, precisely because these two corresponding variants have no common ancestor other than the source text. In this way we are able to reconstruct part of the lost original. One can develop similar logical argumentation for every variant, with the stemma as the guiding pattern.

[92] Heather Windram, Prue Shaw, Peter Robinson, and Christopher Howe, 'Dante's *Monarchia* as a test case for the use of phylogenetic methods in stemmatic analysis', *Literary and Linguistic Computing*, 23(4), 2008, pp. 443–63.

[93] Salemans, *Building Stemma's with the Computer in a Cladistic, Neo-Lachmannian, Way*, p. 4. Similar examples can be found in textbooks on stemmatic philology, such as Walter Greg, *The Calculus of Variants: An Essay on Textual Criticism*, Oxford University Press, 1927, or Vinton Dearing, *Principles and Practice of Textual Analysis*, University of California Press, 1974.

[94] Salemans, *Building Stemma's with the Computer in a Cladistic, Neo-Lachmannian, Way*, pp. 4–5.

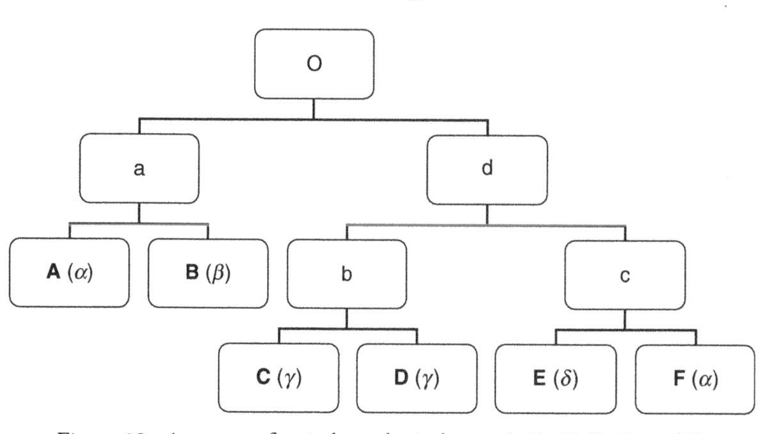

Figure 12. A stemma for six hypothetical texts A, B, C, D, E, and F.

Lachmannian reconstruction thus takes place on the grounds of *logical inference based on the differences between and agreements in the genealogical relationship between texts*. This is why we described Poliziano's genealogical method in the conclusion to chapter 4 as a 'logical formalization', no matter how implicit his method is—it has, rather, the status of a proto-theory. Lachmann's method on the other hand was worked out in sufficient detail in order to go through its life as a 'theory'. However, it turned out to be an enormous task to manually build up a stemma for a substantial text in which all differences and agreements in all versions have to be compared, let alone going on to deduce emendations. It is moreover possible that very little can be emended, and it can even be the case that no genealogical tree can be developed. Usually, though, if there are several versions of a text, they can be organized in a genealogical relationship using Lachmann's method. The stemmatic approach was therefore a giant step forward compared with earlier philological techniques.

Lachmann's impact. Lachmann's philology came as a bombshell. It resulted in his reconstruction of Lucretius, which remains unequalled to this day, and also to a revised version of the New Testament that represented a rejection of Erasmus's *textus receptus*, which had served as the standard for centuries (see 4.1). Lachmann's greatest influence, however, was exerted on the reconstruction of medieval literature, including the poems of Walter von der Vogelweide, the *Hildebrandslied* and the *Nibelungenlied*.[95] Humanistic scholars had ignored the medieval lyric and epic and did not discover or rediscover them until some time in the eighteenth century. For example, the *Nibelungenlied* ('The Song of the Nibelungs') was lost at the end of the sixteenth century but was unearthed again in 1755.[96] Before long there were no fewer than thirty-four manuscripts in circulation, none of which agreed with the others (and which often consisted of fragments). The versions could be put into a

[95] Martin Hertz, *Karl Lachmann: eine Biografie*, Verlag von Wilhelm Hertz, 1851, pp. 100–19.

[96] *Das Nibelungenlied: Song of the Nibelungs*, translated by Burton Raffel, foreword by Michael Dirda, introduction by Edward Haymes, Yale University Press, 2006.

stemma and reconstructed thanks to Lachmann's method.[97] The scope of this discovery and reconstruction is virtually impossible to overestimate.[98] The *Nibelungenlied* was declared to be *the* national German epic and (despite criticism) elevated to the same level as Homer's *Iliad* and *Odyssey*. Passages from the *Nibelungenlied* constantly appeared on posters and during speeches. *Nibelungentreue* ('Nibelung Loyalty'), in which mutual fidelity between vassals was on a higher plane than family loyalty or one's own life, became the cornerstone of German wartime propaganda, with the later National Socialism as the nadir.

Yet we cannot blame Lachmann for this nationalist exploitation of philology. He himself was a largely independent philologist. This emerged all the more when he applied his method, which was considered to be of use primarily for old literature, to contemporary authors too. For a long time it had been assumed that text reconstruction was unnecessary for works that the author himself had had printed. However, after the death of an author a text could soon degenerate if it were reprinted a number of times. New misprints appeared with every edition.[99] Lachmann showed how stemmatic philology could be useful for texts from the recent past. For example, he was responsible for a painstaking edition of the work of Gotthold Ephraim Lessing (1729–1781).

The neo-Lachmannian school: towards complete formalization. Under Lachmann, philology was applied to all periods and his method represented the standard for text reconstruction in Europe and beyond. Lachmann's method also had a following outside philology. It became the cornerstone of Rankean history, which utilized philological source criticism for 'objective' historiography (see 5.1). Historians like Georg Waitz (1813–1886) were pupils of both Ranke and Lachmann and continued to develop the philologization of historiography.[100]

All this success meant that the shortcomings of Lachmann's stemmatology might almost be overlooked. His method was based on a number of assumptions that were not always valid, such as the supposition that every version is derived from exactly one direct ancestor and that a copyist only made new mistakes without correcting the errors of predecessors. Lachmann's theory proved to be flexible enough, though, to be corrected in regard to these assumptions. A more serious problem was that the fundamental concept of a 'common error' was not defined with precision. For instance, are differences in word order errors or not? It was not until the twentieth century that a start was made on formalizing Lachmann's method down to the smallest detail. An important step was taken by Walter

[97] Karl Lachmann, *Der Nibelunge Noth und die Klage nach der ältesten Überlieferung mit Bezeichnung des Unechten und mit den Abweichungen der gemeinen Lesart*, Reimer, 1826 (reprinted by de Gruyter, 1960).

[98] John Evert Härd, *Das Nibelungenepos: Wertung und Wirkung von der Romantik bis zur Gegenwart*, Francke, 1996.

[99] On misprints, see Marita Mathijsen, *Naar de letter: handboek editiewetenschap*, Van Gorcum, 1995, p. 22.

[100] Guido Wölky, *Roscher, Waitz, Bluntschli und Treitschke als Politikwissenschaftler: Spätblüte und Untergang eines klassischen Universitätsfaches in der zweiten Hälfte des 19. Jahrhunderts*, PhD thesis, University of Bochum, 2006, pp. 151–63.

Greg, who gave an unambiguous method in *The Calculus of Variants: An Essay on Textual Criticism* (1927) for constructing a stemma on the basis of variants—although the definitive explanation of Lachmann's method is usually attributed to Paul Maas's *Textkritik* (1960). More recent versions of neo-Lachmannian philology have tried to completely 'mechanize' the assembly of a stemma. This involves building a stemma in two steps rather than one. First of all, a type of underlying structure is set up in the form of a chain, after which the final stemma is deduced in a second stage.[101] This process has since been defined so precisely that it is both reproducible and implementable by using a computer program that automatically works out a stemma from a number of entered variants.[102] Consequently, stemmatic philology appears to be the only humanities discipline to have become a 'normal science'. While the job of stemmatic philology has not yet been finished, the contours have been so clearly defined that the main activity in the field is problem solving. As we saw with Panini's grammar (2.1) and the formalization of organum compositions (3.4), contemporary stemmatic philology embraces the procedural system of rules principle—an unequivocal result is obtained on the basis of a finite number of steps. There is less and less fun to be had by the philologist in manually building up a stemma.

Nevertheless, only a small fraction of textual criticism can be reduced to stemmatology. Stemmatic philology is of little help if one is trying to reconstruct a Chinese poem by combining written bamboo strips or if one is attempting to retrieve a Latin text from carbonized scrolls. The stemmatic method can only be employed when several versions of a source text have been recovered.

A broadening of philology. For centuries the view of philology as textual criticism was dominant (see 2.3 and further). Yet from the early modern era onwards, philology developed branches into literary history, numismatics, epigraphy, palaeography, and other domains. During the course of the eighteenth century these branches could count on burgeoning interest, for example from Gesner (see 4.1) and Vico (see 4.2), and in the nineteenth century from Lachmann's contemporary August Böckh. 'Philology' was no longer taken to mean only textual criticism but the complete study of language and literature in their historical context. We will not discuss this notion of philology in the current section, but we will do so in the appropriate sections on linguistics (5.3) and literary studies (5.6).

Philology: a victim of its own success? Stemmatic philology is possibly the most successful humanistic discipline. It provides a precise method with which texts from *all* periods and regions can be reconstructed. Stemmatic philology is to the humanities what classical mechanics is to natural science. While it is an approximation of reality, it is such a splendid approximation that it is a showpiece for the field as a whole. Yet whereas every natural scientist is instructed in classical

[101] For some recent developments in stemmatic philology, see Pieter van Reenen and Margot van Mulken (eds), *Studies in Stemmatology*, John Benjamins, 1996. See also Pieter van Reenen, August den Hollander, and Margot van Mulken (eds), *Studies in Stemmatology II*, John Benjamins, 2004.

[102] For example, Salemans, *Building Stemma's with the Computer in a Cladistic, Neo-Lachmannian, Way*.

mechanics, stemmatic philology is languishing. It is no longer taught to students of linguistics or literary studies. Stemmatology is still on the curriculum, but only in more specialized sub-disciplines such as paleography or book history. In this regard greater historical awareness can be ascribed to natural science than to the humanities. Text reconstruction philology has been downgraded from the 'queen of learning' (see 4.1) to an auxiliary discipline with a marginal status. Yet because of its resounding success it deserves a place of honour in the canon of the humanities.

5.3 LINGUISTICS AND LOGIC: THE LAWS OF LANGUAGE AND MEANING

The study of language also became historical: people searched for *diachronic* patterns in sound changes between languages that could contribute to finding the Indo-European 'protolanguage'. In the twentieth century, however, the quest for *synchronic* laws gained the upper hand, and the concept of Universal Grammar was also revived. From the nineteenth century onwards, logic developed into a form of meta-mathematics, which comes outside the scope of our historical overview. Like the study of meaning, though, logic remained part of the humanities and formed a duality with linguistics.

The development of comparative linguistics. In Europe it was assumed until far into the eighteenth century that Biblical Hebrew was the ultimate ancestor of all languages, even though this had been convincingly negated by De Laet in the seventeenth century on the basis of American Indian languages (see 4.3).[103] Previously, the Florentine merchant Sassetti had spotted a few striking similarities between Sanskrit and Italian, which were extended by linguists such as Beauzée to other languages. However, these discoveries were not followed up on. This changed in 1784 when William Jones gave his famous lecture *On the Hindus* (see 4.3).[104] He argued on the basis of comparisons between verb stems that European, Indian, and Persian languages must have had a common source language. Grammars of Sanskrit were compiled in a short space of time and Henry Thomas Colebrooke published an English translation of Panini's grammar in 1809.[105] Initially it was primarily British scholars who immersed themselves in oriental languages, but the study of Sanskrit came to Germany via the brothers August and Friedrich von Schlegel.[106] A true quest for the *protolanguage* developed, thanks also to the Romantic goal of the 'original' and the penchant for the 'exotic'. However, the emergence of *comparative linguistics* had more modest beginnings, with a hunt for connections between European languages.

Grimm's Sound Shift Law. Wilhelm and Jacob Grimm are well known as indefatigable collectors of folk tales, but their greatest achievement may possibly have been in the field of comparative linguistics.[107] Of the two brothers, Jacob was the more active in linguistics, and it is his name that is linked to the sound shift law. However, the basic principle had already been postulated by the Dane Rasmus Rask in 1818 when he assumed that if there were agreements between the forms of

[103] On the early modern roots of comparative linguistics, see Toon Van Hal, 'Linguistics "ante litteram": compiling and transmitting views on the diversity and kinship of languages before the nineteenth century', in Bod, Maat, and Weststeijn, *The Making of the Humanities, Volume II: From Early Modern to Modern Disciplines*, 2012, pp. 37–53.

[104] Jones, *Discourses delivered before the Asiatic Society: and miscellaneous papers, on the religion, poetry, literature, etc., of the nations of India.*

[105] Henry Thomas Colebrooke, *Panini*, Calcutta, 1809. See also Frits Staal (ed.), *A Reader on the Sanskrit Grammarians*, The MIT Press, 1972, p. 34.

[106] Friedrich von Schlegel, *Über die Sprache und Weisheit der Indier: ein Beitrag zur Begründung der Althertumskunde*, Mohr & Zimmer, 1808, translated into English by Ellen Millington, Bohn, 1849.

[107] Lothar Bluhm, *Die Brüder Grimm und der Beginn der deutschen Philologie*, Weidmann, 1997.

Greek:	$p\,b\,f$	$t\,d\,\theta$	$k\,g\,x$
	↓	↓	↓
Gothic:	$f\,b\,p$	$\theta\,t\,d$	$x\,k\,g$
	↓	↓	↓
German:	$b\,f\,p$	$d\,z\,t$	$g\,x\,k$

Figure 13. Grimm's Sound Shift Law.

words between two languages such that rules for letter changes could be discovered, there had to be an underlying relationship between these languages.[108] The second edition of Grimm's *Deutsche Grammatik* (1822) contained the well-known law, after Jacob Grimm had read Rask's work.[109]

In its original form Grimm's Law gave the sound changes for only a few European languages, but later it was extended by others to Indian and Persian languages. Grimm found that Greek *voiceless* sounds like *p*, *t*, and *k* had shifted to the Gothic *voiceless* sounds *f*, θ, and *x*, and then to the High German *voiced* sounds *b*, *d*, and *g*. The θ represented the aspirated *t*, which is also written as *th*, and the *x* sounds like the last consonant in *Bach*. There was therefore a visible linguistic historical pattern of voiceless to voiced consonants. The reverse happened with the Greek voiced consonants, which changed to German voiceless consonants (see Figure 13).

Grimm believed that he could derive from these sound changes a cyclical pattern *from voiceless to voiced and back again*. Historical cycles cropped up everywhere in the nineteenth century (see 5.1) and that made the phenomenon of sound shift all the more plausible. At the same time Grimm was aware of the exceptions to the regularity he had found. He contended that his 'law' was only a tendency where the changes occur in many but not all cases. For example the Greek word *treis* (three) complies with Grimm's Law when it shifts via the Gothic *θreis* to the German *drei* (where the shift takes place in accordance with the pattern above from *t* via θ to *d*). The first letter of the Greek *pous* (foot) also complies with Grimm's Law initially when it changes to the Gothic *fotus*. But the last step predicted by the sound change law (from *f* to *b*) did not occur: the German word is *fuss* rather than *buss*.

While Grimm was on the trail of a fascinating trend, it was not an absolute law without exceptions. Nor does he appear to have gone in search of a possibly deeper pattern that was without exceptions (which was undertaken later on by the Neogrammarians—see below). Nevertheless, despite its imperfections, Grimm's sound shift is a pattern that appeals to the imagination. Who can fail to be struck by the many regular historical changes from Greek to German (and other Germanic

[108] Robins, *A Short History of Linguistics*, pp. 188–9. It has been observed that as early as 1710 Lambert ten Kate also investigated and found systematic sound changes between languages—see Dan Brink, 'Lambert ten Kate as Indo-Europeanist', in Jeanne van Oosten and Johan Snapper (eds), *Dutch Linguistics at Berkeley*, The Dutch Studies Program, UC Berkeley, 1986, pp. 125–36.

[109] Jacob Grimm, *Deutsche Grammatik*, 2nd edition, volume 1, Dieterichsche Buchhandlung, 1822, reprinted by Wilhelm Scherer in 1870, p. 503.

languages), such as from *phrater* to *bruder* ('brother') and from *chortos* to *garten* ('garden'). The study of the pattern of sound changes literally offered one a chance to dive into the past of languages, and it was created at a time when millions of Germans (and Britons) were starting to consider themselves to be heirs of ancient Greece.[110] Grimm's discovery gave legitimacy to this Philhellenic world view that had Northern Europe in its grip.

Comparative linguistics takes root: Humboldt's role. Grimm's hypothesis that *languages undergo sound changes in a historically regular way* was widely embraced. It spurred generations of linguists on to an even more systematic study of similarities between languages, and comparative linguistics was to develop into an independent discipline with its own methodology.[111] This result would not have been achieved, though, without the ambition of Wilhelm von Humboldt (1767–1835) to design a new academic structure. Humboldt came from a fabulously rich family and as a high Prussian official he was the driving force behind the academic system in which professors taught but also had the freedom to do research.[112] Humboldt's new university structure gave stability and continuity to scholarship and resulted in a long series of chairs in comparative linguistics.

Humboldt also had innovative views about language.[113] He was a philosopher and linguist and he wrote a three-volume work about the Kawi language on Java (published posthumously in 1836). However, Humboldt became well known primarily because of his speculative ideas about language. Building on the work of Vico and Herder, he argued that a language is not a static product but a dynamic phenomenon or, in Humboldt's Greek terminology, not *ergon* but *energeia*. Moreover, Humboldt distinguished between *interior* and *exterior* form. The exterior form of language consisted of the raw material (the sounds) from which different languages were built up. The interior form was the structure of the grammar that could construct meaning from these sounds. This distinction had a major impact on the later structuralist and generative linguistics (see below). The insight that language users can produce an unlimited number of linguistic utterances using a finite number of sounds has also been attributed to Humboldt, but we already found this in all its glory in the work of Panini around 500 BCE (see 2.1). The notion of a Universal Grammar has also been regularly ascribed to Humboldt, but we came across this centuries before among the Modists (see 3.1).[114] Another of Humboldt's influential theses was the idea that language determines the realm of

[110] Suzanne Marchand, *Down from Olympus: Archaeology and Philhellenism in Germany, 1750–1970*, Princeton University Press, 2003.

[111] Anna Morpurgo-Davies, *History of Linguistics, Volume IV: Nineteenth-Century Linguistics*, Longman, 1998. See also Els Elffers, 'The rise of general linguistics as an academic discipline: Georg von der Gabelentz as a co-founder', in Bod, Maat, and Weststeijn, *The Making of the Humanities, Volume II: From Early Modern to Modern Disciplines*, 2012, pp. 55–70.

[112] Clemens Menze, *Die Bildungsreform Wilhelm von Humboldts*, Schroedel, 1975.

[113] Wilhelm von Humboldt, *Werke in fünf Bänden*, Wissenschaftliche Buchgesellschaft, 2002. See also Elsina Stubb, *Wilhelm Von Humboldt's Philosophy of Language, Its Sources and Influence*, Edwin Mellen Press, 2002.

[114] Pieter Seuren, *Western Linguistics: An Historical Introduction*, Blackwell Publishers, 1998, pp. 109ff.

thought, with which he laid the foundations for the later Sapir-Whorf hypothesis. Humboldt was a great inspirer, and his ideas on language transcend comparative linguistics by far.

Bopp's conjugation system. Franz Bopp (1791–1867) was one of the first to extend Grimm's language studies to Indian and Persian languages. Bopp went further than others—he studied the mutual language relationships not so much on the basis of sound shifts but by examining word structures, in particular those of verb conjugations. In his *Conjugationssystem* (1816) Bopp stated that his main goal was to reconstruct the grammar of a language that had disintegrated step by step into the Indo-European languages. Bopp considered Sanskrit to be the closest to this original language in terms of morphology. Only a comparative quest for the principles of many grammars could produce an understanding of the origin of conjugation. The idea that absolute laws underlie the oldest word forms had been investigated previously by Hemsterhuis for Greek, but without much success (see 4.3). It is possible that Bopp was influenced by Hemsterhuis's approach via Grimm and then applied this to Sanskrit.[115] Bopp did not focus just on concrete language comparisons. In 1833 in his *Vergleichende Grammatik* he also tried to formulate a general 'conjugation law'. This 'law', inspired by Newton's Law of Gravity, contended that a type of attractive force (*Gewichtsmechanismus*) existed between the *vowel of a verb stem and the ending* of that verb. The heavier vowels exerted a greater force on the ending, as a result of which those endings become shorter. On the other hand light vowels permit longer endings to be created.[116]

Although Bopp's conjugation system has not withstood the ravages of time—it emerged that it is stress that is decisive—his method illustrated the formal frame of mind within which nineteenth-century humanities existed.[117] Even during High Romanticism, the formal approach remained unprecedentedly strong. Humanists like Ranke, Lachmann, and Bopp were exponents of German Romanticism, but they employed precise, almost mechanistic principles for source-critical historiography (5.1), stemmatic philology (5.2), and comparative linguistics respectively. There was much more continuity with the *early* modern age than has often been assumed.

One would have expected that Bopp's exploration of Sanskrit would also have resulted in a reassessment of Panini's grammatical formalism with its recursive, algorithmic system of rules (see 2.1). Although Bopp studied Panini's morphological rules closely, he did not understand the latter's formalism, consisting of *context-sensitive* rules. However, like his distant predecessor, Bopp searched for the

[115] Jan Noordegraaf, 'The "Schola Hemsterhusiana" revisited', in Jan Noordegraaf, *The Dutch Pendulum: Linguistics in the Netherlands 1740–1900*, Nodus Publikationen, 1996, pp. 23–55.

[116] Franz Bopp, *Vergleichende Grammatik des Sanskrit, Zend, Griechischen, Lateinischen, Litthauischen, Gothischen und Deutschen*, Part II, 1833–52. Translated by Lieutenant Eastwick, *A Comparative Grammar of the Sanscrit, Zend, Greek, Latin, Lithuanian, Gothic, German, and Slavonic Languages*, 3 volumes, Madden and Malcolm, 1845–50.

[117] Bart Karstens, *'Die Boppsche Wissenschaft'?*, MA thesis, Utrecht University, 2009, p. 37. See also Bart Karstens, 'Bopp the builder: discipline formation as hybridization: the case of comparative linguistics', in Bod, Maat, and Weststeijn, *The Making of the Humanities, Volume II: From Early Modern to Modern Disciplines*, 2012, pp. 104–27.

smallest meaning carriers of Sanskrit's word structure.[118] But Bopp disagreed with the 'Indian grammarians', probably because his goal—comparative grammar—was different.

Schleicher's family tree theory and Proto-Indo-European. Neither Bopp's conjugation system nor Grimm's Sound Shift Law was without exceptions. A generation later, August Schleicher (1821–1868) therefore proposed that language should be seen as a branch of biology, where no absolute laws applied either. Schleicher argued that languages could be depicted in a genealogical family tree, in the same way that texts could be placed in a philological stemma (see 5.2). Such a family tree expresses relationships between languages without professing absolute sound change or conjugation laws. Schleicher first published his graphical family tree in 1853 and caused a sensation in the academic world.[119] His *Stammbaumtheorie* was also criticized, though, primarily because the theory assumed a strict separation between languages, whereas in practice they overlap. Even so, Schleicher's genealogical tree still expresses in broad-brush terms the generally accepted evolution of Proto-Indo-European into approximately two hundred mutually related but very divergent languages, from Persian, Bengali, and Urdu to Croatian, Romanian, and English. With his evolutionary family tree Schleicher was moreover several years ahead of Charles Darwin's *On the Origin of Species by Means of Natural Selection* (1859).[120] Schleicher's family tree received support from a surprising quarter when a similar genetic family tree could be deduced from twentieth-century comparative DNA research.

Schleicher was also the author of a fable—*Das Schaf und die Pferde* ('The Sheep and the Horse', 1868)—which he invented and which he believed he could translate from German into Proto-Indo-European. Although his reconstruction was based more on fiction than fact, it set the tone for a long series of new reconstruction attempts. 'Improved' versions were proposed by linguists until quite recently, by e.g. Frederik Kortlandt in 2007 and Rosemarie Lühr in 2008. While the ultimate goal of historical linguistics remains the reconstruction of proto-languages, such as Proto-Indo-European,[121] it is perhaps more readily achievable to translate into Proto-Indo-European a text of which the lexical proto-forms have been traced with some likelihood. Such a conversion can furthermore be refuted and improved if new historical material surfaces.

[118] Johannes Bronkhorst, 'Panini's view of meaning and its Western counterpart', in Maxim Stamenov (ed.), *Current Advances in Semantic Theory*, John Benjamins, 1992, pp. 455–64.

[119] August Schleicher, 'Die ersten Spaltungen des indogermanischen Urvolkes', *Allgemeine Zeitung für Wissenschaft und Literatur*, August 1853. See also August Schleicher, *Compendium der vergleichenden Grammatik der indogermanischen Sprachen*, Böhlau, 1861–62.

[120] Robert O'Hara, 'Trees of history in systematics and philology', *Memorie della Società Italiana di Scienze Naturali e del Museo Civico di Storia Naturale di Milano*, 27(1), 1996, pp. 81–8. On the influence of comparative linguistics on biology, see Benoît Dayrat, 'The roots of phylogeny: How did Haeckel build his tree?', *Systematic Biology* 52(4), 2003, pp. 515–27. For Schleicher's view, see August Schleicher, *Die Darwinsche Theorie und die Sprachwissenschaft. Offenes Sendschreiben an Herrn Dr Ernst Häckel, Professor der Zoologie und Director des zoologischen Museums an der Universität Jena*, Böhlau, 1863.

[121] Robert Beekes, *Comparative Indo-European Linguistics*, John Benjamins, 1995.

Linguistic laws without exceptions: Verner and the *Junggrammatiker*. The numerous exceptions to the laws of Grimm and Bopp were not accepted by a new generation of linguists who would become known as the *Junggrammatiker* or *Neogrammarians*.[122] They suggested that sound laws that had no exceptions whatsoever had to exist and that it was the task of linguists to track down these deeper and more complex laws. Linguists who were satisfied with 'tendencies' were labelled as unscientific. According to the Neogrammarians, Schleicher's family tree theory was a sort of scientific defeatism because he did not search for absolute laws. The *Prinzipien der Sprachgeschichte* (1880) by Hermann Paul (1846–1921) is usually considered to be the neogrammatical manifesto,[123] in which this new generation distanced itself from their teachers—which was unheard of in those days. In the opinion of the Neogrammarians, as many linguistic elements as possible should be considered in the process of finding a fundamental exception-free law. This was because, in their view, a sound change was not determined just by the sound or the phoneme itself, as in Grimm's 'law', but also depended on other factors in the word, for instance the surrounding phonemes and the location of the stress. These environment-dependent factors led to a new law that became known as *Verner's Law*, named after the Danish linguist Karl Adolf Verner, who discovered it in 1875.[124]

We can illustrate this law using the Indo-European archetypes for *father* and *brother*, namely *pəter* and *bhrater*. If Grimm's Law were to be applied to these words, in their further development these forms would have had to evolve in Old English into *fathar* and *brothor*. However, this was true only for the form of *brother* and not for *father* (which is *fæder* in Old English). Verner succeeded in bringing this 'exception' in the sound change within the scope of a law by including the original stress system of Indo-European and Proto-Germanic. He showed that the sound shift described by Grimm had indeed taken place, but the *fricatives* that had been created were voiced because the stress in the word was on the immediately following syllable. This was the case for *pəter*, where the stress was on the second syllable. In this way it was possible to explain why one word developed differently from another word, depending on the stress and the surrounding sounds (phonemes). Grimm's Law was thus turned from an inexact tendency into an exact law. Putting it another way, the exceptions to Grimm's Law were only ostensible and could be accounted for by the systematic way in which these exceptions came about. Without being aware of it, Verner was using the concept of the *context-sensitive* rule that had been introduced more than two thousand years earlier by Panini (see 2.1). By using this explanation, Verner was, of course, walking on thin ice. A law can always be rescued by using so-called provisos (see also

[122] Eveline Einhauser, *Die Junggrammatiker: ein Problem für die Sprachwissenschafts-geschichtsschreibung*, Trier Wissenschaftlicher Verlag, 1989.

[123] For an English translation, see Hermann Paul, *Principles of the History of Language*, translated by H. A. Strong, McGroth Publishing Company, 1970.

[124] Karl Verner, 'Eine Ausnahme der ersten Lautverschiebung', *Zeitschrift für vergleichende Sprachforschung auf dem Gebiete der Indogermanischen Sprachen*, 23(2), 1877, pp. 97–130.

chapter 6). However, since it was clear that the stress and surrounding phonemes were determining for the pronunciation of a word, Verner's Law was not considered to be just an ad hoc solution, but a deeper linguistic generalization.

To the Neogrammarians, Verner's Law was a model of correct linguistics. Not everyone was convinced, however. In 1885 Hugo Schuchardt pointed out that Verner's Law only applied in a highly idealized context in which a language was seen as a homogeneous system.[125] Such systems might have existed in the form of a corpus of surviving texts in Old English or Proto-German, but according to Schuchardt no single living language found itself in such a situation. However much Schuchardt was right in this regard, Verner's Law was unequalled within the bounds of the assumed idealization.

Linguistics frees itself of its historical yoke: Saussure and structuralism. While the Neogrammarians claimed that they had distanced themselves from the previous historical–comparative linguistics, they actually represent its de facto apex. The beginning of the twentieth century saw the creation of a new school that was to become known as *structuralism* and became associated with the name of Ferdinand de Saussure (1857–1913).[126] Saussure started his career as a comparative linguist with a doctoral thesis on the vowel system of Indo-European languages, but he soon came up with a new approach. It was no longer based on a historical or diachronic method but on a *synchronic* analysis. Saussure's works were published posthumously by his students as *Cours de linguistique générale* (1916),[127] as a result of which it is hard to determine what Saussure himself argued and what was added by his students.

Saussure rejected the notion of absolute laws in language. He maintained that the Neogrammarian quest for exceptionless regularities gave no insights into what a language was for a speaker. What primarily concerned Saussure—and the *structuralists* after him—was the intrinsic structure of language. Forms and meanings were not to be identified with anything external to language.[128] If 'regularities' existed in language, then they only existed by reference to internal, structural differences in the language itself. These structural contrasts define a language system as a whole. For example, if one wants to describe the form of the English word *bat*, then it does not make sense to do this in terms of the sequence of sounds /bat/. Instead, the relation of /bat/ with constrasting words in English should be investigated, such as /kat/, /fat/ and /hat/. One of the goals of linguistics is to search for the set of phonological contrasts of each language. It turns out that there is a no fixed set of phonemes across all languages.[129] Similarly, the meaning of *bat* cannot be identified with a certain species because different languages use different words for biological species.

[125] Hugo Schuchardt, 'Über die Lautgesetze: gegen die Junggrammatiker', in Leo Spitzer (ed.), *Hugo-Schuchardt-Brevier, ein Vademekum der allgemeinen Sprachwissenschaft*, Halle, 1922.

[126] John Joseph, *Saussure*, Oxford University Press, 2012.

[127] For an English translation, see e.g. Ferdinand de Saussure, *Course in General Linguistics*, edited by Charles Bally and Albert Sechehaye, translated by Roy Harris, Open Court, 1983.

[128] Roy Harris, 'Modern linguistics: 1800 to the present day', in Keith Brown (ed.), *Concise Encyclopedia of Philosophy of Language and Linguistics*, Elsevier, 2010, p. 474.

[129] Harris, 'Modern linguistics: 1800 to the present day'.

Saussure further conceptualized his view by distinguishing between *langue* and *parole*. *Langue* was the underlying system (or the structure) of the language's signs, whereas *parole* consisted of the concrete linguistic utterances of individual speakers. According to Saussure linguists should take the *langue* as a research subject and not usage, which was subject to change. If usage (*parole*) was studied, it should only be in relation to *langue*, the token system that consisted of signs that were symbolic and arbitrary. For example, the words *three* and *thirty* are arbitrary signs of the numerical concept *thirty-three*, but the combination of these signs is subject to a precise language system in English (just as *thirty-four, thirty-five*, etc.).

According to Saussure the language itself, or *langage*, consisted of the *langue/parole* relationship. Linguistic communication is nothing more than a message that is coded by the sender in terms of the *parole*, and decoded by a receiver using a general structure, the *langue*. Here Saussure's vision is strikingly similar to the linguistic communication theory of the Indian grammarian Bhartrhari in the sixth or seventh century, in which there was also a message, a conceptualization, and decoding (see 3.1). According to Saussure, linguistics had to be seen as a branch of the general study of signs: *semiology*.

Saussure's linguistics thus differed fundamentally from linguistics before him in that he was not searching for laws but for *relations of differences* between linguistic signs of a specific language. According to Saussure there were only two sorts of relations, which were either *syntagmatic* or *paradigmatic*. A syntagmatic relation means that the meaning of a linguistic token, a word for instance, is determined by its relation to other tokens in a sentence. In a paradigmatic relation on the other hand, the meaning of a word is considered in relation to the words in the same word class. Whereas syntagmatic relations are looked on by some as a horizontal relation (of words in a sentence or phonemes in a word), paradigmatic relations are considered by some to be vertical relations (of words in a word class or phonemes in a phoneme class).

Saussure outlined a research programme that distanced itself entirely from nineteenth-century linguistics.[130] Language was seen as an autonomous system and no longer as a historical object. It is true that we find some of Saussure's ideas in the work of Humboldt, such as the notion of *interior* and *exterior* form (see above), but Humboldt left them largely unelaborated. Although Saussure never used the term structuralism, this became the label attached to his conception of linguistics. Structuralism was eagerly picked up in other disciplines as well, from anthropology to literary studies (see 5.6).[131]

The dominance of structuralist linguistics: from Jakobson to Bloomfield. Saussure's structuralist programme was to hold linguistics in its grip for forty years, from the Prague school with linguists like Roman Jakobson (1896–1982) and Nikolai Trubetzkoy (1890–1938) to the American school with Leonard Bloomfield (1887–1949) as the most important exponent. The Prague school's

[130] Carol Sanders (ed.), *The Cambridge Companion to Saussure*, Cambridge University Press, 2004.

[131] Jean Piaget, *Le structuralisme*, PUF, 1968. For an English edition, see Jean Piaget, *Structuralism*, Harper & Row, 1971.

showpiece was phonology.[132] Rather than simply making a list of a language's sounds, this school investigated in a Saussurean way how these sounds related to one another. They established that the collection of sounds in a language could be analysed in terms of contrasts, or 'distinctive features'. In English the sounds /p/ and /b/ are different phonemes because there are cases (minimal pairs) where the difference between these two sounds represents the distinctive meaning between separate words, for example *pat* and *bat.* The study of such distinctive features in the phoneme systems resulted in new explanations, including, for example, for the fact that speakers of Japanese have problems in distinguishing between the sounds /r/ and /l/ in English. The reason is that in Japanese these sounds have no distinctive features associated with them. The /r/ and the /l/ are interchangeable in Japanese and refer to the same phoneme, like the /v/ and the /b/ in Spanish for instance. This analysis is currently standard practice in linguistics, but in the 1920s the unravelling of phoneme systems offered a wealth of new insights. Phonology became the example of the success of structuralism.[133]

Nevertheless, the analysis of sounds on the basis of distinctive features and minimal pairs was less innovative than the Prague structuralists believed. A grammar of Old Icelandic compiled eight centuries earlier was also based on the technique of minimal pairs (see 3.1).[134] This twelfth-century grammar bore the telling title First Grammatical Treatise. Neither Saussure nor the Prague school knew the treatise. Shortly after the twelfth century, prosperous Iceland entered a period of economic misfortune, after which natural disasters and climate changes did the rest. The country became isolated from Europe, the work remained unpublished until 1818, and for a long time afterwards it was known only in Scandinavia. In the nineteenth century it was solely via Panini that continental Europe came into contact with a comprehensive analysis of phonology, but without the technique of minimal pairs. Knowledge of the First Grammatical Treatise would certainly have been of help to European linguistics, but as it turned out this wheel was reinvented eight centuries later.

In the 1930s a new structuralist movement developed in the United States, which became known as *distributionalism*, developed under the leadership of Leonard Bloomfield.[135] His objective was to make linguistics completely scientific, and it is thanks to him that linguistics was spread far and wide among American universities. Bloomfield became associated with *behaviourism*, which came from the

[132] Geoffrey Sampson, *Schools of Linguistics*, Stanford University Press, 1980, pp. 103ff. See also Josef Vachek, Josef Dubsky, and Libuse Duskova, *Dictionary of the Prague School of Linguistics*, John Benjamins, 2003.

[133] John Joseph, Nigel Love, and Talbot Taylor, *Landmarks in Linguistic Thought, Volume II: The Western Tradition in the Twentieth Century*, Routledge, 2001, pp. 17ff.

[134] Einar Haugen (ed.), *First Grammatical Treatise: The Earliest Germanic Phonology*, 2nd edition, Prentice Hall Press, 1972.

[135] Leonard Bloomfield, *Language*, University of Chicago Press, 1933, reprinted in 1984. See also John Joseph, *From Whitney to Chomsky: Essays in the History of American Linguistics*, John Benjamins, 2002, p. 139.

social sciences and was popular in his day.[136] It was based on the belief that only observable human behaviour should be studied, without speculating about human cognition. To Bloomfield this meant that only sentences that actually occurred were open to linguistic analysis. Theoretically devised sentences were not acceptable for scientific research. According to Bloomfield's distributional method, every sentence was split up into different components, each with its own syntactic function, which were in turn broken up into the parts from which they were composed. This was continued until the 'ultimate constituent' was reached, usually a *morpheme*, the smallest unit of a word with meaning, for example the stem of a verb or an ending. A morpheme consists of phonemes, but these are atomic and are not themselves constituents. The morphemes are distributed across the different constituents and the distributionalist method makes it possible to ascertain the spread of a particular element in a language.

Bloomfield's attempt to expose the underlying distributional system of linguistic elements was in line with the Saussurean paradigm of *langue*. In his adherence to behaviourism, though, he limited himself to observed usage, without making any pronouncements about sentences that were *not* observed. Bloomfieldian linguistics is consequently analytical but not predictive. It can only dissect observed sentences but it cannot define the class of all *possible* sentences. There was thus a crucial difference here between Bloomfield and Panini, whom he greatly admired,[137] and whose finite system of rules could cover an infinite number of linguistic utterances.

The rise of generative linguistics: Harris and Chomsky. The structuralist method was successful in regard to phonology and morphology, in which a finite number of units exist, but it proved to be inadequate for syntax, in which the number of sentences is unlimited. According to Noam Chomsky (born 1928) a different approach was needed to account for the richness of word order in language. In *Syntactic Structures* (1957) he asserted that a system of rules was needed that can correctly generate *all* grammatical sentences in a language rather than a structuralist system of 'relations of differences'. However, these rules are not 'laws' like those of the Neogrammarians; rather, they make up a grammar in the tradition of Panini (see 2.1). Chomsky's approach was therefore not behaviourist but, as he himself said, rationalist.[138] He concentrated not on concrete usage but rather on the knowledge of a language user that can produce an unlimited number of sentences. In line with the approach of Saussure, Chomsky made a distinction to this end between *performance* and *competence*.[139] Performance is usage with all its imperfections (slips of the tongue, hesitations, memory limitations), whereas competence includes the knowledge of the idealized language user in the form of an underlying grammar. The goal of linguistics was to account for competence phenomena and not performance phenomena.

[136] B. F. Skinner, *The Behavior of Organisms*, Appleton-Century-Crofts, 1938. See also John Mills, *Control: A History of Behavioral Psychology*, New York University Press, 2000.

[137] Bloomfield, *Language*, p. 11.

[138] For Chomsky's attack on behaviourism, see Noam Chomsky, 'A review of B. F. Skinner's verbal behavior', *Language*, 35(1), 1959, pp. 26–58.

[139] Noam Chomsky, *Aspects of the Theory of Syntax*, The MIT Press, 1965, p. 3.

S → NP VP
NP → Det N
VP → VP PP
VP → V NP
PP → P NP

Det → the, a
N → man, woman, garden, house, child
V → kissed, saw, was
P → in, behind

Figure 14. A context-free grammar for a small part of English.

Chomsky elaborated the insights of his teacher Zellig Harris (1909–1992), who had started out as a distributional structuralist,[140] but who had also developed a grammatical formalism known as *context-free grammar* with *transformations*. This formalism was worked out in detail by Chomsky in a completely individual way to produce a new vision of language that addressed both structuralism and behaviourism.[141] Figure 14 gives an example of a context-free grammar for a small fragment of English (the rules of which are usually called 'context-free rewrite rules').

The first five rules are the actual grammatical rules, while the last four 'rules' represent the lexicon. The meanings of the symbols (or categories) are as follows: S is 'sentence', NP is 'noun phrase', VP is 'verb phrase', PP is 'prepositional phrase', Det is 'determiner (article)', N is 'noun', V is 'verb' and P is 'preposition'. These rules can be used to generate existing English sentences as well as sentences that have not previously been uttered.

The application of the grammatical rules involves rewriting or replacing one category with one or more other categories *independent* of the context in which this category is (hence the term *context-free* rule in contrast to Panini's *context-sensitive* rule—see the later discussion below). The first rule (S → NP VP) is used to begin the generation of sentences on the basis of a context-free grammar. This rule can be read as follows: 'S is rewritten in an NP followed by a VP'. The next step is to rewrite the NP, for which there is only one option in this grammar: NP → Det N. This results in the sequence Det N VP. Now there are two choices for Det: 'the' or 'a'. If for the sake of the current example we select 'the', our sequence becomes 'the N VP'. We once again have several options for N, and we select 'man': 'the man VP'. There are now two possible ways to rewrite the VP, and we opt for the first (VP → VP PP), which yields 'the man VP PP'. We therefore once again have a VP in the sequence. We could rewrite it once more using *the same* rule (VP → VP PP), resulting in 'the man VP PP PP'. We can make this sequence as long as we want by continuing to use the same rule. We came across this before in Panini's

[140] Zellig Harris, *Methods in Structural Linguistics*, University of Chicago Press, 1951.
[141] John Lyons, *Noam Chomsky (Modern Masters)*, Viking Press, 1970.

grammar and we referred to it as *recursion* (see 2.1). However, we select the other VP rule, VP → V NP, which leads to 'the man V NP PP PP'. If we continue along this line we obtain 'the man V Det N P Det N P Det N', after which we can put in words from the lexicon, by again using the rewrite rules above, to yield for example the sentence *the man kissed a man in the garden behind the house*. This is only one of the unlimited number of sentences that can be generated using the mini-grammar quoted above. For instance, using the recursive rule VP → VP PP this sentence can be made endlessly longer, for example *the man kissed a man in the garden behind the house behind the woman behind the man...*

Although sentences like this are extremely improbable in everyday usage (they may never occur), they belong to the class of grammatical sentences and represent part of the *competence* of an English speaker. Moreover, for every sentence processed using the grammar described above, a *hierarchical tree diagram* can be constructed that shows how every category of the S symbol splits up into underlying categories (in this case NP and VP), which in turn divide up into further categories (Det N and VP PP respectively) until we finally reach the actual words.[142] We also saw a hierarchically stratified structure in stemmatic philology (see 5.2) and it is a recurring theme in the humanities—for instance in Koch's musical phrase structure (see 4.4), in Alberti's *disegno* theory of hierarchical composition of a painting (see 4.5), and even in the *Annales* historiography of hierarchical time layers (see 5.1).

The grammar discussed above can be tested using human judgments, which Chomsky called linguistic intuitions. If the rewrite rules can produce a sentence that is judged ungrammatical by native speakers, the grammar has to be rejected or amended. Now, the grammar in Figure 14 forms only a fragment of English and at best it describes the phenomenon of iterative (or recursive) prepositional phrase attachment. However, it is easy to see that more complex phenomena can also be accounted for using a context-free grammar. Sentence embedding, for instance, can be incorporated by adding one extra rewrite rule. These are sentences that contain other sentences as subclauses such as *the man who saw the woman kissed the child*, and *the man who saw the woman who was in the garden kissed the child*. These sentences can be made longer and longer to our heart's content. Ever more complex sentences can be covered by recursive rules, as was the case in Panini's grammar. Obviously, a complete grammar of present-day written English is much bigger than the rules in Figure 14 and, in spite of the great number of attempts that have been made, so far a complete system of rewrite rules has not yet been produced for any living language.[143] The only complete grammar for a human language appears to be Panini's Sanskrit grammar with nearly 4,000 rules, but Sanskrit no longer has any native speakers to test it.

However, Chomsky's theory of grammar does not consist solely of context-free rewrite rules. In *Syntactic Structures* he showed how movement rules or

[142] Noam Chomsky, *Syntactic Structures*, Mouton, 1957, p. 27.

[143] See e.g. Daniel Jurafsky and James Martin, *Speech and Language Processing*, Pearson Prentice Hall, 2009.

transformations link structurally related linguistic phenomena. For instance, a sentence like *John is hungry* can easily be transformed into the interrogative sentence *Is John hungry?* by a movement rule. Using slightly more complex transformations, active sentences can be converted into passive ones (for example from *the woman kissed a man* to *a man was kissed by a woman*), and so on and so on. In *Aspects of the Theory of Syntax* (1965) Chomsky developed his theory into a general vision of language. One of his most famous, but unproven, contentions is that linguistic competence is innate.[144] According to Chomsky the staggering speed with which children learn languages on the basis of relatively few linguistic stimuli (*poverty of the stimulus*) and the fact that all children learn their native languages in a similar fashion indicates an inherent underlying grammar, a Universal Grammar or UG—a term which has existed since the thirteenth century, but which was given a new interpretation by Chomsky (see 3.1).[145] During children's linguistic development UG is 'instantiated' by the stimuli that children receive from their environment. In his *Lectures on Government and Binding* (1981) Chomsky argued that the differences between languages can be explained by parameter settings in the human brain (for example the *pronoun-drop* parameter that indicates whether a sentence has to contain an explicit subject or not, which is the case in English for instance, but not in Italian). Such parameters have been compared with 'switches' that can be turned on or off. Chomsky's elaboration of the UG hypothesis attracted a following but also criticism,[146] primarily because research into language acquisition revealed an opposite pattern. In fact, children appear to learn languages very gradually rather than by turning 'switches' on or off. There are indications that learning a language does not take place on the grounds of categorical rules but through gradual statistical matching based on wholesale storage of linguistic stimuli.[147]

Chomsky simplified his linguistic theory more and more in his later work until in *The Minimalist Program* (1995) it made use of just a single combination operation, which was called *merge*. By then the original programme of developing testable grammars had largely been abandoned, however. This does not alter the fact that during several stages of his academic career Chomsky initiated new traditions—quite aside from his parallel career as a political activist.[148] For example, his original work in the 1950s launched a quest for possible grammatical formalisms that resulted in the Chomsky hierarchy of formal languages.[149] This hierarchy is useful for the formal study of human languages but also (rather unexpectedly) for the development and analysis of programming languages,

[144] Chomsky, *Syntactic Structures*, 1965, pp. 25ff.

[145] Noam Chomsky, *Rules and Representations*, Blackwell, 1980.

[146] Geoffrey Pullum and Barbara Scholz, 'Empirical assessment of stimulus poverty arguments', *Linguistic Review*, 19, 2002, pp. 9–50.

[147] See e.g. Michael Tomasello, *Constructing a Language: A Usage-Based Theory of Language Acquisition*, Harvard University Press, 2003.

[148] Robert Barsky, *Noam Chomsky: A Life of Dissent*, The MIT Press, 1998.

[149] Noam Chomsky, 'Three models for the description of language', *IRE Transactions on Information Theory*, 2, 1956, pp. 113–24.

which are represented in the form of grammars as well (see also chapter 6).[150] Chomsky's work moreover contributed to the study of human cognition,[151] and his formal approach proved to be very fruitful in other areas of scholarship, such as the study of grammars in musicology (see 5.4), literary studies (see 5.6), and film studies (see 5.7).

Since Chomsky, the Paninian tradition has been alive and kicking once more: with a finite number of rules linguists try to describe an infinite number of sentences. With his cognitive claims, however, Chomsky goes much further than Panini. Yet surprisingly enough—contrary to Panini—he kept semantics mostly outside the object of study. He asserted that syntax is autonomous and that its most important feature is the notion of recursion. We already came across this notion in Panini's formalism. And as we explained in 2.1, in essence Panini's formalism consists of the following type of rewrite rule: $A \rightarrow B/C\_D$, which means that A can be rewritten as B in the context of C and D. This makes Panini's rules *context-sensitive*, whereas the rewrite rules in Figure 14 are *context-free*. However, since Chomsky had shown in *Syntactic Structures* that context-free rewrite rules are not sufficient for explaining the syntax of human language, he, like Harris, extended his formalism with transformations. Since the 1970s, though, it has become clear that transformations are actually too powerful. They can generate all possible sequences of words without any restrictions.[152] It has since been demonstrated that human languages are in any event at least 'mildly context-sensitive', but probably no more than that.[153] Thus context-free grammars are too poor, transformations are too rich (they overgenerate), while context-sensitive grammars lie somewhere in between.

It seems surprising that Panini's 2,500-year-old grammatical formalism is closer to the structure of language than Chomsky's transformation-based formalism. How can this be explained? We must be wary of exaggerating the scope of Panini's grammar. In the first place, it was intended as a system of rules for Sanskrit—not as an overarching theory of human language. It is its surprisingly broad applicability to other languages and linguistic phenomena that makes Panini's *underlying formalism* seem so far ahead of its time. But its importance did not become clear until quite recently. While there was great fascination with Panini's work, the value of his context-sensitive rules was not understood for a long time—even though such rules were used implicitly by the nineteenth-century Neogrammarians (see above). In 1884 William Whitney wrote, 'Panini's work is a miracle of ingenuity but of perverse and wasted ingenuity.'[154] Panini was too hard a nut to crack, even for

[150] See e.g. Martin Davis, Ron Sigal, and Elaine Weyuker, *Computability, Complexity, and Languages: Fundamentals of Theoretical Computer Science*, Academic Press, Harcourt, Brace, 1994, p. 327.

[151] See e.g. J. A. Fodor, *The Modularity of Mind*, The MIT Press, 1983. See also Tomasello, *Constructing a Language: A Usage-Based Theory of Language Acquisition.*

[152] Stanley Peters and Robert Ritchie, 'On the generative power of transformational grammars', *Information Sciences*, 6, 1973, pp. 49–83. Also Chomsky's phonological work used transformations, see Noam Chomsky and Morris Halle, *The Sound Pattern of English*, Harper & Row, 1968.

[153] Stuart Shieber, 'Evidence against the context-freeness of natural language', *Linguistics and Philosophy*, 8, 1985, pp. 333–43.

[154] In Frits Staal (ed.), *A Reader on the Sanskrit Grammarians*, The MIT Press, 1972, p. 142.

seasoned linguists. He is now reasonably accessible, but only thanks to years of exegetic work.[155] And the significance of his grammatical formalism is widely acknowledged. Nevertheless, Panini is not part of the linguistic curriculum. Linguistics students all over the world are still being taught without any knowledge of Paninian linguistics. If someone knows his name, it is mainly because of his phonology, and not his syntax or semantics, let alone his immensely successful grammatical formalism. Even in the recent linguistic literature it is still—but wrongly—assumed that the concept of formal grammar is a Western invention.[156]

The integration of linguistics and logic: from Frege to Montague. As early as in *Syntactic Structures* (1957) Chomsky kept semantics off the agenda. This appears to be justified given his declared goal: to describe and account for syntax. But linguistics without meaning is like a meal without taste. A comprehensive linguistic theory must also be able to derive the meaning of a sentence. Logic proved to be ideal for this. During the nineteenth century logic initially developed in the direction of a meta-mathematical discipline, particularly with the emergence of the algebraic logician George Boole (1815–1864). This logic was to create the foundations of future computer science, but it is outside the scope of our historical overview.

At the end of the nineteenth century, however, a new logic was developed that could 'work out' the meaning of a sentence on the basis of the meaning of its parts. This *predicate logic* was designed by Gottlob Frege (1848–1925), who with the publication of his *Begriffsschrift* ('Conceptual Notation', 1879) became the most important logician since Aristotle in one fell swoop.[157] As we saw in 2.6, by means of *propositional logic* the truth value of a complex sentence, such as 'John is clever and Peter is stupid', can be computed on the basis of the truth values of the elementary propositions that make up the sentence ('John is clever', 'Peter is stupid'). In propositional logic these elementary propositions can only be combined by *connectives* (such as 'and', 'or', 'if... then ... ') and have no internal structure. So a sentence like 'John is clever' is represented by p, for instance, and 'John kisses Mary' by, say, q, so that the information that both propositions are about John is lost. In *predicate logic* on the other hand, they are represented by C(J) and K(J, M) respectively where C stands for the predicate 'clever', J for 'John', K for 'to kiss' and M for 'Mary' (the symbols for these 'predicates' are of course arbitrary). We will call C(J) and K(J,M) meaning representations for the sentences 'John is clever' and 'John kisses Mary'. Meaning representations using predicate logic are independent of their syntactic realizations. For example, 'John kisses Mary' and 'Mary is kissed

[155] A good introduction for understanding Panini's grammar is George Cardona, *Panini: His Work and Traditions*, Motilal Banarsidass, 1988.

[156] See e.g. Adrian Akmajian, Richard Demers, Ann Farmer, and Robert Harnish, *Linguistics*, 6th edition, The MIT Press, 2010, p. 4.

[157] Gottlob Frege, *Begriffsschrift: eine der arithmetischen nachgebildete Formelsprache des reinen Denkens*, Halle, 1879. For an English translation, see Gottlob Frege, *Conceptual Notation and Related Articles*, translated and edited with a biography and introduction by Terrell Ward Bynum, Oxford University Press, 1972. See also Jean van Heijenoort, *From Frege To Gödel: A Source Book in Mathematical Logic, 1879–1931*, Harvard University Press, 1967.

by John' have the same representation K(J,M). It is usually assumed that while the word order varies from one language to another, the underlying meaning representation is independent of the language. Predicate logic formulas can moreover be complex. For instance the sentence 'Peter sees that John kisses Mary' may be represented as S(P, K(J,M)). Aside from the *constants* John, Peter, to kiss, etc., there are also *variables* in predicate logic. The sentence 'He kisses Mary' can be written in predicate logic as K(x, M), where x is an unbound variable, as long as it is not known to whom 'he' refers. A variable can be bound by *quantifiers*: the universal quantifier $\forall$ or the existential quantifier $\exists$. These quantifiers are today (also) part of elementary mathematics, but they were first introduced in logic. A sentence like 'Everyone kisses Mary' can be written in predicate logic as $\forall$x: K(x, M). Predicate logic moreover contains the connectives of propositional logic, as a result of which it really is an extension of the latter (see 2.6).[158]

The idea that the meaning of a sentence can be derived from the meaning of its parts and the rules that combine these parts is known as the *principle of compositionality*,[159] an idea which has also been attributed to Frege. As we have seen, the compositionality principle had been formulated centuries before by the Indian language philosopher Yaska in the fifth century BCE (see 2.1). It was later adopted and defended with fervour by the Nyaya school, which opposed the *semantic holism* of the Sphota school (see 3.1). It was Frege, though, who was the first to give a formal account of the principle. While predicate logic is not rich enough to represent the meaning of all the sentences of a natural language, Frege was able to use it to solve a number of classical problems that had been too much for Leibniz (see 4.3). Predicate logic also proved to be able to form the basis for all kinds of richer logics with more quantifiers (for instance *modal logic*) and a more refined system of meaning references (for example *intensional logic*).[160]

However, a logical language is not yet enough. We also need a system of rules that can *predict* for a certain sentence what its meaning is. It turns out that the syntactic rewrite rules of a grammar are suited for this. To this end each syntactic rewrite rule has to be extended with a semantic enrichment that stipulates *how the compositional meaning of the constituent depends on the meaning of its parts*. For example the rewrite rule S → NP VP (see Figure 14) is semantically augmented to become S[VP(NP)] → NP VP, where the representation between the square brackets indicates how the meaning of the whole sentence S is derived from the separate meanings of the NP and VP. The syntactic word order for an English sentence is thus NP VP, whereas the corresponding meaning representation is VP (NP)—this may be exemplified by the sentence 'John walks' which follows the syntactic pattern NP VP and has a meaning representation 'walks(John)'. The other

[158] See e.g. L. T. F. Gamut, *Logic, Language and Meaning*, 2 volumes, University of Chicago Press, 1991.

[159] Theo Janssen, 'Compositionality', in Johan van Benthem and Alice ter Meulen (eds), *Handbook of Logic and Language*, Elsevier, 1997, pp. 417–73.

[160] Van Benthem and ter Meulen, *Handbook of Logic and Language*.

rewrite rules can be enriched with semantic rules in a similar fashion, for example VP [V(NP)] → V NP. The separate words are not usually enhanced with semantic rules. They often refer to an extralinguistic context, unless they correspond to connectives or quantifiers. (We are making things somewhat easier for ourselves because the semantic structure of a sentence does not always need to coincide with its syntactic structure, and the problems associated with word meaning represent a separate discipline.) When a sentence is generated with such semantically enriched rewrite rules, the predicate-logical meaning of that sentence can be built up at the same time. In this way a linguistic grammar is actually integrated with a logical language—an objective that had been attempted for centuries, among others by the Port Royal linguists (see 4.3), but has never been achieved before. Panini did not make use of compositional semantics either. While he assigned semantic roles to separate words, such as agent, patient, and recipient (see 2.1), he offered no possibility of integrating them into a single representation of meaning.

The integration of linguistics and logic was one of the major attainments of twentieth-century humanities. It has had an enormous effect on other disciplines both within and outside the humanities, from musicology (5.4) to literary studies (5.6) and from cognitive psychology to artificial intelligence.[161] The integration outlined above has been necessarily simplified. For example, a simple substitution is often not enough to combine the sub-meanings of constituents. The combination of predicate-logical functions is regulated by a richer calculus that in itself also represents a rewrite system and which is known as *lambda calculus*.[162] The first person to fully work out the integration of generative linguistics and compositional semantics was Richard Montague (1930–1971). In *English as a Formal Language* (1970) he showed how the three formalisms of predicate logic, generative grammar, and lambda calculus resulted in a new logical grammar that has become known as *Montague Grammar*.[163] Although the Montagovian approach was the first grammar to really go beyond Panini, it has had surprisingly little impact on Chomskyan linguistics, in which semantics is still largely excluded.[164] In the end, Montague Grammar formed a school in its own right.[165]

Shortcomings of generative linguistics and the rise of computational linguistics. Combined with logical semantics, generative linguistics appeared to contain everything needed for a perfect success story. However, during the 1980s and 1990s a number of problems emerged that gnawed away at the foundations of generative and Montagovian thinking.

161 Stuart Russell and Peter Norvig, *Artificial Intelligence: A Modern Approach*, 3rd edition, Prentice Hall, 2009.

162 L. T. F. Gamut, *Logic, Language and Meaning*. See also Barbara Partee, Alice ter Meulen, and Robert Wall, *Mathematical Methods in Linguistics*, Kluwer, 1990.

163 David Dowty, Richard Wall, and Stanley Peters, *Introduction to Montague Semantics*, Kluwer, 1981.

164 Except for the generativist notion of *logical form*, which is however barely formalized.

165 Paul Portner and Barbara Partee (eds), *Formal Semantics: The Essential Readings*, Blackwell, 2002.

(1) ***Categorical versus gradient aspects of language, and the notion of probabilistic grammar.*** The Chomskyan distinction between *competence* (language knowledge) and *performance* (language use) was not embraced by all linguists. Many argued that linguistics should not focus solely on knowledge of language, but on the study and explanation of language use, too. For instance, the concept of grammaticality proved to be too categorical. It emerged from both psycholinguistic and sociolinguistic research that there was a continuum between grammatical and ungrammatical sentences,[166] as well as between standard and non-standard language use.[167] The social variability of language use, and it systematics, were already investigated from the 1960s onwards, in particular by William Labov (born 1927), who is often seen as the founder of sociolinguistics. If we want to account for the continuum in language, we arrive at a theory that is *statistical* in nature. This was supported by early research into regularities in the probability distributions of words and word groups in language by George Zipf (1902–1950).[168] These insights resulted in a vision of language in which generative and statistical methods had to be integrated. This is found in the notion of *probabilistic grammar* where a system of rules is expanded to include a statistical component.[169] Probabilistic grammars are related to linguistic approaches like *usage-based grammar* and *construction grammar.*[170] According to construction grammar new sentences are not produced by combining context-free or context-sensitive rules but by putting together *constructions*—the form–meaning pairs of earlier language observations. Constructions can be derived from fragments of previous linguistic stimuli. The more frequent a fragment, the more stable the resulting construction.[171]

(2) ***Compactness versus redundancy of grammars.*** From the very beginning, the generative linguistic tradition put a strong emphasis on the compactness of grammars. Chomsky and other generativists tended to make their grammars as small as possible. Yet if one linguistic insight has surfaced over recent decades, it is that human language is highly 'redundant'. It has become clear from psycholinguistic, evolutionary, and neuroscientific research that when children learn a language, there is massive and redundant storage of

[166] See e.g. Carson Schütze, *The Empirical Base of Linguistics: Grammaticality Judgments and Linguistic Methodology*, University of Chicago Press, 1996.

[167] William Labov, *Principles of Linguistic Changes: Social Factors*, Blackwell, 2001.

[168] According to Zipf's Law, the frequency of occurrence of a word is inversely proportional to the rank of the word in the frequency table—see George Zipf, *Selected Studies of the Principle of Relative Frequency in Language*, Harvard University Press, 1932.

[169] For an overview, see Rens Bod, Jennifer Hay, and Stefanie Jannedie (eds), *Probabilistic Linguistics*, The MIT Press, 2003.

[170] Charles Fillmore, Paul Kay, and Mary Catherine O'Connor, 'Regularity and idiomaticity in grammatical constructions: the case of *let alone*', *Language* 64, 1988, pp. 501–38. See also Adele Goldberg, *Constructions at Work*, Oxford University Press, 2006.

[171] See e.g. Rens Bod, *Beyond Grammar: An Experience-Based Theory of Language*, CSLI Publications, 1998.

linguistic utterances without much hierarchical structure.[172] It is as though the example-based idea of Sibawayh (see 3.1) is being confirmed in the most extreme sense. Theories that generalize over stored linguistic utterances appear to be quite promising. This is supported by the rise of *computational linguistics* in which the most successful applications, such as automatic speech recognition and machine translation, are based on statistical generalizations from examples.[173] Nevertheless, the Paninian basic principle continues to hold up here, too: the pursuit of infinite productivity on the basis of a finite number of resources.

(3) ***Typology of languages: free versus strict word order.*** There are over 6,000 languages in the world with apparently endless variation. Language universals are still far away. Chomksy's formalism assumes that all languages have a stable syntax in which the word order is determined by precise rules. However, there are languages without any set word order. Latin is close to such a language, but Warlpiri (spoken in Australia) is almost completely free of word order.[174] The relevant rules in Warlpiri appear to be mostly morphological and semantic. Linguistic theories have since been developed that do justice to both fixed and free word order,[175] and that have now been integrated with the construction-based, statistical approach. Moreover, a language has recently been discovered in which recursion appears to play no part at all—Pirahã, which is spoken in the Amazon area.[176] Although further research is necessary, this discovery could completely undermine the concept of Universal Grammar.

Linguistics in India and China. Linguistics outside Europe and the United States came up with strikingly few new themes at the beginning of the modern age. While in the nineteenth century Indian linguists like Ramkrishna Gopal Bhandarkar concentrated on the study of Panini and his commentators,[177] the first Chinese grammar by a Chinese scholar was published at the end of the Qing Dynasty. This *Mashi wentong* ('Basic Principles for Writing Clearly and Coherently by Mister Ma') was written in 1898 by Ma Jianzhong, who fashioned his grammar on the Latin

[172] Stefan Frank, Rens Bod, and Morten Christiansen, 'How hierarchical is language use?', *Proceedings of the Royal Society B*, 297(1747), 2012, pp. 4522–31. See also Stefan Frank and Rens Bod, 'Insensitivity of the human sentence-processing system to hierarchical structure', *Psychological Science*, 22, 2011, pp. 829–34.

[173] For a discussion, see Rens Bod, Remko Scha, and Khalil Sima'an (eds), *Data-Oriented Parsing*, CSLI Publications/University of Chicago Press, 2003.

[174] Kenneth Hale, 'Warlpiri and the grammar of non-configurational languages', *Natural Language and Linguistic Theory*, 1(1), 1984, pp. 5–47.

[175] For example, Lexical-Functional Grammar (LFG) and Head-Driven Phrase Structure Grammar (HPSG). See respectively Joan Bresnan (ed.), *The Mental Representation of Grammatical Relations*, The MIT Press, 1982; Ivan Sag and Thomas Wasow, *Synactic Theory: A Formal Introduction*, CSLI Publications, 1999.

[176] Daniel Everett, 'Cultural constraints on grammar and cognition in Pirahã: another look at the design features of human language', *Current Anthropology*, 46, 2005, pp. 621–46.

[177] Ramkrishna Gopal Bhandarkar, *Second Book of Sanskrit*, Government Central Book Depot, 1870.

model.[178] There were also many other fascinating Chinese linguists. For example Luo Changpei studied the non-Chinese languages in China, and Zhang Binglin revealed the stages in the history of Chinese. However, we have not been able to discover new principles in their work.

It is no exaggeration to assert that from 1900 onwards virtually all over the world linguistics was dominated by the Western (usually colonial) influence. Comparative linguistics had already been exported to all continents, and during the course of the twentieth century linguistic schools developed a 'globalized' character.

The state of contemporary linguistics: between generativism and constructionism. Whereas in the nineteenth century linguistics developed as a historical discipline, from the twentieth century onwards the ahistorical study of language became dominant. It has been argued that in so doing, linguistics moved in the direction of the exact sciences.[179] Yet if the presence of a historical component were an indication of a humanistic or scientific discipline, biology—with its evolution theory—would be one of the humanities, as would astronomy with its cosmology.

It is, however, the case that linguistics is the humanistic field that is ideally suited to the pattern-seeking nomothetic method, which has indeed become common currency. Our overview of linguistic movements, though, is far from complete. We have not, for instance, addressed the twentieth-century generativist approaches that preceded Chomsky.[180] We have similarly not discussed recent developments in the generativist camp (*optimality theory* for example) or in the constructionist or statistical camp (take *exemplar theory*, in which hierarchical structure is relinquished) or in the logical semantic camp (such as *discourse representation theory*).[181] We have barely mentioned or left aside interdisciplinary sub-areas of linguistics such as sociolinguistics and psycholinguistics.

Despite its general pattern-seeking character, present-day linguistics displays a striking lack of unity. There may well be more linguistic theories than universities. However, many of these theories prove to be interchangeable. The question, though, is whether this is a promising result. It is not until we consider present-day linguistic theories from a distance that a clearer picture emerges, namely that broadly speaking one can discern two clusters. In one cluster we see the approaches that champion a rule-based, discrete method, whereas in the other cluster an example-based, gradient method is advocated. The most notable development seems to be the approach that connects both clusters and thus tries to do justice to both rule-based and example-based aspects of language. Yet as happens so often in historiography, we can only evaluate this development if we are able to stand back, not only in terms of space, but also in terms of time.

[178] Victor Mair, 'Ma Jianzhong and the invention of Chinese grammar', *Journal of Chinese Linguistics*, 10, 1997, pp. 5–26.

[179] Jerome Kagan, *The Three Cultures*, Cambridge University Press, 2009.

[180] Such as Categorial Grammar by Kazimierz Ajdukiewicz (1935) and Dependency Grammar by Lucien Tesnière (1954).

[181] For a recent attempt of an overview, see Bernd Heine and Heiko Narro (eds), *Oxford Handbook of Linguistic Analysis*, Oxford University Press, 2010.

5.4 MUSICOLOGY: THE SYSTEMATIC VERSUS THE HISTORICAL

Until the end of the eighteenth century, the study of music in Europe consisted primarily of music theory. The number of historical studies of music could be counted on the fingers of one hand (see 4.4). The balance changed dramatically in the nineteenth century. As in the other humanistic disciplines, the historical approach was given a central position. In 1885 the Austrian musicologist Guido Adler made an influential distinction between *systematic* and *historical musicology* that was to dominate the twentieth-century study of music.[182] The systematic component addressed the 'most important laws that are applicable within the different trends in music',[183] while the historical component focused on 'the history of music', in which there once again had to be a quest for 'regularities' ('gesetzmäßigkeiten').[184] In so doing Adler differed markedly from his contemporary Wilhelm Dilthey, who, in contrast, rejected searching for laws and regularities in the humanities (see 5.1). Compared with Europe, the reverse process was taking place in China: whereas the study of music during the Ming and Qing Dynasties consisted primarily of historical works (see 4.4), there was a return to theoretical approaches in the twentieth century. The dominance of theory continued in India and the Arabic-Ottoman world, and in Africa a surprising unity was revealed in the diversity of musical traditions.

Systematic musicology: a breakthrough in the law of consonances. After centuries of bickering there was a breakthrough in the study of consonant intervals (see 4.4). Humanistic scholars like Gaffurio and Zarlino and natural scientists such as Galileo and Huygens had all been searching for the law of harmonic intervals, but no single pattern seemed to fit the bill. The German physiologist and musicologist Hermann von Helmholtz (1821–1894) seemed to have turned the tide. In *Die Lehre von den Tonempfindungen als physiologische Grundlage für die Theorie der Musik* (1863) Helmholtz focused primarily on the *sensory properties* of dissonances and harmonies.[185] He tried to explain the degree of dissonance on the grounds of the intensity of the beats in the harmonic tones (and the auditory response to them), which he was able to deduce from whether or not *two tones were in phase.* Isaac Beeckman had proposed a similar explanation over two centuries before (see 4.4), but Helmholtz now succeeded where Beeckman had failed. He formulated a mathematical function that proved to be a good yardstick for the observation of dissonance.[186] This function for tonal 'roughness' was to define musicological thinking about dissonance until well into the twentieth century.

[182] Guido Adler, 'Umfang, Methode und Ziel der Musikwissenschaft', *Vierteljahresschrift für Musikwissenschaft, 1*, 1885, pp. 5–20.

[183] Guido Adler, *Methode der Musikgeschichte*, Breitkopf und Härtel, 1919, p. 7.

[184] Adler, *Methode der Musikgeschichte*.

[185] For an English translation, see Hermann von Helmholtz, *On the Sensations of Tone as a Physiological Basis for the Theory of Music*, translated by Alexander Ellis, Longmans, Green, and Co., 1885.

[186] Helmholtz, *On the Sensations of Tone as a Physiological Basis for the Theory of Music*, p. 44.

Yet even Helmholtz's law is nothing more than an approximation of an extremely complex phenomenon. It slowly became clear that several factors play a role in the perception of consonance and dissonance. These factors are acoustic on the one hand and sensory on the other—not to mention the cultural–historical aspects. If we want to get to grips with the perception of consonance it is important to exclude as many external factors as possible. But which factors are 'external'? For a long time both humanities and science scholars believed that the simple monochord (the one-string instrument) was suitable for conducting controlled experiments (4.4). However, acousticians soon found out that the tone of a string was complex and consisted of *partials*: a *fundamental tone* plus *overtones* (see 3.4 and 4.4).[187] In the nineteenth century it was furthermore discovered that complex tones could be mathematically described as a combination of sine functions or sine waves, using *Fourier analysis*.[188] Every partial corresponded to a sinusoid, and only a tone that consists of just one single sinusoid generates a sound without overtones. The problem of consonance could perhaps be unravelled on the basis of such simple tones. And if this was possible for simple tones, it could also be done for complex tones by combining sine waves.

The Dutchmen Reinier Plomp and Willem Levelt were among the pioneers in this field. In their famous study in 1965 they asked Western test subjects to rank pairs of simultaneous simple tones on the basis of how good they sounded on a scale of one to seven.[189] What emerged was that when two tones varied very little in terms of their vibration frequency, beats were observed but the sound was nevertheless evaluated as consonant ('pleasant', 'nice'). If the two tones had a slightly larger difference in vibration frequency, the beats became faster and the sound was judged to be disagreeable (and dissonant). The two tones were not considered to be harmonic again until the difference in vibration frequency had become sufficiently large. Plomp and Levelt discovered that the consonance assessments of test subjects coincided with a physiological property of hearing—the *critical bandwidth*. This is the frequency range within which a tone tends to block another tone in the human auditory organ. The tones did *not* sound dissonant if they differed by more than one critical bandwidth or if they differed by only a small fraction in bandwidth. *Maximum dissonance*, on the other hand, was created when the tones were separated by about a quarter of the critical bandwidth, which corresponded to about three or four per cent frequency difference, or in other words a semi-tone.[190] For the first time the degree of dissonance was expressed in terms of the properties of both hearing and the physical phenomenon.

[187] William Forde Thompson, *Music, Thought, and Feeling: Understanding the Psychology of Music*, Oxford University Press, 2008, pp. 46ff.

[188] Joseph Fourier, 'Mémoire sur la propagation de la chaleur dans les corps solides', *Nouveau Bulletin des sciences par la Société philomatique de Paris I*, Bernard, 1808, pp. 112–16.

[189] Reinier Plomp and Willem Levelt, 'Tonal consonance and critical bandwith', *Journal of the Acoustical Society of America*, 38, 1965, pp. 548–60.

[190] For a sound example of maximum dissonance, as well as the transition to minimal dissonance, see <http://en.wikipedia.org/wiki/File:Dissonance-M2-to-unison.ogg>.

Plomp and Levelt subsequently developed a method that could calculate consonance for complex tones on the basis of the simple tone combinations of which the complex tones consisted. Their calculation summed the dissonances of all combinations of adjacent partials, after which the consonance was inversely proportional to the total dissonance.[191] This calculation proved to agree extremely well with the progressive subjective evaluations. Furthermore, the special cases of unison and octave emerge from the calculations as the most consonant, followed by the fifth and the fourth, then the third, and so on. Plomp and Levelt's approach thus explains why simpler numerical relationships (for example 2:1 for octave and 3:2 for fifth) are perceived as more consonant. Their method has been successfully tested on multiple occasions, including with non-Western test subjects.[192] Yet their work can still not explain all aspects of consonance and dissonance. For instance, the theory says nothing about consonance perception of two *successive,* non-simultaneous tones. This form of consonance perception has been explained as a memory phenomenon, in which a tone is memorized for some period of time. Similarly, Plomp and Levelt's theory does not say anything about the historical shift in the concept of consonance, for example the fact that in Europe the sixth was not considered to be consonant until after centuries of musical practice (see 4.4). Some musicologists ascribe this phenomenon to a learning process. Ultimately any combination of tones can be perceived as 'consonant'. There is only one exception to this: if the tones are separated by a quarter of the critical bandwidth, following Plomp and Levelt the pairs of tones are *always* experienced as dissonant—independent of time, place, or culture.

Plomp and Levelt's impressive result was largely conducted in a discipline that is currently not looked on as being the exclusive preserve of the humanities, i.e. psychoacoustics. Nevertheless, this work comes under Adler's definition of systematic musicology (see above). As we have already seen with regard to modern historiography, linguistics, and philology, it is almost impossible on the grounds of principle to categorize a discipline as belonging to the humanities or (social) science.

Hierarchical analysis and structuralism in music: from Riemann and Schenker to Lerdahl and Jackendoff. Inspired by Quintilian's rhetoric, as early as the sixteenth century Dressler and Burmeister introduced a *hierarchical analysis* of music in which a piece was split up into ever smaller parts, such as phrases and segments (see 4.4). At the end of the eighteenth century Heinrich Koch was even talking about a 'natural law' of musical phrase structure (4.4). The smallest unit of meaning was a segment consisting of no more than one bar. Segments joined together to form a phrase, and phrases in turn made up a period. The hierarchical description of music also dominated the nineteenth century. Such musical forms as the sonata and symphony were broken down into layered structures, and more than once musicologists believed they were on the trail of 'universal musical laws', but these were often based on the philosophical flavour of the month. In 1853, for

[191] Plomp and Levelt, 'Tonal consonance and critical bandwith'.
[192] Diana Deutsch (ed.), *The Psychology of Music*, Academic Press, 1999.

instance, Moritz Hauptmann argued that the basic units of all music consist of patterns with precisely two elements, which in the Hegelian tradition he designated as *thesis*.[193] A longer pattern, comprising three elements for example, forms an *antithesis*, which together with the previous one was resolved in a *synthesis*. Other musical concepts too, such as Koch's antecedent phrase and consequent phrase (see 4.4) were analysed in Hegelian-dialectic terms.

Hugo Riemann (1849–1919) introduced a new musical concept—the *motif*, which he defined as the indivisible musical unit that represented the 'life force' of a piece of music and which according to him, in the best Herderian tradition, went through a rise, peak, and decline.[194] Riemann published in-depth phrase and motif structure editions of Bach, Mozart, Haydn, and Beethoven.

Although analysis in phrase structures was one of the cornerstones of nineteenth-century musicology, a desire for finding precise rules that underlie this structuring process did not arise for a long time. We see a quest for a system of rules that could predict the phrase structure of a piece of music for the first time with the emergence of *Gestalt psychology*. In 1890 a pioneer in this discipline, Christian von Ehrenfels, showed in *Über Gestaltqualitäten* ('On the Qualities of Form') that a melody's phrase structure remained the same if it was transposed to another pitch.[195] Although every note is moved, the form or *Gestalt* remained unchanged. The perceived structure did not depend on the absolute pitch, but the *relative* pitch and the time between notes in a melody, as well as the similarity between tones. For example, if a sequence of notes is separated at a particular point by a relatively large time or tone interval, the musical listener tends to register a phrase boundary at that point. And if several equal notes occur after each other, the listener tends to ascribe a group or phrase to those notes. According to Gestalt theorists these basic 'laws of perception' represented the universal principles of all perception and were designated by them as the 'principle of proximity' and the 'principle of similarity' respectively.[196] During the first half of the twentieth century these and other principles were expanded further by Max Wertheimer, Kurt Koffka, and Wolfgang Köhler, and tested in terms of visual and auditory perception.[197] However, the Gestalt principles did not embody a formal grammar with which the phrase structure could be unambiguously predicted. A first attempt to achieve this was to be undertaken in the second half of the twentieth century.

[193] Moritz Hauptmann, *Die Natur der Harmonik und der Metrik: zur Theorie der Musik*, Breitkopf und Härtel, 1853. Translated into English as *The Nature of Harmony and Metre*, Swan Sonnenschein, 1893, reprinted by Da Capo Press, 1991.

[194] Hugo Riemann, *Musikalische Syntaxis: Grundriß einer harmonischen Satzbildungslehre*, Breitkopf und Härtel, 1877. See also Alexander Rehding, *Hugo Riemann and the Birth of Modern Musical Thought*, Cambridge University Press, 2003.

[195] Christian von Ehrenfels, 'Über Gestaltqualitäten', *Vierteljahrsschrift für wissenschaftliche Philosophie*, 14, 1890, pp. 249–92. See also Reinhard Fabian, *Christian von Ehrenfels: Leben und Werk*, Rodopi, 1986.

[196] David Hothersall, *History of Psychology*, McGraw-Hill, 2003, pp. 207ff.

[197] See e.g. Max Wertheimer, 'Untersuchungen zur Lehre von der Gestalt', *Psychologische Forschung*, 4, 1923, pp. 301–50.

In Heinrich Schenker (1868–1935) we find one of the most original musicologists of the modern era. Although Schenker's musical analysis, like those of his predecessors, was hierarchical, he broke through the primacy of the melody. In *Der freie Satz* (1935) Schenker asserted that all tonal pieces of music could be reduced to harmonic triads, such as do mi sol (C-E-G).[198] Musical compositions were nothing other than transformations and elaborations of basic chords. This did not mean that a piece of music is just a succession of chords, but that the underlying structure of music can be understood as such. Schenker thus abstracted from the notes of a musical composition itself, and he applied his method within Saussure's structuralist thinking in linguistics (see 5.3). The two mechanisms that Schenker introduced for expansion from a triad to a melody—*transformation* and *prolongation*[199]—were explained in such detail that it was only a small step to mould them into a verifiable system of rules.

This was actually done by Fred Lerdahl and Ray Jackendoff in their influential work *A Generative Theory of Tonal Music* (1983). Based on the insights of Chomskyan linguistics, they described a system of rules within which the Gestalt principles and Riemann's notion of motif discussed above, and also Schenker's theory of harmonic analysis, were integrated into one theory. Lerdahl and Jackendoff gave an impressive grammatical synthesis of centuries of musicological research in which methods taken from linguistics were employed for musicology. It should be remarked that there is an important difference between a grammar for language and one for music. Whereas Chomsky's language grammar attempts to define the correct (grammatical) sentences for a language (and in which a hierarchical phrase structure can be produced for every sentence—see 5.3), Lerdahl and Jackendoff's music grammar defines not so much the 'correct' musical compositions as the hierarchical phrase structures that are ascribed to pieces of music by listeners. To this end Lerdahl and Jackendoff used two sorts of rules: *well-formedness rules*, which specify all well-formed phrase structures of a piece, and *preference rules*, which predict the phrase structures actually assigned by an 'experienced listener'.[200] This multi-stage system of rules makes it possible to test their music grammar against listeners' perception. Such testing was indeed conducted—on the basis of, among other things, a collection of Western folk songs (the Essen Folksong Collection of 20,000 pieces). These folk songs were enriched by students of the Essen conservatory with their (perceived) phrase structures. It turns out that a computer implementation of Lerdahl and Jackendoff's theory correctly predicts around seventy-five per cent of the phrases perceived by this group of

[198] Heinrich Schenker, *Der freie Satz: neue musikalische Theorien und Phantasien*, part 3, Universal Edition, 1935. For an English translation, see Heinrich Schenker, *Free Composition (Der freie Satz): Volume III of New Musical Theories and Fantasies*, translated and edited by Ernst Oster, Longman, 1979.

[199] For an extension of Schenker's method to medieval, renaissance, and modern music, see Felix Salzer, *Structural Hearing*, Dover, 1962.

[200] Fred Lerdahl and Ray Jackendoff, *A Generative Theory of Tonal Music*, The MIT Press, 1983, pp. 9ff.

students.[201] Of course, this does not mean much more than that seventy-five per cent of the phrases predicted by the theory agreed with the intuitions of conservatory students, who were obviously far from average listeners. But it did demonstrate that some musical theories can now be precisely tested and replicated. Moreover, the seventy-five per cent accuracy suggested that there was substantial scope for improvement.

One of the possible improvements relates to the *example-based* aspects of music (see 3.4 and 4.4), which Lerdahl and Jackendoff neglected, even though it is well known that listeners remember musical pieces (or parts thereof) fairly easily—from excerpts of a melody to entire songs. If certain sequences of notes occur more frequently in a musical culture than other sequences they form a more stable group (cf. the notion of construction in language—5.3). As in linguistics, the integration of a rule-based approach with an example-based one could be one of the most promising options (5.3). When such an integrated method was tested with the same Essen Folksong Collection, using a new computer implementation, over eighty-seven per cent of the phrases were correctly predicted.[202] This was significantly higher than the seventy-five per cent of phrases predicted correctly by the rule-based method. Moreover, it emerged that this improvement could be attributed in full to the example-based component. After centuries of developing theories, music analysis has become a testable discipline.[203]

Clearly the testing of computational models of music perception on purely tonal music says nothing about *atonal*, twelve-tone music. Whereas Arnold Schoenberg gave the first foundation of twelve-tone compositions in (the last chapter) of his *Harmonielehre* (1911), it was Allen Forte who developed an analytical model in 1973 with which hierarchical structures could also be derived in atonal music.[204] By means of pitch classes Forte managed to derive three levels of structure. Hierarchical stratification is therefore an unbroken strand in atonal music too.

From music history to new musicology. As well as searching for patterns in musical compositions, musicologists have also hunted for regularities and laws in music *history*. The classification into the well-known musical styles and periods (e.g. baroque, classical, romantic, etc.) is the result of such quests. We saw in 4.4 that the periodization by Giovanni Martini broadly speaking followed the biblical pattern of

[201] David Temperley, *The Cognition of Basic Musical Structures*, The MIT Press, 2001, pp. 73ff. For testing Lerdahl and Jackendoff's theory, Temperley had to fix some open ends, where he had to leave the theory at some points.

[202] Rens Bod, 'Memory-based models of melodic analysis: challenging the gestalt principles', *Journal of New Music Research*, 31, 2002, pp. 27–37. For an overview of probabilistic, example-based models of music analysis, see David Temperley, *Music and Probability*, The MIT Press, 2007. See also Rens Bod, 'A unified model of structural organization in language and music', *Journal of Artificial Intelligence Research*, 17, 2002, pp. 289–308.

[203] See e.g. Henkjan Honing, 'On the growing role of observation, formalization and experimental method in musicology', *Empirical Musicology Review*, 1(1), pp. 2–6. The application of computational techniques to musicology has also led to the new field of *computational musicology*.

[204] Allen Forte, *The Structure of Atonal Music*, Yale University Press, 1973.

St Augustine. In the nineteenth century a multitude of historical classifications was used. They were based on Herder's cyclical, Hegel's dialectic, or Comte's positivist model.[205] Darwinian ideas were also applied to music: in 1896 in *The Evolution of the Art of Music* the composer and musicologist Hubert Parry compared the 'embryonic music of the primitive savage' with European medieval music. According to Parry, only Western music had been through all stages of human development to maturity. His music history is comparable to Buckle's world history (see 5.1), in which progress was elevated to a law of nature and Western Europe represented the apex of historical development. We find Parry's racist world view in all nineteenth-century humanistic disciplines.

In *Der Stil in der Musik* (1911) Guido Adler criticized his contemporaries and championed an emphasis on the concept of *style*.[206] Adler gave a number of criteria for determining musical style, such as rhythmic properties, tonality, song, polyphonic construction, instrument use, and performance practice. According to him it was essential that composers be brought together in groups with the same style characteristics. Hero worship was altogether wrong. Adler rigorously analysed the Viennese classical style, and his adherents investigated Beethoven's personal styles. His research was extended to other composers, from Palestrina to Wagner, but the notion of style was initially only applied to relatively short periods. Adler compiled his ideas in his *Handbuch der Musikgeschichte* (two volumes, 1929), which is still being used. Although Adler was highly productive in his scholarly life—he more or less developed the whole of modern musicology—he appears to have been a man of few words in his private life. As the composer Gustav Mahler was wont to say, 'If I want to be alone, I go for a walk with Guido Adler.'[207]

Adler only applied the idea of musical style to short periods, but in 1919 the musicologist Curt Sachs identified longer musical style periods when he used the art historical term *baroque* for music between 1600 and 1750.[208] To do this Sachs systematically applied to music the five principles of art analysis introduced by Heinrich Wölfflin (see 5.5). Style periods from the visual arts and literature were deemed one by one to be applicable to music history, often after lengthy debate. In broad-brush terms this classification of Western music corresponded to the Middle Ages (*ars antiqua, ars nova*), Renaissance, mannerism, baroque, rococo, classicism, romanticism, and impressionism. These were followed by parallel styles such as neo-romanticism, dodecaphony (twelve-tone music), *musique concrète*, and serialism—let alone the many styles of popular music that have been ignored by traditional musicology.

As well as this concentration on European music, the study of music from other regions also arose. Jaap Kunst is seen as one of the founding fathers of

[205] Warren Dwight Allen, *Philosophies of Music History*, Dover Publications, 1962.

[206] Adler's work appears not to have been translated into English. On Adler's science of music, see Kevin Karnes, *Music, Criticism, and the Challenge of History: Shaping Modern Musical Thought in Late Nineteenth Century Vienna*, Oxford University Press, 2008.

[207] Friedrich Engel-Jánosi, *Aber ein stolzer Bettler: Erinnerungen aus einer verlorenen Generation*, Verlag Styria, 1974, p. 30.

[208] Curt Sachs, 'Barokmusik', *Jahrbuch der Musikbibliothek Peters*, 1919, pp. 7–15.

ethnomusicology, a term that he coined. In *De toonkunst van Bali* ('The Music of Bali', 1925) his underlying assumption was that music can only be studied successfully in conjunction with other cultural expressions, such as dance and theatre.[209] Although ethnomusicology is often seen as the study of 'non-Western' music, its method is equally applicable, and has been applied, to Western music.

After 1945, methods from other humanistic disciplines were also used in music history. These included structuralism (taken from linguistics, see 5.3) and the more pattern-rejecting movements like narrativism, critical theory, and deconstructivism (see 5.1). As in historiography, critical theory and deconstructivism cast doubt on every unity and coherence, in this case in musicology.[210] Claims about laws and universality were rejected, but cultural tendencies were investigated. The recently developed *new musicology* is an example of the latter.[211] Like *new cultural history* (see 5.1), it is characterized by the concepts that it uses, such as power, race, gender, ideology, and identity, as well as by the influences of feminism and postcolonial studies. The goal of new musicology appears to be the creation of a new view and criticism of music rather than a desire to increase knowledge about music. For example, the application of Western yardsticks to the study of non-Western music is criticized, as is the fact that academic musicology ignores the study of pop music.

Ottoman Empire and the Arab world: the quarter tone controversy. Nineteenth-century Ottoman and Arabic musicology has remained largely underexposed. The study of the quarter tone scale appeared to be one of the few constants in this era. First introduced by the Lebanese musicologist Mikhail Mishaqa (1800–1880),[212] the twenty-four-tone scale became the subject of fierce debate, particularly about the tuning of this scale.[213] Two camps emerged—the Egyptian camp, which proposed an equal distance between the twenty-four quarter tones, and the Turkish camp, which rejected an even division. Arguments were underpinned by obscure mathematical reasoning. Nowadays the controversy looks like pedantry, but in the Ottoman–Arabic world it was considered to be so important that two major congresses were organized, in 1959 and 1964, in order to settle the difference of opinion. The theoretical controversy did not die down and make way for studies of musical practice until the advent of *electronic* simulations of scales. A comparison with the study of consonance and dissonance in Europe springs to mind. Ultimately it emerged that the arguments about 'natural' tones were clouded, and electronically generated tones led to new insights with which the controversy was settled.

Africa: unity in diversity. The study of music in Africa was dominated for a long time by European colonial powers (see 4.4). However, after the Second World War

[209] Jaap Kunst, *Indonesian Music and Dances: Traditional Music and its Interaction with the West: A Compilation of Articles (1934–1952)*, Royal Tropical Institute, 1994.

[210] Alistair Williams, *Constructing Musicology*, Ashgate, 2002.

[211] David Beard and Kenneth Gloag, *Musicology: The Key Concepts*, Routledge, 2005, pp. 92–3.

[212] Shireen Maalouf, 'Mikhii'il Mishiiqa: virtual founder of the twenty-four equal quartertone scale', *Journal of the American Oriental Society*, 123(4), 2003, pp. 835–40.

[213] Robert Günther (ed.), *Musikkulturen Asiens, Afrikas und Ozeaniens im 19. Jahrhundert*, Gustav Bosse Verlag, 1973.

African musicology gained momentum, in particular thanks to Kwabena Nketia, who identified and analysed African musical traditions in *The Music of Africa* (1974). Over and above the huge diversity, he also revealed the striking unity in Africa. For the first time it became clear how much interchange there must have been between the musical traditions in the hundreds or even thousands of African empires. Nketia's analysis covered descriptions of musical traditions, music groups and their place in the community, and he also presented a detailed melodic, polyphonic, rhythmic, and harmonic examination of African music. The close relationship between music and language in Africa was described by David Rycroft in his study of Zulu, Swazi, and other languages. Rycroft developed a new, circular form of analysis with which he was able to describe the *overlapping question and answer structure* in African music.[214] In terms of the phrase structure we have discussed, this music can be represented in the form of overlapping phrases. The Gestalt principles of 'proximity' and 'similarity', which are deemed to be universal (see above), appear to apply to African music too. Although Nketia contends that phrase structures in Ghanaian music were *not* subject to 'rules',[215] it has meanwhile become clear that these phrase structures can still be predicted fairly accurately by the well-known Gestalt principles.[216] Research into African music did not really start flourishing until recent decades, and it promises a wealth without equal.

China and India: towards globalized musicology. Compared with the early modern age, nineteenth-century China and India produced few, if any, new trends in the study of music. The historiographical tradition of music was maintained, and in India the centuries-old custom of establishing musical systems of rules was continued. Musical practice in China changed radically after 1911, and Western influences became dominant. Even so, there was no musicology in the People's Republic for a long time. This did not change until a couple of decades ago in the form of 'globalized' musicology, which aside from historiography also focuses more and more on the analysis of Chinese music.[217] In twentieth-century India, on the other hand, we see a reverse shift from the dominance of theoretical treatises to a more historical approach. Indian music history was neglected for centuries (like all other types of history in India), but currently it is being studied far and wide.[218]

The state of musicology. For centuries, musicology was the outstanding example of an exact humanistic discipline, where the interaction between theory and empiricism, studying consonance for instance, rose to great heights. In Europe, historiography of music had a marginal existence (but not in China—see 4.4). The balance changed in the nineteenth century—music history was given a huge shot in

[214] David Rycroft, *Zulu, Swazi en Xhosa: instrumentale en vocale muziek*, Koninklijk Museum voor Midden-Afrika, 1969. See also David Rycroft, 'Nguni vocal polyphony', *Journal of the International Folk Music Council*, 19, 1967, pp. 88–103.

[215] Kwabena Nketia, *African Music in Ghana*, Northwestern University Press, 1963, p. 80.

[216] David Temperley, *The Cognition of Basic Musical Structures*, The MIT Press, 2001, pp. 286–90.

[217] See e.g. Sinyan Shen, *Chinese Music in the 20th Century*, Chinese Music Society of North America Press, 2001.

[218] See e.g. Swami Prananananda, *A History of Indian Music*, volume 1, Ramakrishna Vedanta Math, 1963.

the arm, while consonance research moved more towards (psycho)acoustics. It was only music analysis that continued at the same pace. For an extended period musicology was under the influence of different movements—Herderian, Comtean, Hegelian, Rankean, or Darwinian—but under Adler and Schenker musicology became a discipline with its own methodology. During the course of the twentieth century, musicology came under the influence of other movements once again. With some justification it can be called the most interdisciplinary humanistic discipline (with the possible exception of archaeology). It has an exact side (music analysis), a social-science side (music cognition), and a humanistic side (music history). The humanistic side has both pattern-seeking and pattern-rejecting movements. On the one hand there is empirical musicology, which is becoming progressively more dominant, and on the other there is the rise of the *new musicology*, which criticizes the empirical–nomothetic approach—although here other types of patterns (ideological and postcolonial) are being sought.

5.5 ART HISTORY AND ARCHAEOLOGY: TOWARDS A VISUAL PHILOLOGY

As in musicology, European art history saw the development of both historical and systematic components. The historical component developed in the universities, whereas initially the systematic, style-oriented component flourished outside academia. The two components were integrated during the early twentieth century, after which an iconological method for working out the underlying meaning of art was developed. At the end of the twentieth century, art history consisted of a multiplicity of approaches that were pattern-seeking, as well as some that were pattern-rejecting. Archaeology, on the other hand, was primarily pattern-seeking and developed in the direction of the social and natural sciences.

The historical component: steady liberation from classicist art theory. The early modern historiography of Vasari and Bellori was largely founded on classical ideals, but during the eighteenth and nineteenth centuries art history steadily freed itself from the classicist straitjacket (see 4.5 for a description of this process). The discipline was given academic foundations thanks to its institutionalization at the University of Berlin in 1834, with Franz Kugler as its first professor of art history.[219] The most acute question seemed to be the relationship between art history and cultural history.[220] The first nineteenth-century art historical works followed in the footsteps of Hegel and his philosophical history of the Spirit. In his posthumously published *Ästhetik*, Hegel argued that the history of art went hand in hand with the stages in his proposed development of the Spirit (see also 5.1 under Marx): the so-called *symbolic* era in which architecture was dominant, the *classical* era when sculpture was in the forefront, and the *romantic* era (which, according to Hegel, started as far back as the Middle Ages), when painting became central.[221]

Hegel's model was adopted in Franz Kugler's *Handbuch der Kunstgeschichte* ('Handbook of Art History') published in 1842. Although this work was no longer under the classical yoke, it was now squarely in the Hegelian world view. Art history was divided into four periods: pre-Greek, classical, 'romantic' (medieval), and modern (from the early modern age to the nineteenth century). Kugler proposed the Germans as the successors of the Greeks and considered medieval art to be the true expression of the people's feelings. He thought that what we nowadays call the Renaissance was derivative. As the person with overall responsibility for Prussian art policy, however, Kugler had progressively less time for his scholarly work, and he was able to rope in his pupil Jacob Burckhardt (1818–1897) to produce a new, drastic reworking of his *Handbuch*. In Burckhardt's hands the Hegelian world view

[219] Udo Kultermann, *Die Geschichte der Kunstgeschichte*, Ullstein Sachbuch, 1981. Translated into English as Udo Kultermann, *The History of Art History*, Abaris Books, 1993, pp. 89–91.

[220] Marlite Halbertsma, 'De geschiedenis van de kunstgeschiedenis in de Duitssprekende landen en in Nederland van 1764 tot 1933', in Marlite Halbertsma and Kitty Zijlmans (eds), *Gezichtpunten: een inleiding in de methoden van de kunstgeschiedenis*, SUN, 1993, p. 52.

[221] Georg Hegel, *Vorlesungen über die Ästhetik, Berlin 1820/21, Eine Nachschrift*, edited by H. Schneider, Peter Lang, 1995. For an English translation, see Georg Hegel, *Aesthetics: Lectures on Fine Art*, translated by T. M. Knox, Oxford University Press, 1975.

disappeared and in the second edition, published in 1848, the concept of 'Renaissance' was introduced.[222] In the third edition 'romantic' and Germanic were replaced by medieval and Gothic respectively. We have some justification for putting Jacob Burckhardt forward both as the 'discoverer' of the Renaissance and as the originator of cultural history (see 5.1). Burckhardt, who was Swiss, abhorred the nationalist agenda of the German scholars and concentrated on the rebirth of the classics which, he said, led to a new European culture.[223] At the University of Basel he focused on the history of forms of expression at the time of the Renaissance. In Burckhardt's hands the visual arts were looked on primarily as a source for historiography and less as a subject for independent study. Many of his works continue to be read, for instance *Die Kultur der Renaissance in Italien* (1860) and *Geschichte der Renaissance in Italien* (1867).[224] The scope of these works is breathtaking: all forms of Renaissance expression were addressed, from poetry and music to natural science, and from social etiquette and morality to religion. Burckhardt's works have something timeless, perhaps because they were devoid of the nationalist and philosophical fashions of the day.

Burckhardt was the first in a long and illustrious tradition of Swiss art historians. Impressive as his work was, however, it did not provide a set of methodical principles for *analysing* works of art. After his death Burckhardt was succeeded in Basel by Heinrich Wölfflin, then only twenty-eight, who was able to elevate art history from a historical to a more analytical, stylistically-oriented discipline. We will first discuss the creation of this stylistic component, which happened almost completely outside the universities.

The stylistic component: the visual philology of Morelli. In so far as we have talked about an *early* modern stylistic analysis (4.5), this existed primarily in normative treatises. In the fifteenth century Alberti described how the hierarchical stratification of a work of art was structured, but upon close scrutiny he only analysed *one* style: the 'ideal selection' (4.5). The comparison of several art styles is to be found in early modern China, but it did not receive much theoretical attention in the West. Comparative stylistic analysis did not really get going in Europe until the nineteenth century, and it was stimulated by the very lively trade in old art. A need developed for reliable attributions, which prompted the rise of 'visual philology'. Like textual philology, it tried to establish the date and maker (the 'authenticity') of a work using methodical rigour.

The first person to attempt to develop methodical stylistics was the Italian Giovanni Morelli (1816–1891). Trained in comparative anatomy, he applied his taxonomic classification principles to painting. The basic idea behind Morelli's method was that every artist has a personal style which betrays itself in the smallest,

[222] Franz Kugler and Jacob Burckhardt, *Handbuch der Kunstgeschichte*, Ebner & Seubert, 1848.

[223] John Hinde, *Jacob Burckhardt and the Crisis of Modernity*, McGill-Queen's University Press, 2000.

[224] For an English translation of Burckhardt's best known work, *Die Kultur der Renaissance in Italien*, see Jacob Burckhardt, *The Civilization of the Renaissance in Italy*, edited by Peter Murray, translated by S. G. C. Middlemore, with an introduction by Peter Burke, Penguin Classics, 1990.

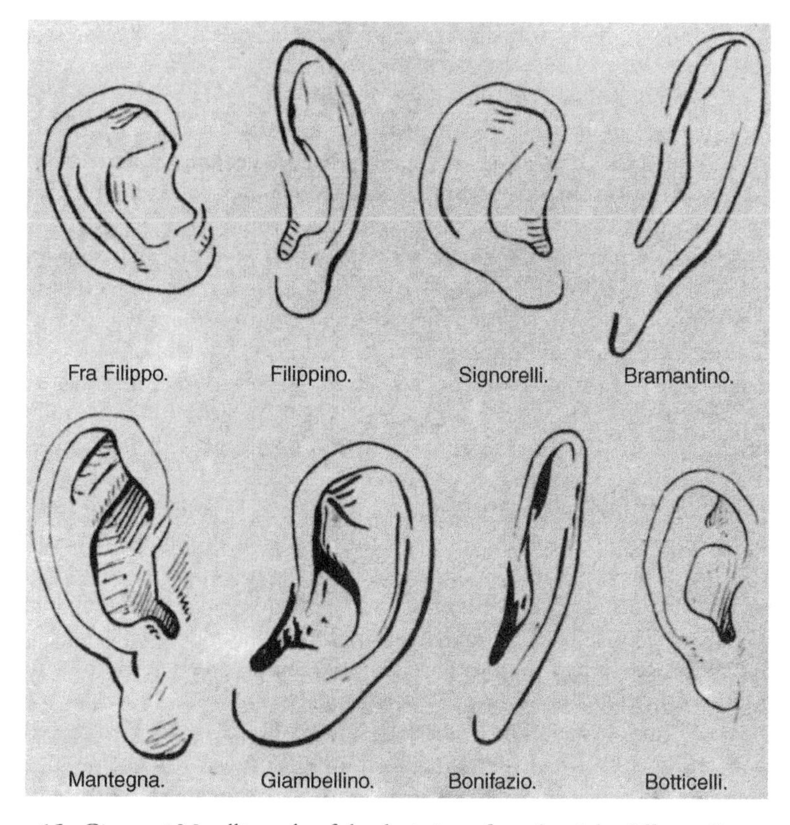

Figure 15. Giovanni Morelli, study of the depiction of ears by eight different Renaissance artists, in *Kunstkritische Studien.*

often insignificant details of a painting over which the artist has no control.[225] Even when an artist tried to paint in another style, his characteristic 'hand' could be recognized by comparing and classifying the painterly representation of ears, noses, hands, and other parts of the body, as well as clouds, leaves, folds, and individual brushstrokes (see Figure 15).[226]

Morelli also compared the poses of figures and the use of colour, but in his view these did not betray the hand of the artist and were barely usable for practical attribution. Morelli analysed works of art from collections in Italy and Germany, which resulted in hundreds of new attributions.[227] The pattern he found—that *the depiction of painterly details remains constant during an artist's career*—was widely embraced, and both connoisseurs and art historians were trained in the Morellian

[225] Richard Wollheim, *On Art and the Mind: Essays and Lectures*, Harvard University Press, 1972, chapter 9 ('Giovanni Morelli and the origins of scientific connoisseurship'), pp. 177–202.

[226] Giovanni Morelli, *Kunstkritische Studien über italienische Malerei*, 3 volumes, Brockhaus, 1890–3.

[227] Ivan Lermolieff (pseudonym of Giovanni Morelli), *Die Werke italienischer Meister in den Galerien von München, Dresden und Berlin*, Verlag von E. A. Seemann, 1880.

model. Morelli's method was also used in archaeology to classify Greek vases and reliefs.

The precision of Morelli's method gives rise to the question as to whether we are dealing with a system of rules here. This proves not to be the case. Morelli's visual philology is not defined in terms of rules, either procedural or declarative. His method has a strongly subjective component. Morelli applied taxonomy to painterly representations of noses, ears, hands, etc., and would use it to try to 'match' depictions in other paintings as closely as possible. However, this matching was not described in detail. Morelli's method is therefore based more on examples or analogies than on rules. In the same way as the classical Alexandrian philologists searched for textual similarities at word level (see 2.3), Morelli looked for visual resemblances at the limb—or even more detailed—level. And like the Islamic linguist Sibawayh, whose description of Arabic was realized not with rules but with examples (see 3.1), in his account of an artist's style, Morelli gives collections of detailed instances, not rules.

Morelli's attribution method achieved massive success.[228] It is in part thanks to his work that there is currently an art historical corpus of tens of thousands of accepted attributions. His attributions could moreover be corroborated or refuted if historical documents, such as deeds of sale, were discovered. Morelli's method also exerted influence outside art history circles. His idea that hidden 'meaning' could be found in the details was approvingly cited by Sigmund Freud and also by Arthur Conan Doyle through the words of Sherlock Holmes.[229]

Heyday of connoisseurship: from Berenson to Hofstede de Groot. Bernard Berenson (1865–1959) had met Morelli in 1890 and was to become his most famous follower. He was one of the first to set up a large archive of photographs of paintings, through which connoisseurship could be tackled even more systematically.[230] It had to be possible to organize works of art, like plants, on purely formal grounds without historical sources. Berenson built up a huge reputation as an art expert, in part because of his manuals about Italian paintings and drawings, of which *Drawings of the Florentine Painters* (1903) is unsurpassed in terms of both scope and the quality of his opinions as a connoisseur. The application of Morelli's method proved to be a very lucrative activity. Astronomical sums were being paid in the United States for Italian masters. Berenson's attribution could make or break a painting, and his five per cent commission made him an immensely rich man. However, accusations of fraud against Berenson resulted in the discrediting of connoisseurship in the twentieth century.[231] All the same, Berenson's photographic archive is still being used and kept in his magnificent Tuscan home *Villa I Tatti.*

[228] Wollheim, *On Art and the Mind: Essays and Lectures*. See also Halbertsma and Zijlmans, *Gezichtpunten: een inleiding in de methoden van de kunstgeschiedenis*.

[229] Carlo Ginzburg, 'Morelli, Freud, and Sherlock Holmes: clues and scientific method', in Umberto Eco and Thomas Sebeok (eds), *The Sign of Three: Dupin, Holmes, Peirce*, Indiana University Press, 1983, pp. 81–118.

[230] Ernest Samuels, *Bernard Berenson: The Making of a Connoisseur*, Belknap Press, 1979.

[231] Mary Ann Calo, *Bernard Berenson and the Twentieth Century*, Temple University Press, 1994.

Photographic archives were to play an increasingly important role in the study of art. In the Netherlands the art connoisseur Cornelis Hofstede de Groot (1863–1930), director of the national print room, the Rijksprentenkabinet, built up a huge private archive of illustrations of Dutch works of art, catalogues, and documentation.[232] His attributions, though, led to fierce controversies on more than one occasion, and even to disgrace when he declared a painting to be an authentic work by Frans Hals even though chemical analysis had shown that twentieth-century pigments had been used in it (which Hofstede de Groot ascribed to later additions).[233] Currently the panel concerned is thought to be one of Han van Meegeren's first forgeries. Slowly but surely it became clear that art connoisseurship was inadequate without historical or iconological support (see below). In addition, natural scientific methods started to play a progressively greater part. Nevertheless, photographic archives remained indispensable and Hofstede de Groot's private collection formed the basis for the Netherlands Institute for Art History (RKD), which opened its doors in The Hague in 1932. It is the largest art historical documentation centre in the world.

Integration of historical and stylistic components: Wölfflin's structuralism and the Vienna School. Although Heinrich Wölfflin (1864–1945) was a pupil of Burckhardt, his work must also be seen in the context of Morelli and Berenson.[234] Despite Morelli's success, his method did not prove usable for a stylistic analysis of a painting as a whole. It was largely Morelli's emphasis on the tiniest style units that was to blame.[235] After all, no single visual element can be seen as separate because it is linked to the other components of a painting. Contrary to Morelli's detail-based approach, Wölfflin developed an analytical method in which it was not only all the separate parts of the work that were examined, but also their relationship with the whole, as well as the use of light and colour. This method appeared to dovetail with Alberti's fifteenth-century *disegno* theory, which also focused in the part–whole relationships in a painting (see 4.5). But where Alberti used his *disegno* as a normative, prescriptive tool, Wölfflin was interested in a descriptive analysis. In so doing, Wölfflin 'discovered' new style periods in art, which gave his stylistic work a historical component that was missing in Morelli.

True to the tradition of Herder and Hegel, Wölfflin still looked on style periods as a pattern of rise, zenith, and ultimately decline. However, he was the first to introduce the term *baroque* as a designation for the art style that developed in Europe after the Renaissance. Initially Wölfflin also considered the sixteenth-century movement currently known as mannerism as baroque. His predecessors

[232] Jan Emmens and Simon Levie, 'The history of Dutch art history', in Jan Emmens, *Kunsthistorische opstellen*, collected works, volume 2, Van Oorschot, 1981, pp. 35–50.

[233] Cornelis Hofstede de Groot, *Echt of onecht? Oog of chemie? Beschouwingen naar aanleiding van het mansportret*, Van Stockum, 1925.

[234] Meinhold Lurz, *Heinrich Wölfflin: Biografie einer Kunsttheorie*, Werns, 1981. See also Joan Hart, *Heinrich Wölfflin: An Intellectual Biography*, PhD thesis, UC Berkeley, 1981.

[235] Gerrit Willems, 'Verklaren en ordenen: over stijlanalytische benaderingen', in Marlite Halbertsma and Kitty Zijlmans (eds), *Gezichtpunten: een inleiding in de methoden van de kunstgeschiedenis*, SUN, 1993, p. 127.

Burckhardt and Berenson had dubbed these movements as degenerate, but Wölfflin decided not to make normative comments. In line with Saussure's structuralism (see 5.3), in his *Kunstgeschichtliche Grundbegriffe* (1915),[236] Wölfflin introduced a gamut of new stylistic concepts that he grouped in five pairs of opposites in order to characterize the style transition from Renaissance to baroque:

1. From a *linear* to a *painterly* method (where the linear technique defined the contours sharply whereas the painterly approach was based on blurred transitions).
2. From a *flat* (two-dimensional) to a *deep* (three-dimensional) composition.
3. From a *closed* to an *open* form.
4. From *multiplicity* to *unity* in the parts of the composition.
5. From *clear* to *diffuse* in the pictorial representation.

As in structuralist linguistics, the use of these opposites corresponds to the *relations of differences principle* (see 5.3).

Wölfflin's concept of style periods profoundly influenced the study of such periods in the other branches of the humanities. His notion of baroque matched the historiography of architecture and sculpture, and also the historical analysis of music (see 5.4), literature (5.6), and theatre (5.6). Alongside the broad impact, Wölfflin's method was also criticized, particularly by Walter Benjamin, who in his essay *Strenge Kunstwissenschaft* (1933) argued that Wölfflin ignored the *social* and *cultural* backgrounds that launch a style change.[237] The Vienna School of Art History, which came into the ascendant in the 1920s, added a social–cultural perspective to Wölfflin's formalist approach.[238] Art historians like Alois Riegl and Franz Wickhoff, who were trained in the Morellian model, continued Wölfflin's work and were able to rehabilitate other neglected style periods—mannerism, early Christian art, and late Roman art.

From form to meaning: Panofsky and iconology. During the 1920s a new school developed in both Germany and the Netherlands that focused on the meaning rather than the form of the subject depicted. Godefridus Hoogewerff initiated this tradition in the Netherlands, and in Germany it was instigated by Aby Warburg, whose immense library was to form the core of the world-famous Warburg Institute in London.[239] The most important exponent, though, was Warburg's pupil Erwin Panofsky (1892–1968), who put iconology on the map as an independent art-historical method.[240] In 1933, driven by Hitler's racial laws,

[236] For an English translation, see Heinrich Wölfflin, *Principles of Art History: The Problem of the Development of Style in Later Art*, translated by M. D. Hottinger, Dover Publications, 1932.

[237] For an English translation, see Walter Benjamin, 'Rigorous study of art', translated by Thomas Levin, *October*, 47, 1988, pp. 84–90.

[238] Christopher Wood (ed.), *The Vienna School Reader: Politics and Art Historical Method in the 1930s*, Zone Books, 2003.

[239] Ernst Gombrich, *Aby Warburg: An Intellectual Biography*, The Warburg Institute, 1970.

[240] Michael Holly, *Panofsky and the Foundations of Art History*, Cornell University Press, 1985.

Figure 16. Jan van Eijck, *Arnolfini Wedding*, 1434.

Panofsky fled to the United States, where he made a very fruitful impact on the country's emerging art history. Panofsky was fascinated by the assumed deeper symbolic meaning in art. For instance, he asserted that Jan van Eyck's famous *Arnolfini Wedding* painted in 1434 (see Figure 16) depicted more than a wedding ceremony and was also a visual contract of the act of marriage.[241] Panofsky believed he could unravel a large number of hidden symbols that referred to the marriage sacrament—the little dog in the foreground represented fidelity, the oranges on the table symbolized purity, and the cast-aside shoes indicated that the pair were standing on sacred ground. Some of these interpretations seem far-fetched, but Panofsky was able to substantiate most of them on the basis of historical sources

[241] Erwin Panofsky, 'Jan van Eyck's Arnolfini Portrait', *The Burlington Magazine for Connoisseurs*, 64(372), 1934, pp. 117–19 and 122–7.

that established that these symbolic meanings were fairly generally known in the fifteenth century.

Panofsky systematically explained his interpretation method in *Studies in Iconology* (1939). He defined three levels of analysing the meaning of a work of art:[242]

1. *Primary or natural subject matter.* Broadly speaking, this corresponds to Wölfflin's form analysis. First and foremost the description required practical experience (for example, familiarity with objects and events, which seems like the 'knowledge of all visible things' that we came across in early modern art treatises, see 4.5). This pre-iconographic description moreover had to be supported by knowledge of the style history that had been initiated by Wölfflin.
2. *Iconographic analysis.* This included determining the subject of the painting in terms of the figures, stories, and allegories. The requirements for this are knowledge of literary sources on the one hand and art historical typology (the history of the themes and motifs in pictorial tradition) on the other.
3. *Iconological interpretation.* This contained the deeper significance of the painting, including the symbolic references. This calls for art historical expertise, but also, and above all, the 'synthetic intuition' that springs from psychological insight and a thorough knowledge of relevant world views *from the period when the work of art was created.*

Panofsky emphasized that these levels cannot be considered separately from one another. Only the exhaustive analysis of the three levels can satisfactorily reveal the meaning of the work. Naturally Panofsky did not offer any formal system of rules for analysing and interpreting art, but his methodical principles have kept several generations of art historians employed.

For example, the Dutch art historian Eddy de Jongh argued that Dutch seventeenth-century *genre painting*, which apparently depicts aspects of everyday life, represented a form of 'seeming realism'. In the realistic looking representation of an everyday scene there were moralizing meanings that could be correlated to contemporary texts.[243] An apparently true-to-life scene in an inn by Jan Steen acquires a moralizing meaning if we spot a boy blowing bubbles next to a skull. In the well-known seventeenth-century emblem books (like those by Roemer Visscher and Jacob Cats) the soap bubble can be a symbol for the transience of life that can burst at any moment. Jan Steen's painting therefore serves, as the popular seventeenth-century saying has it, 'to teach and delight', a view which dates back to Horace. However, critics of the iconological method argue that De Jongh's and Panofsky's method sets no limits for symbolic interpretation. Every depicted object

[242] Erwin Panofsky, *Studies in Iconology: Humanistic Themes in the Art of the Renaissance*, Oxford University Press, 1939, reprinted by Harper & Row, 1972, pp. 5ff.

[243] Eddy de Jongh, 'Realisme en schijnrealisme in de Hollandse schilderkunst van de zeventiende eeuw', *Rembrandt en zijn tijd*, Tentoonstellingscatalogus Brussel, 1971, pp. 143–94. Translated into English as Eddy de Jongh, 'Realism and Seeming Realism in Seventeenth-Century Dutch Painting', in Wayne Franitz (ed.), *Looking at Seventeenth-Century Dutch Art: Realism Reconsidered*, Cambridge University Press, 1998.

can be interpreted symbolically, whereas much Dutch painting probably had no goal other than to record the visible reality (see for example the concept of 'picturesque' in seventeenth-century art treatises in 4.5).[244] Even so, Panofsky's method changed art history. In so far as symbolic interpretations can be supported by emblem books, his method had put the modern viewer on the track of *meaning patterns* that were not picked up by previous generations of art historians. Although these meaning patterns are culture-specific, the method used to find them is not.

Panofsky also applied his iconographic method to longer periods. He was pioneering in his comparison of rebirths or 'renascences' in the history of the visual arts, such as Carolingian Renaissance, the tenth-century Ottonian Renaissance, the twelfth-century Renaissance, and the fifteenth-century Italian Renaissance. In *Renaissance and Renascences in Western Art* (1960) Panofsky showed that only the fifteenth-century Renaissance was comprehensive and lasting. While classical themes and elements were used in earlier rebirths, it was temporary and testified to nostalgia for Antiquity. According to Panofsky, only the fifteenth-century Renaissance displayed a unity of form and content.[245] This arose from the awareness that Antiquity was definitively in the past and that a literal return was pointless.

Towards a cognitive-historical approach: Gombrich. As it did in linguistics and musicology, the psychological–cognitive approach also blossomed in art history. Now art analysis could be tackled on the basis of the psychology of the viewer rather than from a purely historical perspective. This technique was initiated in the Vienna School we discussed above, in which Ernst Gombrich and Rudolf Arnheim were some of the leading exponents. In the same way as Lerdahl and Jackendoff tried to get a grip on musical perception (see 5.4), Gombrich and (in a more psychologically-oriented way) Arnheim attempted to get a hold on how the visual arts were observed by the viewer.[246] However, the units of visual stimuli are much more difficult to pin down than those of musical or linguistic stimuli. In music and language the basic units consist of a relatively small number of notes and phonemes, whereas in the case of a visual stimulus, almost all things visual can be basic units and there is no prospect of a definition, despite Morelli's earlier attempts. Although Gombrich convincingly explained in his *Art and Illusion* (1960) how the interpretation of a work of art is influenced by the automatisms with which a two-dimensional projection surface is regarded as a three-dimensional space, his work did not result in a precise method for art analysis.[247] In 1950, however, Gombrich did assimilate his ideas into what many consider to be the most approachable

[244] See e.g. Svetlana Alpers, *The Art of Describing: Dutch Art in the Seventeenth Century*, University of Chicago Press, 1983.

[245] Erwin Panofsky, *Renaissance and Renascences in Western Art*, Harper and Row, 1972, p. 113.

[246] Rudolf Arnheim, *Art and Visual Perception: A Psychology of the Creative Eye*, University of California Press, 1954.

[247] Ernst Gombrich, *Art and Illusion: A Study in the Psychology of Pictorial Representation*, Phaidon, 1960. See also Ernst Gombrich, *The Image and the Eye: Further Studies in the Psychology of Pictorial Representation*, Phaidon, 1982.

introduction to art history, *The Story of Art*, which he continued to update until shortly before his death in 2001.

Computational and natural scientific methods in art analysis. When the digital, computational analysis of art began to develop in the second half of the twentieth century, a formal theory of visual stylistics appeared to be within reach. Algorithms from *computational image analysis* were used to anatomize the use of light and colour, brushstrokes, and perspective from different periods.[248] Although the computational approach to more complex concepts such as composition and interpretation is as yet underdeveloped, the computational technique already has several achievements to its name.[249] Using extremely detailed digitized scans of paintings, computers can statistically analyse the texture and the use of light and perspective with greater precision than a trained art expert or artist can. Insights have emerged about the use of perspective by Jan van Eyck and the application of light by Caravaggio that have disproved earlier theories.[250] No matter how much the digital, computational approach is in its infancy, it provides a powerful tool for testing hypotheses relating to painting. We saw the potency of this technique earlier in the development of algorithms for stemmatic philology, linguistic analysis, and music analysis. The computational approach in the humanities has led to a new interdisciplinary field that is usually referred to as *digital humanities*, though the term *computational humanities* may be more appropriate here (see also 5.6). Theories and ideas from the humanities can now be tested faster and more precisely than ever, and new patterns are being discovered that would probably never have been found manually.

The natural science approach to art is much older. As long ago as the beginning of the twentieth century a start was made on using X-ray analyses of paintings in order to make underdrawings visible. This was soon followed by infrared photography, analysis with ultraviolet light and other methods,[251] including panel investigations into tree ring dating. Meanwhile, natural science techniques have become indispensable to studying underdrawings or underlying paint layers of art works, although it continues to be people who have to interpret underdrawings. It is fascinating to see how natural science expertise is contributing to more reliable attributions or to a better informed interpretation of a work of art. Scientific methods can be decisive in major research projects like the Rembrandt Research Project in the attribution of a painting to Rembrandt—although these methods have been at loggerheads with more traditional art history more than once.[252]

[248] See e.g. *Proceedings of Computer Vision and Image Analysis of Art*, SPIE, 2010–12.

[249] David Stork, 'Computer image analysis of paintings and drawings: an introduction to the literature', *Proceedings of the Image Processing for Artist Identification Workshop*, Van Gogh Museum Amsterdam, 2008.

[250] Stork, 'Computer image analysis of paintings and drawings: an introduction to the literature', 2008.

[251] See e.g. B. Keisch, R. Feller, A. Levine, and P. Edwards, 'Dating and authenticating works of art by measurement of natural alpha emitters', *Science*, 155, 1967, pp. 1238–41.

[252] Ernst van de Wetering, 'Thirty years of the Rembrandt Research Project: the tension between science and connoisseurship in authenticating art', *IFAR Journal*, 4(2), 2001, pp. 14–24.

A multiplicity of approaches and the *new art history*. As is the case in musicology, art history benefits from a multiplicity of methods and techniques. Aside from the stylistic and iconological approach (which may be specific to art history) and the methods already discussed from cognitive psychology, natural science, and information technology, there is also an anthropological–Marxist (Meyer Shapiro), psychoanalytical (Sigmund Freud), and sociological (Arnold Hauser) study of art. It is not entirely surprising that—parallel to *new musicology* and *new cultural history*—*new art history* is now also on its way.[253] This is an alliance of common interests and schools rather than an independent discipline. Criticism is an important goal, and the criticism is justified. As Edward Said showed in *Orientalism* (1978), it is extremely dubious to apply Western standards to other cultures, including its artistic expressions. *New art history* looks at art from a number of different perspectives—a Marxist approach but also a deconstructivist, feminist, and postcolonial method. This form of pluralism, though, is a very long way from the computational and natural scientific techniques, possibly because of the alleged universalist claims of the latter. Perhaps we can state that there are two art histories at the beginning of the twenty-first century, but it is not easy to label them as pattern-seeking or pattern-rejecting.[254] For example, pattern-seeking methods of analysis are also used in new art history; one such is the use of *narratology* in art interpretation, which has its roots in literary theory (more about this in 5.6).[255]

China and other regions. At the beginning of the Qing Dynasty (1644–1912) art and style criticism was still being produced on an impressive scale (see 4.5), but in the late Qing era the number of art literati appeared to gradually decrease.[256] The literati level reached a nadir when the gates of the Qing Dynasty were violently opened during the Opium Wars (1839–1860). The arts flourished in 'free' cities like Shanghai, but their scholarly study seemed not to have taken place. Art theory did not receive a new impulse until the age of Republican China (1912–1949) and above all the People's Republic, but it was soon forced into a Marxist straightjacket of socialist realism. There was a de facto reduction in the number of art theoreticians in China to one—Mao Zedong. The absolute low point was the Cultural Revolution (1966–1976), when the repression and intimidation instigated by Mao's fourth wife Jiang Qing made the study of art completely impossible. Until well into the twentieth century the scholarly study of Chinese art took place primarily outside China. The first historical overview of Chinese art, which appeared in 1912, was written by the American art historian Ernest Fenollosa.[257]

[253] Jonathan Harris, *The New Art History: A Critical Introduction*, Routledge, 2001.

[254] Charles Haxthausen (ed.), *The Two Art Histories: The Museum and the University*, Yale University Press, 2003.

[255] See e.g. Mieke Bal, *Reading Rembrandt: Beyond the Word-Image Opposition*, 2nd edition, Amsterdam University Press, 2006.

[256] See e.g. Susan Brush, *China, Painting Theory and Criticism*, Oxford Art Online, Oxford University Press, 2004; Sirén, *The Chinese on the Art of Painting: Texts by the Painter-Critics, from the Han through the Ch'ing Dynasties*.

[257] Ernest Fenollosa, *Epochs of Chinese and Japanese Art*, Heinemann, 1912.

Other regions were also studied. At the University of Calcutta the Austrian Stella Kramrisch became the first professor of Indian art history. In 1924 she laid the foundations for the systematic study of Indian art in her book *Principles of Indian Art*. She was able to rescue the complex systems of rules in the *Vishnudharmottara* dating from about 400 CE (see 2.5) from obscurity.[258] The history of art in other regions was also recorded, initially from a strongly colonial point of view, but by the end of the twentieth century a more local perspective had become increasingly frequent.

The rise of archaeology. The early modern history of archaeology is fragmentary, as we outlined in our discussion of Flavio Biondo as the first early modern archaeologist (and as the discoverer of the material source principle). There were archaeological activities as far back as ancient China and Greece, as well as in Islamic civilization (see the discussion in 4.2). During the Renaissance, Raphael and Michelangelo descended into Nero's *Domus Aurea* in Rome to study the classical frescos,[259] and Pirro Ligorio identified and analysed the remains of the Villa Hadriana in Tivoli.[260] The search for historical artefacts and works of art, though, was generally speaking nothing more than foraging, which continued up to the eighteenth-century excavations of Herculaneum and Pompeii. The only exception might have been Thomas Jefferson's meticulous digging up of a burial mound on his Virginia estate (1784).[261] Greater continuity and more systematic foundations for the discipline did not emerge until the nineteenth century and much more so in the twentieth century. As such, archaeology became related to history, philology, linguistics, and above all art history because of its initial focus on works of art.

Where and when did archaeology develop into a field with *methodical principles*? When he discovered Troy and Mycenae between 1870 and 1880, Heinrich Schliemann, to the great irritation of his colleagues, asserted that all he had done was follow Homer's *Iliad* step by step. Be that as it may, Schliemann's procedure is seen as the first archaeological method of the modern age: *text-based archaeology*.[262] This approach resulted in new successes. The Mesopotamian city of Ur was discovered as a result of a close reading of the Bible. Yet most archaeological finds were still unearthed by coincidence. Moreover, all nineteenth-century archaeologists were amateurs.[263] This was also true of the most important people in the field around 1900, from Howard Carter, who laid bare Tutankhamen's tomb, to Arthur Evans, the discoverer of the Minoan civilization on Crete. A new practice developed during the course of the twentieth century after the institutionalization of archaeology at the modern universities. It became obligatory for infrastructure

[258] Barbara Stoler Miller, *Exploring India's Sacred Art: Selected Writings of Stella Kramrisch*, University of Pennsylvania Press, 1983.

[259] Ida Sciortino and Elisabetta Segala, *Domus Aurea*, Electa Mondadori, 2006.

[260] C. David Coffin, *Pirro Ligorio: The Renaissance Artist, Architect, and Antiquarian*, Penn State Press, 2003.

[261] William Kelso, *Archaeology at Monticello*, UNC Press Books, 2002.

[262] Bruce Trigger, *A History of Archeological Thought*, Cambridge University Press, 1996, pp. 40ff.

[263] Trigger, *A History of Archeological Thought*.

construction projects to be preceded by an archaeological site investigation. Aerial photography was also used to systematically search for archaeological patterns. Even *Google Earth* has been used for archaeological air surveys, and amateur archaeologists have once again had a significant role in the discovery of previously unknown Roman ruins.[264]

Once an archaeological site had been found, the standard method involved an excavation. The digging up of Herculaneum in the early eighteenth century was nothing short of destructive, but in fact so is any excavation. They are based on an assumption derived from *stratigraphy*, that is to say where there is one layer on top of another, the deeper layer is older than the one above it.[265] This assumption is not without exceptions, and dating techniques are therefore essential. For a long time dating was based primarily on philological sources (see 4.1 and 4.2), but these were of no use when it came to prehistoric cultures. It was established as early as 1878 that clay layers corresponded to the ice layers of glaciers. This enabled new dating that went back to the last ice age, some 12,000 years ago. Dendrochronology (tree ring dating) emerged at the beginning of the twentieth century, and this provided the possibility of making more refined estimates.[266] However, the most spectacular breakthrough came with the discovery of the carbon-14 dating method in 1949 by Willard Libby. It uses the principle of radioactive decay.[267] After its death, a plant or animal stops absorbing carbon from the atmosphere, after which the unstable carbon-14 (C-14) decays at a very precise and regular rate. The age of vegetable or animal material that is found in or on an archaeological artefact can then be calculated by establishing the quantity of C-14 that is still present. In this way, artefacts can be dated up to 80,000 years ago. This method resulted in a revolution in archaeology and drastically improved our knowledge of the oldest human cultures in a spectacular way. It causes a media sensation whenever an unearthed statue proves to be tens of thousands of years old.[268]

Natural science techniques are useful for dating, but they provide little to go on when it comes to the interpretation of cultures and artefacts. The most widely-used interpretation methods are squarely in the humanities tradition. As in historiography, linguistics, and musicology, the approaches of virtually all nineteenth-century movements have been utilized in archaeology, from the Herderian cycles and the Marxist ages to frankly racist evolutionary construals. Until well into the twentieth century the Western belief in the inferiority of African peoples was so strong that all impressive archaeological finds south of the Sahara were attributed to influences from the north (see also 5.1 under Africa). A typical example is the

[264] See e.g. Declan Butler, 'Enthusiast uses Google to reveal Roman ruins', *Nature*, NatureNews 14 September 2005.

[265] Edward Harris, *Principles of Archaeological Stratigraphy*, 2nd edition, Academic Press, 1989.

[266] Colin Renfrew and Paul Bahn, *Archaeology: Theories, Methods and Practice*, 4th edition, Thames & Hudson, 2004, pp. 144–5.

[267] Sheridan Bowman, *Interpreting the Past: Radiocarbon Dating*, University of California Press, 1990.

[268] See for example 'Full-figured statuette, 35,000 years old, provides new clues to how art evolved', in *New York Times*, 13 May 2009.

nineteenth-century discovery, or rediscovery, of the monumental city of Great Zimbabwe, which flourished from the fifth to the sixteenth centuries.[269] Until 1980 the colonial Southern Rhodesian government banned any reference to the black origins of this city, yet over two hundred of these archaeological finds are known in Zimbabwe alone.[270]

Nonetheless, patterns that have withstood the test of time were also found in the nineteenth century. One of the most influential of these is the system of three prehistoric ages—the Stone Age, the Bronze Age, and the Iron Age—that was introduced in 1848 by the Dane Christian Jürgensen Thomsen.[271] Thomsen based these categories on the materials used to make the artefacts that were found. Nevertheless, these ages are regularly misused, for example when some contemporary cultures, particularly in the popular media, are relegated to the 'Stone Age' because of their technological situation. During the course of the twentieth century a need arose for a more scientific basis for archaeological interpretations, which in around 1960 resulted in an adoption of an anthropological method. This *processual archaeology* is based on testing specific hypotheses, as is normal in the social sciences.[272] Because of its alleged positivist approach this archaeology has been criticized by what is currently known as *post-processual archaeology*.[273] This latter movement is under a postmodernist influence and, as in the case with *new art history*, it is not so much one method but a conglomerate of them with a common goal of creating a comparative archaeology. Post-processual archaeology has in turn been criticized for a lack of scientific character.

Although archaeology is normally assigned to the humanities faculties, in the meantime, many of its methodical principles have been put under natural or social science. Currently archaeology is considered to be an interdisciplinary field where knowledge of natural sciences, social sciences, and humanities are combined in order to reconstruct the past.

Art history in the modern period: from prescriptive to descriptive. Several aspects of art history run parallel to musicology, for example in the creation of the historical–stylistic approach, the cognitive–computational movements and not least in the ascent of the *new musicology/new art history*. The most important development, however, has been the return to a more descriptive history of art. Until far into the eighteenth century, art historians searched for exact proportions and universal rules for the 'right' art. In the latter part of the eighteenth century and above all in the nineteenth, people began to realize how damaging this quest had been. The baroque and other styles were largely overlooked, and many artists were

[269] See e.g. Peter Garlake, *Early Art and Architecture of Africa*, Oxford University Press, 2002.

[270] Innocent Pikirayi, *The Zimbabwe Culture: Origins and Decline of Southern Zambezian States*, Rowman Altamira, 2001.

[271] Bo Gräslund, *The Birth of Prehistoric Chronology: Dating Methods and Dating Systems in Nineteenth-Century Scandinavian Archeology*, Cambridge University Press, 1987, pp. 17ff.

[272] Gordon Willey and Philip Phillips, *Method and Theory in American Archeology*, University of Alabama Press, 2001.

[273] Michael Shanks and Christopher Tilley, *Re-Constructing Archaeology: Theory and Practice*, Routledge, 1987.

brushed aside because they did not keep to the classical canon. If there was a significant development in art history, then it was not so much liberation from the humanistic as from the prescriptive yoke.

The present-day status of art history is diffuse. Scholars are using physical, chemical, computational, and of course historical, stylistic, and iconographic methods for attributing, interpreting, and reconstructing works of art.[274] In its quest for attribution and reconstruction, the discipline of art history somewhat resembles *forensic science*, which is at home in all markets and employs all available resources to identify the 'perpetrator' and reconstruct the 'offence'. Even so, the natural scientific methods do not have the same status in art history as they do in musicology. A number of phenomena in musicology actually are physical or physiological (such as dissonances, see 5.4). It is much more difficult to find such physical phenomena in the history of art, and we should rather talk about the use of natural scientific methods as 'tools'. There is moreover a strongly anti-positivist movement in art history, the *new art history*, which has become especially well established in the United States. There appears to be no collaboration between the two traditions.

[274] See for example the range of techniques and methods used in the reconstruction of one work of art: Machtelt Israëls (ed.), *Sassetta: The Borgo San Sepolcro Altarpiece*, Villa I Tatti Primavera Press, 2009, pp. 161–203.

5.6 LITERARY AND THEATRE STUDIES: THE CURIOUS DISAPPEARANCE OF RHETORIC AND POETICS

While (stemmatic) philology was searching for the rules of text reconstruction (5.2) and linguistics for regularities in languages (5.3), literary theory was seeking patterns in literary productions like poetry and novel-writing. To an extent, literary theory was the heir of classical philology as well as poetics—a position that it shared with theatre studies. The once flourishing rhetoric, a 'science of everything', was absorbed in argumentation theory and continued to live on in part in literary studies.

Literary historiography: from positivism to formalism. The damage done by early modern classicist poetics was immense (see 4.7). All literature, poetry, and theatre had to comply with prescriptive, mechanical rules in order to claim universal assumed beauty and clarity. By contrast, the flourishing baroque literature and poetry remained completely unstudied. During the course of the eighteenth century this vision became untenable and we come across attempts to study literature from a broader and more equal perspective (see 4.7). During the era of Romanticism there was great interest in everything associated with the national past. Medieval texts were rediscovered and, after reconstruction, published (see 5.2). A need developed to place these texts in a historical overview so that the cultural heritage could be seen in perspective.

The first literary histories in modern Europe were primarily bibliographic. They gave an endless list of writers and their works, with many divisions and subdivisions, of which the *Grundriss* ('Outline', 1795–8) by Erduin Koch is an example.[275] A generation later, literary historiography looked very different. For example in 1835 Georg Gervinus wrote a nationalist-leaning history of German literature (*Geschichte der poetischen National-Literatur der Deutschen*). Instead of a bibliographic summary, the reader was served up with a story that began with a glorification of German medieval literature, which was slowly but surely corrupted by the clergy, was then controlled by scholars during humanism, and was finally liberated in the eighteenth century by critics.

What we see is that nineteenth-century literary histories, like histories of art and music, were under the influence of the philosophical fashion of the day. They were Herderian (see 4.2), Hegelian (see 5.3), or Darwinian, but in almost all cases positivist.[276] These histories wanted to show how literary works could be *explained*, either based on the spirit of the age or the life of the author. Francesco de Sanctis's influential history of Italian literature (*Storia della letteratura italiana*, 1870) is an interesting example.[277] His historical overview runs as

[275] Erduin Koch, *Grundriss einer Geschichte der Sprache und Literatur der Deutschen von den ältesten Zeiten bis auf Lessings Tod*, 2 volumes, Königliche Realschulbuchhandlung, 1795–8. Erduin Koch should not be confused with the musicologist Heinrich Koch—see 5.4.

[276] See e.g. Michael Batts, *A History of Histories of German Literature, 1835–1914*, McGill-Queen's University Press, 1993.

[277] For an English translation, see Francesco de Sanctis, *History of Italian Literature*, translated by Joan Redfern, 2 volumes, Oxford University Press, 1930.

one causal chain from the early Sicilian and Tuscan poets to Manzoni, by way of Dante and the Renaissance. The fact that de Sanctis's literary history was a typical product of his time was tellingly expressed by his loyal pupil Benedetto Croce, who commented that all history was contemporary history (see 5.1).

We find an example of a Darwinian literary history in the work of Ferdinand Brunetière, who in 1890 described the evolution of French literature from the tenth to the nineteenth centuries.[278] Brunetière looked on the novel as a genre that developed at the expense of other genres (tragedy, lyric poetry), and consequently reached a progressively higher level. Hippolyte Taine (1828–1893) went the furthest with his *naturalist* description of literary history in his *Histoire de la littérature anglaise* ('History of English Literature').[279] According to Taine every piece of writing was completely causally defined by three factors: *race, milieu*, and *historical moment*. If one knew these three factors, one could explain and possibly even predict the writer and his work. In Taine's opinion all human expression was the result of these three factors, which in turn were nothing other than the outcome of chemical and physical processes. Although this last outlook is not unheard-of in present-day natural science, it ignored the existence of autonomous levels of explanation: in order to understand a work, it is not always necessary—let alone possible—to reduce it to physics or chemistry (see chapter 6).

While nineteenth-century literary histories differed from one another in many ways, they were based on one common methodological principle—*a work can be explained causally* (Darwinian, naturalistic or by way of the Herderian *Zeitgeist* and *Volksgeist*—see 4.2 and 5.1). The 'causal explanation principle' is a typical nineteenth-century phenomenon. Present-day literary theory does not consider explanations of literary works to be feasible. On this point Wilhelm Dilthey appears to have been right when he contended that the humanities are not concerned with explaining an expression of the human mind (*erklären*) but understanding it (*verstehen*). At the same time, though, Dilthey criticized the notion of searching for patterns in such human expressions as literary works. He was proved to be less correct on this point. Pieces of writing from one and the same period or region normally display common features, and revealing these has been, and is still being, successfully undertaken by humanities scholars. Searching for patterns in works of literature does *not* imply that they are (or have to be) causally explained.

This last point was recognized by a new school of literary theorists—the *formalists*—who were active in Russia between 1915 and 1930.[280] Viktor Shklovsky and Yury Tynyanov were among their most important representatives. In line with Saussure's structuralist programme (5.3), they searched for *internal* regularities in literary works, such as form characteristics and their effects, and not for *external* laws that could clarify the creation of the works. The efforts of these

[278] Ferdinand Brunetière, *Études critiques sur l'histoire de la littérature française (1849–1906)*, 8 volumes, 1880–1907.

[279] For an English translation, see Hippolyte Taine, *History of English Literature*, translated by H. van Laun, 2 volumes, T. Y. Crowell, 1873.

[280] Peter Steiner, *Russian Formalism: A Metapoetics*, Cornell University Press, 1984.

formalists resulted in a new vision of literature that reflects how literary works have developed over time in terms of style and content. Tynyanov, for instance, argued that a literary style period was determined by a system of dominant forms, rules, and genres. This did not make other movements and genres impossible, but they were not at the centre of the period concerned.[281]

Of course, Russian formalism, like any theory, had shortcomings, but the net effect on literary theory was that literary historiography became more stylistic and less positivist. The formalists were opposed to psychological explanations, interpretations based on biographies of authors, and Marxist literary views. However, during the 1920s the formalists became less and less tolerated under Stalin, after which their method was continued by the Prague structuralists. They included the linguist Roman Jakobson (see 5.3), who had started his career with the Petersburg formalists. The influence of the formalists was greater than is often surmised. The concept of a piece of writing as an autonomous work can be traced back to them, and it is not uncommon to date literary theory proper from the formalists.

We thus find a transition from a philosophical–positivist to a style-oriented approach to the history of literature. We previously found this transition in art and music historiography. We can therefore conclude that *the historiography of the arts (music, painting, literature) went through a process of change from a nineteenth-century philosophical–positivist approach to a twentieth-century style-focused one*. This parallel shift to style-driven historiography took place in music with Adler, in the visual arts with Wölfflin, and in literature with the Russian formalists. During the course of the twentieth century the limitations of a style-oriented approach were also recognized, and we see the advent of a more pluralistic view, where the social–cultural context of the writer and the work's reception are also considered alongside style. Attempts to produce deductive explanations seem not to occur anymore.

A grammar for narratives: Propp. The Russian formalists were even more important to literary theory than they were to literary history. The most formal of the formalists was Vladimir Propp (1897–1970), who concentrated on the analysis of Russian folk tales. His theory of literature contains the first attempt to create an exact model of analysis for narratives. Propp's goal was extremely ambitious. Could a system of rules be designed to which an entire literary genre was subject? In his *Morphology of the Folk Tale* (1928) he used a structuralist method to analyse 100 Russian folk tales.[282] He divided these stories into ever smaller parts until he reached the smallest narrative units, which he called *narratemes*. 'The villain collects information about the victim' is an example of a narrateme. Narratemes have *functions* that serve as stable, constant elements in a story, independent of how and by whom they are accomplished.

[281] Yury Tynyanov, *Formalist Theory*, translated by L. M. O'Toole and Ann Shukman, *Russian Poetics in Translation*, volume 4, University of Essex, 1977. See also Steiner, *Russian Formalism: A Metapoetics*, pp. 115ff.

[282] Propp's work was translated into English as late as in 1958: Vladimir Yakovlevich Propp, *Morphology of the Folktale*, University of Texas Press, 1958.

Propp found a total of thirty-one different functions for the 100 Russian folktales. Damage, mediation, start of the response, departure of the hero, recognition, punishment, solution, and marriage are examples of such functions. Propp also found seven character types, or *dramatis personae*. These were the hero, the villain, the helper, the donor (who provides a magic spell), the dispatcher (who sends the hero on an adventure), the false hero, and the royal personage (the princess and her father). These thirty-one functions and seven character types are the building blocks in Propp's theory. By combining and recombining them it is possible to generate *all* Russian folk tales, and more besides.[283] The sequencing of narratemes into a story is comparable to the word order in a sentence. We can consider this sequencing as the syntax or grammar of storytelling. Propp assigned a symbol to every narrateme and used them to develop a formal system that shows schematically how these symbols can be threaded together to form or represent fairy tales. Propp's formalism also permits iterations of symbols, which means there is no limitation to the number of folktales that his system can cover.

Although Propp limited himself in accordance with the structuralist tradition to a finite corpus (see 5.3), his system can be tested using new fairy tales or even fairy tales that have yet to be 'conceived'. This means that Propp's theory is of the same rank as Paninian and Chomskyan linguistics, where in principle an infinite number of phenomena can be covered with a finite number of resources.[284] Something like this had never previously been demonstrated in literary theory, either in the rule-based poetics of Aristotle or the formal poetics of India, the *Natya Shastra* (see 2.8). Of course, Propp made it relatively easy by confining himself to fairy tales, but it enabled him to unravel the structure of a complete genre.

In terms of subject, Propp's work appears to be part of the same tradition as Wölfflin's art analysis (5.5) and Lerdahl and Jackendoff's music analysis (5.4). The primary emphasis here is on an analysis of the form and not the meaning. In fact, the French anthropologist Claude Lévi-Strauss (1908–2009) criticized Propp precisely for the mere focus on form rather than meaning and proposed his own method of finding binary oppositions in the narrative instead (see below).[285] Although Lévi-Strauss contrasted his structuralist approach against Propp's formalist one, both can be seen as exponents of structuralism—Lévi-Strauss by searching for binary oppositions like the phonologists had done, and Propp by splitting a story into ever smaller parts as the distributionalists had done—see 5.3. What is more, Propp's underlying method turned out to be unprecedentedly powerful and

[283] Alan Dundes, 'On computers and folktales', *Western Folklore*, 24, 1965, pp. 185–9.

[284] It should be emphasized, though, that Propp's approach still requires a manual assignment of functions and character types to parts of the narrative discourse before a formal representation of the narrative can be created. It has been shown that this assignment is much more subjective than often assumed and can lead to different representations by different annotators. This is in contrast with formal linguistics where the words of the sentences are directly used as input to the system of rules. See Rens Bod, Bernhard Fisseni, Aadil Kurji, and Benedikt Löwe, 'Objectivity and reproducibility of Proppian annotations', in Mark Finlayson (ed.), *The Third Workshop on Computational Models of Narrative*, Cambridge, MA, 2012, pp. 17–21.

[285] On the debate between Propp and Lévi-Strauss, see Alan Dundes, 'Binary opposition in myth: the Propp-Lévi-Strauss debate in retrospect', *Western Folklore*, 56, 1997, pp. 39–50.

influential—it was suitable for analysing virtually every narrative phenomenon: theatre, film, television, and even games (see 5.7). Propp's exact approach has influenced several generations of literary theorists and anthropologists. It resulted in new schools, from the Prague structuralists to the French narratologists. Even the opponents of Propp's work have ushered in new movements, such as the Marxist analysis of Michail Bachtin and the post-structuralist school of which Roland Barthes was a founding father (see below).

The integration of Propp's formalism with later structuralism: *narratology*. After Propp had been translated into French and English in the 1950s, there was a quest to augment literary analysis with meaning within the structuralist framework. The best known exponent was Claude Lévi-Strauss, who analysed myths on the basis of opposites such as life versus death, non-life versus non-death, and high versus low, not high versus not low.[286] However, Lévi-Strauss's method did not yet provide a system of rules with which these assignments of meaning could be derived.

It was fairly straightforward to combine the structuralist and earlier formalist principles, which resulted in a new literary theory movement that became known as *narratology*, a term that was coined by Tzvetan Todorov (born 1939) but that is also associated with Algirdas Greimas (1917–1992) and Gérard Genette (born 1930).[287] Narratology seeks to have a complete narrative analysis that reveals not solely the constituent parts and their functions and relations, but the themes, motives, and plots, too.[288] The concept of plot brings us back to Aristotle, who in his *Poetics* defines the plot as the arc of tension that runs from the beginning via the middle to the end, where the highest level of tension coincides with the actual middle of the story (see 2.8). Yet stories can contain the most diverse plot lines that do not comply with Aristotle's theory or Propp's formalism. One of the criticisms of Propp's work was his limitation to literary genres with a linear plot structure. Genette and others showed that a profound, narratological analysis is also possible for much more complex narrative structures, as in Marcel Proust's *À la recherche du temps perdu*. In most literary genres the constituent parts do not form a linear structure but a hierarchical or even a more complex 'rhizomatic' one in which the ordered parts can in turn be connected to one another (see 5.7).

Narratological analysis has been formalized to such an extent for some media that it can be executed algorithmically wholly or in part. For example, recent work has shown that the narrative structure of the popular TV series *CSI: Crime Scene*

[286] Claude Lévi-Strauss, *Anthropologie structurale*, Plon, 1958. For an English translation, see Claude Lévi-Strauss, *Structural Anthropology*, translated by Monique Layton, Penguin, 1978.

[287] For English translations of these narratologists, see Tzvetan Todorov, *The Poetics of Prose*, translated by Richard Howard, Oxford University Press, 1977; Algirdas Greimas, *Structural Semantics*, translated by Daniele McDowell, Ronald Schleifer, and Alan Velie, University of Nebraska Press, 1984; Gérard Genette, *Narrative Discourse: An Essay in Method*, translated by Jane Lewin, Cornell University Press, 1983.

[288] For an introduction to narratology, see for example Mieke Bal, *Narratology: Introduction to the Theory of Narrative*, 2nd edition, University of Toronto Press, 1997.

Investigation can be derived computationally.[289] Surprisingly, it emerges that all the episodes that have been investigated consist of not more than eight building blocks (narratemes). The use of computational and digital methods in literary studies has become part of the overarching discipline of digital humanities.[290] It has, among other things, resulted in quantitative methods for style recognition and authorship attribution, similar to those developed in musicology and art history (see 5.4 and 5.5).[291] Narratology is also being used in the visual arts to reveal the narrative structure of paintings, but this approach is extremely controversial among art historians.[292] Finally, narratology has found its way into feminist and postcolonial literary criticism.

From narratology to poststructuralism and deconstructivism. One of the features of narratology that drew criticism is that it is based primarily on the text rather than the reader. During the late 1960s this criticism resulted in what is currently called *poststructuralism*.[293] It is not easy to summarize this movement in a few sentences, on the one hand because it is still in a phase of rapid development and on the other because it is more of a conglomerate of ideas and critiques than a clearly demarcated theory. We can state that *under the poststructuralist approach, the reader is given the place of the author as the primary subject of research*. This shift is usually designated as the 'decentring' of the author. Poststructuralists are investigating other sources for analysis and attribution of meaning, such as cultural norms, other literature and, of course, readers. These alternative sources do not guarantee consistency. In many cases the goal of a poststructuralist analysis is to demonstrate that many meanings can be ascribed to a text and that these meanings can conflict with each other depending on the reader. This same idea occurred in a rudimentary form as long ago as the sixteenth century in the *Essays* of the humanist Michel de Montaigne (see also 4.7), but it was not worked out systematically until the poststructuralists did so in the twentieth century. One of the most significant (and nowadays widely accepted) results of poststructuralism, consequently, is that it has shown that *the meaning of any text changes depending on the identity of the reader*.

This was what lay behind Roland Barthes's famous 1967 essay 'The Death of the Author', in which he announced the metaphorical death of the author as an authentic source for interpretation.[294] The death of the author was at the same time the birth of the reader as the source of the many meanings of a text. Barthes illustrated his method in detail in his book *S/Z* (1970), in which he analysed

[289] Benedikt Löwe, Eric Pacuit, and Sanchit Saraf, 'Identifying the structure of a narrative via an agent-based logic of preferences and beliefs: formalizations of episodes from *CSI: Crime Scene Investigation*™', in Michael Duvigneau and Daniel Moldt (eds), *Proceedings of the Fifth International Workshop on Modelling of Objects, Components and Agents, MOCA'09*, 2009, pp. 45–63.

[290] Though computational approaches to literature are also explored in computational linguistics and artificial intelligence—see e.g. Scott Turner, *The Creative Process: A Computer Model of Storytelling and Creativity*, Erlbaum, 1994.

[291] See e.g. David Hoover, *Stylistics: Prospect & Retrospect*, Rodopi, 2007.

[292] Bal, *Reading Rembrandt: Beyond the Word-Image Opposition*, 2006.

[293] Stefan Münker and Alexander Roesler, *Poststrukturalismus*, Metzler, 2000. See also Catherine Belsey, *Poststructuralism: A Very Short Introduction*, Oxford University Press, 2002.

[294] Barthes's essay was originally published in English in the journal *Aspen*, no. 5–6 in 1967.

Balzac's story *Sarrasine*.[295] Still under the influence of structuralism, Barthes dissected this novella into 561 reading units ('lexies'). His analysis is based on a hierarchical stratification of a story and, like the analysis of music and art, it followed the 'hierarchical analysis principle' (see 5.4 and 5.5). He then analysed the units in terms of different meaning attributions. It emerged that Balzac's realistic text was full of symbolic and other connotations, which could be interpreted in various different ways by the reader.

Poststructuralism has had enormous influence on philosophy. Currently the two are so intertwined that poststructuralism can be called both a philosophical and a literary theory movement. Jacques Lacan, Roland Barthes, and Michel Foucault (see also 5.1) are among the most important figures in poststructuralism. The work of Jacques Derrida (1930–2004) and Julia Kristeva (born 1941) is also considered to be an example of poststructuralism, although thanks to his *La Voix et le phénomène* (1967),[296] Derrida has to be seen first and foremost as the architect of *deconstructivism*. We discussed the deconstructivist method in some detail under historiography (see 5.1). However, Derrida did not look on deconstructivism as a 'method'. According to him 'texts deconstruct themselves' as a result of internal opposites and contradictions. A deconstruction of a text reveals to what extent these contrary concepts are interwoven.[297] Also postcolonial literary criticism can be considered part of deconstructivism. Colonial discourses are deconstructed and colonial concept pairs like Western/Eastern are criticized.[298] Deconstructivism as a movement got a foot in the door at many universities, but it also got into people's bad books because of its extreme relativism. It was the Belgian Paul de Man (1919–1983) who was able to establish the deconstructivist vision in American academia after he emigrated to the United States. However, De Man's work was disgraced shortly after his death with the discovery of many anti-Semitic articles he had written during the Second World War and published in a collaborationist Belgian newspaper. It looked as though De Man had a personal interest in his deconstructivist qualification of texts and the role of the author.[299] Derrida's relativizing of De Man's anti-Semitic writings did not do the movement any good either.

Both deconstructivism and poststructuralism are sometimes upbraided for not being scholarly,[300] but Barthes's method demonstrates the opposite. His

[295] For an English translation, see Roland Barthes, *S/Z: An Essay*, translated by Richard Miller, with a preface by Richard Howard, Farrar, Straus, and Giroux, 1975.

[296] Translated into English, as *Speech and Phenomena: And Other Essays on Husserl's Theory of Signs*, translated by David Allison, Northwestern University Press, 1973.

[297] Jacques Derrida, *De la grammatologie*, Les Éditions de Minuit, 1967. For an English translation, see Jacques Derrida, *Of Grammatology*, translated by Gayatri Spivak, Johns Hopkins University Press, 1998.

[298] Edward Said, *Orientalism*, Vintage Books, 1978. See also Gayatri Spivak, 'Can the subaltern speak?', in Cary Nelson and Lawrence Grossberg (eds), *Marxism and the Interpretation of Culture*, MacMillan, 1988.

[299] David Lehman, *Sign of the Times: Deconstruction and the Fall of Paul de Man*, Poseidon, 1991.

[300] This was particularly highlighted during the Sokal-affair. In 1996 the physicist Alan Sokal submitted an article full of nonsensical reasoning and pseudo-scientific jargon to the 'leading' journal *Social Text*. Sokal's intention was to show that postmodernist journals publish everything as long as the

conscientious *syntactic* analysis is carried out in an empirical way on the basis of methodical principles that we also came across in art theory, musicology, and linguistics. It is mainly the *semantic* analyses, with the many meaning attributions, that are not unambiguous. Yet this is exactly what the poststructuralists have in mind. Although they are pattern-rejecting with regard to the 'universal validity' of patterns, at the same time they point to an extremely stable pattern, namely the result *that every text has a multiplicity of interpretations*, which they also observed for every historical event (see 5.1).

Parallel movements: hermeneutics and the anticipatory 'method'. It is possible to identify lines in the history of literary theory other than our story from positivism via structuralism to poststructuralism outlined above. *Hermeneutics*, or the study of interpretation, is one of these developments. It existed before positivism reigned supreme. Friedrich Schlegel (1772–1829) initiated this movement and coined the term Romanticism at the same time, which he based on the then new literary genre—the *roman* (novel). According to Schlegel, 'intuition' about the whole is necessary for an understanding of a text.[301] However, an understanding of the whole cannot be obtained simply by bringing together the understanding of the separate parts. The reader therefore has to juggle two levels—the parts and the whole—simultaneously.

Friedrich Schleiermacher (1768–1843) took this idea further.[302] He argued that since any form of interpretation is context dependent, the parts have to be understood in the context of the whole. Yet the whole consists of nothing other than the constituent parts. So how is it possible to break out of this vicious circle, which is known as the *hermeneutic circle*? Interpretation can only take place on the grounds of a 'divinatory' or guessing method that is based on empathizing with the writer's state of mind. An understanding of the text can gradually be built up on the basis of the circular movement described above. Wilhelm Dilthey (1833–1911) was on the same wavelength as Schleiermacher with his concept of *verstehen* (see 5.1).[303] He regarded hermeneutics as a process of re-experiencing the author's world. The central concern of hermeneutics was to understand the author's inner world and historical life. For Dilthey, history consisted of world views without any firm standard for taking one world view as superior over another. Thus, interpretations of texts are historically relative and dependent on the cultural context and the perspective of the interpreter.[304]

editors are appeased by ideologically correct views. The publication of the article, followed by Sokal's disclosure, caused great indignation in the academic world. See Alan Sokal and Jean Bricmont, *Fashionable Nonsense: Postmodern Intellectuals' Abuse of Science*, Picador USA, 1998.

301 Friedrich Schlegel, *Literarische Notizen, 1797–1801*, edited by Hans Eichner, Ullstein, 1980. Translated into English as *Literary Notebooks, 1797–1801*, edited by Hans Eichner, University of Toronoto Press, 1957.

302 Martin Redeke, *Schleiermacher: Life and Thought*, Fortress, 1973. See also Kurt Nowak, *Schleiermacher: Leben, Werk und Wirkung*, Vandenhoeck & Ruprecht, 2001.

303 Wilhelm Dilthey, *Selected Works, Volume IV: Hermeneutics and the Study of History*, translated and edited by Rudolf Makkreel and Frithjof Rodi, Princeton University Press, 1991.

304 See also Xing Lu, *Rhetoric in Ancient China, Fifth to Third Century* BCE: *A Comparison with Classical Greek Rhetoric*, University of South California Press, 1998, pp. 19–20.

Nineteenth-century hermeneutics delivered admirable interpretations but it contained a strongly subjective component: guessing the author's intention. Twentieth-century hermeneutics, on the other hand, puts the idea of *Vorverständnis*—pre-understanding or the entirety of prejudices, preconceptions, and prejudgments that have been determined historically and with which every interpretation begins—at the centre.[305] The philosopher Martin Heidegger (1889–1976) contended that the understanding of a text continues to be influenced by our a priori judgment, but according to him the most important thing is that the interpreter is aware of his position.[306] According to the most prominent twentieth-century hermeneuticist, Hans-Georg Gadamer (1900–2002), an interpretation is created as a result of a fusion of the horizons from the past and the present.[307] What Gadamer meant with these high-flown words is that roughly speaking an understanding of a text is created when it is applied to the reader's own situation. This brings hermeneutics into the vicinity of poststructuralism. Yet with its anticipatory starting point, hermeneutics is in essence outside the scope of our quest for methodical principles, whereas this is not the case with the poststructuralist method. While it is true to say that the latter is based on sources of often obscure and very diverse natures, it does not commit itself to 'premonitions'.

Other twentieth-century movements stemming from formalism and structuralism. A large number of other literary theory movements emanated directly or indirectly from formalism, structuralism, and poststructuralism. Most of them were pattern-seeking, although their methodical principles were not always precisely defined. We discuss a few below.

New criticism and new historicism. *New criticism* blossomed primarily between 1930 and 1950, and like the formalists it had the autonomy of literary works at its centre. The literary theorist's most important activity was the *close reading* of a text, divorced from its context. The purpose of this close reading was to reveal the complexity and the ambiguity as well as the interaction between form and content. However, new criticism did not develop any formally defined working principles. Although the movement no longer exists, considerable importance continues to be attached to close reading. Later on *new historicism* arose out of new criticism, and since 1990 it has become influential. The new historicism appears to represent a return to an understanding of a work in its historical context, while at the same time one tries to comprehend the cultural history by means of literature.[308] Here literary analysis runs over into cultural history (see also 5.1).

Empirical literary theory and psychonarratology. The psychological study of people's literary behaviour developed in response to the rather subjectivist, non-

[305] Jean Grondin, *Introduction to Philosophical Hermeneutics*, Yale University Press, 1994.

[306] Martin Heidegger, *Sein und Zeit*, 1927, in *Gesamtausgabe*, volume 2, edited by F. von Herrmann, Klostermann, 1977. For an English translation, see Martin Heidegger, *Being and Time*, translated by Joan Stambaugh, State University of New York Press, 1996.

[307] Hans-Georg Gadamer, *Wahrheit und Methode: Grundzüge einer philosophischen Hermeneutik*, Mohr, 1960. For an English translation, see Hans-Georg Gadamer, *Truth and Method*, translated by J. Weinsheimer and D. G. Marshall, Crossroad, 2004.

[308] Harold Veeser (ed.), *The New Historicism*, Routledge, 1989.

empirical analysis of the poststructuralists. This movement studies how people mentally represent and process texts and what they do with them.[309] Psychonarratology is concerned with the extent to which readers attribute mutually consistent analyses to stories, such as the subdivision into narrative units and plot lines, as well as the recognition of the characters and the narrator.[310] This interdisciplinary area lies between the humanities and the social sciences.

Rhetorical analysis and argumentation theory. Although rhetoric saw barely any new developments in the nineteenth century (see also 4.6), it made a comeback in the twentieth century in the exposure of rhetorical arguments in texts that are analysed in terms of their persuasiveness. Stephen Toulmin's study *The Uses of Argument* (1958) was the most influential example. Nevertheless, rhetoric has had a marginal existence within literary theory. However, the interdisciplinary field known as *argumentation theory* has been flourishing and is an important auxiliary subject in such areas as the study of jurisprudence.[311]

Study of orality. In the twentieth century the study of oral literature expanded enormously.[312] An unexpected treasure trove of oral stories from many parts of the world came to light, and at the same time the mechanisms that enable oral poets to improvise poetry were also studied. One of the first important insights in this area—based on the study of Serbian oral epic poems—is that poets have stored an enormous amount of poetic 'formulas' in their memory.[313] The concept of poetic formula in this discipline is comparable to the notion of 'construction' in linguistics or 'narrateme' in Propp's formalism. These formulas can be combined or recombined in a conventional fashion, which results in extremely fast (oral) poetry production. As in music, oral poetry proves to be both rule-based and example-based, in which the formulas are example-based and the procedure for combining them is rule-based (see 5.4).

Chinese literary theory: from Ruan Yan to revolutionary realism. Chinese literary theory was still flourishing during the early part of the Qing Dynasty (see 4.7), but in its later phases the level of activity dropped sharply.[314] One of the most important exponents of Chinese poetics was Ruan Yuan (1764–1849), who argued that only texts written in rhyme and parallel forms (such as alliterations) can be designated as 'literature' (*wen*) as opposed to 'standard texts' (*pi*).[315] The study of Chinese classics was the dominant tone in the work of Ruan Yuan, whose influence lasted until after the fall of the Qing Dynasty.

309 See e.g. Richard Gerrig, *Experiencing Narrative Worlds: On the Psychological Activity of Reading*, Yale University Press, 1993.

310 Marisa Bortolussi and Peter Dixon, *Psychonarratology: Foundations for the Empirical Study of Literary Response*, Cambridge University Press, 2003.

311 See e.g. Frans van Eemeren and Rob Grootendorst, *A Systematic Theory of Argumentation: The Pragma-dialectical Approach*, Cambridge University Press, 2003.

312 Jack Goody, *The Interface between the Written and the Oral*, Cambridge University Press, 1987.

313 Albert Lord, *The Singer of Tales*, Harvard University Press, 1960.

314 James Liu, *Chinese Theories of Literature*, University of Chicago Press, 1975, pp. 103ff.

315 Betty Peh-T'i Wei, *Ruan Yuan, 1764–1849: The Life and Work of a Major Scholar-official in Nineteenth Century China before the Opium War*, Hong Kong, University Press, 2006.

During the Republican Era (1912–1949) the most significant movement was the *New Cultural Movement.* Literary reformers like Hu Shi and Chen Duxiu declared the classical language to be dead, and asserted that a dead language cannot produce a living literature.[316] The use of written vernacular Chinese and vernaculare literature was vigorously promoted. In 1942 Mao Zedong gave his famous series of lectures in which he argued that art and literature should be subordinate to politics.[317] These so-called 'Talks at the Yan'an Forum on Art and Literature' were to form the national guideline for all culture in the People's Republic. Literary theory could only proclaim socialist realism combined with 'revolutionary realism' and 'revolutionary romanticism'. The whole of literary theory came to a standstill during the Cultural Revolution (1966–1976).

Critical literary theory in present-day China is still hard to find in the universities. Literary theory can barely get off the ground in a country where books are heavily censored and regularly banned. However, there is underground 'literary criticism': some 4,000 illegal publishers are operating in China and the number of banned books has been estimated at one third of the total number of books published.[318] These works cover novels as well as subjects from the humanities, particularly relating to fields where censorship is the strictest, for example history and literary criticism. It will probably never be possible to adequately describe the history of such underground humanities. The government regularly conducts public book burnings. In the eyes of the government such books are 'spiritual pollution'—although more and more illegal books are also finding their way abroad.[319]

The rise of theatre studies: in the shadow of literary theory? Modern theatre studies is often considered to be a discipline that stemmed from literary theory. From a historical point of view, though, the opposite is closer to the truth. For centuries poetics was more about the theatre than literature. We saw this with Aristotle, Bharata Muni (2.8), and Castelvetro's three unities of time, place, and action (4.7). As a result of the eighteenth-century rise of the novel and the nineteenth-century importance that was attached to a country's own national literature, the European study of literature was concerned not so much with its theatre as with its literary history. It was not until the twentieth century that the study of theatre was taken up again, which led to the creation of modern theatre studies. This was largely due to Max Herrmann (1865–1942), who from 1919 held the first chair of theatre studies.[320] In his *Forschungen zur deutschen Theatergeschichte des Mittelalters und der Renaissance* ('Researches on German Theatre

[316] Chang-tai Hung, *Going to the People: Chinese Intellectuals and Folk Literature, 1918–1937*, Harvard University Press, 1985.

[317] Junhao Hong, 'Mao Zedong's cultural theory and China's three mass-culture debates:a tentative study of culture, society and politics', *Intercultural Communication Studies*, 4(2), 1994, pp. 87–103.

[318] Shen Yuan, 'The secrets of China's most profitable sectors', *China Rights Forum*, 3, 2005, pp. 34–8.

[319] Haig Bosmajian, *Burning Books*, McFarland & Company, 2006, pp. 32ff, pp. 178ff.

[320] Erika Fischer-Lichte, 'From text to performance: the rise of theatre studies as an academic discipline in Germany', *Theatre Research International*, 24(2), 1999, pp. 168–78. See also Stefan Corssen, *Max Herrmann und die Anfänge der Theaterwissenschaft*, Niemeyer, 1998.

History of the Middle Ages and the Renaissance', 1914) Herrmann stated that he had an extremely ambitious goal, which was the reconstruction of past theatrical performances such that 'lost performances' were brought back to life again. This historicizing approach was clearly influenced by the neo-Rankean tradition of historical writing (see 5.1). However, Herrmann did not confine himself to the study of drama texts. He also used as many other sources as possible from theatrical practice in order to reconstruct the past—performance reports, minutes of meetings, yearbooks, reviews, journals, letters, pamphlets, novels, and posters.

The problems associated with such a reconstruction soon led to controversy in Herrmann's lifetime, but with his 'all possible sources principle' he set the standard for theatre studies in the twentieth century. Even though a complete reconstruction is currently thought to be utopian, Herrmann's approach nevertheless resulted in a huge quantity of historical material that gives us a vivid idea of Greek, medieval, and Renaissance theatre. In 1923 Herrmann became the first director of the renowned *Theaterwissenschaftliche Institut* in Berlin, but ten years later, when Hitler's racial laws were introduced, he was forced into retirement. Herrmann carried on working on his grand overview of drama in the most difficult circumstances—for example, in the Berlin State Library he was only allowed to read manuscripts while standing up—but in 1942 the *éminence grise* was transported to Theresienstadt concentration camp, where he died two months later.

Currently Herrmann's method has been set aside in academic theatre studies as being positivist, and it is the poststructuralist approaches that are calling the tune.[321] Nevertheless, these days theatre reconstruction is still being practised in all its glory, for instance in the form of computer-based multimedia simulations of historical European theatres, such as the *Theatron Project*.[322] By far the greatest theatre reconstruction of its kind was the recent rebuilding of Shakespeare's *Globe Theatre* in London, where architects and theatre historians worked together to try to reconstruct the famous playhouse as accurately as possible.[323] It appears that Herrmann's dream has come true after all.

As regards *theatre analysis*, most of the approaches in literary theory have also been employed for drama.[324] Methods from art history have moreover been used, in particular Panofsky's iconological procedure for the interpretation of visual sources (see 5.5). The study of acting techniques is one of the more theoretical theatrical approaches. This is as old as Aristotle's *Poetics* and Bharata Muni's *Natya Shastra* (see 2.8). During the nineteenth century acting theory in Europe blossomed in the work of the Frenchman François Delsarte, who tried to get the

[321] See e.g. Christopher Balme, *The Cambridge Introduction to Theatre Studies*, Cambridge University Press, 2008. See also Phillip Zarrilli, Bruce McConachie, Gary Jay Williams, and Carol Fisher Sorgenfrei, *Theatre Histories: An Introduction*, Routledge, 2010.

[322] Susan Schreibman, Raymond Siemens, and John Unsworth, *A Companion to Digital Humanities*, Blackwell, 2004, p. 123.

[323] J. R. Mulryne and Margaret Shewring (eds), *Shakespeare's Globe Rebuilt*, Cambridge University Press, 1997.

[324] Balme, *The Cambridge Introduction to Theatre Studies*.

inner experience to correspond with gestures and movements.[325] He based this on his observations of human interaction. In the twentieth century there was a fascinating sequel to this empirical approach to acting techniques in the study of *intercultural acting*, which is known as *theatre anthropology*. This movement was launched by the Italian Eugenio Barba, and it endeavours to discover the universal acting principles.[326] In theatre anthropology, theatrical traditions from diverse cultures are compared, and centuries-old Indian, Chinese, and Japanese texts are consulted. Although this technique is still in its infancy and strongly related to theatre performance, the intercultural study of theatre practices is striking a sympathetic note.

[325] Nancy Lee Chalfa Ruyter, 'The Delsarte Heritage', *Dance Research: The Journal of the Society for Dance Research*, 14(1), 1996, pp. 62–74.

[326] Eugenio Barba and Nicola Savarese, *A Dictionary of Theatre Anthropology: The Secret Art of the Performer*, Routledge, 2005.

5.7 THE STUDY OF ALL MEDIA AND CULTURE: FROM FILM STUDIES TO NEW MEDIA

Literature, poetry, theatre, painting, and music have existed from Antiquity and have been studied continuously ever since. Film, television, and digital media, on the other hand, have only existed since the twentieth century and the study of them is less than a century old, and in some cases less than a few decades. Initially these studies employed methods from other humanistic disciplines, in particular literary studies. While the study of film, TV, and new media was recently combined in media studies, another discipline has evolved with overarching aspirations: *cultural studies*.

Film and television studies: from structural to cultural patterns. The history of film is barely a century old, and as an academic discipline, film studies date from the 1970s. However, the first studies in this field are as old as the medium itself. The French philosopher Henri Bergson (1859–1941) is often considered to be the forerunner of film theory. In 1896 in his *Matière et mémoire* he described the development of the medium as the 'moving image' and the 'time-image'.[327] The first study of film as a form of art was by the Italian Ricciotto Canudo (1879–1923) who worked primarily in France. In his essay *La Naissance d'un sixième art* (1911) he looked on the new medium as a new art but also as a synthesis of five old arts: architecture, sculpture, painting, music, and poetry, to which he later added dance.[328]

It was not until the emergence of formal film theory in the first two decades of the twentieth century that a precise method was developed that focused on an analysis of the technical elements of film, such as lighting, screenplay, sound, shot composition, and colour.[329] In this tradition, however, no quest was launched for the underlying film structure. We do, though, find it in structuralist film theory, in particular in the work of Christian Metz (1931–1993) who applied Saussure's theory of semiology (5.3) to film. In *Film Language: A Semiotics of Cinema* (1974) Metz developed his 'Grande Syntagmatique' in which he called the 'universal' building blocks of film 'syntagmas'. He provided a hierarchical organization for them so that the cinematic narrative structure could be visualized. Metz's Syntagmatique was elaborated by others, particularly by Michel Colin, into a system of rules with which in principle all possible structures of all possible films could be generated.[330] Analogous to generative linguistics and musicology, a cinematic narrative structure is interpreted as a tree diagram (see 5.3 and 5.4) where the

[327] For an English translation, see Henri Bergson, *Matter and Memory*, translated by N. M. Paul and W. S. Palmer, Zone Books, 1990.

[328] For an English translation see Ricciotto Canudo, 'The birth of the Sixth Art', in Richard Abel (ed.), *French Film Theory and Criticism: A History/Anthology, 1907–1939*, Princeton University Press, 1993, pp. 58–66.

[329] See e.g. Robert Stam, *Film Theory: An Introduction*, Wiley, 2000.

[330] Michel Colin, 'The Grande Syntagmatique revisited', in Warren Buckland (ed.), *The Film Spectator: From Sign to Mind*, Amsterdam University Press, 1995, pp. 45–86.

leaves of the tree represent film scenes and the branched structure reflects the relationships between the scenes.

As a procedural system of rules, though, Colin's work does not come up to the mark. The grammatical building blocks used by Colin have categories like shot, shot sequence, and scene, but they are not linked to *concrete* shots in the way that a language grammar is linked to concrete words of a language or a music grammar is linked to concrete notes of a musical idiom. This leaves cinematic grammar dangling at a rather abstract level, and it is difficult to test it. The Grande Syntagmatique therefore belongs more to the tradition of semiology than the tradition of formal linguistics.

On the other hand, there are other methods, in particular the *narratological* approach (see 5.6), that have been developed to the point that they can even be tested using computational techniques for the derivation of the narrative structure of TV series. However, for the time being this method only works for relatively simple films and series,[331] and not yet for the more complex films that Metz and Colin focused on.

While film studies in the 1970s were dominated by linguistic and semiotic principles with a universalist perspective, we see increasing (and renewed) interest in film history in the 1980s. One of the great contributions was the systematic historiography of the classical Hollywood film by David Bordwell, Janet Staiger, and Kristin Thompson (1985), *The Classical Hollywood Cinema: Film Style and Mode of Production to 1960*. The development of technical filmmaking has also been historically investigated, which has resulted into a strongly empirical approach to film studies.[332] Similar to the quantitative analysis of music, art, and literature (see 5.4, 5.5, and 5.6), stylometric methods are applied to films in order to identify the individual style of a director. This is done by collecting formal parameters of films, such as duration of shots, shot scales, camera movements, and angles of shots, which are next statistically analysed.[333]

Besides these heavily empirical approaches, film has also been studied from the perspectives of critical theory (see 5.1), poststructuralism (see 5.6), and feminist and postcolonial perspectives. These methods depart from searching for 'universal' (and even 'empirical') patterns, but they do find cultural patterns in film, particularly with regard to the portrayal of male–female relationships and power structures. Film has also been investigated using approaches from outside the humanities. For example, Jacques Lacan uses a Freudian psychoanalytical method, Gilles Deleuze (see below under New Media) connects film to the brain, and Slavoj Žižek analyses the philosophical notion of the 'real' in film.

We also see the absence of a single specific methodology in the study of television.[334] Nevertheless, patterns are being revealed here too. In *Television: Technology and Cultural Form* (1974), for instance, Raymond Williams discovered

[331] See e.g. Löwe, Pacuit, and Saraf, 'Identifying the structure of a narrative via an agent-based logic of preferences and beliefs: formalizations of episodes from *CSI: Crime Scene Investigation*™'.

[332] Barry Salt, *Film Style and Technology: History and Analysis*, Starword, 2003.

[333] Warren Buckland, 'What does the statistical style analysis of film involve?', *Literary and Linguistic Computing*, 23(2), 2008, pp. 219–30.

[334] See e.g. Robert Allen and Annette Hill (eds), *The Television Studies Reader*, Routledge, 2004.

a remarkable shift in the development of the medium. In the 1970s, programmes were no longer made with separate, successive blocks like news, quiz, and film; instead everything flowed virtually seamlessly into everything else. The natural breaks were now commercials and announcements about films and quizzes on the following day. This phenomenon, which Williams called *flow*, resulted in a non-stop stream of information, advertising, entertainment, and trailers with the aim of keeping the viewer tuned to a particular channel. In addition to this critical analysis, television has also been studied using other methods, such as the narratological analysis of TV series described above. Like film, television has been studied using virtually all approaches from the humanities and social sciences, with the aim both of criticism and of unravelling trends in the history and structure of the medium.[335]

The rise of media studies. The general study of film, TV, journalism, and digital media is denoted using the umbrella term *media studies*, which has grown into an academic discipline.[336] One of the doyens in this field was Walter Benjamin (1892–1940), who was also one of the founders of the Frankfurt school of critical theory (see 5.1). Although Benjamin's work was brought up earlier in connection with art history (5.5), we have not yet mentioned his most influential contribution—*Das Kunstwerk im Zeitalter seiner technischen Reproduzierbarkeit* ('The Work of Art in the Age of its Technological Reproducibility') written in 1936.[337] This essay is about both art theory and media theory. According to Benjamin the new technologies for reproducing and distributing works of art bring about a radical change in the relationship in regard to art and artists. Film, photography, and the gramophone alike rejected the traditional concept of the *authenticity* of a work of art. This is because the distinction between an original and a copy is meaningless in film and photography. The same film can be shown in different locations simultaneously, so no single viewer has a privileged position vis-à-vis other viewers. The work of art has become part of mass culture and has lost its mystical character or *aura* that has surrounded it for centuries. Benjamin believed he could see a *pattern of democratization* in the new arts and—as a child of his age—he analysed the use of art in fascist and communist regimes. The patterns observed by Benjamin have largely determined the critical movement in media studies.

The analyses conducted by Marshall McLuhan (1911–1980) were of a rather different nature. In *Understanding Media: The Extensions of Man* (1964) McLuhan argued that media, like the telephone and television, were an extension of human senses. He analysed the relationship between man and medium as a complex interrelationship between body and world. According to this analysis a shoe or a bicycle is just as much part of the media as television or telephones. People can therefore only fathom out the world by getting to the bottom of the media (i.e. the man–world relationship). McLuhan is famous primarily for his slogan 'The

[335] John Corner, *Critical Ideas in Television Studies*, Oxford University Press, 1999.

[336] Robert Kolker, *Media Studies: An Introduction*, Wiley-Blackwell, 2009. See also Sue Thornham, Caroline Bassett, and Paul Marris (eds), *Media Studies: A Reader*, 3rd edition, NYU Press, 2010.

[337] For an English translation, see Walter Benjamin, *Illuminations: Essays and Reflections*, edited by Hannah Arendt, translated by Harry Zohn, Harcourt, 1968, pp. 217–52.

medium is the message', by which he meant that the medium itself is the message that deserves our attention and extends the senses. This is why the eye acquired a bigger role than the ear when writing largely replaced the oral culture. Although McLuhan's analyses fundamentally shaped media studies,[338] in recent years the discipline has become increasingly influenced by *cultural studies*.

Cultural studies: eclecticism *in extremis*. The humanistic movement in which methodical eclecticism is not considered to be a problem and in fact has been elevated to a paradigm is *cultural studies*. Originally, cultural studies was a continuation of critical theory (see 5.1), but it was soon augmented by economics, communication studies, sociology, linguistics, literary studies, cultural anthropology, philosophy, art history, poststructuralism, and gender studies (among other things).[339] Cultural studies attempts to combine all these disciplines in order to study *cultural practices and their relationship with power*. A distinction is no longer made between high and low culture—everything is culture. The term 'cultural studies' was coined by Richard Hoggart in 1964 upon the foundation of the Birmingham Centre for Contemporary Cultural Studies (CCCS), although the discipline is currently primarily associated with Stuart Hall, Hoggart's successor.[340]

Like the school of critical theory, neo-Marxism was the initial source of inspiration for cultural studies. Under that heading, it associated itself closely with the thinking of Antonio Gramsci (1891–1937), who introduced the distinction between hegemonic and subaltern (see 5.1). The subsequent embracing of a multitude of methods from the humanities and social sciences has resulted in a discipline that makes no further demands on a coherent method and has become a patchwork of approaches. Is this the ultimate form of interdisciplinarity, or is it a case of anything goes? No matter how many methods have been adopted by cultural studies, we do not find the procedural system of rules principle among them—while we do come across it in the other humanistic disciplines. Despite (or perhaps because of) the absence of an unambiguous, coherent method, cultural studies has exposed trends and patterns in cultural material that were not found previously. For example, new relationships between power and knowledge have surfaced, and persistent myths about the accessibility of the digital world and the oneness of national identity have been negated. This 'boundless diversity of present-day culture'[341] is precisely what cultural studies is aiming at.[342] As a hybrid discipline, however, it lacks a clear basis that guarantees any controllability. Consequently,

[338] Paul Levinson, *Digital McLuhan: A Guide to the Information Millennium*, Routledge, 1999. See also Philip Marchand, *Marshall McLuhan: The Medium and the Messenger*, The MIT Press, 1998.

[339] Richard Johnson, Deborah Chambers, Parvati Raghuram, and Estella Tincknell, *The Practice of Cultural Studies*, SAGE Publications, 2004, chapter 2 ('Multiplying methods: from pluralism to combination'), pp. 26–43.

[340] Stuart Hall, 'Cultural studies: two paradigms', *Media, Culture and Society*, 2, 1980, pp. 57–72. See also Simon During, *The Cultural Studies Reader*, Routledge, 2003.

[341] From the cover text of Jan Baetens, Joost de Bloois, Anneleen Masschelein, and Ginette Verstraete, *Culturele studies: theorie in de praktijk*, Vantilt, 2009.

[342] In a related approach known as *Cultural Analysis*, Mieke Bal calls this form of practice an *interdiscipline*—see Mieke Bal, *The Practice of Cultural Analysis: Exposing Interdisciplinary Interpretation*, Stanford University Press, 1999.

cultural studies runs the risk of degenerating into an easy-going field in which everything can be claimed. Are we back with classical rhetoric, which changed from a discipline of everything into a discipline of nothing (see 4.6)? Time will tell.

New media: a discipline in a state of flux with a 'rhizomatic' structure. The study of new media is the latest in the humanities, and one of the most fascinating problems in this field is the definition of its own subject of study.[343] Currently new media is primarily taken to mean *digital media*, such as the internet, the web, virtual reality, video games, mobile telephony, digital film, and interactive television. A generation earlier, though, around 1975, new media referred mainly to *audio-visual media*, which included film, TV, and video. A generation before that people considered new media to be analogue telephony, radio, and photography. In short, the concept 'new media' itself is in a state of flux. Among other things the discipline therefore investigates to what extent new media are 'new', and what this means in a historic and conceptual sense. In *Remediation: Understanding New Media* (2000), Jay Bolter and Richard Grusin identified as a pattern the fact that every 'new' medium in history absorbs the content of earlier media. For instance, film contains narrative structures that already existed on the stage, and a computer game uses narrative structures from films.

Another quest in new media studies is to reveal the underlying structure of products like hypertext, websites, and video games. The traditional hierarchical structures that are used in linguistics, literature, and film studies (such as tree structures) do not appear to be fit for purpose. The very first explorations in this field were visionary—they investigated a theoretical form of media that did not yet exist at the time of publication. In 1941 in *El jardín de senderos que se bifurcan* ('The Garden of Forking Paths'), for example, Jorge Luis Borges described the idea of a novel that can be read in several ways via a labyrinthine structure.[344] Borges's idea had great influence on the development of hypertext in the 1960s. Another important visionary was Vannevar Bush, who was also at the birth of the atomic bomb. In his essay *As We May Think* (1945) he foresaw the possibility of 'wholly new forms of encyclopedias, ready-made with a mesh of associative trails running through them'. Such structures underlie present-day digital media like websites and games, in which labyrinthine interweaving is a recurring element.

The concept of an interwoven structure was articulated further by Gilles Deleuze and Félix Guattari, who introduced the term 'rhizome' to this end in *Mille plateaux* (1980).[345] This word is taken from botany, where it refers to an underground, usually horizontal, stem that often bends upwards again and thus creates a new

[343] Martin Lister, Jon Dovey, Seth Giddings, Iain Grant, and Kieran Kelly, *New Media: A Critical Introduction*, Routledge, 2009. See also Noah Wardrip-Fruin and Nick Montfort, *The New Media Reader*, 2003.

[344] The English translation of Borges's essay is included in Wardrip-Fruin and Montfort, *The New Media Reader*, pp. 30–4.

[345] For an English translation, see Gilles Deleuze and Félix Guattari, *A Thousand Plateaus*, translated by Brian Massumi, Continuum, 2004.

plant. A *rhizomatic structure* in the humanities does not assume a basis with branches. Rather it is a network that fans out sideways and consists of a multiplicity of junctions without a clear beginning or end. Such a structure is more complex than a hierarchical tree structure in stemmatic philology and linguistics. In a rhizome the different parts that are split up hierarchically in a tree structure can also be directly connected to one another. In mathematics and information technology, a rhizomatic structure is covered by the concept of *graph*.[346] Thus the structure of a website or a video game cannot normally be represented as a linear or hierarchical structure but as a rhizomatic one. The levels in a game and the pages in a website can be linked to one another such that one can jump freely, and not in a hierarchical fashion, to and fro between levels or pages. Virtually all exploration of the structure of new media results in a rhizomatic pattern. The idea of a rhizomatic structure is not limited to new media, though—it was devised in literary studies and has been used to analyse writers like Marcel Proust (see 5.6).

In *The Language of New Media* (2001) Lev Manovich also adopted the idea of a distributed, rhizomatic platform as the structural basis of new media. Manovich described this structure as a *relational database*, which he compared with the more traditional narrative forms. While traditional forms have a beginning and an end, databases create a structure without a specific path to be followed, just like rhizomes.

Over and above this quest for underlying structures, the new media discipline is also involved in new media art, cyberculture, interactivity, digital politics, and more. In terms of its subject of study, new media studies is probably the most innovative field in the humanities. Nothing ages faster than the concept of 'new', so the discipline is continuously concerned with redefining itself. As regards methods, however, new media studies is just as hybrid as the other media areas.

The hybridity of new humanistic disciplines. It may seem strange that our history of the humanistic quest for principles and patterns ends with cultural studies and new media. Yet nowhere is the search for methodical underpinning so topical as it is in these disciplines (so too in the even more recent subfields that we have not considered here, such as visual culture studies and performance studies).[347] The hybridity of methods is a phenomenon that is found in all 'beginning' disciplines—even in the nineteenth century and before.[348] Early literary studies, art history, and musicology were based on methods drawn from nineteenth-century Hegelian philosophy and Comtean positivism; at the beginning of the twentieth century theatre studies was based on methods from literary studies, whereas in the latter part of the century cultural and media studies looked to the

[346] See e.g. Gary Chartrand, *Introductory Graph Theory*, Dover, 1985.

[347] See e.g. Nicholas Mirzoeff, *An Introduction to Visual Culture*, Routledge, 1999; and Soyini Madison and Judith Hamera (eds), *The Sage Handbook for Performance Studies*, SAGE, 2006.

[348] See Karstens, 'Bopp the builder: discipline formation as hybridization: the case of comparative linguistics', 2012, for the hybridity of nineteenth-century linguistics.

social sciences. A discipline does not create its own method until it has detached itself from the fields where it originated. In the case of literary studies this process took the entire nineteenth century and longer. There is no reason to assume that such a process will take place any faster with the recent humanistic disciplines. Yet one should remain apprehensive about interdisciplinary mules: they may prove to be usable for a while, but in the end they are not fertile.[349]

[349] I owe this metaphor to Wijnand Mijnhardt.

CONCLUSION: IS THERE A BREAK IN THE MODERN HUMANITIES?

Common patterns in the modern humanities. Before addressing the question of whether the modern humanities signify a break with the early modern age, we will unfold a number of common patterns that we have found in the nineteenth- and twentieth-century humanities. During this period the humanities outside Europe were largely under Western influence, with African historiography as the most striking exception (see 5.1).

The emancipation of the past: the humanities became historical. Whereas the history of music, art, literature, and theatre received sporadic attention in the early modern humanities, starting in the nineteenth century the historical approach was given a central place in every humanistic discipline—from Europe to India (in China, on the other hand, art historiography had occupied a central place for centuries). The history of the arts was subdivided into style periods, where each period displayed a pattern of rise, peak, and decline. We have moreover seen that essentially all twentieth-century humanities were studied from cognitive, computational, and postmodern perspectives. Barring the occasional exception (linguistics and possibly philology), all humanities have both a pattern-seeking and a pattern-rejecting component, where the latter only rejects 'universal' patterns but not local or culture-specific ones.

The trend from prescriptive back to descriptive continued. This trend was initiated in the eighteenth century and continued into the modern age. Initially most Western humanities were still under the romantic or nationalist yoke, when the influence of Hegelian, Comtean, and Herderian movements was dominant, but in the twentieth century the descriptive approach seemed to have returned in the study of music, art, literature, and theatre. This began in art history with the anti-nationalist attitude of Burckhardt and Wölfflin, and spread to the other humanities. We recall that in Antiquity there was a trend from descriptive to prescriptive (see the conclusion to chapter 2). If we try to survey the history of the humanities as a whole, there appears to be a cyclical tendency from descriptive to prescriptive and back to descriptive.

Hierarchical stratification as a recurring theme. This stratification came from rhetoric, but it cropped up in all other humanistic disciplines—a text, piece of music or piece of art was split up into its constituent parts, which were subdivided into ever smaller units until one ended up with the 'atomic' parts, such as phonemes in language, notes in music, constituents in painting, and scenes in film. We could assert that analysis in the form of a hierarchical structure was a constant in European humanities—from the rhetoric of the ancient world with its hierarchical classification (see 2.7), and Vitruvius's analysis of architecture (see 2.5) by way of medieval Modist linguistics (see 3.1) and humanistic art theory (4.5) to nineteenth-century stemmatic philology and musicology and twentieth-century linguistics, history, and literary studies. The recent new media studies show a further quest for even more complex patterns, such as rhizomatic structures (5.7).

Grammars as a recurring theme: rules versus examples. Hierarchical stratification indicated the existence of a grammar or procedural system of rules for humanistic expressions. The constituent parts of the stratification can be denoted by categories and the atomic units as words, notes, or scenes. These are then linked together by a system of rules, and—behold—a grammar. However, practice is more obstinate. No system of rules is perfect, and a rule-based method only seemed to bear fruit if it was extended by an *example-based* component—simply because so many linguistic, musical, literary, cinematic, and artistic expressions were based on conventions. While the strict rule-based tradition originated in India (Panini), the example-based tradition came primarily from the Arab world (Sibawayh), while both traditions also existed in Greece (Apollonius Dyscolus). The integration of the two stemmed from twentieth-century Europe (see 5.3).

Cross-fertilization with the sciences (and social sciences) continued. While the humanities developed into a separate branch of disciplines in the modern period, the interaction with the sciences remained particularly strong. Dating methods from physics found their way in historiography and archaeology. X-ray and infra-red analysis of paintings became well established in art history, and mathematical analyses were used in (economic) history and linguistics. And there were influences from the humanities to the sciences as well: Morelli's stylistic approach inspired forensic science, Chomsky's formal linguistics came to be used in computer science and biology, and Saussure's structuralist approach was adopted in almost all social sciences. In fact, there was a continuous interaction between the social sciences and the humanities. For example, Comte influenced Marx's historical work, Weber had a major influence on historiography, and Geertz's anthropological and Labov's sociological work influenced and were influenced by respectively cultural history and linguistics. Perhaps the strongest interaction between the sciences and humanities is currently happening in the upcoming field of digital humanities. Digital approaches have led to computational explorations of all of the humanities. Thousands of literary works, musical pieces, and works of art can be compared in one go and have, among many other things, resulted in new techniques for authorship attribution and style recognition (see e.g. 5.5 and 5.6).

Break or continuity with the early modern humanities? The transition from early modern to modern disciplines is usually seen as a conceptual break or even as a revolution in intellectual history.[350] According to this vision the early modern humanities were still in step with the other sciences as regards method and results, whereas in the course of nineteenth century, and especially in the twentieth century, two types of disciplines appeared on the stage, each with its own method—natural science, which searched for 'nomothetic' patterns and was

[350] See e.g. Reinhart Koselleck, 'Einleitung', in O. Brunner, W. Conze, and R. Koselleck (eds), *Geschichtliche Grundbegriffe: Historisches Lexikon zur Politisch-Sozialen Sprache in Deutschland*, volume 1, Klett-Cotta, 1972. Johan Heilbron, Lars Magnusson, and Björn Wittrock (eds), *The Rise of the Social Sciences and the Formation of Modernity: Conceptual Change in Context, 1750–1850*, Kluwer, 1998. Foucault, *The Order of Things*. Ian Hacking, *Historical Ontology*, Harvard University Press, 2002.

concerned with *erklären* (explaining), and the humanities, which studied 'idiographic' unique events and was involved with *verstehen* (understanding)—see chapter 1 and 5.1.

Our historical quest showed a rather different picture. Contrary to the constitutive recommendations of Dilthey, Windelband, and others, there has been a continuous humanistic tradition from Antiquity to the present day that focuses on the quest for patterns and rules (with alongside it a parallel tradition that concentrates on the *rejection* of patterns—to which we shall return at the end of this conclusion). There was no clear-cut conceptual break. As we have seen, all nineteenth-century disciplines and methods were part of a longer process that started already in the eighteenth or seventeenth century, and often earlier. The 'innovations' of Lachmann's philology, Ranke's historiography, and Grimm's linguistics, to name a few, were often systematizations of methods and practices developed long before, such as can be found in Poliziano's philology, Scaliger's historiography, and Hemsterhuis's or Ten Kate's linguistics. What is novel in the nineteenth century is thus the systematization and the proliferation of these practices thanks to their institutionalization in the newly founded universities. This resulted in a disciplinary stability that had never been seen before. If there was anything like a revolution, it was on an institutional rather than on a conceptual level.[351]

Cumulative progress in the modern humanities as well? The issue of continuity in the humanities gives rise to the question of whether there was cumulative progress in the modern period, as well. We dealt with this question in the conclusion to chapter 4 with regard to the early modern humanities in terms of Kuhn's notion of 'problem-solving capacity'. Below we will describe briefly for each humanistic discipline what the continuity with its early modern predecessor consisted of, and check whether or not it included progress in solving the problems discussed in the conclusion to chapter 4. We will not consider the newer disciplines of film studies, cultural studies, and new media because they have no early modern counterpart to compare them with.

Historiography. Ranke's source criticism (5.1) brought about a philologization of modern history. Although Ranke changed the general practice of history writing, his philological approach did not represent a break. His source criticism was a conscious continuation and extension of the humanistic historiography of Francesco Guicciardini, Joseph Scaliger, and others, who also employed strict philological criticism (see 4.2). We have moreover seen how African oral history (between the sixteenth and nineteenth centuries) reached European history by way of anthropologists and in so doing brought about a degree of continuity between different regions. As regards cumulative progress, we have seen advances in the problem-solving capacity when we consider, as we did in the conclusion to chapter 4, the problem of dating historical events. Here we found ever greater refinement thanks to the inclusion of all possible sources by the *Annales* school and

[351] See also Rens Bod, 'Introduction: the dawn of the modern humanities', in Bod, Maat, and Weststeijn, *The Making of the Humanities, Volume II: From Early Modern to Modern Disciplines*, 2012, p. 17.

others, and also thanks to the use of new techniques for establishing the age of historical artefacts, such as dendrochronology and the C-14 method (see 5.5).

Philology. Nowhere was the continuity greater than in stemmatic philology. Lachmannian text reconstruction built on the humanistic methods of Poliziano and Bentley, which were in the tradition of the Arab *isnad* philologists and can ultimately be traced back to the Alexandrian scholars. There was also cumulative progress on the problem of text reconstruction. As we saw in 5.2, Lachmann was the first philologist to formalize the method as a system of rules, after which philologists in the nineteenth and twentieth centuries gave the text reconstruction method an increasingly formal and more mathematical basis. Currently, stemmatic philology appears to be a 'completed' discipline, like classical mechanics or geometrical optics in physics. However, (non-stemmatic) philology is also dependent on the reading and hermeneutic interpretation of texts, and this reconstruction is less subject, if at all, to cumulative progress.

Linguistics. There was also a high degree of continuity to be seen in linguistics. The historical linguistics of Grimm and Bopp built on Beauzée's earlier work and Hemsterhuis's ideas about an original language without exceptions. The Neogrammarians continued down the same path as Grimm, Bopp, and their colleagues. With their focus on synchronic rather than diachronic linguistics, the structuralists appeared to display a break, but in terms of method, they still employed a system of rules and there was no interruption in the search for patterns. With Chomsky we saw an integration of structuralist ideas with the work of Panini. The more example-based (data-oriented) linguistics built on the traditions of both Panini and Sibawayh. Progress was easiest to see in testing the grammars developed in computational linguistics. The cumulative progress here is not just present—it is quantifiable too.

Logic. The greatest change seemed to have taken place in logic. A new system of logic was developed for the first time since Aristotle—Frege's predicate logic. Yet this logic continued to build on the early modern attempts of Leibniz and others, and Frege's predicate logic even generalized the centuries-old propositional logic. *Montague grammar* was also a solution to a classical problem, that is the integration of logic and linguistic systems. Moreover, Frege and Montague advanced the centuries-old Indian concept of compositionality, even though they were probably not aware of it. Thanks to Frege's predicate logic, progress with determining the validity of a logical argument (see conclusion to chapter 4) was also demonstrable. For the first time, entire sentences were converted into predicate-logical formulas, which were then combined into complex arguments of which the validity could be computed exactly. It was an unparalleled result in the history of logic.

Musicology. In musicology both the early modern quest for consonance and the early modern development of musical grammars continued. However, there was much greater emphasis on the historical study of music, although this was primarily in Europe (the opposite happened in China). For the first time there was cumulative progress in musicology: while Plomp and Levelt built on the work of Helmholtz and earlier musicologists and acousticians, they solved the problem of dissonance versus consonance to a degree of satisfaction. In any event, they reduced the problem to a partly physiological and partly historical subject. Centuries-old musical grammars, with their hierarchical stratification, were formalized and tested against listeners' intuitions through Schenker and the work of Lerdahl and Jackendoff. As in linguistics, these musical grammars were enhanced by an example-based component, after which there was also detectable progress in the accuracy of predicting the observed musical phrase structure (5.4).

Art history. The style-focused art analysis of Morelli, Berenson, and Wölfflin was largely absent in the early modern era, yet such a tradition existed in China during the Ming Dynasty (4.5). However, there was a high degree of continuity between older approaches like Alberti's *compositio* and Wölfflin's method, namely in the hierarchical part–whole analysis of works of art (5.5). The transition from early modern to modern art historiography consisted largely of a shift from prescriptive to descriptive. As far as progress was concerned, the problem of depicting three-dimensional objects on a two-dimensional surface was as good as solved in the early modern age (see the conclusion to chapter 4). We therefore see that the greatest advances were in the problem of attributions to specific masters. Initially they were still made in an intuitive way, but Morelli developed a taxonomy of painterly details, which was expanded into a general style analysis by Wölfflin. The ongoing development of natural scientific methods has meant that a painting can be attributed to a particular painter, or not, with greater precision.

Archaeology. It is difficult to talk about long-term continuity in archaeology because excavations did not take place until the eighteenth century. There was, however, some continuity that can be identified since Antiquity in the study of archaeological artefacts and in the form of epigraphy and antiquarianism. However, if we want to talk about cumulative progress, we must confine ourselves to the eighteenth, nineteenth, and twentieth centuries. Here we find progress in detecting archaeological sites and dating artefacts. Initially sites were located haphazardly or through textual sources, but since archaeological soil surveys became a regular part of infrastructure projects, the site problem has been highly systematized and the chance of finding one has risen dramatically. The introduction of aerial photography has increased the detection options even more. Dating has also improved significantly, initially through the study of clay layers, then by examining tree rings and more recently thanks to the C-14 method (see 5.5).

Literary and theatre studies. We came across nineteenth-century literary historiography in Europe in the early modern work of Bembo and Huet (see 4.7). The development of narrative grammars by Propp, Todorov, and others appeared entirely new, but in a manner of speaking we also found them in classical poetics, with its prescriptive narrative and theatrical structures. While early modern poetics is completely prescriptive, however, modern literary analysis is primarily descriptive. The use of a system of rules in Russian formalism and narratology, on the other hand, is different from all other early modern systems. Here there was a break in formalization. Cumulative progress was visible in the problem of deriving the underlying narrative structure. Propp inferred the building blocks in his narrative grammar ('narratemes') subjectively, whereas in narratology these structural components can be very largely predicted, and the process can even be executed automatically in some computational approaches. When it comes to the attribution of meaning and interpretation, though, progress is difficult or even impossible to establish. And what applies to literary studies is also valid for theatre studies. Theatre histories were written early on, by J. C. Scaliger and others, but Herrmann's reconstruction of historical theatre was new. Similarly, modern acting theory is as old as Aristotle and Bharata Muni. Progress was visible mainly in the accuracy of theatre reconstructions, for example the Globe Theatre (see 5.6).

Progress in all humanistic disciplines. The discussion above shows that, as in the early modern humanities, there was also progress in the modern humanities (in terms of problem-solving capacity). However, we should once again emphasize that this progress was visible with regard to only one or two problems in each discipline. Yet contrary to the commonly held view, there was progress in all fields.

Dichotomy in the post-war humanities? Although there is no break between the early modern and the modern humanities, there is a divide in the post-war humanities. We see it primarily in the rise of the deconstructivist and poststructuralist movements. While it is true to say that the quest for universal patterns remained, alongside it a tradition arose that rejected this search, even though culture-specific patterns continued to be identified (see 5.1, 5.5, and 5.7). The two traditions do not appear to be reconcilable—the stylistic and computational analysis of art, music, and literature is a long way from the poststructuralist and deconstructivist approach, and vice versa. There are two art histories, two musicologies, and two literary studies that operate without any real contact, usually ignorant of each other's attainments. As if history is repeating itself, in the present-day humanities there is on the one hand the 'analogist' movement, which goes back to the Alexandrians in the third century BCE, and on the other we find an 'anomalist' movement, which has its origins in Pergamon around the same time—see 2.3.

It is striking that the pattern-seeking and pattern-rejecting activities are not equally vigorous in every discipline. For example, pattern-rejecting movements, which repudiate universal patterns, are far less common in linguistics and logic than in literary studies, musicology, and art history. We also find this dichotomy in the third century BCE, and sometimes even in one and the same person. In his philological opinions, Chrysippus of Soli maintained a pattern-rejecting view (which nevertheless resulted in fascinating text interpretations, see 2.3), but at the same time he was the designer of the most systematic logic in Antiquity—propositional logic (see 2.6). Such a scholarly split is an exception and it would have remained unnoticed if we had not approached the history of the humanities in a comparative fashion.[352] However, the seed of the thinking that rejects patterns germinated in Antiquity and it is a puzzle why poststructuralists do not seek their spiritual forebears among the anomalists of Pergamon. Or are they afraid, as Marx said, that everything in history occurs twice—the first time as tragedy and the second time as farce?[353]

[352] This kind of 'mental' split is also found in the history of science, e.g. in figures like Ibn Sina (Avicenna) and Descartes for whom natural philosophy and mathematical knowledge of nature were entirely separated areas. I thank H. Floris Cohen for pointing this out.

[353] Karl Marx, 'Der 18te Brumaire des Louis Napoleon', *Die Revolution: eine Zeitschrift in zwanglosen Heften, erste Hefte*, 1852.

6

Conclusions: Insights from the Humanities that Changed the World

In this book I have uncovered a line in the history of the humanities—the quest for methodological principles and empirical patterns that has lasted from Antiquity to today. By summing up in this chapter only those insights from the humanities that 'changed the world', I do not do justice to the rich and broad history of the humanities. Nevertheless, it is precisely these world-changing insights that were ignored for too long and often unjustly credited to the natural sciences. In this conclusion I will therefore begin by shining the spotlight on the impact of the humanities. After that I will discuss the relationship between the humanities and the sciences and outline a few future prospects.

If there is one picture that emerges from this conclusion, it is that the quest for principles and patterns in language, music, art, theatre, and literature exists all over the world. The history of the humanities indicates that precise methods are not limited to the natural sciences,[1] and that 'world-changing discoveries' are made in all disciplines.

The invention of source criticism was at the birth of the Reformation and brought about the Enlightenment. The invention of source criticism led to one of the greatest changes in the Western world view, and is still hugely influential to this day, particularly in historical truth-finding. In fact, we are talking here about two discoveries that were amalgamated during the nineteenth century into one overarching method—*source criticism* and *source reconstruction*. *Source criticism* was created in a rudimentary form in Greece (see 2.2), and via the formal *isnad* method (3.2) it resulted in a precise scholarly method among early modern humanistic scholars (4.1). In the hands of Valla, Scaliger, and others, it became a powerful weapon with which centuries-old sources could be exposed on the basis of purely grammatical, historical, and logical analyses. Now one individual could make mincemeat out of a source that had been deemed unimpeachable. The most famous example is Valla's refutation of the *Donatio Constantini*, dating from 1440, and the subsequent *de jure* lapsing of the legitimacy of the papal state (see 4.1). Initially Valla's result had little or no effect, but during the Reformation it was used by Martin Luther and others as potent evidence against the secular power of the Catholic Church. Through Poliziano, Valla sowed the seed of formal

[1] A similar thesis was defended by Frits Staal, but only for logic and linguistics—see Frits Staal, *Universals: Studies in Indian Logic and Linguistics*, University of Chicago Press, 1988.

text reconstruction, which could be used to reconstruct the original source version from surviving copies by means of a genealogical derivation (4.1). In the late sixteenth century Joseph Scaliger established on the basis of the reconstruction of Egyptian lists of kings (by Manetho) that world history had to be older than biblical history (4.2). Although Scaliger tried to save the Bible by introducing a notion of 'mythological' time, during the course of the seventeenth and eighteenth centuries his result led to fierce criticism of the Bible that ushered in the Enlightenment. Ranke and Lachmann perfected source criticism and source reconstruction in the nineteenth century, which even today forms the basis of much historical research (5.2). In addition, international institutions as well as national governments often use source criticism and philological reconstructions to help establish the reliability of documents. It is not just in the legal system that source criticism is part of the toolkit of auxiliary sciences. It is also part of everyday practice in, among other things, the historical truth-finding with regard to genocide.

The humanistic discovery of the interaction between theory and empiricism formed the basis for the Scientific Revolution. In fifteenth-century philology, art theory, and musicology there was intriguing interaction between theory and empiricism, where empiricism had the last word, no matter how solid the theory might have seemed. As we saw above, this interplay began in philology with Valla and Poliziano, and via Scaliger and Casaubon it resulted in an early modern discipline with its own methodology (4.1). An example of the impact of this discipline is the fact that many sixteenth- and seventeenth-century 'natural scientists' were also philologists, from Galileo and Kepler to Newton. Philology was the first early modern discipline to show how hypotheses and even theories could be brought down on the basis of new observations (for instance, newly discovered texts). We find this interaction in art theory too. Although Alberti created a splendid mathematical structure for linear perspective, it emerged that *empirical* perspective did not comply with it. Leonardo left no stone unturned in order to unravel the underlying theory of this empirical perspective, but apart from a few rules of thumb he did not produce theoretical underpinning (4.5). Leonardo's experiments, in which he investigated the effect of the light source, colour, and position of an object very systematically, were a source of inspiration for the New Scientists. Meticulous string experiments and their mathematical underpinning by humanistic musicologists were similarly important. Through their studies of the laws of consonance, they conveyed the interaction between hypothesis development and experiment virtually directly to the new generation of scholars, which was literally the case between father and son Galilei (4.4). The quest for the underlying patterns in philology, art theory, and music theory revealed a continuity with the later search for the underlying laws of falling bodies, planetary movements, and their mathematical foundations, where some of the results from the first hunt formed the starting point for the second (see 4.4 and the conclusion to chapter 4).

The invention of grammar favoured imperialism and laid the basis for computer science. The scope of grammars has been multifaceted. First and foremost the pedagogic grammars of Dionysius Thrax, Donatus, and Sibawayh enabled the spread of, respectively, Greek in the Hellenistic world, Latin in the post-Roman world, and Arabic in Islamic civilization. Without these practical grammars no language would have developed into a lingua franca and a Hellenistic, Roman, or Arab empire would not have been created, or in any event not in the same way. In addition to the imperialist function of grammars, their underlying *formalism* changed the world as well, albeit in a completely different way. Panini's formal system of rules was used nearly 2,500 years later as the underlying formalism for higher programming languages (see 2.1). The person who actually established this connection between linguistics and computer science was Noam Chomsky, who referred to Panini as his spiritual father (see 5.3). It was Chomsky's precise notation for defining grammars that was applied to programming languages by computer scientists like John Backus.[2] Chomsky's syntactic definition of a language served as the pattern for the structure of the whole compiler for ALGOL—the first higher programming language. Thus, the development of modern programming languages was initiated through formal linguistic work that laid the foundations of information technology with its many impressive applications. This unexpected application of the study of language is rarely if ever mentioned in the historiography of linguistics, while it is widely acknowledged in (the history of) computer science.[3] The disregard of applications of humanistic insights seems to be symptomatic for the humanities.[4] This has also happened with the unexpected application of philological reconstruction methods to the fields of genetics and DNA analysis.[5]

Other insights: Indo-European revealed the relationship between peoples, ancient texts fuelled nationalism, accessibility of the digital world shown to be a myth. The examples above are the tip of the iceberg. We can remind ourselves how the nineteenth-century discovery of the Indo-European language family (5.3) defined our view of the relationships between peoples, for better and worse. Among other things, this discovery led to the hypothesis of the existence of a 'pure' Aryan race, a theory which would be hijacked by the National Socialists.[6] This shows that the impact of the humanities, like that of the sciences, is not necessarily positive. The claim that the humanities are important for democracy and developing critical citizens,[7] thus deserves a more nuanced discussion. For nineteenth-century scholars like Max Müller and Christian Lassen it was crystal clear that the linguistic evidence

[2] Erol Gelenbe and Jean-Pierre Kahane (eds), *Fundamental Concepts in Computer Science*, Imperial College Press, 2009, p. 99.

[3] See e.g. Edwin Reilly, *Milestones in Computer Science and Information Technology*, Greenwood Publishing Group, 2003, pp. 43ff. See also Martin Davis, Ron Sigal, and Elaine Weyuker, *Computability, Complexity, and Languages: Fundamentals of Theoretical Computer Science*, Academic Press, Harcourt, Brace, 1994, p. 327.

[4] In a recent collection of 24 essays on the relevance of the humanities commissioned by UK's Arts and Humanities Research Council, there is no contribution that takes on the case for the practical applications of the humanities—see Jonathan Bate (ed.), *The Public Value of the Humanities*, Bloomsbury Academic, 2010.

[5] For details, see Rens Bod, 'Discoveries in the humanities that changed the world', *Annuario 53, 2011–2012*, Unione Internazionale degli Istituti di Archeologia, Storia e Storia dell'Arte in Roma, 2011, pp. 189–200.

[6] Stefan Arvidsson, *Aryan Idols: Indo-European Mythology as Ideology and Science*, University of Chicago Press, 2006, pp. 241ff.

[7] Nussbaum, *Not for Profit: Why Democracy Needs the Humanities*.

for an ur-language meant that there also had to be a pure Aryan race and that some other races were endlessly mixed and impure.[8] Many of the most critical philologists and linguists accepted this view. A similar negative impact followed the unearthing and reconstruction of old literary texts, paintings, and archaeological artefacts in the nineteenth and early twentieth century: these had a vast influence on national self-image and identity construction, which contributed more than a little to the growth of nationalism and racism during the twentieth century (as happened in particular with the rediscovery of the *Nibelungenlied* which was used as the cornerstone of later German war propaganda—see 5.2).

To add some recent examples of the impact of the humanities, we should not forget the post-war 'discoveries' in film studies, television studies, and media studies, such as the debunking of the myth about the accessibility of the digital world (see 5.7).[9] Those discoveries also include an analysis of the medium of television indicating that viewers are captured through 'flows'—i.e. non-stop streams of information, advertising, entertainment, and trailers—whose purpose is to keep the viewer tuned to a particular channel (see 5.7). And what to think about the disturbing discovery that the TV series *Crime Scene Investigation*, which has dragged on for years, consists of only eight narrative building blocks (5.6)? Time will tell whether these insights and discoveries are going to change the world, but they are in any event sensational in all respects.

No radical dichotomy between humanities and science. Nowhere in our history of the humanities did we come across an acute divide between the humanities and science. Both humanists and scientists search for underlying patterns, which they try to express in logical, procedural or mathematical formalizations. There is, moreover, a continuity between humanistic and scientific disciplines as regards the 'nature' of the patterns. Whereas patterns in the humanities appear to be less absolute and subject to changes, in the sciences they seem to be absolute—in any event in physics. However, in the humanities themselves there is also a gradual shift from virtually absolute sound shift laws to less absolute harmonic rules to changeable culture-specific patterns (see 5.5 and 5.6). Yet there is also a gradual shift like this to be found in natural science—from the absolute laws of theoretical physics to the more approximate laws in chemistry to the local and variable patterns in biology. The eminent biologist Ernst Mayr contended that universal patterns do not exist in biology.[10] Mayr admitted that the laws of physics and chemistry, of course, apply to biological systems at a molecular level. In a complex system, though, no biological regularity has ever been observed that complies with the rigorous definition of a 'law' in theoretical physics. According to Mayr, what biologists mean by a 'law' is a pattern that is usually local and not universally valid and is moreover often statistical. These regularities are widely used in

[8] Georges Vacher de Lapouge, 'Old and new aspects of the Aryan question', *American Journal of Sociology*, 5(3), 1899, pp. 329–46.

[9] Jan Baetens, Joost de Bloois, Anneleen Masschelein, and Ginette Verstraete, *Culturele studies: theorie in de praktijk*, Vantilt, 2009, pp. 131ff.

[10] Ernst Mayr, *This is Biology: The Science of the Living World*, Harvard University Press, 1997, p. 62.

explanations of biological phenomena, without their being reduced to deeper physical or chemical laws.

The philosopher of science Philip Kitcher agreed with this when he stated that there are *autonomous levels* of biological explanation.[11] In biology the set of concepts and explanations used at cell level is different from that used at an ecological level, for instance. This does not exclude the reduction—sooner or later—of complex biological processes to physical ones. However, it does not make sense to reduce a biological phenomenon to elementary particle physics in order to understand it. In line with Kitcher we argue that there are also *autonomous levels of analysis and understanding in the humanities.* Obviously the laws of physics also apply to the human brain, and therefore also indirectly to the products of that brain. Yet it is not the case that we need to consult biology or physics for the analysis of a human expression like a literary work or a piece of music. The cognitive and neurosciences have produced important insights into the study of language and music,[12] among other things, but it becomes impossible and even senseless if we try to understand Greek vase painting or Renaissance architecture in terms of the sum total of all brain activities relevant at the time. It proves to be the case that *autonomous* analysis levels—from the art-historical analysis of paintings to the narratological analysis of texts—deliver the most insightful patterns and interpretations.

The concept of 'exception': a difference between humanities and science after all? There might, however, be a difference between the humanities and natural science in the notion and treatment of exceptions. The statement that 'the exception proves the rule' seems unthinkable in natural science—although we should stress here that in the humanities this pronouncement is only used in the prescriptive tradition of secondary school grammars (see 2.1). All the same, there are most certainly exceptions in the humanities. However, they are not solely to be found in the humanities, but in the natural and social sciences too.

Theoretical physics, with its universal laws, is sometimes referred to as the only exceptionless discipline. This may represent a possible demarcation. Yet this demarcation characterizes not so much the difference between science and the humanities, as between theoretical physics and other scientific fields. While theoretical physics permits no scope for exceptions, applied physics is full of ad hoc corrections, phenomenological constants, normalizations, and so-called provisos. And although the universal laws of nature are considered to be exception-free, in mathematical derivations and explanations of specific phenomena, ad hoc approximations and corrections are used more than once.[13] Here too it is not possible to assert anything other than that there is a gradual scale from disciplines with the fewest exceptions to those with the most. While theoretical physics reflects an ideal

[11] Philip Kitcher, '1953 and all that: a tale of two sciences', *Philosophical Review*, 93, 1984, pp. 335–73.

[12] See e.g. Aniruddh Patel, *Music, Language and the Brain*, Oxford University Press, 2008.

[13] For an overview, see Nancy Cartwright, *How the Laws of Physics Lie*, Oxford University Press, 1983.

picture, it is not feasible for most natural sciences (see our discussion of biology above), let alone for other areas of scholarship.[14]

The history of the humanities as liberation from the biblical, classical, and nationalist yokes. One could look on the history of the humanities as an ongoing search for principles and patterns and their formalizations—as we have done in this book—but also as an ongoing liberation from imposed ways of thinking. While the Greek humanities of Antiquity were based primarily on Aristotle, in the Middle Ages in Europe there was a pursuit of biblical coherence (in which the Aristotelian system of thought was integrated). In the early modern era, on the other hand, there was the endeavour to attain humanistic classicism that slowly but surely distanced itself from both the Bible and Aristotle, whereas the modern age showed above all a historicization of the humanities, which was then employed for nationalist or colonial goals. To oversimplify, we can assert that after an initial Aristotelian phase, Western humanities found itself under the yokes first of the Bible, then of classicism, and finally that of nationalism. Originally the Bible, the classics, and nationalism were useful frameworks in which new ideas, methods, and patterns were swiftly created. If the boundaries of such ways of thinking were reached, though, which often happened all too soon, they represented a straightjacket that could only be cast off with great difficulty. The super-fast embrace of frames of mind followed by a laborious deliverance from them is a recurring pattern not just in the humanities: we find it in all disciplines and in all regions.[15]

Supra-disciplinary humanities and humanists. This book has put a number of unexpected humanities scholars in the limelight. The fact that they did not emerge earlier as influential thinkers is because most humanists have little notion of the history of the humanities as a whole. It was predictable, of course, that Aristotle would have an eminent place, like Leibniz and Chomsky. But who could have suspected that Chrysippus of Soli from the third century BCE would go down in history as one of the most important figures in the humanities? While it is true that Chrysippus is well known as the founder of propositional logic (see 2.6), as the instigator of the *anomalist* tradition, he only enjoys some fame among philologists (see 2.3). Yet, with his anomalist approach, Chrysippus launched the entire pattern-rejecting tradition, which has represented an ongoing line in the history of the humanities alongside the *analogist*, pattern-seeking movement (see the conclusion to chapter 5). This line can only be recognized if the humanities are considered in their totality.

Scholars who were long considered to be marginal also receive a significant place in our history for the first time. The seventeenth-century researcher William Holder is a prime example. While Holder's work in the *separate* disciplines of musicology and linguistics was snowed under by the work of others, if his

[14] For further discussion, see Rens Bod, 'Towards a general model of applying science', *International Studies in the Philosophy of Science* 20(1), 2006, pp. 5–25.

[15] William Dampier, *A History of Science and Its Relation to Philosophy and Religion*, Cambridge University Press, 1966. See also H. Floris Cohen, *How Modern Science Came into the World: Four Civilizations, One 17th-century Breakthrough*, 2010.

work is looked at as a whole, he acquires a very different status. His study of musical consonances and microintervals (see 4.4), his in-depth investigation of sign languages (see 4.3), and his discoveries in the field of articular phonetics—which disappeared from view in the nineteenth century, see 4.3—reveal a brilliant supradisciplinary scholar with a striking unity of research. Holder concentrated primarily on the human production of sound (language and music) as well as on its absence (the deaf). Holder's contribution is therefore greater than the sum of its parts. Holder only remained marginal in the histories of the disciplines because the subjects he worked on did not develop into a discipline ('the study of the human production of sound'). We can snatch scholars like this from obscurity if we no longer address the history of the humanities on a discipline-by-discipline basis but as one *overarching* field. I readily admit that my book also uses a classification based on 'disciplines'—otherwise I would have had no handholds or footholds at all. By addressing these disciplines alongside one another and continually making comparisons, we spot 'possible' disciplines that did not make it. A future book on the history of the humanities would do best to drop these disciplinary designations, which, after all, were only created in nineteenth-century Europe. The modern compartmentalization of the humanities should not stand in the way of its history.

Aside from scholars like William Holder who were almost forgotten, it also emerges that well-known names in the humanities were often active on a much broader front than that for which they are known. Leon Battista Alberti, for example, emerges as one of the most influential humanistic scholars of all time. While he is best known for his art theoretical innovations (such as linear perspective and the *disegno* theory), he has major achievements to his name in archaeology, philology, musicology, and linguistics (see chapter 4). The same can be said of Lorenzo Valla, whose work was not limited to philology, but extended to logic, linguistics, and rhetoric, too. Similarly we should not forget Joseph Scaliger, whose discerning philological, historical, and linguistic work changed the European world view (although he was well below par as a mathematician—see 4.2). St Augustine was likewise not just a theologian or philosopher—he played a decisive role in the medieval humanities as a rhetorician, historian, and poeticist (see chapter 3).

The Islamic humanistic scholars also had unprecedented breadth. Al-Farabi was a linguist, musicologist, logician, and much more besides, like Avicenna (Ibn Sina) and Averroës (Ibn Rushd), who in fact covered the humanities, natural, and social sciences. Al-Biruni, for example, proved to be an Indologist, linguist, historian, astronomer, mathematician, and more too. Take Sima Qian (see 2.2), as well as the Chinese exponents of the Empirical School (see 4.1 and 4.3), who were at home in historiography, philology, linguistics, and poetics. The Indian Bharata Muni (see chapter 2) is famous primarily in the history of theatre studies, but he was also one of the most important musicologists and literary theorists. The insight that Panini (see 2.1) would play such an overpowering role in the rule-based tradition in the humanities is largely new—while not forgetting the example-based tradition that started with Sibawayh (see 3.1).

Nevertheless, in our umbrella history of the humanities there are also people who are not given the position they deserve. The fifteenth-century scholar al-Suyuti was both a linguist (see 3.1) and a fascinating historian (3.2), but because of our focus on principles and patterns, he only gets a brief mention—like many others. Some brilliant scholars do not appear in this book at all, or are discussed only indirectly, such as Max Weber (see 5.1), who, besides being a sociologist, was also a cultural historian, historian of religion, musicologist, economist, and jurist. However, the sociological aspect of his work was so predominant that he does not fit well into our history of the humanities. Similar reasoning almost led me to decide to treat Auguste Comte the same way, but I could not get around his positivism. We have, of course, come across such overwhelming breadth before—the icons of natural science proved to be humanistic scholars too. Galileo was one of the most important early modern music researchers, Kepler was an eminent philologist, and Newton spent the greater part of his life on theology and historiography (in which he far from excelled, however—see 4.2). The modern age can also boast supra-disciplinary, integrated humanities-science scholars, with Hermann von Helmholtz (5.4) and Noam Chomsky (5.3) as shining examples. Only an overarching history of all fields of learning could do justice to them.

Unexpected influences back and forth. The contacts between India, China, Europe, Africa, and the Arab world have been plentiful and fruitful. Many of these influences are known, but have still not been recognized in all cases. It has emerged, for instance, that Indian linguistics from the fifth century BCE became known first in China and then in Islamic civilization, and later it overran Europe and the United States, after which it was the turn of Western linguistics to influence the other regions. We have also seen that Arab logic, historiography, linguistics, and musicology played a decisive role in the development of European humanities in the later Middle Ages and the Renaissance.

It is less easy to demonstrate other influences. For the time being, for example, there is no concrete evidence that the Arab *isnad* method of *oral* source reconstruction (see 3.2) influenced the European method of *written* source reconstruction (see 4.2). Yet the methods are so similar and the flow of knowledge from Islamic to Christian Europe was of such a magnitude (see chapter 3) that influence cannot be ruled out. We have moreover seen parallel, sometimes almost identical developments in different regions, in regard to which direct effects are equally difficult to establish, such as the fundamental laws of logic in China and Greece (see the conclusion to chapter 2), the rules of art history in China and India (see 2.5), and the laws of harmony in India, China, and Greece (see 2.3).

This book has also exposed an unexpected influence—the line than runs from Africa to Europe. This line was largely already known, but it has seldom been referred to by name. Here we give four examples of the effects that Africa has had on European humanities:

(1) The lists of kings by the North African historian Manetho (see 2.2) were used by Joseph Scaliger to construct his revolutionary chronology, the net effect of which was to overthrow the Christian vision that the age of the earth could be derived from the Bible (see 4.2). In the long term this resulted in a new, secular view of man and society.

(2) The Christian revolution (see the conclusion to chapter 3) was unleashed by North African, often Berber, historians such as St Augustine, Sextus Africanus, and Orosius and led to the redefinition and reinterpretation of virtually all activities in the humanities in medieval Europe (see chapter 3).

(3) The cyclical model with cultural propagation that was developed by the Tunisian Ibn Khaldun influenced the vision of the Western world among European historians such as Spengler and Toynbee (see 5.1).

(4) The centuries-old African oral history, and especially the nineteenth-century African historiography which emerged from it, seems to have brought about European oral history via Western anthropologists (see 5.1). It was thanks to this influence that 'non-elitist' oral history went on to play a significant role in twentieth-century historiography.

Future research: from Japan to pre-Columbian America. Although I have tried to write this book from a rather global perspective, Western humanities nevertheless receives disproportionate attention. The reasons for this were discussed in chapter 1—a very great many manuscripts from other regions are not yet accessible. Hundreds of thousands of early modern Arabic, African, and other manuscripts are waiting to be unlocked (see 4.2). I have moreover had to limit myself to written sources, whereas in many regions the study of literature, art, and music exists in oral forms too. Yet even within these constraints I have left many areas untouched.

The illustrious Japanese historiography, for instance, has not been addressed. It would have been fascinating to investigate whether the medieval *shogun* histories, with their cyclical pattern,[16] would fit into the metapattern we found in 3.2 that the time structure of a culture's canonical texts corresponds to the time structure of its historiography. In general, Japanese humanistic activities have barely been explored. It is commonly assumed that Japanese science and learning did not begin to blossom until the country opened up to the West in the late nineteenth century. Prior to this, from 1641 to 1853, the only access that Japan had to Western learning was through the Dutch trading post on the island of Deshima. The literal meaning of the Japanese word for 'science' at the time, *rangaku*, was 'Holland studies' ('Holland' is *O-ran-da*). For centuries, all external scientific knowledge in Japan came from books that had been translated into Japanese via Dutch, or that were even read directly in Dutch. These thousands of documents were eagerly snapped up in Japanese cities, where seventy to eighty per cent of the population was literate. However, these works were restricted to natural science and technology—fields in which Japan had a lot to learn from the West. In the domain of the humanities, though, Japan had its own centuries-old tradition. In addition to historiography, poetics also flourished, with one of the highlights being the famous fourteenth-century treatises of Zeami Motokiyo about the *No* drama, which found

[16] Markus Völkel, *Geschichtsschreibung: eine Einführung in globaler Perspektive*, Böhlau Verlag, 2006, pp. 181ff.

their way to the West.[17] What was the status of the other humanistic disciplines in Japan? And, starting in the seventeenth century, was there interaction between the humanities and the *rangaku*, or did these two activities represent separate worlds? In the latter case, Japan would be an exception to our generalization that the humanities and natural science in all regions were studied jointly and mutually reinforced each other.[18]

We have not discussed pre-Columbian America in our book either, yet here we find one of the most spectacular forms of historiography—the pictorial histories of the Aztecs.[19] The *Codex Mendoza*, for example, consists of a combination of text and images dating from 1541 which describes the conquests of the Aztec rulers, together with a list of the taxes paid by the conquered provinces and accounts of daily life. If such a thing is possible, the *Aubin Codex* is even more impressive. In beautifully coloured pictures, it portrays the departure of the Aztecs from Aztlán as a result of Spanish domination. It has eighty-one pages and covers one of the most dramatic moments in Aztec history—evidence of the bloodbath and the devastation of the temple in Tenochtitlan in 1520. Although we have not yet investigated these documents for principles and patterns, it is obvious that the personal experience principle plays a significant role.

Now that the Mayan script has also been virtually completely decoded, it has become clear that there is an extensive Mayan historiography, including dynastic chronicles, biographies, descriptions of political controversies, and battles.[20] We also find musicological texts in Mayan culture, parts of which are to be found in the historical–mythological book *Popol Vuh* of the Quiché in Guatemala. And there is more.

We have also bypassed the humanities in Korea, Vietnam and the once so flourishing Khmer Empire. And Africa may well offer the most promise for the future. There is no other continent with as many different cultures and languages. The staggering number of hidden manuscripts in and around Timbuktu alone appeals to the imagination. Meanwhile 20,000 of the estimated 700,000 manuscripts have been collected, but have barely been studied as yet.

However, one thing has become clear in this book. The quest for principles and patterns in language, texts, music, art, and literature is of all times and regions. There is no reason whatsoever to exclude any region as a source of inspiration.

[17] See Masakazu Yamazaki (ed.), *On the Art of the No Drama: The Major Treatises of Zeami*, translated by Thomas Rimer, Princeton University Press, 1984.

[18] We know that at least for the development of Japanese linguistics, the study of Dutch grammar was highly influential. Since Japanese scholars of *rangaku* all knew Dutch, Japanese linguistic terminology was determined by Dutch grammar. Here we thus have a clear though rather unique example of how the study of science (*rangaku*) influenced the study of language. See Frits Vos, 'The influence of Dutch grammar on Japanese language research', in Sylvain Auroux, E. F. K. Koerner, Hans-Josef Niederehe, and Kees Versteegh (eds), *History of the Language Sciences: An International Handbook on the Evolution of the Study of Language from the Beginnings to the Present*, volume 1, Mouton de Gruyter, 2000, pp. 102–4.

[19] Elizabeth Hill Boone, *Stories in Red and Black: Pictorial Histories of the Aztec and Mixtec*, University of Texas Press, 2000.

[20] Dennis Tedlock, *2000 Years of Mayan Literature*, University of California Press, 2010.

Linguists can refer both to Panini, Dionysius Thrax, and Sibawayh as their spiritual fathers. Art historians should not just look to Vasari but also to Xie He as the 'first' art historian. Literary and theatre theorists do not need to be influenced by Aristotle alone. Bharata Muni and Liu Xie would do equally well and bring in different perspectives. And Herodotus is not the only doyen of historiography—there are Sima Qian and Ibn Khaldun too.

What about the future of the humanities? Although speculations about the future rapidly become gratuitous, it is possible to outline a few trends in the humanities. Despite the prejudices about their supposed uselessness—which we hope to have effectively refuted—the humanities are doing better than ever. Methods and approaches from different fields are being integrated and are leading to new interpretations of the past, languages, works of art, literature, music, films, new media products, and other cultural artefacts. The quest for patterns represents an uninterrupted constant in humanistic research and is being investigated increasingly often with the aid of cognitive and digital approaches. If trends are taking shape, they are thus at least the following:

(1) *The cognitive approach* to investigating humanistic material has resulted in new psychologically motivated testing methods in language, music, literature, and art. This approach has delivered new patterns in all fields of the humanities and represents one of its most active branches (see 5.3 to 5.7).

(2) *The digital, computational approach* to humanistic material has led to new comparisons and methods of analysis. Without digitized sources it would have been impossible to compare thousands of texts or paintings with one another in one fell swoop or to find historical patterns in the abundance of sources. The digital humanities are bringing not just new understanding but also new questions that have never been asked before (see 5.2 to 5.7). This movement is drastically changing humanistic practice.

(3) *The integration of supra-disciplinary methods* from the humanities, sciences, and social studies has produced new disciplines (for example, cultural studies and media studies), and these methods are now also being employed in more traditional humanistic areas (see e.g. 5.4 and 5.5). No matter how great the danger that is lurking in the uncritical combination of different scholarly fields, this trend is proving to be extremely productive and irreversible.

To assert that these trends will continue in the future is, of course, crystal gazing. No matter how inspiring and innovative they seem now, if we look back after twenty years we may discover that entirely different trends have surfaced.

However, what we can safely say is that the palette of the humanities in the past and now—and probably in future too—covers the full spectrum of methods, from the most subjective approach to the most mathematical one, and from the most relativist approach to the most universalist one. This multicoloured palette is both the strength and the weakness of the humanities. Diversity is not always appreciated. There is ever growing pressure to streamline research, to publish in the same leading journals, using the same methods and techniques. The coexistence of mathematical modelling (for example in cliometrics and linguistics) and subjective narration (for instance in narrativism and poststructuralism) alone makes the humanities the most unlikely of all human doings. I started this book with a quest for pattern-seeking activities in the humanities, but towards the end it emerges that the pattern-rejecting tradition is at least as fascinating. We would be better advised not to just put up with the versatility of the humanities, but to embrace it.

APPENDIX A

A Note about Method

In writing this book I have searched through around five hundred humanistic treatises for underlying principles and empirical patterns (see chapter 1). These works are about the study of such subjects as language, music, art, the past, literature, poetry, and theatre, and are referred to in the main body of this book. They cover the period from roughly speaking 600 BCE to 2000 CE and concern China, India, the Arab world, Africa, and Europe/the United States. I studied these treatises in their original language only if they were written in German, Dutch, French, Italian, Spanish, or English, together with some fragments in Latin. I read the other works—written in, for instance, Sanskrit, Arabic, Chinese, Ge'ez, Fulani, Greek, Turkish, and Russian—in the form of translations. I also examined around a thousand secondary sources, most of which are referred to in the notes to this book. I realize that I have ignored possibly hundreds of untranslated treatises (see also chapter 1).

I put the following three (a priori) basic questions when I studied each humanistic discourse:

1. *Which methods and/or principles* did the humanities scholar develop and use for studying the material (language, art, the past, music, literature, and theatre), and what questions did he/she ask?
2. *How did the humanities scholar apply these methods and principles?* How consistently did he/she use them, and why these methods and not others (in so far as this can be answered)?
3. *What answers and patterns did the humanities scholar find in the material studied?* Were the methodical principles employed of use in deriving patterns?

I then put a number of follow-up questions. However, these did not come to the surface until my investigation was in progress because at the outset I could not have known which principles and patterns could be found and which additional questions about them were possible. Without guaranteeing completeness, these were the following (a posteriori) follow-up questions:

(a) What is the *nature of the method and the patterns found with it*: a procedural system of rules, a declarative system of rules, analogies or something else?

(b) Is the method used *rule-based or example-based*, and is the method *pattern-seeking or pattern-rejecting*?

(c) Is there any *formalization of the patterns found* and, if so, what is nature of this formalization (procedural, logical, or mathematical)?

(d) What is the *structure of the patterns found*? Is it linear, hierarchical, or something else (interwoven, rhizomatic)?

(e) Is there an observable *process from descriptive to prescriptive* (or vice versa) in the treatment of the humanistic material?

(f) What are the *parallel or unique developments in different regions*?

(g) Are there *analogies between different humanistic disciplines* with regard to the questions asked, the methods used, and the patterns found (within and between regions)?

(h) What are the *commonalities and differences between the principles or patterns in the humanities versus the sciences*?

(i) What are the *central problems in the humanities* (in different periods and regions)?

(j) Is there (cumulative) *progress in terms of problem-solving capacity*?

(k) Is there *continuity or a break* in the history of the humanities?

(l) To what extent are the principles and patterns under an *ideological yoke* (Pythagorean, Buddhist, Christian, classicist, nationalist, or other)?

(m) Are there *metapatterns* (on top of the patterns found)?

(n) Is there *interaction between theoretical principles and empirical patterns*? In other words, is synergy whipped up between theory and empiricism?

(o) To what extent have the patterns found had *societal impact*, and can we speak of '*humanistic discoveries*'?

I have woven the answers to these questions into a narrative whole, the exact method of which is still a mystery to me—even after reading dozens of works about literary composition—but of which this book is the result.

If my method, or the narrative resulting from it, evokes in someone an association with Whig history, I would like to refer them to my own explanation of it in 5.1. The concept of Whig history has recently gone through an enormous meaning inflation, as a result of which it has become a sort of vague standard reproach to anyone who wonders how something that we know now, originated and developed previously.[1]

[1] With many thanks to H. Floris Cohen, p.c. I wouldn't mind if my narrative be called an anti-anti-Whig history (Michiel Leezenberg, p.c.).

APPENDIX B

The Most Important Chinese Dynasties

Xia Dynasty	2100–1600 BCE
Shang Dynasty	1600–1045 BCE
Zhou Dynasty	1045–256 BCE
Qin Dynasty	221–206 BCE
Han Dynasty	
Western Han Dynasty	202 BCE–9 CE
Eastern Han Dynasty	25–220
Period of Disunity	221–589
Jin Dynasty	265–316
Eastern Jin Dynasty	317–420
Wei Dynasty	386–534
Sui Dynasty	589–618
Tang Dynasty	618–907
Period of Five Dynasties and Ten Kingdoms	907–960
Song Dynasty	960–1279
Yuan Dynasty	1272–1368
Ming Dynasty	1368–1644
Qing Dynasty	1644–1911
Republic of China	1912–present (since 1919 in Taiwan)
People's Republic of China	1949–present

Index

Abbasids 76, 92, 94
Abelard, Peter 124, 126, 140
Abi Ishaq, Ibn 76
Abul Fazl 182
acting theory 337, 351
actio 61
Adler, Guido 301, 307, 310, 328
Adorno, Theodor 264
Aelfric of Eynsham 82
Aelianus, Claudius 21
Aeschylus 65
Affan, Uthman Ibn 106
Africa 5, 6, 8, 74, 84–6, 88, 94, 96, 102, 125, 135, 129, 142, 161–2, 179–81, 198, 208–9, 238, 243, 251, 269–70, 301, 309–9, 323, 346, 348, 359–61
Agamemnon 44
Agathias 90, 93
Agricola, Rudolf 152, 191, 230, 248
Aitzema, Lieuwe 252
Akhfash, al- 78
Al-Andalus 81, 98, 128
Alberti, Leon Battista 47, 51, 163, 184–5, 193, 204, 211–19, 222–5, 227–8, 232, 241, 247, 259, 292, 312, 315, 350, 353, 358
Alcuin of York 79, 87, 104, 130
Alenio, Giulio 196
Alexander the Great 18, 24, 43, 45, 57, 61, 95
Alexandria, Alexandrian 31–5, 40, 60, 70, 78, 85, 87, 105, 148, 246, 272, 314, 349, 351
Alfred the Great 88
Alhazen (Ibn al-Haytham) 213
allegorical, allegory 33, 34, 86, 102, 119, 131–3, 141, 145, 218, 318
Amalric, Arnaud 133n
Ambrose, St 130
anachronism 8, 9, 11
Analects, see *Lunyu*
Analogists, analogistic 31–5, 40, 60, 77, 105, 148, 351, 357
anatomical, anatomy 46–8, 119, 240, 312, 330
Ankersmit, Frank 264
Annals, annalistic 25–7, 34, 35, 84, 86, 88, 98–100, 166, 245, 267
Annals of the Old Kingdom 24
Annius of Viterbo 154, 169
Anomalists, anomalistic 10, 32–5, 40, 54, 71, 72, 131, 236, 351, 357
anthropology 2, 9, 10, 78, 95, 132, 254, 257n, 266, 270, 288, 321, 324, 329–30, 338, 342, 347–8, 360
Antigonus of Carystus 44
antiquarianism 163, 164, 350
Apelles of Kos 45
Apollonius Dyscolus 10, 18–19, 40, 77, 245, 347
Apuleius 144
Aquinas, Thomas 120, 121, 131–2, 145
Arab, Arabic 6, 10, 11, 74–8, 79, 81, 84, 86, 93–8, 101, 102, 103–7, 112, 114–15, 116, 117, 125, 127–9, 131, 133–5, 139–40, 150, 154, 162, 169, 179–80, 181, 185, 198, 208–9, 213, 238, 242, 245, 248, 262, 275, 301, 308, 347, 349, 354, 359, 360, 364
Aramaic 154, 169
archetype 31, 274–5, 286
argumentation 53, 56–63, 126, 130, 136, 191–2, 196, 248, 276, 326, 335
Aristophanes of Byzantium 31–2, 72
Aristotelian 19, 40, 55–6, 60–3, 72, 79, 121, 124–30, 133–4, 140, 195, 209, 219, 230, 233–5, 238, 240–1, 248–9, 357
Aristotle 6, 14, 18, 24, 31, 38–9, 44, 52–5, 57–67, 69, 71–2, 79, 80, 85, 107, 121, 124–7, 129, 133–4, 137, 191–2, 195, 197, 201, 229, 230, 233–5, 237, 241, 243, 248–9, 295, 329–30, 336–7, 349, 351, 357, 362
Aristoxenian 72, 109, 200, 209, 215
Aristoxenus of Tarentum 19, 38–42, 46, 60, 66, 69, 72, 111, 157, 198–9, 203, 246–7
Armenia 94
Arnauld, Antoine 187–8, 192
Arnheim, Rudolf 319
ars antiqua 111–13, 307
ars combinatoria 125
ars dictaminis 130, 137
ars historica 168
ars mnemonica 230
ars nova 113, 307
ars poetica 67, 233
ars praedicandi 130
ars prosandi 130
artes liberales 3, 34, 37, 41, 52, 61, 103, 109, 118, 145, 184, 244
Asconius 146
Ashanti 269
astronomical, astronomy 1, 2, 24, 27, 34, 37, 43, 61, 87, 95–6, 107, 154–5, 169, 300, 314, 358
Asturlabi, Al- (al-Saghani) 98
Augustine, St 1, 34, 84–6, 103, 120, 130–2, 141, 145, 307, 358
Augustus, emperor 26
Aurangzeb 183

Aurelius, Cornelius 167n
Averroes (Ibn Rushd) 125, 128, 133–4, 162, 358
Avicenna (Ibn Sina) 115, 125, 127–8, 133–4, 139, 191, 248, 358
axiom, axiomatic 18–19, 39–41, 54–5
Aztec 361

Babylon, Babylonian 24, 37n, 42, 84, 103, 169, 170, 245, 262, 273
Bach, Johann Sebastian 157, 205, 206, 304
Bachtin, Michail 330
Bacon, Francis 193, 230–1, 240
Bacon, Roger 20, 80, 104, 184, 213
Baghdadi, Al- (Abd-al-latif) 163
Bakcheios 41
Bakr, Abu 106
Balzac, Honoré de 332
Bamboo Annals 26, 35
Ban, Gu 28–9, 30, 100
Ban, Zhao 28–9
Barba, Eugenio 338
Baronio, Cesare 148
baroque 203, 227, 234, 236, 242, 306–7, 315–16, 324, 326
Barthes, Roland 330, 331–2
Barzizza, Gasparino 211
Basque 190
Battuta, Ibn 123
Baumgarten, Alexander 223
Bayt al-Hikma 76
Beauzée, Nicolas 190, 281, 349
Bede, Venerable 86–8, 93, 94, 169, 245
Beeckman, Isaac 201–2, 301
Beethoven, Ludwig van 304, 307
behaviourism 289–91
Behistun inscriptions 273
Belisarius 90
Bellori, Giovanni 219–20, 311
Bembo, Pietro 237, 350
Benedetti, Giovanni Battista 199, 216
Bengali 285
Benjamin, Walter 316, 341
Bentley, Richard 155–7, 246, 349
Berber 84, 360
Berenson, Bernard 314–16, 350
Bergson, Henri 339
Bernard of Clairvaux 120
Berossus 23–4, 26, 154, 169, 245
Berr, Henri 259
Bessarion of Trebizond 151
Bharata Muni 3, 41–2, 69, 71, 116, 206, 336–7, 351, 358, 362
Bhartrhari 75, 288
Bhatta, Kumarila 136
Bible 86, 93, 99, 102, 103–4, 119–20, 130–2, 141, 145, 149, 153, 160, 170–2, 178, 186, 190, 233, 237–8, 240, 246, 322, 353, 357, 359
biology 2, 4, 285, 300, 347, 355–7
Biondo, Flavio 163–4, 322
Biruni, Al- 1, 17, 29, 74, 78, 93, 95–6, 106, 123, 139, 358
Bismarck, Otto von 252
Bisschop, Jan de 222
Bloch, Marc 258–60
Bloomfield, Leonard 288–90
Boccaccio, Giovanni 144–5, 237
Böckh, August 279
Bodin, Jean 167–8
Boethius, Anicius Manlius Severinus 41, 55, 109, 124, 127, 130, 198–9
Boetius of Dacia 80
Boileau, Nicolas 235, 249
Bolter, Jay 343
Bonet, Juan Pablo 190
Boniface, St 103
Book of Changes (*Yijing*) 56, 158
Book of Documents (*Shujing*) 21, 26–7, 159
Book of Rites (*Liji*) 43, 137
Book of Songs (*Shijing*) 27, 43, 137, 158–9, 196
Boole, George 295
Boolean logic 195
Bopp, Franz 157, 284–6, 349
Borges, Jorge Luis 343
Borromini, Francesco 227
Boswell, James 236
Boyle, Robert 210
Bracciolini, Poggio 145–6, 163, 184, 229
Brahe, Tycho 72
Braudel, Fernand 258–60
Bredero, Gerbrand Adriaensz 220
Brethren of the Common Life 152
Brouwer, Adriaen 222
Brunelleschi, Filippo 211
Brunetière, Ferdinand 327
Bruni, Leonardo 145, 162–3
Bruno, Giordano 230
Bryennius, Nikephoros 90
Buckle, Henry Thomas 254, 307
Buddhism 69–70, 96, 101
Burckhardt, Jacob 263, 264, 266, 311–12, 315–16, 346
Buridan, Jean 125, 126–7, 139, 191, 248
Burke, Edmund 68, 223, 228, 237, 238, 253
Burmeister, Joachim 204, 303
Burney, Charles 206
Bush, Vannevar 343
Byzantium, Byzantine Empire 5, 62, 84, 90–1, 93, 101, 102, 104–5, 118–21, 124, 135, 151, 181

Caesar, Julius 26, 33, 58, 86, 164
Cai, Yuanding 117, 207
calculus ratiocinator 195
calendar, calendar systems 28, 87–8, 89–90, 95–6, 169–72, 245–6
Camden, William 167n

canon:
architectural 226–7
of ideal selection 216–19, 325
of mathematical proportions 71
of picturesque (*schilderachtig*) 220–3
of Polykleitos (classical canon) 45–8, 122
of Tala proportions 49
Cantemir, Dimitrie 182, 183, 209
Canudo, Ricciotto 339
Cao, Cao 70
Cao, Pi 70
Caravaggio, Michelangelo Merisi da 220, 320
carbon-14 dating 323
Carolingian Renaissance, *see* Renaissance
Carracci, Annibale 220
Carter, Howard 322
Cartesian 173–4, 192, 202–3, 230
Casaubon, Isaac 155, 160, 167, 246, 353
Cassiodorus 79, 109n
Cassius Dio, Lucius 26
Castelvetro, Lodovico 233–4, 236, 249, 336
Castiglione, Giuseppe 228
categorial grammar 300n
catharsis 65–8, 233, 236
Catiline conspiracy 30
Cats, Jacob 318
Catullus, Gaius Valerius 143
catuskoti 129
causal explanation 177, 263, 264, 327
Celtic 79, 82, 88, 190
Celtis, Conrad 167n
Cennini, Cennino 218
century of discoveries 145
Chalcondyles, Laonicus 91, 181
Champollion, Jean-François 272–3
Charlemagne 79, 88, 104–6
Charles IV 147n
Charles the Bald 109
chemistry 2, 4, 327, 355
Chen, Di 158, 196
Chen, Kui 136–8, 139, 140, 239, 249
China 5, 6, 8–9, 26–9, 30, 35–6, 43, 48–51, 56–7, 62–4, 70–1, 75–6, 94, 98, 100, 102, 107–8, 116–17, 121–2, 129, 136–8, 139, 159–60, 177–8, 183, 196–7, 207, 227–8, 238–9, 266–8, 269, 299–300, 309, 312, 321–2, 336, 346, 349–50, 359
Chomsky, Noam 20, 82, 193, 290–5, 298, 300, 305, 347, 349, 354, 357, 359
Chomsky hierarchy 293
Christianity 85, 86, 89, 96, 103, 130, 159, 162
chronicle, chronicle writing 24–7, 84–9, 91, 99, 100–2, 116, 121, 161, 162, 177, 179, 180–3, 208, 236, 267–9, 361
chronology 23–4, 26, 34, 87, 88, 95, 101, 154–5, 169–72, 359
Chrysippus of Soli 18, 32, 54, 351, 357
Chrysoloras 151
Cicero, Marcus Tullius 34, 55, 61–2, 67, 84, 105, 130, 137, 145, 146, 150, 154, 157, 165, 168, 191, 218, 229, 275
Cimabue 118, 219
Cinnamus 90
Circumscriptio 216
classicism 204, 220, 227, 233–7, 307, 357
Cleionides 41
Clement of Alexandria 85
cliometrics 261–3, 363
close reading 322, 334
Clovis 86, 94
Cluny, Odo of 110
cognition, cognitive approach 290, 294, 299, 310, 362
Colin, Michel 339–40
Collegium trilingue 153
comma Johanneum 152–3
Como, Martino da 164
compositionality 17, 75, 296–7, 349
computational humanities 320
computational linguistics 299, 349
computational musicology 306n
computer science 20n, 295, 347, 353–4
Comte, Auguste 10, 174, 251, 254–6, 258, 317, 347, 359
Comtean 310, 344, 346
concinnitas 244
Condorcet, Nicolas 174–5, 183, 253
Confessor, Theophanes 91
Confucianism 28, 101
Confucius 10, 14, 18, 21, 35, 43, 63, 70, 76, 159–60, 177, 267, 268
Congo 208
conjugation law, *see* law
connectives 54, 125, 295–7
connoisseur 227, 313–15
consequentiae 124, 126, 140
consonance (of interval) 27–9, 42, 112, 114–15, 198–202, 203, 215–16, 207, 209–10, 212, 215–16, 224–5, 227, 232, 242, 246–7, 257, 301–3, 308, 309–10, 349, 353, 358
Constantine the Great 26, 90, 147–8
construction grammar 298
context free 291–2, 294, 298
context sensitive 16, 20, 284, 286, 291, 294, 298
contrapposto 45
Coptic 88, 102, 272
Corpus Hermeticum 155, 218, 246
Cotto, Johannes 112
counterfactual method 148, 261
Crates of Mallos 32
Creation 26, 29, 84–6, 93, 161, 169–71, 177–8, 183, 189
Cremona, Gerard of 79, 162
Crete 322
critical bandwidth 302–3
Critical school, critical theory 264–5, 308, 340–2
Critobulus 91, 181
Croatian 285
Croce, Benedetto 258, 260, 327
Croesus 23
cultural history 262, 265–6, 311–12, 334, 347

Cultural Revolution, *see* Revolution
cultural studies 339, 342–3, 344, 348, 362
cuneiform 273
Cusa, Nicholas of (Cusanus) 147, 152

Dalgarno, George 82, 190–1, 194–5, 272
dance 42, 69, 308, 339
Dante Alighieri 118, 121, 132, 165, 237, 327
Daoism, *see* Taoism
Darius 23
Dark Ages 161–2
Darwin, Charles 285
Darwinian 307, 310, 326–7
dating 30, 34, 86–8, 169–72, 242, 245–6, 268, 320, 323, 347, 348, 350
De Morgan's laws, *see* law
Deacon, Paul the 88
decentring 265, 331
decipherment 272–3
declarative system of rules, *see* rule system
deconstructivism, deconstructivist 10, 308, 321, 331–2, 351
deduction, deductive 39, 41, 52, 56–7, 59, 72, 128, 140, 192, 196, 203, 209, 248, 254, 262, 328
Deleuze, Gilles 340, 343
Delsarte, François 337
Demetrius of Scepsis 33
Denmark 89
dependency grammar 77n, 300n
Derrida, Jacques 265, 332
Descartes, René 172, 192, 201, 202, 231, 240
descriptive 6, 15, 18, 40, 66, 69, 71, 78–80, 82, 111, 116, 140, 146, 155, 188, 190, 204, 206, 208, 211, 216, 222, 227–8, 234, 242, 315, 334, 346, 350, *see also* pattern; prescriptive; tendency
Deshima 360
Diamond Sutra 99
Diderot, Denis 203, 223, 237
diesis 39
digital humanities 320, 331, 347, 362
Dignaga 129, 135
Dilthey, Wilhelm 7, 174, 256–7, 260, 263–4, 301, 327, 333, 348
Dimashqi, Al- 134
Din, Rashid Al- 98, 181–2
Din, Safi Al- 115, 209
Dinawari, Al- 6, 93–4
Diocletian 87
Dionysius the Areopagite 120–1
Dionysius Thrax 18–19, 32, 40, 76, 245, 353, 362
Dionysius of Halicarnassus 66–7, 68, 70, 72, 137, 168
Diop, Cheikh Anta 270
Dioscorides 107
discovery 5, 10, 14, 32, 35, 43, 45, 57, 59, 83, 99, 113, 114, 129, 137, 139, 143, 145, 146, 165, 169, 170–1, 190, 229, 232, 234, 240, 270, 272–3, 278, 283, 299, 323–4, 332, 353–5, *see also* century of discoveries
  of archaeological sites 323, 331
  of Bamboo Annals 35
  of Behistun inscriptions 273
  of Bells of Zenghouyi 43
  of canon 45–8
  of consonance, *see* consonance
  of cycle of rise, peak, and fall 5, 23, 27–8, 57, 97, 183
  of exceptions, *see* exception
  of formalization, *see* formalization
  of grammar, *see* grammar
  of harmony, *see* harmony
  of hierarchy, *see* hierarchical structure
  of hybridity of symbols 273
  of Indo-European language family 284–5, 354
  of interaction between theory and empiricism, *see* interaction between theory and empiricism
  of laws, *see* law
  of logical rules 53, 57, 63–4, 125–6
  of manuscripts 137, 145–6, 229
  of meta-patterns, *see* meta-pattern
  of method, methodic principles, *see* principle
  of overtones 113–14
  of patterns, *see* pattern
  of perspective, *see* perspective
  of regularities in textual corruptions 145, 149–51
  of Rosetta Stone 272
  of rule systems, *see* rule systems
  of style periods, *see* style period
  of Troy 10, 322
  of two Plinys 143
discovery procedure 83
disegno 216–17, 219–20, 254, 292, 315, 358
dispositio 61
dissonance (of interval) 37, 39, 198, 200–2, 301–2, 308, 349
distributionalism 289–90, 329
Dominici, Giovanni 145
Donatello 211, 219
Donatio Constantini 147–9, 164, 352
Donatus 19, 353
Doryphoros 45, 47
Doyle, Arthur Conan 314
dramatis personae 329
Dressler, Gallus 203–4, 226, 247, 303
Droysen, Johan Gustav 253
Duhem, Pierre 9
Dunstaple, John 198
Dürer, Albrecht 216
Dutch 1n, 11, 156, 160, 189, 206, 216, 221–3, 228, 234, 235, 238, 242, 253, 315, 318–19, 360

Easter Island 274
Eastern Roman Empire 90, *see also* Byzantine Empire
ebced notation 209
economic, economics 2, 56, 95, 100, 255–6, 258–61, 263–4, 268–9, 289, 342, 347
Egypt 22, 24, 61, 88, 163, 166, 168–72, 182, 262, 272, 273, 308, 353
Ehrenfels, Christian von 10, 304
Elamite 273
ellipsis 15, 186–7, 245
elocutio 61
emendatio 275–7
emendation 105, 149, 151, 156, 200, 277
Emmius, Ubbo 170
Empirical School 158–60, 178–9, 246, 358
Engels, Friedrich 256
Engerman, Stanley 261
English 1, 11, 15, 77, 82–3, 90, 168, 182, 230, 234, 236, 253, 255, 281, 285–6, 287–93, 296–7, 327, 330
Enkyklios paideia 3, 61
Enlightenment 142, 171–2, 174–6, 206, 243, 253, 271, 352–3
enthymeme 58–62, 133, 248
Ephorus of Cyme 23
epigraphy 164, 279, 350
equal temperament 199, 207
Erasmus, Desiderius 147, 149, 151, 152–4, 157, 172–3, 230, 246, 277
Eratosthenes 24, 41
Erfurt, Thomas of 80, 81, 184, 188
erklären 7, 257, 327, 348
Ernesti, Johann 157
Erpe, Thomas van (Erpenius) 78
Ethiopia 6, 84, 88, 101–2, 140, 171, 180
ethnomusicology 308
ethos 60
Etruscan 274
Euclid 20, 39, 41, 53, 192, 196, 248
Euler, Leonhard 196, 201, 203
Euripides 65
Eusebius 24, 26, 74, 84, 91, 245
Evans, Arthur 273, 322
examinatio 275–6
example-based 19, 75–8, 83, 112, 116–17, 139, 140, 155, 157, 187, 231, 299, 300, 306, 335, 347, 350, 358, *see also* rule-based versus example-based
excavation 322–3, 350
exception (on patterns) 7, 9, 25, 32, 45, 77, 79, 112, 115, 118, 121, 140, 155, 156, 166–7, 187, 208, 230, 249, 251, 269, 282, 285, 286–7, 303, 323, 346, 349, 351, 356–7, 361
exceptionless 287, 356
exegesis 34, 69, 131–2
exemplar theory 300
Exiguus, Dionysius 87
experiment, experimental 41, 68, 114, 193, 199, 210, 215, 247, 302, 353
Eyck, Jan van 216, 317, 320

falsification 30, 68, 72
Fang, Zhongtong 196
Farabi, Al- 114–15, 125, 127–8, 133–5, 139, 206, 209, 358
Farra 78
Fazang 76
Fazl, Abul 182
Febvre, Lucien 259
feminism, feminist 226, 308, 321, 331, 340
Ferrières, Lupus de 105, 143
feudalism 148, 255, 268–9
Feuerbach, Ludwig 256
Ficino, Marsilio 151, 155, 216, 218
fifth (musical interval) 37, 39, 42–3, 71, 110–11, 115, 117, 140, 198–9, 203, 205, 303
Filelfo, Francesco 151, 184
film studies 2, 294, 339–40, 343, 348, 355
fiqh 128
First Grammatical Treatise 83, 289
Flaccus, Valerius 146, 150
Flaccus, Verrius 33
Fogel, Robert 261
folk song 176, 305
folk tale 281, 328–9
forensic science 325, 347
foreshortening 48–9, 122–3, 139, 228, 247
forgery 35, 147–9, 153, 157, 169
Formalism (literary studies) 326–8, 334, 350
formalism (of rule system) 14–15, 17, 20, 284, 291, 293–5, 297, 299, 329–30, 354
formalization 69, 92, 109, 128, 135, 140–1, 241–2, 243, 274, 278–9, 350, 357
  logical 241–2, 263, 277, 355
  mathematical 241–2, 247, 263, 355
  procedural 241–2, 247, 355
Forte, Allen 306
Foucault, Michel 4n, 265, 332
Francesca, Piero della 214–15, 241
Franco of Cologne 112
Frankfurt school 264–5, 271, 341
Frederick III of Brandenburg 156
Frege, Gottlob 295–6, 349
Freising, Otto von 88
French 11, 95, 144, 164, 167, 170, 172, 174, 176, 182, 188, 202, 223, 234–5, 237, 243, 245, 251, 253–5, 258–9, 265–6, 272, 327, 329–30, 337, 339, 364
French Revolution, *see* revolution
Freud, Sigmund 314, 321, 340
Fruin, Robert 252
Frumentius 101
Fu, Sheng 35
Fulani 11, 179, 364
Fulda, Rudolf of 88

Fulgentius 132
Furtado, Francisco 196
Fux, Johann 204–5

Ga 269
Gadamer, Hans-Georg 5, 7, 334
Gaffurio, Franchino 199, 210, 301
Galen 47, 107
Galilei, Galileo 1, 41, 172n, 173, 200–2, 210, 240, 243, 301, 353, 359
Galilei, Vincenzo 41, 198–200, 210
Gallicanus 148
gamaka 206–8
Gangesa 129
Gansfort, Wessel 152
Garlandia, Johannes de 113
Gautama, Aksapada 55, 63
Gautama, Medhatithi 55
Ge'ez 10, 11, 111, 180, 364
Geisteswissenschaften 1n, 7, 9n, 257
Gellius, Aulus 105, 168
gender studies 342
genealogical method/theory 149–51, 153, 155, 157, 159, 200, 242, 246, 275–7, 285, 353
genealogy 29, 89, 179–80, 182, 269
generalization:
  in historiography 165–6, 175, 183, 258–61, 263
  in linguistics 19, 83, 287, 299
  in musicology 115
  in poetics 66
  in rhetoric 63
generative grammar, *see* grammar
genetics 276, 285, 354
Genette, Gérard 330
geographical, geography 9, 22, 24, 33, 94, 135, 139, 162, 171, 190, 259–60, 267
George of Trebizond 229
Gerard of Cremona, *see* Cremona, Gerard of
Gerbert, Martin 206
German, Germany 1n, 11, 105, 146, 156, 157, 164, 182, 183, 185, 189, 190, 206, 216, 223, 231, 234, 235, 237, 251, 253, 261, 267, 278, 281–2, 283–7, 301, 311–13, 316, 326, 336, 355, 365
Gervinus, Georg 326
Gesner, Johann, Matthias 156–7, 279
gesta 88
Geyl, Pieter 262n
Ghazali, Al- 128
Gibbon, Edward 176, 182
Giorgio, Francesco di 224
Giotto 118, 218, 219
Goeree, Willem 222, 223
Golius (Jacob van Gool) 162
Gombrich, Ernst 319
Gongsun, Long 56
Gorgias 58
Gothic 120, 190, 282, 312
Gottsched, Johann 231, 235–6
Gracián, Balthasar 234
grammar:
  cinematic 339–40
  context-free 291–4, *see also* context free
  context-sensitive 20, 294, *see also* context sensitive
  declarative 71
  descriptive 18, 79, 185, 188
  example-based 76–8, 139–40, *see also* example-based
  film, *see* cinematic
  formal 6, 295, 304
  generative 16n, 193, 297
  logical 297
  melodic 157, 203, 249
  Montague 297, 349
  music 46, 60, 203–4, 305, 340, 349, 350
  narrative 328–9
  prescriptive 18
  probabilistic 298
  rhetorical 230–1
  school 14, 18, 356
  Speculative Grammar 75, 80–1, 83, 184, *see also* modists
  Universal Grammar 20, 78, 80–2, 83, 188, 190, 281, 283, 293, 299
  vernacular 82, 83, 184
grammatical (of a sentence) 16, 77, 232, 290, 292, 305
grammaticality 298
Gramsci, Antonio 261n, 269, 342
Great Zimbabwe 324
Greece, Greek 5–6, 8, 11, 13–14, 18–19, 21–5, 32, 34, 37, 39–40, 42–4, 46–7, 49, 52–3, 55, 57, 59, 61–2, 65, 68–73, 75–6, 77, 79, 89–90, 93n, 97, 101–4, 106–7, 111, 114, 121, 123, 127–8, 131, 134–5, 139–40, 145, 151–3, 155–7, 162–3, 168–9, 172, 175, 182, 187, 189–91, 193, 198, 223, 227–9, 233, 237, 245–7, 249, 257, 272–4, 276, 282–4, 311, 314, 337, 353, 356–7, 364
Greg, Walter 279
Gregorian 109–11, 141
Gregory I 118–20
Gregory of Tours 85–7, 93, 103
Grimm, Jacob 281–6, 348, 349
Grimm, Wilhelm 281
Gronovius 156
Grote, Geert 152
Grotius (Hugo de Groot) 154, 160, 189
Grusin, Richard 343
Gu, Kaizhi 50–1
Gu, Yanwu 158, 178, 246
Guarino da Verona 151, 184
Guattari, Félix 343
Guha, Ranajit 269
Guicciardini, Francesco 164, 166–7, 168, 348

Guido d'Arezzo 110
Guo, Ruoxu 122
Gusdorf, Georges 4n
Gutenberg Bible 99

hadith 6, 91–2, 106, 123n, 128
Hall, Stuart 342
Halle, Morris 294n
Hals, Frans 220–1, 315
Han dynasty 5, 27–9, 43
Händel, Georg Friedrich 206
harmonia mundi 37
harmonists 38, 40–1
harmony:
in architecture 224–5
in art theory 46–7, 68, 71
in musicology 3, 9, 37–9, 41–3, 57, 71, 109, 113–15, 117, 198, 203, 205, 209, 241, 359
Harris, Zellig 290–1, 294
Hauser, Arnold 321
Hawkins, John 206
Haydn, Joseph 304
He, Qiutao 267
head-driven phrase structure grammar 299n
Hebrew 93, 103, 131, 153–4, 169, 185, 189–90, 281
Hegel, Georg Wilhelm 256, 307, 311, 315
Hegelian 304, 310, 311, 326, 344, 346
Hegius, Alexander 152
Heidegger, Martin 334
Heinsius, Daniel 156, 235
Hellenism, Hellenistic 3, 18, 20, 24, 31, 34, 36, 57, 61, 67, 71, 223, 353–4
Helmholtz, Hermann von 201, 301–2, 349, 359
Hempel, Carl 262
Hemsterhuis, Tiberius 156–7, 284, 348, 349
Henry IV 155
Herder, Johann Gottfried 174, 176, 183, 190, 250, 251–2, 262, 283, 307, 315
Herderian 304, 310, 323, 326–7, 346
Hermagoras of Temnos 61
hermeneutics 33, 333–4, 349
Hermes Trismegistus 155
hermetic 155, 246
Hermogenes 58, 62, 134, 229–30
Herodian, Aelius 19
Herodotus 5, 21–4, 28, 30, 33, 66, 71, 92, 94–5, 97, 101, 140, 161, 165, 168, 175, 245, 271, 362
Herrmann, Max 336–7, 351
Hesiod 21, 32
heuristic, heuristics 23, 60, 71, 133, 192
hexachord 110, 113
Heyne, Christian Gottlob 157
Hezarfen, Huseyn 181
hierarchical structure:
in architecture 217, 346–7
in art 216–19, 222, 312, 346–7, 350
in film 339–40, 346–7
in history 258–60, 346–7
of knowledge 220
in language 78, 81, 292–3, 299, 300, 346–7
in literature 330–2, 346–7
in music 204, 303–6, 346–7, 350
in new media 343–4, 346–7
in rhetoric 217–18, 229–30, 346–7
of the world 120–1
hieroglyphs 207, 272–3
Higden, Ranulf 89–90, 94, 96–7
hijra 246
Hildebrandslied 277
Hindu 17, 29, 49, 69, 116, 281
Hippocrates 107
historicism, historicist 167, 173, 257–8, 260
historism 251–3
history from below 269
Hitler, Adolf 316, 337
Hobbes, Thomas 230–1
Hofstede de Groot, Cornelis 314–15
Holder, William 4, 190–1, 201, 357–8
holism 75, 296
Holland 222, 360
Homer 18, 22, 31–3, 45, 59, 65, 67–8, 85, 131, 237, 272, 278, 322
Homiletics 130, 134
Hoogewerff, Godefridus 316
Hoogstraten, Samuel van 216, 220, 222–3
Horace 6, 66–7, 233, 249, 318
Horkheimer, Max 264
Hornius, Georgius 172n
House of Wisdom, see *Bayt al-Hikma*
Hrabar, Chernorizets 82
Hu, Shi 336
Hu, Yinglin 238–9
Huang, Zongxi 177
Hucbald 109–10, 112
Huet, Pierre Daniel 172, 237–8, 350
Huizinga, Johan 262–3, 266
human sciences 4n, *see also* social science
Humboldt, Wilhelm von 252, 283–4, 288
Hume, David 176, 183
Huygen van Linschoten, Jan 208
Huygens, Christaan 200–2, 241, 301
Hyginus, Julius 33
hypertext 343

iconographic 318–19, 325
iconological, iconology 311, 315, 315–18, 321, 337
ideal selection 193, 216, 218–20, 223, 224, 228, 232, 238, 312
idiographic 7, 257–8, 260, 348
illusionism 44–8, 119, 211, 219, 247
imitatio 66–8, 157, 222, 232
Imola, Benvenuto da 118
India 5–6, 8–9, 14, 17, 24, 29, 35–6, 41–2, 48–50, 55–7, 61–2, 71, 73, 75, 78, 83–4,

94–5, 98, 101–2, 122–3, 128–9, 135–6, 139–40, 182, 189, 197, 206–8, 227–8, 238, 248, 266, 268–9, 299, 301, 309, 329, 346–7, 359, 364
Indo-European 17, 157, 190, 281, 284–7, 354
induction 63, 262
Industrial Revolution *see* revolution
interaction between theory and empiricism 151, 202, 203, 209–10, 240, 309, 353
interpretation 2, 9, 16–17, 33, 40, 69, 75, 85–6, 117, 119–20, 128, 131–2, 136, 139, 141, 145, 166, 168, 174, 176, 189, 193, 220, 227, 235, 249, 253, 256, 265, 275, 293, 317–19, 320–1, 323–4, 328, 331, 333–4, 337, 349, 351, 356, 360, 362
  allegorical 131–2, 141, 145
  anagogical 132
  decentred 331–3
  hermeneutic 333–4, 349
  literal 119
  moralistic 132
  symbolic 317–19
interval 37–9, 72, 109, 111, 114, 115, 117, 198–203, 205, 209, 215, 246–7, 260, 301, 304, 358, *see also* consonance; dissonance
inventio 59, 61, 64, 70
Ireland 79
Isfahani, Al- 115
Ishaq, Abi Ibn 76
Ishaq, Hunayn Ibn 91, 106–7
Isidore of Seville 103
Islam 5, 6, 8, 61, 74–9, 81, 84, 89–90, 91–8, 102–3, 106–8, 114–15, 116–17, 123, 125, 127–30, 133–6, 139–40, 142, 162–3, 174, 181–3, 197, 206, 238, 246, 248–9, 272, 314, 322, 354, 358–9
isnad 6, 84, 91–8, 102, 128, 139–40, 150, 181, 242, 246, 252, 271, 275, 349, 352, 359
Isocrates 58
Italian, Italy 1n, 11, 79, 131, 142, 144–6, 151, 164, 167–8, 176, 189, 199, 206–8, 211, 215–16, 220, 226, 233–7, 258, 273, 281, 293, 312, 314, 319, 326, 338–9, 364
Italicus, Silius 146

Jackendoff, Ray 303, 305–6, 319, 329, 350
Jacques de Liège 113
Jahan, Shah 182
Jahiz, Al- 185
Jaimini 69–70
Jain 129
Jakobson, Roman 288–9, 328
Jansenists 188
Jefferson, Thomas 322
Jerome, St 84, 103, 149, 153
Jesuit 159–9, 188, 196, 207, 228, 234, 267
Jewish 24, 79, 89, 103, 169, 245
Jiang, Yong 159
Jin Dynasty 35
Jinni, Ibn 78
Johnson, Samuel (historian) 269
Johnson, Samuel (literary critic) 236–8
Jones, William 182, 188, 190, 281
Jongh, Eddy de 318
Jonson, Ben 235
Josephus, Flavius 24
Jugurthine war 30
Junggrammatiker, see Neogrammarians
Junius, Franciscus 222
Junius, Hadrianus 167n
Jürgensen Thomsen, Christian 324
Jurjani, Al- 78
Justin Martyr 85
Justinian, emperor 74, 90, 118
Juzheng, Xue 100

Kalhana 30, 101
Kant, Immanuel 223
Kashi, Al- 123
Kashmir 30, 84, 101–2, 268
Kati, Mahmud 179, 183, 270
Katyayana 17
Kebra Nagast 101–2, 180
Kempis, Thomas à 152
Kepler, Johann 1, 37, 72, 156, 202, 240–1, 353, 359
Khaldun, Ibn 5, 96–8, 100, 102, 139, 161–2, 168, 179, 183, 208, 242, 260, 262, 360, 362
Khmer 361
Khoikhoi 208
Kilwa 180
Kindi, Al- 114
Kircher, Athanasius 207, 272
Kitcher, Philip 356
Ki-Zerbo, Joseph 181n
Koch, Erduin 326
Koch, Heinrich Cristoph 203, 205, 292, 303–4
Kolb, Peter 208
Komnene, Anna 91
Koran 76, 102, 106, 123, 128, 134, 237
Korea 15, 159, 361
Kristeva, Julia 332
Kugler, Franz 311–12
Kuhn, Thomas 244, 348
Kulmus, Luise Adelgunde 236
Kulturwissenschaften 257n
Kunst, Jaap 307
Kurdish, Kurds, 94, 115
Kurr, Ibn 209

Lacan, Jacques 332, 340
Lachmann, Karl 7, 150, 274–9, 284, 348–9, 353
Ladhiqi, Al- 209
Laet, Johannes de 188–90, 197, 242, 281
Lairesse, Gerard de 222–3, 225
Lamy, Bernard 230–1, 248

Lancelot, Claude 187–8
language family 354
langue 288, 290
Las Casas, Bartolomé de 169
Lassus, Orlandus 204
Latin 11, 16, 19, 33, 61, 79, 82–3, 87, 90, 103–5, 110, 114, 116, 124, 132, 140, 146–7, 151–6, 162, 184–91, 193–4, 232, 251, 279, 299, 353, 364
  Classical 146–7, 184–5
  Late 146
  Medieval 146–7
  Neo-Latin 147
  Silver 146
Latin school 147, 152
law:
  absolute, universal 287, 355–6
  Bopp's conjugation 284
  consonance, *see* law of harmonic intervals
  De Morgan's laws 125–6
  of excluded middle 53, 57, 64
  Grimm's laws, *see* sound shift laws
  of harmonic intervals 37, 198–202, 301–3, 353, 359
  historical laws 165–7, 174, 259
  of human progress 254
  of melody 38–40, 203
  of nature 224
  of non contradiction 53, 57, 64
  perception laws 304
  of perspective 211–13
  sound shift laws 9, 190, 281–3, 285, 355
  Verner's 286–7
  Zipf's 298n
  see also pattern
Le Clerc, Jean 172–3
Le Roy Ladurie, Emmanuel 266
LeBrun, Charles 220
Leibniz, Gottfried Wilhelm 82, 125, 156, 190, 194–6, 197, 230, 241, 248, 296, 349, 357
Lenin, Vladimir Ilyich 261n
Lennep, David Jacob van 157
Leo Africanus 179
Leo III 119
Leonardo da Vinci 48, 214–16, 219, 225–6, 228, 238, 247, 353
Leoninus 112
Lerdahl, Fred 303, 305–6, 319, 329, 350
Lermolieff, Ivan, *see* Morelli, Giovanni
Lessing, Gotthold Ephraim 118, 157, 237, 278
Levelt, Willem 302–3, 349
Lévi-Strauss, Claude 10, 329–30
lexical-functional grammar 299n
Li, Si 27, 35
Li, Zhi 177, 183, 243
Libby, Willard 323
Ligorio, Pirro 322
Linacre, Thomas 184–5
linear pattern, *see* pattern
Linnaeus, Carl 194
Lipsius, Justus 154
Liu, An 3, 43, 71, 73
Liu, Fenglu 267
Liu, Xie 64, 70, 136, 239, 362
Liu, Zhiji 99–100, 260
Livy (Titus Livius) 25–6, 30, 84, 105, 144, 165
Locke, John 156
logica nova 124
logica vetus 124, 127
logos 60
Lomazzo, Gianpaolo 219
Longinus 67–8, 70, 72, 223, 237
Lord, Albert 335n
Louis the Pious 121
Louis XIV 187–8, 235
Lovati, Lovato 143
Lucretius 105, 146, 191, 275, 277
Lull, Ramon 125, 195
Lunyu (Analects) 63
Luo, Changpei 300
Lütfulla, Ibn 181–2
Luther, Martin 148, 152, 352
Lyotard, Jean-François 265

Ma, Jianzhong 299–300
Maas, Paul 279
Mabillon, Jean 172–3
Macaulay, Thomas 253
McLuhan, Marshall 341–2
Machaut, Guillaume de 113
Machiavelli, Niccolò 164–8, 183
Macrobius 34, 105, 109n, 131
macrohistory 260
Mahavamsa 101n
Mahler, Gustav 307
Makdisi, George 93n
Makeda, Queen of Sheba 101–2
Ma'mun, Al- 76
Man, Paul de 332
Mander, Karel van 220
Manetho 23–6, 30, 154, 170, 245, 353, 359
Manilius 146, 154, 156
mannerism 307, 315–16
Manovich, Lev 344
mantiq 127
Manuzio, Aldo 151–2, 229, 237
Manzoni, Alessandro 327
Mao, Zedong 268, 321, 336
Marcellinus, Ammianus 26, 146
Marini, Giambattista 234
Martianus Capella 34, 41, 79, 109
Martin of Tours 86
Martini, Giovanni Battista 206, 306
Marx, Karl 174, 251, 255–6, 258, 260, 268, 311, 347, 351
Marxism, Marxist 261n, 264–9, 321, 323, 328, 330
Masaccio 212, 219

Masoretes 103
Masudi, Al- 6, 94–5
Matamba 208
Mathémata 37
mathematics 1, 2, 20, 80, 87, 113, 172, 192, 195, 199, 210, 215, 246, 281, 296, 344
Matociis, Giovanni de 143
Mattheson, Johann 205
Maurus, Rabanus 105
Maya 272, 274, 361
Mayr, Ernst 355
means of production 255–6, 261
mechanical, mechanics 125, 217, 237, 279–80, 349
Medici, de 165
medicine 34, 56, 93, 107, 128
Meegeren, Han van 315
Megasthenes 24, 95
melodic law, *see* law
memoria 61
Mencius 35, 63
Menelik I 101–2, 180
Merolla, Girolamo 208
Mersenne, Marin 114, 200–1, 206
mesohistory 260
Mesomedes of Crete 40n
Mesopotamia 14, 24, 273, 322
metahistory 264
meta-pattern 6, 72
metaphysics 53, 80
metarule 15, 78
method, methodic principle, *see* principle
Metz, Christian 339–40
Michael, emperor 121
Michelangelo Buonarroti 219, 224–5, 322
Michelet, Jules 174, 254
microhistory 260
Milton, John 180
Mimamsa 69, 132, 136
mimesis 44, 65–7, 134, 233–8
Ming Dynasty 36, 101, 108, 158, 177, 198, 207, 227–8, 239, 246, 301, 350
minimal pairs, technique of 83, 289
Minoan 322
miracle 85–7, 93
Mishaqa, Mikhail 308
Misri, Dhul-Nun al- 272
modal logic 296
Modern Devotion 152
Modism, Modists 78, 80–2, 126, 139, 184, 188, 193, 245, 259, 283, 346
modus ponens 124
modus tollens 124
Mohism, Mohist 56–7, 62–3, 71–2, 129, 156, 196–7
Molinier, Guilhem 132–3
Mommsen, Theodor 252
Mongols 76, 115
monochord 41, 199, 302
Montague, Richard 295, 297, 349
Montague grammar, *see* grammar
Montaigne, Michel de 235, 331
Montecassino 106, 144
Montesquieu, Charles de 25n, 175, 182–3
Monteverdi, Claudio 201, 237
Morelli, Giovanni 312–16, 319, 347, 350
morpheme 290
morphology 16, 83, 234, 290
motif 208, 304–5, 318
Mozart, Wolfgang Amadeus 304
Mozi (Mo Tzu) 56–7
Mubarrid 78
Mughal 98, 142, 161, 182–3, 228, 242, 268
Muhammad 91–3, 106
Mulerius, Nicolaus 170
Müneccimbaşi, *see* Lütfulla, Ibn
Muris, Johannes de 113
musica:
  activa 112
  enchiriadis 110–12, 117, 204–5, 247
  ficta 204n
  humana 41
  instrumentalis 41
  mundana 41
  poetica 198, 204
  recta 204n
  speculativa 112
musical notation 39, 109–10, 112–13, 207, 209, 247
music grammar, *see* grammar
Music of the Spheres 37–8, 41
musique concrète 307
Mycenae, Mycenaean 273–4, 322
myth 21, 29, 35, 70, 96, 162, 173, 190, 262, 265, 330, 342, 353

Nadim, Ibn al- 135
Nafis, Ibn al- 128
Nagesa 16
Naima, Mustafa 182
narrateme 328–9, 331, 335, 350
narrative structure 65–6, 330–1, 339, 340, 343, 350
narrativism 264–5, 308, 363
narratology 94n, 321, 330–1, 334–5, 340–1, 350, 356
nationalism, nationalist 176, 250–2, 266, 278, 312, 326, 346, 354–5, 357
natural science, *see* science
naturalist 220, 327
Natya Shastra 42, 69, 116, 136, 329, 337
Navya-Nyaya 129, 135
Nennius 88
neoclassicism 227
Neo-Confucianism 101
Neogrammarians 282, 286, 287, 290, 294, 349
Neo-Latin, *see* Latin
neo-Marxism 264, 269, 342

Neoplatonism 55, 131, 216, 218
Neo-positivism 98, 262–3
neo-Rankean 257–8, 260, 337
neo-romanticism 307
Nero, emperor 147n, 322
neume 109, 247
neuroscience 298, 356
new art history 321, 324–5
new criticism 334
new cultural history 266, 270, 308, 321
New Cultural Movement 336
new historicism 334
new humanism 156
new media 343–4, 348, 362
new musicology 306, 308, 321, 324
New Science 156, 172, 192, 200–2, 206, 209–10, 230–1, 234, 247
  Ibn Khaldun's 96–7
  Vico's 173–4
New Testament 103, 131–2, 152–3, 238, 275, 277
Newton, Isaac 1, 156, 172, 202, 210, 241, 284, 353, 359
New World 168–9, 172n, 193, 254
Nibelungenlied 277–8, 355
Nicole, Pierre 187, 192
Nietzsche, Friedrich 149
Nigeria 180, 269
Nketia, Kwabena 309
No drama 360
nominalism, nominalist 81–2, 126
nomothetic 7, 257, 260–2, 264–6, 300, 310, 347
normal science 244, 279
normative 6, 14, 18, 62, 67, 71, 76, 128, 131, 140, 183, 218, 222, 228, 230, 233, 241, 242, 312, 315–6, *see also* prescriptive
North Africa 76, 85–6, 96, 102, 359–60
nouvelle histoire 266
numismatics 164, 279
Nyaya 17, 55–6, 62–3, 72, 75, 128–9, 296

objective, objectivity 21, 28, 90, 100, 166, 178, 203, 216, 219, 251–2, 256, 258, 260, 278, 229n
oblique perspective, *see* perspective
Occitan 83, 133, 147, 249, 266
Ockham, William of 81, 125–7, 139
octave 37, 42–3, 71, 110–11, 117, 198–9, 201, 203, 205, 303
Odysseus 45
Old Bulgarian, *see* Old Church Slavonic
Old Church Slavonic 82
Old English 82, 286–7
Old Icelandic 83, 289
Old Norse, *see* Old Icelandic
Old Persian 273
Old Testament 103, 131
optics 56, 349
optimality theory 300
oral history 266, 269–70, 348, 360
oral source, *see* source
oral tradition 22, 33, 36, 179, 269
ordinatio 47
Oresme, Nicole 113–14, 117, 201
organology 206
organum 110–13, 139–40, 198, 201, 204, 279
Origen 85, 103, 132
Orosius 84–5, 93n, 360
Ostrogoths 90, 124
Ottoman 5, 123, 161, 181–3, 198, 208–9, 242, 301, 308
Ouyang, Xiu 99–100
overtone 113–14, 117, 201–3, 205, 302
Oyo 269

paleography 280
Palestrina, Giovanni Pierluigi da 204–5, 307
Pali 101n
Palladio, Andrea 225–6
Pan, Lei 196–7
Panini 5–6, 10, 14–20, 22, 30, 32, 38, 40, 42, 55, 59–60, 66, 69, 72–3, 75, 77–8, 83, 111, 186–7, 193, 195, 203, 244, 249, 279, 281, 283–4, 286, 289–92, 294–5, 297, 299, 329, 347, 349, 354, 358, 362
Paninian 55, 66, 75, 78, 249, 294–5, 299, 329
Panofsky, Erwin 120, 261n, 316–19, 337
paradigmatic 288
paradox 56, 127, 191
parallel developments/discoveries 57, 71, 139, 242
Pareia, Ramis de 199, 210
parole 288
Parrhasius 44
Parry, Hubert 307
Patanjali 17
pathas 36, 41
pathos 60
Patrizi, Francesco 168
pattern:
  cultural/culture-specific 23, 339–40, 351, 355
  cyclic 5, 23, 25, 27–8, 57, 87, 97, 102, 114, 116, 135, 139, 168, 173, 175–6, 183, 198, 209, 220, 242, 282, 323
  from descriptive to prescriptive 6, 71, 140, 242, 346
  hierarchical, *see* hierarchical structure
  linear 26, 68, 85, 102, 175, 183
  from prescriptive to descriptive 6, 242, 324, 346, 350
  of progressive deterioration 178
  rhizomatic 330, 343, 344, 346
  from rules to examples 75, 112, 116–7, 129, 140
  spiral 174–6, 183
  universal 263, 346, 351, 355

pattern: (*cont.*)
from voiced to voiceless 282
from voiceless to voiced 282
world-historical 263
see also law
pattern-rejecting 54–5, 251, 258, 260–1, 263, 265, 268, 308, 311, 321, 333, 351, 357, 363, 364, *see also* Anomalistics, anomalistic
pattern-seeking 7–8, 260–3, 265–6, 300, 310–11, 321, 334, 346, 351, 357, 363, 364, *see also* Analogistics, analogistic; pattern
Paul, St 130–2, 149
Pausanias 163
Peçevi, Ibrahim 181
pedagogical, pedagogy 82, 152, 156, 187, 192, 203, 208, 211, 253
Peloponnesian wars 22–3
Penelope 45
performance (linguistics) 290, 298
performance studies 344
Pergamon 10, 32–3, 35, 40, 131, 351
Pericles 22
periodization 8, 84–5, 161, 163, 254, 267–8, 306
permutatio litterarum 189
Perotinus 112
Perrault, Charles 235
Persia, Persian 17, 21, 23–4, 43, 74, 76–7, 89–90, 93–5, 97, 106, 123, 127, 169, 174, 182, 190, 245, 273, 281–2, 284–5
Persian wars 21
perspective:
empirical 214, 216, 232, 247, 353
linear/mathematical 214, 216, 232, 247, 353, 51, 123, 211–16, 224, 228, 241, 247, 253, 358
oblique/parallel/ruled-line 49–51, 228
Peter of Spain 125
Petrarch, Francesco 142–6, 148, 161–3, 179, 183, 229, 237, 242
Peyrère, Isaac la 170–1, 189–90
Phidias 44–5
Philo Mechanicus 47
Philo of Megara 54
Philodemus of Gadara 35
philosophy:
of art 44, 223
Chinese 27, 43, 101, 177, 268
and empiricism 10, 176
Greek 18, 37, 41, 44, 52, 55, 58–9, 63, 65, 67, 72, 85, 90
Hermetic 155, 218–19
of history 173–4, 255–6, 258, 311
of humanities 7, 256–7, 260
Indian 57, 69, 296
Islamic 133–5
Jansenist 187
of language 10, 14, 17–18, 75–6, 190, 283
of logic 52, 56
medieval 79, 81, 106, 113, 121
moral 145
natural 240, 243, 351n
philosophes 174–6, 183
philosophical materialism 256
and poetics 68, 70
political 171, 255
Ramist 192
Roman 34, 62
of science 243, 356
see also Critical school; positivism; postmodernism; poststructuralism
phoneme 83, 286–90, 319, 346
phonetics 78, 190–1, 194, 358
phonology 16, 78, 83, 158, 197, 289–90, 295
Photius 104–5, 135
phrase structure, *see* hierarchical structure
physics 2, 4, 93, 327, 347, 349, 355–6
Piccolomini, Ennea Silvio, *see* Pius II
Pictor, Fabius 24–5
picturesque (art theory) 220, 222, 319
pinyin transcription 26n
Pirahã 299
Pisistratus 23, 31
Pius II 148, 164
Plato 10, 14, 18, 23, 37, 41, 44, 52, 58, 63, 65, 67, 72, 74, 85, 90, 124, 137, 151, 155, 218
Plethon 151
Pliny 3, 6, 8, 44–5, 47–8, 66, 71–2, 118, 122, 143, 211, 218–19, 236, 247
Plomp, Reinier 302–3, 349
plot 66, 330, 335
Plutarch 26, 61, 164
Poliziano, Angelo 149–53, 155, 157, 159, 169, 172, 200, 232, 241–2, 246, 275, 277, 348–9, 352–3
Polybius 24–5, 30, 68n, 72, 90, 179
Polykleitos 45–7, 218
polyphony 109–10, 112, 117, 204
Pompeii 322
Pope, Alexander 235–6
Porphyry 124, 197
Port Royal 187–8, 191–3, 231, 241, 245, 297
positivism, positivist 3–4, 251, 254–6, 262, 264, 307, 324–5, 326, 328, 333, 337, 344, 359, *see also* Neo-positivism
post-colonialism 266
postmodernism 265, 268, 324, 332n
post-processual archaeology 324
poststructuralism 33, 331–5, 337, 340, 342, 351, 363
poverty of the stimulus 293
Praetorius, Michael 206
Prague school 288–9
Praxiteles 44
pre-Adamites 170
pre-Columbian America 5, 360–1
predicate logic 55, 295–7, 249
premise 52–3, 56, 59–60, 128–9, 133, 195, 236

Presbyter, Theophilus 118
prescriptive 6, 18, 40–1, 48–9, 61–2, 66, 69–71, 111, 136, 140, 146, 204, 208, 211, 216, 218, 222, 227–8, 233, 238, 241–2, 249, 315, 324–6, 350, 356, 364, *see also* descriptive; normative
presentism, presentist 8, 11, 258
Prez, Josquin des 204
principle:
  all possible sources 99–100, 260, 337
  analogy 32, 78
  analysis of means of production 255
  anomaly 32, 45, 48, 68, 122, 132, 258
  basic Confucian virtues 28
  biblical coherence 84–6, 90, 102, 104, 119–20, 130–2, 139–40, 170
  chronological consistency 147, 157, 159
  compositionality 296, *see also* compositionality
  declarative system of rules 40, 46, 66, 117, 122, 197, *see also* rule system
  eliminatio 150, 200, *see also* oldest source principle
  example-based description 78, 83, *see also* example-based
  eyewitness account 22, 24, 30, 48, 90–2, 106, 140
  hierarchical analysis 332, *see also* hierarchical structure
  linguistic consistency 147–8, 157, 159
  logical consistency 147–8, 157, 159, 192, 241
  logical inference 277
  material source 163–4, 322
  mathematical proportions 45–6, 48, 115, 117, 122
  mimesis 234–6
  most probable source 22, 30, 33, 92, 140
  oldest source 149–50, 152–5, 169–70
  oral transmission 181, 243, 270
  original language 152–3
  personal experience 24–5, 30, 90, 94, 179, 181, 243, 269n, 361
  philologically underpinned source 252, *see also* oldest source principle
  procedural system of rules 16, 59–60, 81, 83, 185, 195, 197, 241, 279, 342, *see also* rule system
  proximity 304, 309
  relations of differences 288, 290, 316
  similarity 304, 309
  sociologically analyzed source 97
  verum factum 173
  visual source 48
  written source 23–5, 27, 30, 48, 86, 89, 91, 93, 96, 98, 150, 165, 251
Priscian 19, 79, 184
problem-solving capacity 243–4, 246–9, 348, 351, 365
Probus, Flavius 87
Probus, Valerius 33
procedural system of rules, *see* rule system
processual archaeology 324
Procopius 90, 118
progress 3–4, 28, 48, 174–6, 183, 219, 240, 243–9, 253–4, 262, 264–5, 207, 348–51, 364
Propertius 143, 150
propositional logic 53–5, 124–5, 129, 295–6, 349, 351, 357
Propp, Vladimir 7, 328–30, 335, 350
Protagoras 58
Protogenes 45
Proto-Indo-European 285
Proust, Marcel 330, 344
Prussian 156, 283, 311
Prussian Academy 156
Psellos 41
psychoacoustics 303
psychology 2, 9, 10, 113, 259, 297, 304, 318–19, 321, 328, 334, 362
psychonarratology 334–5
Ptolemy 3, 41, 61, 114, 198–9
Ptolemy II 31
Ptolemy V 272
Punjab 34, 61
Puranas 29, 102
pyknon 39
Pythagoras 1, 3, 37, 40, 42, 45–6, 48, 71, 114, 118, 199, 246
Pythagorean 23, 37–8, 41, 43, 72, 109, 114–15, 117, 198–9, 202, 215, 224–5, 238, 247, 365
Pythagorean comma 43, 247

Qalanisi, Ibn al- 98
Qasim, Abu 'l- 123
Qian, Mu 268
Qin Dynasty 27, 35, 57, 73, 177
Qing Dynasty 99–100, 177, 207, 239, 267, 299, 321, 335
quadrivium 3, 34, 127
Quarrel of the Ancients and the Moderns 235
Quechua 190
Quintilian, Marcus Fabius 41, 62, 137, 146, 157, 191, 229, 303
Quintilianus, Aristides 41
Quran, *see* Koran

racism, racist 261n, 307, 323, 355
raga 42, 116, 198, 206–8
Raleigh, Walter 172n
Rameau, Jean-Philippe 202–3
Ramism 188, 192
Ramus, Peter 191–2, 197, 230
rangaku 360–1
Ranke, Leopold von 251–3, 257, 278, 284, 348, 353

Rankean 252, 254, 264, 268–71, 278, 310
Raphael 229, 322
rasa 69, 136
Rashiq, Ibn 135
Rask, Rasmus 281–2
realism, realist:
 in art theory 44, 119, 220, 318, 321
 in historiography 167
 in literary studies 332, 336
 in philosophy 81
recensio 275–6
receptio luminum 217
Reconquista 81, 98
recursion 15–17, 55, 203, 284, 292, 294, 299
reductio ad absurdum 52
reduction 80–1, 321, 356
Reformation 148, 352
regularity 9, 137, 157, 175, 259, 282, 355, *see also* pattern
Reindorf, Carl Christian 269
relativism, relativist 23, 96, 177, 183, 243, 258, 332, 362
Rembrandt van Rijn 220, 223, 320
Renaissance:
 Carolingian 19, 79, 87–9, 104–5, 130, 143, 319
 fifteenth-century/Italian (the 'Renaissance') 6, 113, 124, 127, 137, 142, 151, 155, 165, 203–5, 207, 229, 307, 311–13, 315–16, 319, 322, 327, 337, 356, 359
 Ottonian 319
 twelfth-century 76, 79, 116, 143, 319
replicable 30, 35, 72, 306
Reuchlin, Johannes 152, 185–6
Revius, Jacob 170
revolution:
 Christian 141, 360
 Cultural 268, 321, 336
 French 250, 254–5, 259
 Humanistic 244
 Industrial 255, 263
 religious 140
 Scientific 1, 142, 172, 210, 244, 254, 263, 353
rhizomatic structure 330, 343–4, 346, 364, *see also* hierarchical structure
rhythm 36, 38, 112–16, 135, 198, 209, 307, 309
Ricci, Matteo 159, 196
Rickert, Heinrich 257
Ricoeur, Paul 264
Riegl, Alois 316
Riemann, Hugo 303–5
Rigveda 36, 55
Robortello, Francesco 155, 168
rococo 307
Roman, Rome 3, 18, 24–6, 30, 33–4, 44, 46–8, 50, 55, 61–2, 67, 72–5, 78–9, 82, 84–6, 89–90, 93n, 97, 103, 105, 109, 118, 130, 132, 137, 139, 143–5, 147–9, 151, 153, 155, 159, 161–6, 169, 176, 182, 185, 187, 189, 199, 228–9, 232–3, 245–8, 252, 255, 260, 261n, 262, 272, 316, 322–3, 353–4
Romantic, Romanticism 190, 281, 284, 306–7, 311–12, 326, 333, 336, 346
Rongorongo 274
Rosellini, Ippolito 273
Rosetta Stone 272–3
Rousseau, Jean-Jacques 175, 190
Royal Society 191, 194, 230
Ruan, Yuan 335
Rudolf IV 147n
rule system (system of rules):
 declarative 40, 42, 46, 48–9, 60, 66, 71, 111, 115–17, 122, 135, 157, 166, 197, 203, 205, 207, 209, 222, 225, 314, 364
 heuristic 60, 71, 133
 procedural 16, 18, 30, 40, 59–60, 71, 81, 83, 111, 117, 124, 129, 132–3, 135, 157, 185, 195, 197, 203, 205, 241, 263, 279, 314, 340, 342, 347, 364
 see also grammar
rule:
 absolute 249
 context-free 291
 context-sensitive 16, 284, 286, 291, 294, 298
 exception 187
 grammar 15, 17, 66, 111, 291–3
 hidden 212
 logical 66, 191
 preference 305
 recursive 17, 292, *see also* recursion
 well-formedness 305
rule-based 19, 30, 55, 69, 78, 83, 115–17, 132, 134, 157–8, 187, 207, 219, 231, 300, 306, 335, 347, 358, 364, *see also* rule-based versus example-based
rule-based versus example-based 83, 116–17, 140, 219, 231, 300, 316, 335, 347, 350, 358, 364
rule-breaking 139
rule-free 121–2
Rushd, Ibn, *see* Averroes
Russian 10–11, 94, 267, 327, 328–9, 350, 364

Saccheri, Giovanni 196
Sacchi, Bartolomeo (Platina) 164
Sachs, Curt 307
Sadanga 48–9, 71, 122, 247
Sadi, Al- 179
Saenredam, Pieter 216
Safavids 182
Saghani, Al-, *see* al-Asturlabi
Said, Edward 269, 321
Sakatayana 17
Salemans, Ben 275
Sallust (Sallustius), Gaius 26, 30, 165

Salutati, Coluccio 145–6, 150, 162, 229
salvation history 89, 93, 140, 161, 169, 176, 255–6, *see also* Universal History
Samarkand 74, 94
Samaveda 41
Sanctis, Francesco de 326–7
Sanctius, Franciscus 185–8, 193, 195, 197, 199, 219, 241, 245
Sandrart, Joachim von 220
Sanskrit 10–11, 14–18, 20, 22, 36, 40, 42, 55, 69, 72, 75–8, 95, 101, 112, 116, 129, 189–90
Sapir-Whorf-hypothesis 284
Sappho 67
Sardis, Melito of 85
Sarngadeva 116
Sarpi, Paolo 168
Sarton, George 3, 95
Sassetti, Filippo 188–9, 281
Saumaise, Claude 171n
Saussure, Ferdinand de 7, 287–90, 305, 316, 327, 339, 347
Scaliger, Joseph Justus 24, 30, 154–7, 160, 167, 169–73, 189–90, 233, 235, 245–6, 252, 348, 352–3, 358–9
Scaliger, Julius Caesar 154, 184–5, 233, 235, 237, 351
Scamozzi, Vincenzo 226
Schenker, Heinrich 7, 303, 305, 310, 350
schilderachtig, *see* picturesque
Schlegel, August von 281
Schlegel, Friedrich von 281, 333
Schleicher, August 285–6
Schleiermacher, Friedrich 333
Schliemann, Heinrich 10, 322
Schoenberg, Arnold 306
Schola Hemsterhusiana 157, 284n
scholastic 125–6, 184, 191, 197, 248
Schuchardt, Hugo 287
science 1, 2, 3, 4, 7, 9, 10–11, 14, 20, 31, 34, 37, 59, 61, 71–4, 78–9, 93, 96–8, 114, 127, 151, 156, 172–3, 175, 192, 195, 198–202, 206, 209–10, 230–1, 234, 240–1, 243–4, 247, 250, 254, 257, 262–3, 279–80, 300, 310–12, 320–1, 323–4, 327, 347, 352, 354–62, 365, *see also* astronomy; biology; chemistry; forensic science; genetics; mathematics; medicine; physics; social science
Scientific Revolution, *see* revolution
Scotland 176, 183
Scotus, Johannes 121
script system 272–4
Second Punic War 161
secularization 161, 243
Seleucus 61
semantics 16, 19, 75, 294–5, 297
semiotics 339
Semler, Johann 176n
scenario 200
seeming realism 318
Selden, John 168
Seneca, Lucius Annaeus 144, 146, 149–50
Senegal 180, 270
Septuagint 103
Serenus of Marseilles 119
Serlio, Sebastiano 225
Sextus Africanus, Julius 26, 84–5, 91, 360
shadja 42
Shakespeare, William 236–7
Shang Dynasty 21, 267
Shapiro, Meyer 321
Sharma, Ram 269
Shiites 92
Shklovsky, Viktor 327
Shogun 360
Shu, Xi 35
Shun, emperor 21
Sibawayh 75–8, 81–3, 139–40, 185–7, 245, 299, 314, 347, 349, 353, 358, 362
Sicily 34, 61, 74–6, 79
Sikkit, Ibn al 107
siku quanshu 178
Sima, Biao 29
Sima, Guang 100–1
Sima, Qian 5, 26–8, 30, 35, 43, 70–1, 73, 97–9, 116, 121–2, 161, 163, 168, 178, 252, 267, 358, 362
Sima, Tan 27
Sina, Ibn, *see* Avicenna
sira 91
sixth (musical interval) 37–8, 198, 200, 205, 303
slave, slavery 184–5, 236, 255, 261
Snow, C. P. 1
social-economic history 256, 258, 261
social science 1–11, 254, 290, 303, 310–11, 324, 335, 341–2, 345, 347, 356, 358, 362, *see also* anthropology; economics; geography; pedagogy; psychology; sociology
sociology 2, 9–10, 96, 254, 342
Socrates 58
Sokal-affair 332n, 333n
solmization 110
Solomon, king 101
Solon 31
Song dynasty 49, 100–1, 107–8, 116–17, 122, 136, 207, 227
Songhai Empire 142, 179, 208
sophists 37, 56, 58–9
Sophocles 65
sound shift laws, *see* law
source:
  material 159, 163–4, 322
  oldest 149–50, 152–5, 169–70
  oral 22, 25, 27, 158, 181, 252, 259, 269, 359
  original 31, 35–6, 172, 246, 353
  visual 48, 337

source: (*cont.*)
written 23–5, 27, 30, 48, 85–6, 89, 91, 93, 96, 98, 150, 165, 359–60
source criticism 21, 91, 93, 95–7, 169–70, 173, 178, 251–2, 268–9, 271, 278, 348, 352–3
source reconstruction 6, 34–6, 91–2, 103–8, 143–4, 146, 149–50, 153–7, 157, 200, 232, 243–4, 257, 272, 274–8, 280, 326, 349, 352–5, 359
South Africa 208, 309
Spain 74–6, 79, 81, 125, 128, 168, 179, 185, 234
speech (rethoric) 53, 58, 60–2, 68, 130, 146, 162, 187, 190, 217, 229, 231, 278
Spengler, Oswald 262, 266, 360
Speroni, Sperone 168
Sphota school 75, 296
Spinoza, Baruch 169, 171, 238
Sprat, Thomas 230
Spring and Autumn Annals 26–7, 137, 159
Sri Lanka 94
sruti 116
Staal, Frits 352n
Stalin, Joseph 328
stasis theory 61–2
Steen, Jan 318
stemma 274–9, 285
stemmatic philology 274–80, 284, 292, 320, 326, 344, 346, 349
Stoic, Stoics 32, 35, 53–5, 68
stratigraphy 323
structuralism:
in art history 315–16
in film studies 339–40
in linguistics 283, 287–91, 349
in literary studies 327–31, 333–4
in musicology 304, 308
studia humaniora, see studia humanitatis
studia humanitatis 3, 145, 184, 240
style:
in art history/theory 47, 50, 120–1, 220, 227–8, 243, 311, 312–16, 318, 321, 324, 346, 350
in film studies 340–1
in historiography 27, 90, 94, 99–100, 164, 172, 267
in logic 72
in musicology 39, 111–13, 117, 203–5, 247, 306–7
in philology 32, 157
in poetics/literary studies 67, 70, 137, 239, 249, 328, 331, 347
in rhetoric 60–1, 137, 232
style, definition of 113, 117
style period 307, 315–16, 328, 346
Su, Song 107–8
subaltern 266, 269, 342
sublime 67–8, 72, 223, 237
substitution (linguistics) 77–8, 82, 139–40, 186, 195, 297
Suetonius, Gaius 26, 143
Suger, abbot 118, 120
Suli, al 107
Sulla, Lucius Cornelius 61, 67
Sunnis 92
suppositio 126, 278
Suyuti, Al- 78, 98, 358
Swahili 180
swara 116
Sybel, Heinrich von 253
syllogism 52–3, 59–60, 126, 128–9, 133, 191–2, 195–6
Sylvester I 147–8
symbolic logic 194–5
symmetria 47
sympathetic kinaesthesia 230–1, 248
syntagmatic 288
syntax 16, 18–19, 73, 77, 83, 185, 187–8, 204–5, 225–6, 245, 290, 293–5, 299, 329
Syria, Syriac 61, 76, 92n, 101–3, 106, 127, 154, 169, 227, 273
system of rules, *see* rule system

Tabari, Al- 6, 8, 93–4, 245
Tacitus, Publius Cornelius 26, 144, 165, 168
Taine, Hippolyte 327
takhyil 134
Tala 48–9, 116, 122
Tamasheq 179
Tamil 17, 20, 75
Tang Dynasty 29, 51, 94, 99, 107–8, 116, 121, 136, 227
Taoism 28, 101
taxonomy 314, 350
technological, technology 94, 321, 324, 340–1, 344, 354, 360
tempus prolepticon 170
tendency:
from descriptive to prescriptive 140, 346, *see also* pattern
from prescriptive to descriptive 242, 346, *see also* pattern
to example-base systems 140
to rule-based systems 113
towards global view 266
towards order and simplification 197
towards theoretical foundation 200
Terence 150, 233
terministic logic 125–6
Tertullian 85
Tesauro, Emanuele 234
tetrachord 39, 114
text reconstruction, *see* source reconstruction
textus receptus 227
Thabit, Zaid Ibn 106
Tha'lab (Al-Kufah) 134–5
Theagenes of Rhegium 131
theatre anthropology 338
theatre reconstruction 337, 351

Theodulf of Orleans 104
Theodora, empress 90
Theodoric the Great 124
theology 2, 10, 56, 80, 87, 92, 127, 134–5, 145, 172, 220, 241, 359
Theophrastus 38, 65
theory of celestial spheres 121, 241
third (musical interval) 27–9, 198, 200, 205, 303
Thucydides 5, 21–4, 28, 30, 90, 92–3, 97, 140, 161, 163, 165, 168, 245, 252
Tibetan 17, 20, 75
Timaeus of Tauromenium 24
Timanthes 44
Timbuktu 179–80, 208, 270, 361
time layers/spans 258–9, 292, *see also* hierarchical structure
time reckoning 86–8
time structure 102, 139, 161, 183, 360
Todorov, Tzvetan 330, 350
total history 154, 259
Toulmin, Stephen 335
Toynbee, Arnold 360, 362, 365–6
transformation (linguistics) 291, 293–4
translation school 106
transmission theory, *see isnad*
Trattatisti 168, 182–3
trivium 34, 52, 127, 184
Trubetzkoy, Nikolai 288
truth table 54, 124
Tuareg 179
Tunisia 96, 162, 360
Turgot, Anne Robert Jacques 174–5, 253
Turkish 11, 91, 181–2, 209, 308, 364, *see also* Ottoman
Tuscan 147, 185, 193, 211, 314, 327
Tutankhamen 322
twelve-tone music 43, 247, 306–7
Tynyanov, Yury 327–8
Tzimiskes, Johannes 105

Umayyads 92
unity of time, place and action 233–7, 249
Universal Grammar *see* grammar
Universal History 24–6, 84–94, 96, 101, 169, 181–2, *see also* salvation history
universal language 141, 193–4, 230
universals 81, 299
universities 9, 79, 89, 127, 145, 250, 252, 254, 289, 300, 311–12, 322, 332, 336, 348
Urdu 285
ur-language 355
usage-based grammar 298
Ussher, James 170

Valckenaer, Lodewijk Caspar 157
validity (of an argument) 126, 243, 247–8, 349
Valla, Lorenzo 146–9, 152–3, 157, 159, 171–3, 184, 191–2, 197, 229, 232, 246, 248, 261, 352–3, 358
Vandals 79, 86, 90
vanishing point 123, 213
variants (of a text) 36, 106, 275–6, 279
Varro, Marcus Terentius 19, 25, 30, 33–4, 118, 184
Vasari, Giorgio 219–20, 223, 311, 362
Vatsyayana 48
Vaugelas, Claude de 188
Vedas 17, 36, 41, 55, 69, 132, 136, 249
Ventris, Michael 273–4
Venustas 47
vernacular grammar, *see* grammar
Verner, Karl Adolf 286–7
verstehen 7, 257–8, 264, 327, 333, 348
verum factum 173–4
Vickers, Brian 229
Vico, Giambattista 3, 173–4, 176, 183, 190, 231, 237, 242, 250–1, 258, 262, 279, 283
Vietnam 159, 361
Villard de Honnecourt 118
Vinsauf, Geoffrey of 131
Virgil 132, 150
virtual reality 343
Visscher, Roemer 318
visual culture 344
Vitruvius, Marcus 3, 6, 47–8, 73, 105, 217, 224–5, 346
Vitry, Philippe de 113
Vogelweide, Walter von der 277
Volksgeist 176, 327
Voltaire 174, 182
Vondel, Joost van den 235
Vossius, Gerardus 12, 156, 168, 170–1, 230, 235, 237
Vossius, Isaac 171, 189–90
Vulgate 103–4, 149, 153, 246

Wade-Giles transcription 26n
Wagner, Richard 307
Wahshiyya, Ibn 272
Waitz, Georg 252, 278
Wallis, John 191, 201
Wang, Fuzhi 178
Wang, Shizhen 227, 239
Wang, Yi 227
Warburg, Aby 316
Warlpiri 299
Weber, Max 10, 263, 347, 359
Wei Dynasty 70
Wei, Xi 50
Wei, Yuan 257
Wertheimer, Max 304
West Francia 89, 109
Western Roman Empire 78, 85, 90, 147–8, 161
Weyden, Rogier van der 216
Whig history, Whig interpretation of history 253–4, 365
White, Hayden 264

Whitney, William 294
Wickhoff, Franz 316
Widukind 89
Wilkins, John 82, 193–5, 272
Willem III 253
William III 253
Williams, Raymond 340–1
Winckelmann, Johann Joachim 157, 223, 227, 242
Windelband, Wilhelm 7, 256–7, 260, 263, 348
Wölfflin, Heinrich 7, 307, 312, 315–16, 318, 328–9, 346, 350
Wood, Robert 227, 242
word forms 18, 22, 60, 83, 187, 243–5, 284, *see also* morphology

Xenocrates of Sicyon 44
Xenophon 24, 90
Xia Dynasty 163
Xia, Wenyan 122
Xie, He 49–50, 121–2, 217, 227, 239, 247, 362
Xu, Guangqi 196
Xuanzang 75, 129
Xuanzong, emperor 116

Yan, Ruoqu 159–60
Yao, emperor 21
Yao, Zui 121
Ya'qubi, Al- 94
Yaska 17, 75, 296
Ying, Shao 107
Yoruba 269
Young, Thomas 272–3
Yuan Dynasty 227
Yugas 29

Zarlino, Gioseffo 199–200, 202, 210, 301
Zeitgeist 252, 258, 327
Zeng, Gong 107
Zeno of Citium 32
Zeno of Elea 52
Zenodotus of Ephesus 31, 103, 148
Zeuxis 44–5, 218
Zhang, Binglin 300
Zhang, Xuecheng 177–8
Zhang, Yanyuan 21
Zhao, Zhixin 238–9
Zhou Dynasty 43
Zhu, Jingxuan 121
Zhu, Quan 207
Zhu, Xi 100–1
Zhu, Zaiyu 207
Zimbabwe 324
Zipf, George 298
Žižek, Slavoj 340
Zuccari, Federico 219–20
Zumpt, Carl 275
Zuo, Mister 26n